The

Adirondack

Atlas

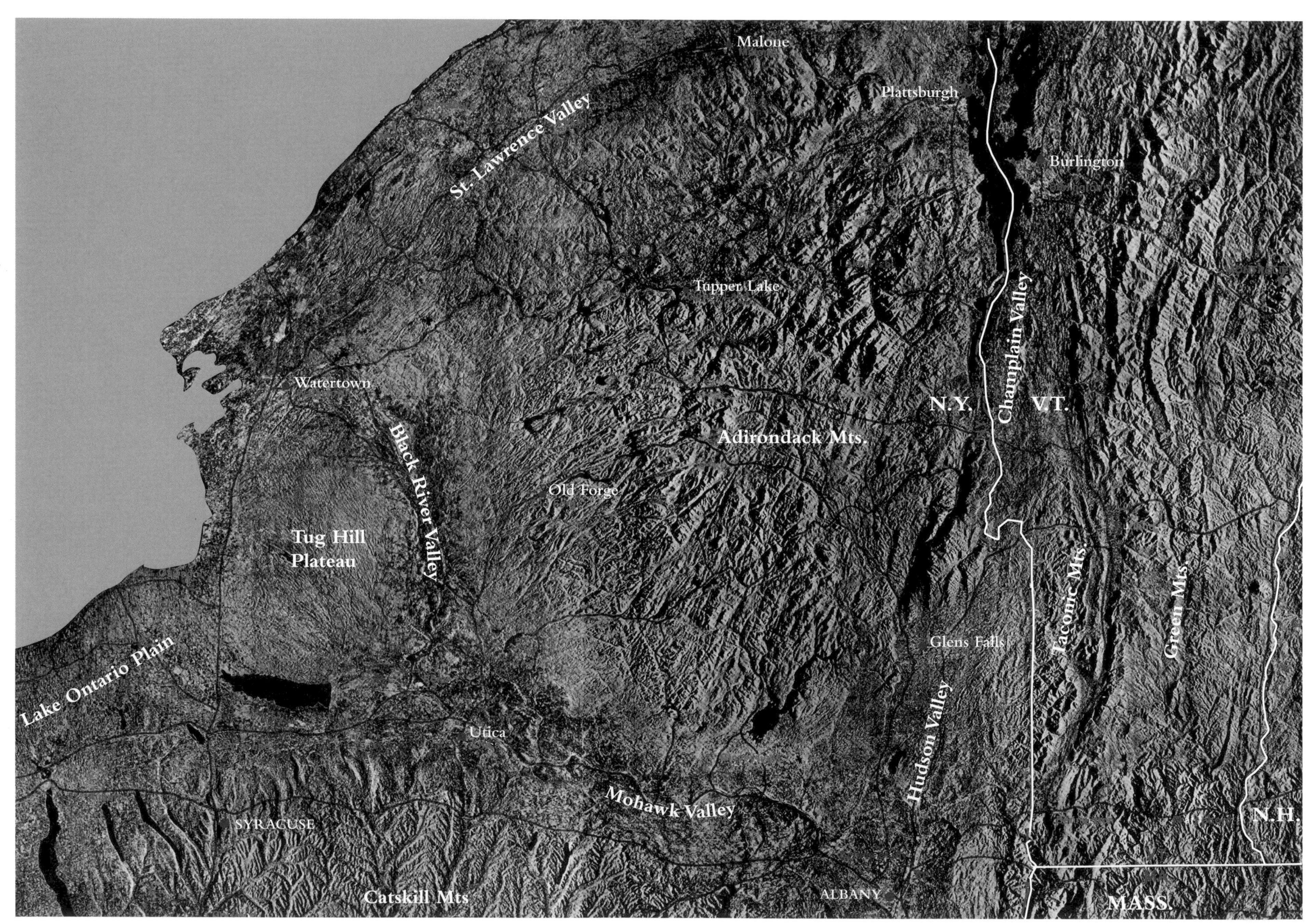

SIDE-LOOKING AIRBORNE RADAR IMAGE OF NORTHERN NEW YORK

THE ADIRONDACK ATLAS

A GEOGRAPHIC PORTRAIT OF THE ADIRONDACK PARK

JERRY JENKINS

WITH

ANDY KEAL

SYRACUSE UNIVERSITY PRESS & THE ADIRONDACK MUSEUM

A PROJECT OF THE WILDLIFE CONSERVATION SOCIETY

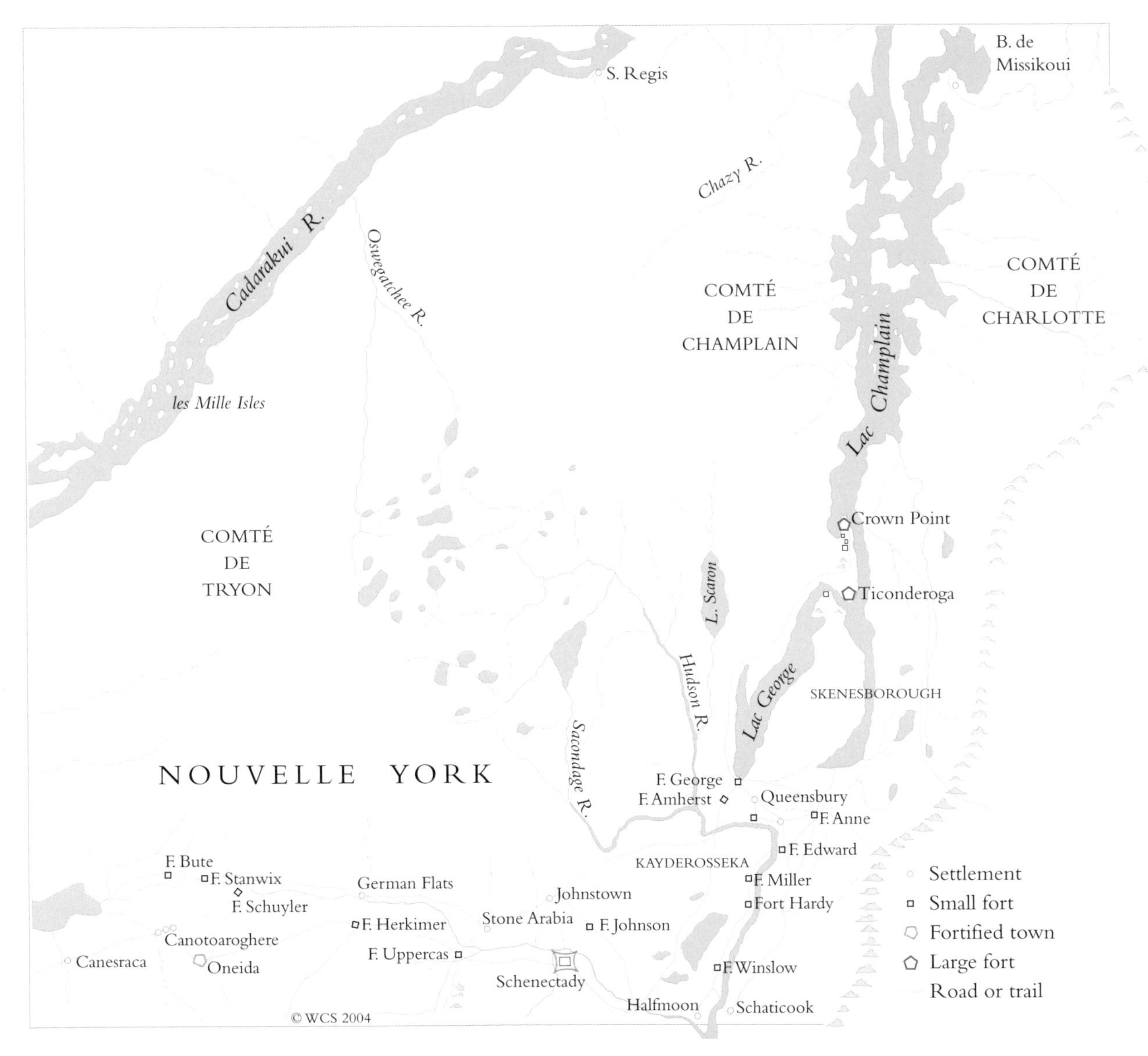

Northern New York in 1777, from a map by Louis Brion de la Tour. Only four of the fifteen main rivers draining the Adirondacks are shown, and the headwater lakes and the streams connecting them are mostly imaginary. Two of the three counties shown are, in a sense, imaginary as well. Champlain County was a real county. Charlotte County was the New York name for the parts of the New Hampshire Grants that it claimed but Vermont controlled. Tryon County, except for a narrow fortified corridor in the Mohawk Valley, was a large area of unceded Indian lands in which there was no European presence, civil or military, at all.

Adirondack Atlas Staff

Director
Bill Weber

Text, Illustrations, Final Maps, & Design
Jerry Jenkins

Research, Preliminary Maps
Andy Keal

Research
Leslie Karasin, Heidi Kretser, Zoë Smith

This book was produced with the assistance of a grant from the Ford Foundation, and published with the assistance of a grant from the John Ben Snow Foundation and from Furthermore: a program of the J.M. Kaplan Fund. It was printed by Friesens in Altona, Manitoba.

Library of Congress Cataloging-in-Publication Data
Jenkins, Jerry.
The Adirondack atlas: a geographic portrait of the Adirondack Park / Jerry Jenkins with Andy Keal
p. cm.
Includes bibliographical references and index.
ISBN 0-8156-0757-1 (alk. paper)
1. Adirondack Park (N.Y.)—History. 2. Adirondack Mountains (N.Y.)—History. 3. Natural history—New York (State)—Adirondack Park. 4. Natural history—New York (State)—Adirondack Mountains. 5. Adirondack Park (N.Y.)—Geography. 6. Adirondack Mountains (N.Y.)—Geography. 7. Adirondack Park (N.Y.)—Maps. 8. Adirondack Mountains (N.Y.)—Maps. I. Keal, Andy. II. Title.

F127.A2J46 2004
974.7'5—dc22 2004041714

Printed in Canada

Contents

Foreword

THE *Adirondack Atlas* is one of those public services so unexpected and so large that it is hard to know how to greet it with more than a quiet "thank you." Before the year is out, it will be difficult for many to imagine how they did without it for all these years—it will quickly emerge as the most-thumbed book on the shelf of anyone concerned with the past, present, and future of these six million acres, and soon it will seem as inevitable and permanent as the cliffs above Chapel Pond. But now, at its first edition and in its first printing, it is worth reflecting for a moment on what it means that such a book has finally emerged.

For most of the decades of the twentieth century, the Adirondacks slumbered along in pleasant obscurity. As the forest slowly recovered from the depredations of the first wave of industrial logging, as the farms on the periphery of the core wilderness grew back into pine, the region only sporadically captured the attention even of the rest of the state. Then, with the construction of the Northway, the proposals for a national park, and the eventual creation of the Adirondack Park Agency, the region went through a necessary but traumatic passage, one that made immediate and hard-fought conservation battles the priority and tended to prevent a clear focus on the larger picture. Now, as generational change has lessened the political temperature a bit, the time is finally ripe for sitting back and seeing what more than a century of conservation has left us with, and then using that base to plan for the future. The *Adirondack Atlas* thus marks an important moment: along with a few other developments, like the rise of Paul Smith's as "the college of the Adirondacks" and the emergence of North Country Public Radio as the first parkwide media, it demonstrates that the region has reached a certain kind of maturity—that we are now able to take a long, close look at ourselves and our institutions. This *Atlas* serves as a kind of mirror.

And when we gaze into that mirror, much that we see is beautiful. The park, as any casual visitor knows, is a place of great and growing beauty. As any biologist could tell you, and as this volume helps make clear, it is also a place of great and growing ecological integrity. Though assailed by acid rain, and now by climate change, it is nonetheless one of the few regions on the face of the earth that can legitimately claim to be growing more whole with each passing year—it is the earth's single great example of successful ecosystem restoration.

But if that is all that the visitor notices, then they miss what may be the park's real glory—the fact that all this wilderness coexists with human settlement. That this is not Yellowstone—that it is something far more real, and hence far more useful as a model. Planners and conservationists from around the globe have started to sense this in the last decade. Struggling to set up "biosphere reserves" and the like in their own lands, they have come in increasing numbers to the Adirondacks, visiting the park agency, the colleges, and so forth in an effort to understand how they might manage the same trick of letting people and nature make their livings in more or less the same place.

Of course, just like the forest, this human world is in flux. And as the landscape must cope with acid rain and global warming, so must communities cope with economic change. As the *Atlas* makes clear, a great many problems persist—ill health, crime, and unemployment daunt the Adirondacks, as they have for generations. But here too the *Atlas* offers ground for more than a little hope. At every turn, new institutions are arising to make life a little easier. One hopes that the *Atlas* itself, by turning a light on those problems still troubling us, will help launch more such efforts at reform.

So far I have written as a resident and a partisan of this wonderful park. But let me add a few words also from my perspective as a writer. This book has significantly raised the bar for anyone who will write about these mountains in the future. It is a triumph not only of mapmaking, but even more of analysis and interpretation. That thinking process obviously started early on, with the crucial decisions about what data to seek out. It bears rich fruit in the modest, straightforward, and thoughtful conclusions that the author draws from the data he has gathered.

It is a great gift. It is the kind of gift that marks a coming of age. The Adirondacks and its residents have reached the point where this depth of calm reflection is possible—only a region that in some ways has reached a certain maturity could produce and make use of such a volume. With any luck, its successor editions for generations to come will chronicle the steady consolidation of the beauties—both human and natural—that make this one of the most interesting and most important corners of the entire planet.

BILL MCKIBBEN
Johnsburg, N.Y.

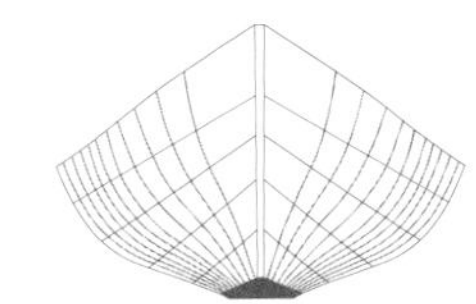

Adirondack guide boat by H. Dwight Grant, 1905.

Preface

FROM its geologic origins to its contentious conservation history, the Adirondack Park has occupied a distinctive place among the world's great natural areas. The surface rocks of the Adirondacks are some of the oldest in North America, yet powerful forces continue to push its mountains higher. The region's inhospitable environment was avoided by Native Americans and colonial Europeans alike until an explosion of resource extraction opened the Adirondacks to settlement in the 1800s. When uncontrolled clearing threatened the forests, the Adirondack Park was created in 1892 and then expanded over the next century to encompass an area as large as Yellowstone, Glacier, and Yosemite National Parks combined. Wildlife species once hunted to extinction found new homes in the healing mix of forest, wetlands, and waterways, only to encounter new threats to their existence. At least 150,000 permanent residents strive to adapt to ever changing economic challenges, while sharing their environment with millions of seasonal and short-term visitors. As it enters the twenty-first century, the park's 9,375 square-mile landscape is almost equally, if irregularly, divided between public and private ownership.

Conflict—public versus private, residents versus outsiders, resource use versus recreation, and various other forms of us versus them—is a recurrent theme in the Adirondack legacy. Yet beneath the conflict is a more fundamental theme, a cycle of change, recovery, and renewal that characterizes the region's history. Whether the subject is economy, culture, habitat, or wildlife, change and recovery are both constants of the Adirondack experience and the cause of many of its conflicts.

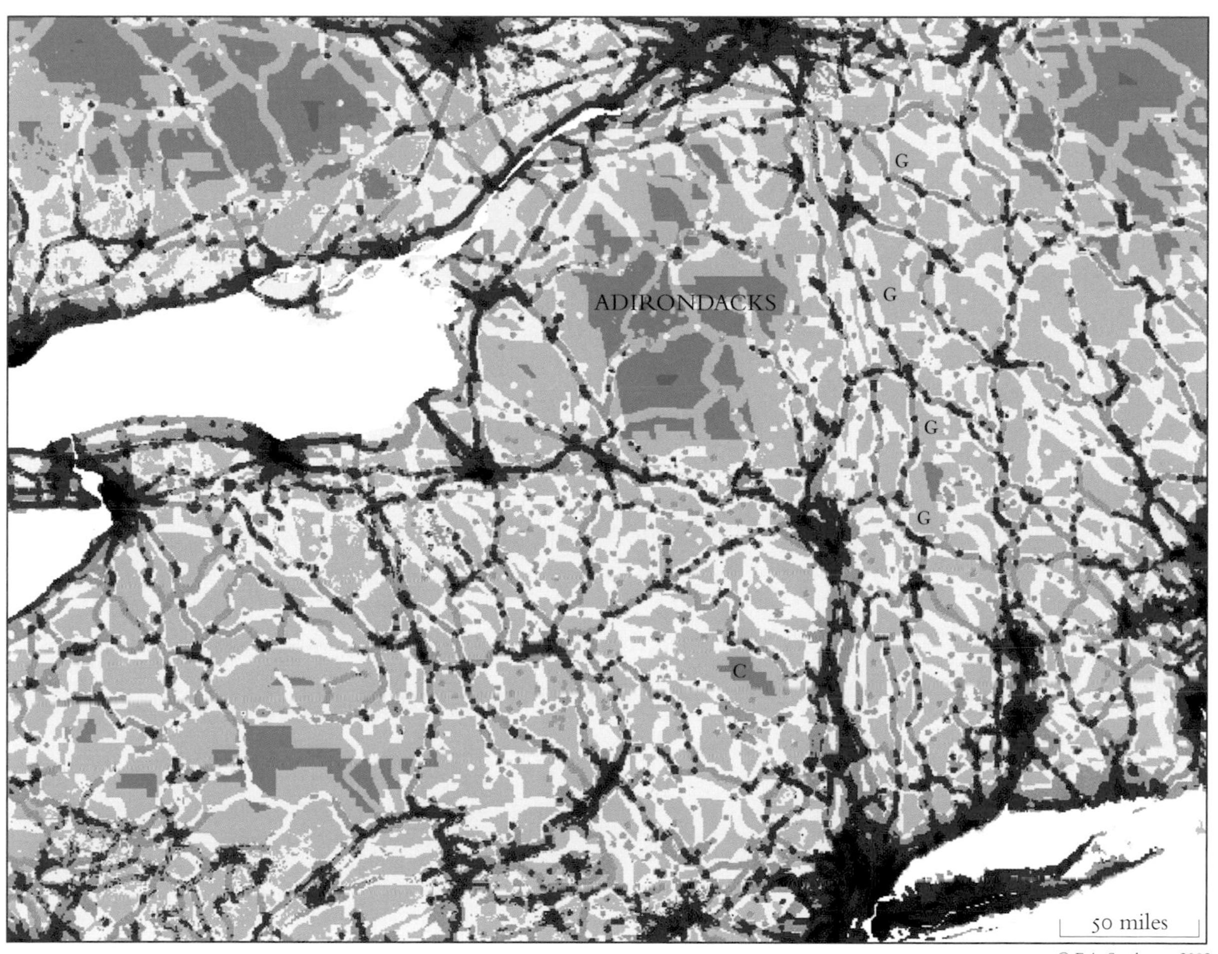

The Human Footprint in New York and New England. This map, by Eric Sanderson of the Wildlife Conservation Society and his associates, estimates the intensity of human activity and the amount that humans influence ecosystems and natural processes. The estimate is based on the human population density, the extent to which natural vegetation has been altered, the presence of settlements and roads, and the level of power consumption, measured by the intensity of lights at night. The map shows developed and urban areas and the transportation corridors connecting them very clearly. It weights the effects of roads heavily and so considers large and relatively roadless forests like the Adirondacks wilder than smaller or more divided ones like the Green Mountains (G) or or Catskills (C).

Relative intensity of human influence

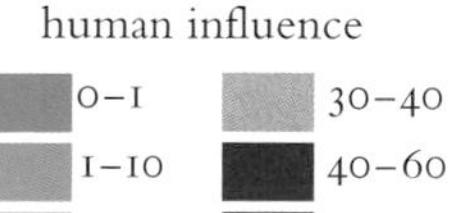

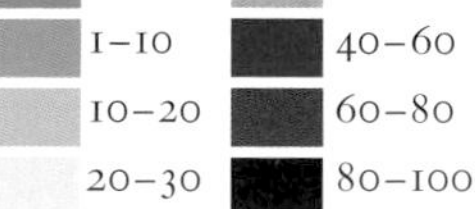

The Adirondack forest itself has experienced dramatic change. Of its six million acres, perhaps no more than 20% escaped the blades of the plow and ax at the peak of clearing. Yet today, nearly three million acres are returning to old-growth conditions under the protection of the Forest Preserve. More than 90% of all wildlife species found in the northeastern United States now call this forest home. Among them are many that had once nearly or completely disappeared, including moose, marten, fisher, beaver, peregrine falcon, and bald eagle. And perhaps the cougar too.

Industries too have come and gone. For almost two centuries, trapping fueled the colonial ambitions of France and England and enriched their Native allies, until first the prey and then most of their pursuers were destroyed in the process. Tanning hides required trees and the Adirondack periphery was settled first by those who mined the hemlock. When the hemlock was consumed, the other trees became the resource. Great wealth and considerable employment were generated, but lasted only as long as the trees themselves. The twentieth century saw vibrant towns develop to service mining operations, then decline in line with production.

Of all the regional industries, only tourism has avoided boom-bust cycles. In the mid-nineteenth century, wealthy outsiders began to seek out the pleasures of the Adirondacks, much as modern, well-to-do westerners seek ecotourism experiences in less developed countries. Today a much larger and more egalitarian base of visitors still seeks many of those same pleasures, and in the process generates the largest single source of employment in the Park.

Over the past two centuries, Adirondack residents and communities have adapted remarkably well to the dramatic changes resulting from these business cycles. Individuals and entire towns have suffered when a sole employer leaves; some, indeed have not recovered. But in general, human communities too have demonstrated a remarkable capacity for renewal.

The *Adirondack Atlas* provides a portal to the past, a mirror of the present, and a framework for the future of a remarkable land and its people. It brings to life the rich mix of history, culture, economics, and wilderness that characterizes the Adirondack region, including its vast capacity for adapatation and recovery.

The Wildlife Conservation Society is committed to an information-based, cooperative approach to conservation at more than 300 projects in fifty-three countries. In the Adirondacks, we have complemented our traditional focus on wildlife research with attention to the needs of local human communities and constituencies. With core support from the Ford Foundation, our Adirondack Communities and Conservation Program (ACCP) works with local partners to promote the healthy integration of conservation and development interests across the region. The idea of the *Adirondack Atlas* emerged from the ACCP experience. In particular, we felt that few residents perceive the Adirondacks as a cohesive landscape, except perhaps as the region enclosed by the Blue Line—which some choose to see as a noose.

While WCS is a conservation organization, environmental and conservation issues do not predominate in the *Atlas*. Rather, the goal is to produce a multifaceted image of the Adirondacks that is both accurate and thought provoking, to increase understanding and inform the debate over the future of this fascinating region. Our target audience includes full-time and seasonal residents, interested visitors, local and state decision makers, educators, researchers, activists, and others with a strong personal or professional interest in the Adirondacks. There is also an audience beyond the region among the many groups seeking to understand the relationship between their communities and the surrounding landscape. In its successes and failures, its advances, declines, and recoveries, the Adirondack experiment is increasingly relevant in a world where people, wilderness, and wildlife must find ways to coexist.

The *Adirondack Atlas* is a WCS project coordinated by Jerry Jenkins. Jerry has written the text, produced the final maps and graphics, and invested the better part of four years in its production. He is a long-time North Country resident and expert botanist and ecologist. Andy Keal is a specialist in geographic information systems. He generated many of the data sets and preliminary maps on which the final *Atlas* images were based. He has been assisted in this effort by the staff of the Adirondack Communities and Conservation Program led by Heidi Kretser, Zoë Smith, and Leslie Karasin. It has been my personal pleasure to oversee, advise, and interact with this dedicated group to help bring this *Adirondack Atlas* to life.

BILL WEBER
Director of the North America Program
of the Wildlife Conservation Society

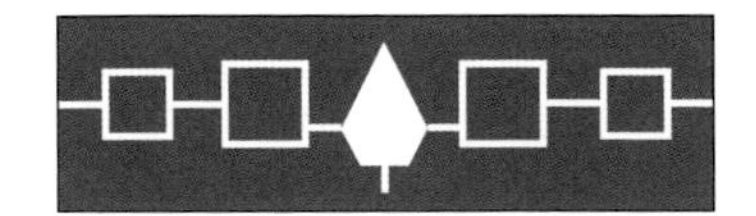

The *Tree of Peace* of the Five Iroquois Nations, in the center of a path that runs from the Western Door on the Genesee to the Eastern Door on the Schoharie.

Acknowledgments

THE *Atlas* would not have been possible without the kindness, effort, and support of many people:

At the Wildlife Conservation Society, our director Bill Weber, his assistants Ingrid Li, Sarah Ward, Stacey Low, and Dawn Greene, and researchers Eric Sanderson and Gillian Woolmer,

At the WCS Adirondack Communities and Conservation Program, which assembled the social and cultural data for the *Atlas,* the indispensable Leslie Karasin, Heidi Kretser, and Zoë Smith.

At the Adirondack Museum, without whose constant help and priceless collections the *Atlas* would have been much poorer, Jane McIntosh, Carolyn Welsh, and especially Jerry Pepper, the man who knew everything we didn't. And directing the museum and chairing its board and thus critical to the publication of the *Atlas,* John Collins and Bob Worth.

At the Department of Environmental Conservation and the Adirondack Park Agency, which not only manages the park but keeps track of a lot of what goes on in it, at least twenty-six people who provided essential information: John Banta, John Barge, Tom Beschle, Al Breisch, Mark Brown, Rangers Jim Giglinto and Steve Ovitt, Ray Curran, Leo Demong, William Dora, Fred Dunlap, Vance Gilligan, Al Hicks, Bob Inslerman, Captains Andy Jacobs and John Streiff, Marie Kautz, Brian Primeau, Ann Rice, Karen Roy, Peg Sauer, Kurt Schwartz, Bob Senior, Dave Smith, Ted Smith, Dan Spada, and Dave Winchell.

In other agencies, and equally helpful and generous: Mary Davis, Linda Foglia, William Heidelmark, Debbie Legree, and Caroline Maloney.

At Huntington Forest, a quiet and deeply committed group of researchers: Charlotte Demers, Stacy McNulty, Scott Haulton, and Ray Masters.

At several conservation groups, a number of extremely knowledgeable Adirondack hands who have put up, always generously, with three years of our questions: Tim and Claire Barnett, Peter Bauer, Bill Brown, Todd Dunham, Dave Gibson, Ross Morgan, and Rich MacDonald.

At large in the Adirondacks, Nina Schoch, Tim Holmes, and our occasional and always valuable researchers Matthew Gillman, Holly Howard, Nadia Korths, and Dominick Ruggeri. Also the equally valuable Adirondackers Cali Brooks, Larry Dennis, Jim Frenette, Greg Frohn, Tom Helmes, Malinda and Glen Chapman, and Don Potter; the town clerks of some sixty towns; and six noble reviewers who waded through drafts of this *Atlas* on our behalf: John Banta, Tim Barnett, Ray Curran, Barbara McMartin, Alec Reid, and Mary Thill.

At Paul Smith's, Caroga Lake, and Bowling Green, Ohio, three inspiring scholars whose works taught us much about the Adirondacks and served as the basis for many of our maps: Michael Kudish, Barbara McMartin, and Philip Terrie.

Scattered through North America, other researchers who shared field days and ideas and information with us: Charlie Canham, Ted Chapin, Charlie Cogbill, Sarah Cooper-Ellis, Matt Gompper, Steve Hale, Lee Harrington, Daniel Harrison, Angie Hodgson, Roland Kays, Ed Ketchledge, Jim McLelland, Justina Ray, Barry Rock, Nancy Slack, Sue Williams, John Weaver, David Wilkie, and Steve Zack.

And finally, gone and missed, Greenleaf Chase, Yngvar Isachsen, and Dick Sage, friends with fine minds and wonderful eyes, from whom we learned much that we could have learned from no one else.

BREEDING SEASON RECORDS OF BALD EAGLES, 2000–2001

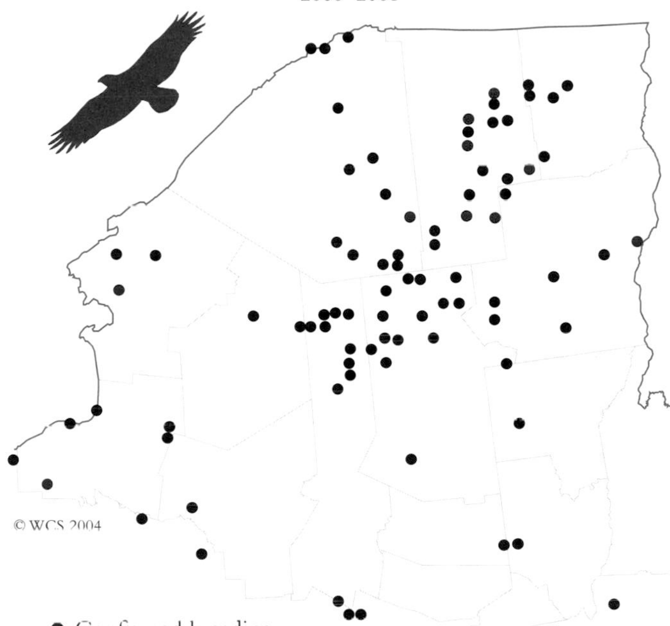

● Confirmed breeding
● Possible or probable breeding

Eagles Return to New York. Sometime about 1960, bald eagles, which like other birds of prey had toxic concentrations of DDT in their tissues, ceased to breed in New York. They did not go extinct; migrant and wintering birds were still seen, and a few individual pairs continued to try to nest. One pair, for example, nested at the same site from 1965 till their deaths in 1981 and 1984 but only laid fertile eggs once in this time.

In 1976 Cornell University and the Department of Environmental Conservation began to reintroduce eagles by hacking (hand-rearing) them in artificial nests. In the next few years, 166 birds were hacked. In 1985, two pairs of hacked birds had bred successfully; by 2002, in one of the most impressive recoveries of an endangered species every recorded, eagles were breeding at thirty-nine sites in New York and had been observed in the breeding season at another 140.

PUBLIC ROADS IN NORTHERN NEW YORK

The map shows 23,018 miles of public roads, 5,285 miles in the Adirondack Park and 17,733 miles outside. The road density is about 1.8 miles of road per square mile of land outside the park and 0.6 miles per square mile within.

Almost three-quarters of the roads are rural roads. Most of them were built early in the nineteenth century to connect farms to towns and are evenly spaced because the early farms, limited to what one family could clear and work, tended to be even-sized. Wherever a meshwork of short roads occurs, even within the Adirondacks, there were farms.

Within many of the apparently roadless areas there are extensive networks of private roads that do not show on this map. Some are abandoned and impassable, some in daily use. *Map 16-7*, for example, shows over a hundred miles of private roads in an area north of Cranberry Lake that appears empty on this map.

NORTHERN NEW YORK ROAD MILEAGE

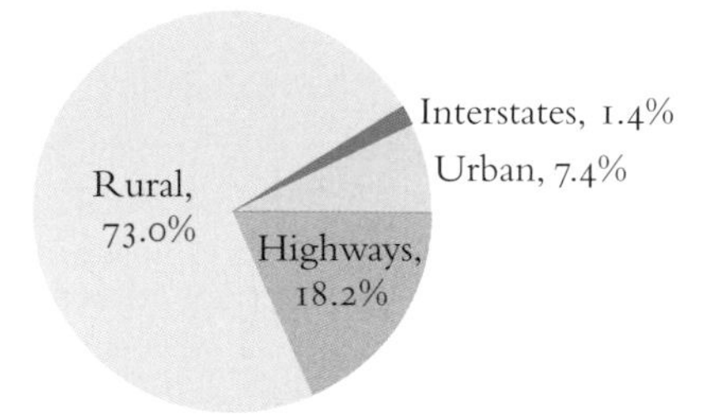

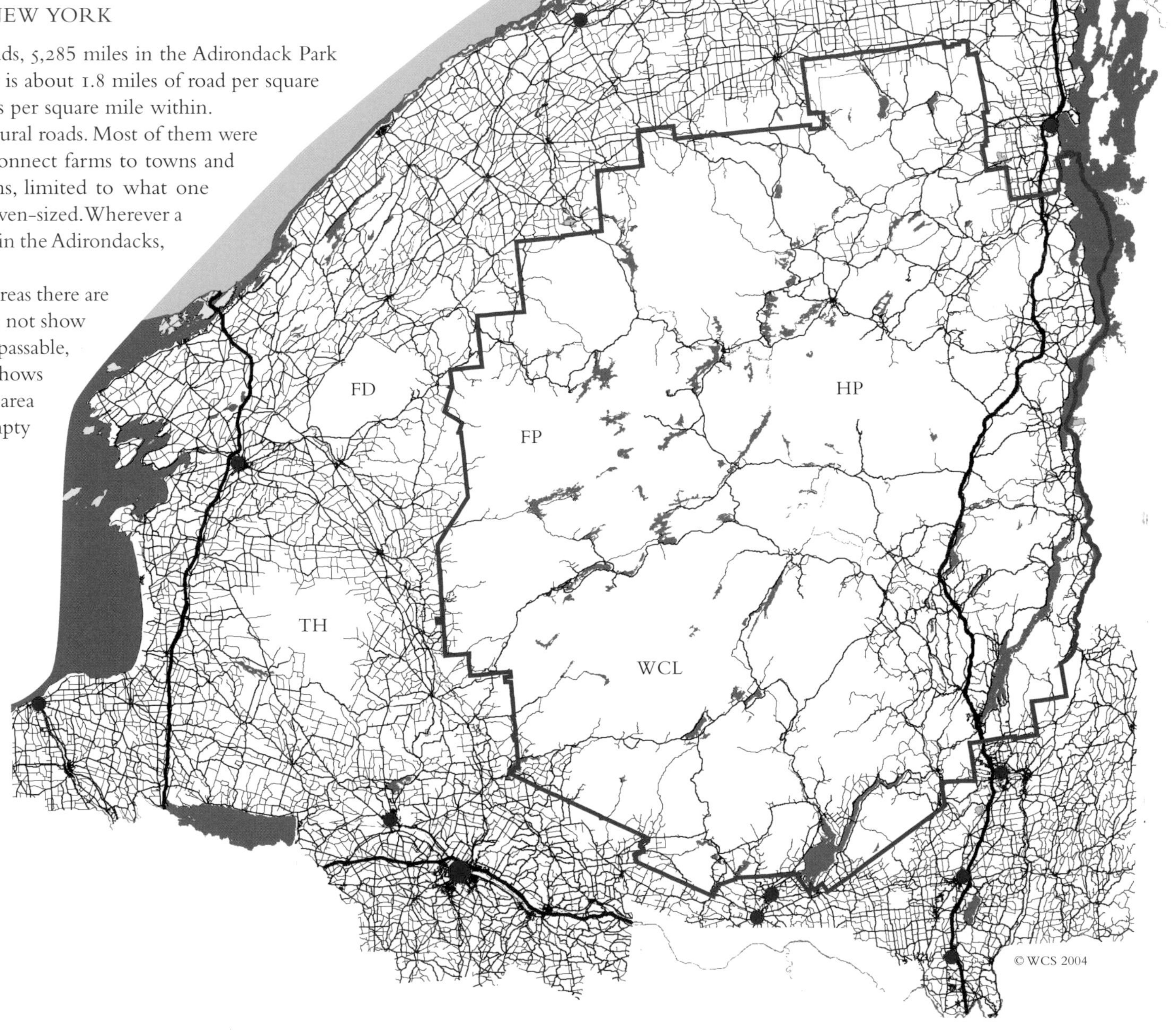

- ● Urban area
- FP Five Ponds Wilderness
- FD Fort Drum
- HP High Peaks Wilderness
- TH Tug Hill Plateau
- WCL West Canada Lakes Wilderness

1 *About the Adirondacks & the Atlas*

EIGHTEENTH-CENTURY maps of eastern North America, for example Louis Brion's graceful *Carte du Theatre de la Guerre Entre les Anglais et les Americans* of 1777 (p. iv), invariably show a large blank space between Lake Champlain and Lake Ontario. Sometimes it is marked *Couchsachrage,* sometimes *The Beaver Hunting Country of the Six Nations,* sometimes *Tryon County.* The first two names were anachronistic because both the beaver and the Six Nations of the Iroquois who had hunted them were largely gone. The third was meaningless because it implied a royal authority which had never existed. Sometimes the blank space is decorated, tentatively and also meaninglessly, by a few unnamed rivers and lakes. But most often it is just empty.

The blank space is of course the Adirondacks. This book is about their history, geography, condition, and future; about, in other words, why there was a blank at all, how it first was filled, how it has changed since then, and what other changes may be ahead.

Geographically, the Adirondacks are a 9,000 square-mile oval of highlands, located between Lake Champlain and the Black River Valley. The land is cold, wet, rocky, and barely farmable, and so remained forested and unsettled long after the surrounding valleys had filled with roads and farms. Even in the mid-eighteen hundreds, when New York State had a population of four million people and canal towns like Glens Falls and Utica were becoming prosperous small cities, the Adirondacks were a genuine wilderness with wolves and cougars and three million acres of virgin forest.

Legally, the Adirondacks are a state park, regulated by land-use legislation and administered by a special agency. Their legal existence began in 1884 when New York State declared that several hundred thousand acres of land it had acquired in tax sales constituted a forest preserve and were to be "forever kept as wild forest lands." In 1892 they gained a name and a boundary when the state drew a blue line on a map to show where it intended to acquire more lands and called the land within it the Adirondack Park. In 1894 they gained permanency when the state, alarmed at destructive logging and logging-related fires, passed a startling constitutional amendment prohibiting any sale or removal of timber at all.

The park that these acts created was unique in 1894 and remains so today. First, it was the biggest park in the contiguous U.S.: 3.1 million acres then, grown to 5.9 million acres now. Second, because the state could sell neither the timber nor the land without a constitutional amendment, it was the best protected park ever created in the U.S. And third, because the state only owned about a sixth of the land in the park and had made no plans and appropriated no money for getting the other five-sixths, it was certainly one of the most incomplete parks, and perhaps also one of the most visionary, ever created anywhere.

THE ADIRONDACKS & THEIR NEIGHBORS

The Adirondacks are separated from the adjacent highlands by the St. Lawrence, Mohawk, and Hudson-Champlain Valleys. The St. Lawrence and Mohawk Valleys are the only two lowland routes to the continental interior between Labrador and Georgia. The Hudson and Champlain Valleys connect them. Lake George, for two centuries the most important warpath in North America, connects the Hudson and the Champlain.

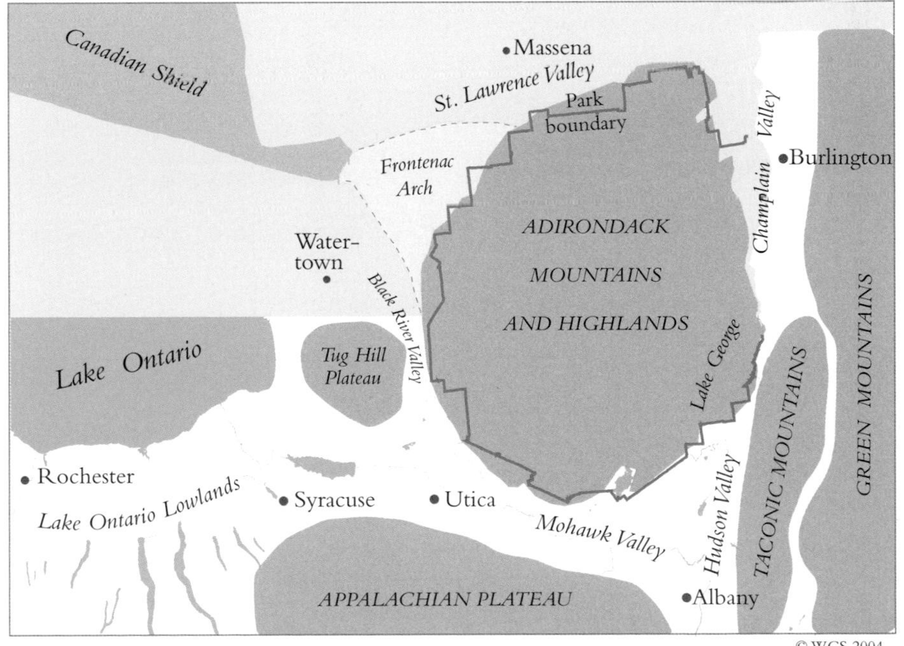

Today it remains large, incomplete, well protected, and visionary. Part of the original vision, that the Adirondacks would become a three-million acre public forest, is now close to coming true. But another part of the vision, that the Adirondacks would preserve a continuous piece of the original wilderness, has not and will not come true. Instead the current Adirondack Park remains, like the original one, a mosaic park, private and public, settled and wild.

The mixture of public and private lands makes the Adirondacks special, sad, and complex. Special because the park belongs to its residents in a way that, say, national parks never belong to their visitors. Sad because we have lost a significant amount of the wilderness that the founders of the park sought to protect. And complex because the park now has three constituencies—those who see it as a home, as a playground, and as a wilderness—and each has a vision of the future that to some extent threatens and excludes the others.

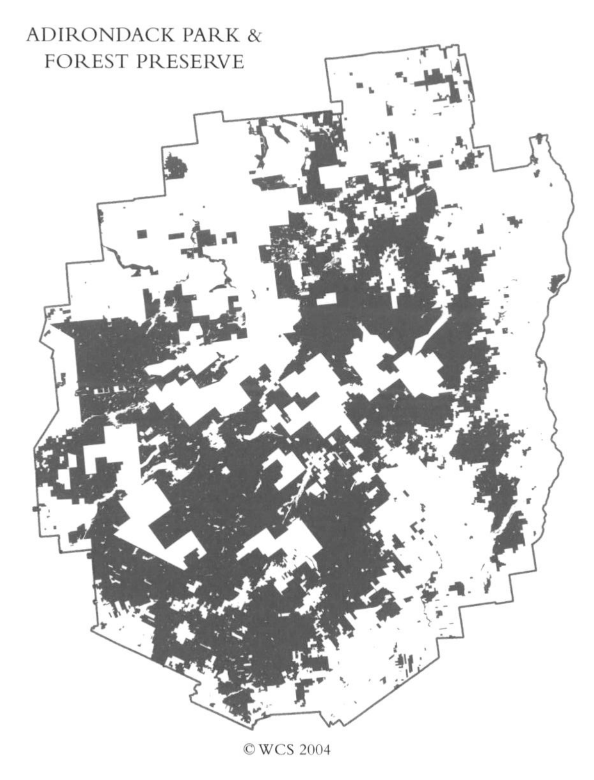

Some History. The roots of the public-private mosaic are historical and, like much Adirondack history, derive from a basic geographic fact: the three great valleys that shaped colonial American history all border the Adirondacks. Because these valleys were geographically important, they attracted much human energy; because they were energetic, they began, very early on, to develop and exploit the wilderness between them.

The importance of these valleys was both economic and geopolitical. From 1600 to 1800 they were both the major American fur-trade routes and the front lines of an intermittent but very grim war about the fate of the Indians and the control of the continent. Peter Stuveysant and Edmund Andros made their fortunes by trading for beaver pelts in them. When the pelts were gone, William Johnson traded land and intrigued and built his brief multicultural empire in them. And at a moment when everything was in doubt, George Washington said that whoever controlled them would control the continent. By 1777, when Louis Brion drew his map, Washington did, and a new phase in North American history had begun.

After 1814, the valleys' importance was largely economic. The East was industrializing and needed iron and timber and water power. The continental interior was turning into farms, and needed to get its grain and animals to eastern markets. The Mohawk and St. Lawrence Valleys were the only water routes from the Atlantic to the continental interior. The Adirondacks, next to them, had huge, unexploited reserves of iron and timber and abundant water power.

By 1820 this combination of resources, industry, and access to the continental interior had triggered an economic explosion in northern New York. It began with canals, lumber, ship-building, and hides and continued with railroads, meat, and iron. By the 1850s New York was calling itself the Empire State, and the name was a much a fact as a boast.

Because the Adirondacks were an essential province of New York's empire, many of the powerful families of the New York peerage—Rockefellers, Morgans, Vanderbilts, Webbs, Whitneys, Huntingtons, Pruyns—were involved in the Adirondacks from early on. They bought and sold Adirondack lands, built mills and railroads, cut and sold timber, patronized Adirondack resorts, and eventually built large country estates there.

Where they went many others followed, inventing ecotourism in the process and creating what had to be one of the oddest landscapes in America—part wilderness, part hinterland, and part resort, simultaneously wild and productive and rough and genteel. In a belt along its edge, which we call the *old industrial zone* (map on p. 5), was the working landscape and the working population: logging camps, forges and furnaces, mill towns, tannery towns, entrepots, halfway houses, junctions, depots; Yankee farmers, Canadian teamsters, Irish loggers, Polish mill hands, Scots ironmasters, Chinese miners, Lithuanian peddlers, black stablemen, and even, as a result of Gerrit Smith's experiment in North Elba, black homesteaders.

Deeper in the woods, in what we call the *Adirondack interior* or *old wilderness,* were the baronies and playgrounds of those for whom the working population worked: private parks and clubs of ten thousand acres or more; rustic "camps," some now considered architectural landmarks, that could accommodate a hundred guests at a time; six-story hotels on wilderness lakes; and of course the squads of cooks, carpenters, waiters, maids, and guides that built and ran the camps and resorts and served their owners.

Odd or not, it was an extraordinarily active landscape, and this activity had a great influence on the kind of park that was created. In retrospect, it seems probable that, despite the founders' hopes, the human energy and diversity in the Adirondacks in 1892 were already so great that there was no way that anything except a mosaic, public-private park could have developed here.

For better or worse—and this is still argued passionately today—a mosaic is what we got. Today the park contains about 2.8 million acres of forest preserve land and 3.1 million acres of private land. The public lands contain 1.1 million acres of legally designated wilderness where, by administrative rule, "the imprint of man's work is substantially unnoticeable." The private lands contain about 2.6 million acres of woods and sparsely settled land and 0.5 million acres of developed land. On the developed lands, where the works of men are making up for their shyness elsewhere, there are ballparks, dams, factories, hotels, mines, museums, marinas, mobile homes, prisons, ski lifts, schools, theaters, and the 80,000 homes of perhaps 150,000 permanent residents and at least an equal number of seasonal ones.

Beneath the obvious mosaic of wilderness and settlement, the park also contains a less obvious but even more complex mosaic of naturalness and disturbance. In the last hundred years there has been a remarkable rewilding of former agricultural and industrial lands. Trees have grown, structures

crumbled or been removed, polluted rivers have cleaned themselves, and many animals returned or arrived. But there has also been an equally notable civilizing of the developed lands: since the park was created in 1892, about 5,000 miles of road have been paved, 16 major dams constructed, and at least 60,000 houses built. And if the rivers and lakes are cleaner, the air now carries acids and greenhouse gases and threatens us with deep changes in geochemistry and climate.

Clearly a park this energetic and inconsistent poses questions. How, for example, do original and a recreated wildernesses compare? Are the new ones we have gained as good as the old ones we lost? How long can the park grow and still keep its wildness and beauty? If it must stop growing, what will happen to its towns and residents? And even if its towns prosper, how long can the forests themselves prevail against climate change and pollution, especially when isolated from the continuous forests they were once a part of?

The Atlas. These questions bring us to the *Adirondack Atlas* project. It began simply. Three years ago we were trying to think about the state and future of the Adirondacks. We found much valuable information but only a few good maps. Andy Keal and I took the information and made some maps of our own, and then, because we were learning too much to stop, made several hundred more. The *Atlas* is a compilation of these maps, with brief stories to set them in context and some drawings to make them less abstract.

The *Atlas* is arranged topically and does not need to be read in sequence. The text is minimal: reading maps is almost as much fun as making them, and we didn't want to spoil your fun. Sometimes when we had the space we mention why we think a map is important or how it relates to other maps. But often we did not have the space or have preferred to let the maps speak for themselves.

Geographic Patterns. Even though we have provided no overall narrative, it is our feeling, after making nearly five hundred Adirondack maps, that there are recurring and instructive patterns in Adirondack geography. These patterns, while not determining what has happened or will happen, do determine what Fernand Braudel called the *limits of the possible*—that which, at any moment, is within a society's cultural and biological reach and that which is beyond it. They suggest, for example, why many of the returning animals refused to return to the old forests; why some of the former industrial towns have had so much trouble converting to tourist towns; and why the unchecked contemporary commercialization of recreation may be as dangerous to the park as the unchecked commercialization of timber was a century ago.

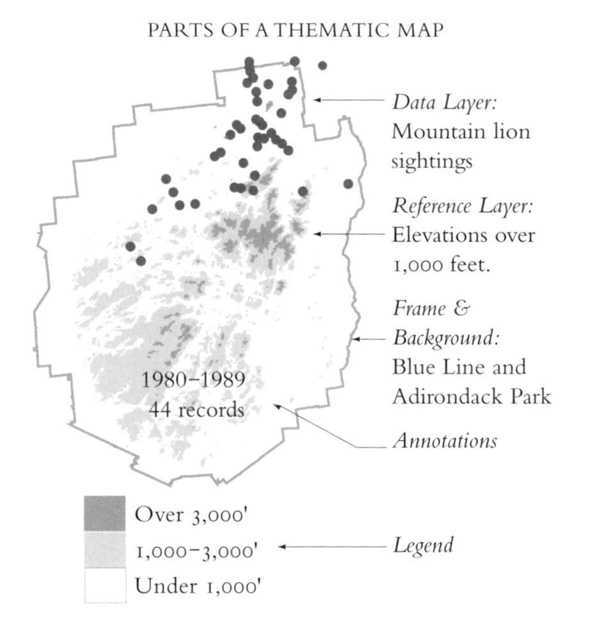

At another time, and perhaps in a more scholarly setting, we may write about Adirondack geography in more detail. Here we simply note four general patterns that seem to underlie many maps in the *Atlas.*

- A hundred years ago the Adirondacks divided into an industrial periphery and a wilderness core, mapped on the opposite page. The industries have largely vanished and the periphery turned into a zone of settlements and young forests, some of which are well on their way to becoming a new wilderness. But many biological and social traces of the old division still persist. Many of the recently reported sightings of mountain lions, for example, were in the periphery where the woods are younger and deer populations higher.
- The Adirondack interior has always been economically and geographically tied to the mill towns and agricultural valleys just outside it. Essentially, interior and valley are a functional unit: without the interior the valley towns could not have prospered; without the valleys, many of the interior towns would not have existed.
- The Adirondacks passed, in a very short time, through both a period of intense and destructive industrial use and a period of deindustrialization and rewilding. They have now entered a third period of increasing use, which both profits from and threatens the wilderness that was recreated in the second period.
- In the old Adirondacks there were two radically different kinds of land, public and private. In the contemporary park, conservation easements are

rapidly blurring the distinction between them and interrelating them in unexpected ways.

The Maps. Most of the maps in the *Atlas* are thematic maps, in which historical, biological, or social information is added to a base map of geographical boundaries and reference points. Structurally, thematic maps are graphical layer cakes, in which the base map frames and organizes the thematic layers and the thematic layers annotate the base map.

The layers in the cake come from a variety of sources. Many of the base maps come from electronic files. Frequently, as in the map of Moriah (*16-8*), we redrew the base map to get the level of detail and the quality of line that we wished. When no electronic base maps existed, we had to create one. If a paper base map existed, as it did for the map of the *Forests in 1885* (*6-1*), we usually scanned it, fitted the scans as best we could to some modern base map, and then used them as a template to draw the map. When no paper map existed, or when the task of scanning and assembling the pieces of a large old paper map was more than we could face, we made a reconstruction using reference points and modern base maps. The map of *Early Forest Preserve Acquisitions* (*6-3*), which took two weeks to create, is a reconstruction that fits historical lotting grids to a modern base map of town lines. The map of *Mineville in 1920* (*16-9*) is an adventurous but we think basically accurate reconstruction based on triangulating with lines-of-sight taken from historical photographs.

The sources of the themes we place over these base maps are equally various. Perhaps ninety percent of the maps are based on publicly available information, taken from books, reports, institutional and on-line databases, and the files of organizations and individuals. The rest come from private information, either gathered by the *Atlas*

In the 1880s, the Adirondacks consisted of a lightly logged, nonindustrial *interior,* a heavily cut but only sparsely farmed industrial *periphery*, and a continuously farmed and much more densely settled *edge* with well-developed water and rail transportation. The current Blue Line is shown; the original Blue Line of 1892 (*Map 3-1*) excluded the industrial periphery.

staff or shared with us by other researchers. We are indebted to the many people who collected this information, and to the many years of their effort which it represents. Large data sets, in our view, are remarkable human creations, with some of the compression and liveliness of poetry. We hope we have presented them with the clarity and dignity their richness deserves.

The maps vary in originality. A few of them, like the Brion map on page iv or the map of *Euro-American Settlement* (5-5) are redrawings of classic maps. In these we have tried to be as true to the originals as we could. Many others are original maps or original combinations of maps, in which we had great freedom to choose what to include and how to present it.

The presentation varies with the topic. Where the data were simple and the variation discontinuous, we use traditional cartographic techniques, often guided by the splendid examples in the *Historical Atlas of Canada* and the recent *Atlas of Oregon*. Where the data were continuous, as in the *Earthquake* and *Forest Cover* maps (*2-2, 4-1*) we use computer visualizations. Where the data were multivariate, as in the *Children's Health* map (*9-3*), or needed to be summarized statistically, as in the *Mortality Rate* maps on page 132, we have not hesitated to use more complicated techniques.

As a result, the *Atlas* contains both easy maps and, like the flowers or thorns in a hedge, more difficult ones. The demographic curves of people and maples (*7-2, 17-5*) are plots of age distributions. The *Income* graph (*8-4*) compares an income distribution to a population distribution. The *Mortality* and *School District* maps (p. 132, *10-1*), which compare quantities with dissimilar variances, are scaled in the quintiles of the distribution of ranks. The climate maps (*2-7*, p. 244) use locally-weighted smoothing and trends obtained from linear regressions. The *Children's Health* and *Well-being & Wilderness* maps (*9-3, 16-10*), the most intensely multivariate graphics, show the amounts by which local indices differs from the regional mean. The map of *BTI Applications* (*4-16*) shows the relations between four different quantitative variables, and allows some fairly complicated questions (for example, do towns that spend more money per square mile also treat more heavily per stream mile?) to be answered graphically.

ADIRONDACK LAKES & RESERVOIRS

The more complex maps took some thought to make and likely will take some thought to understand. We made them this way deliberately: the data they represent are inherently complex and would have been distorted by a simpler presentation.

Scales, Projections, Reference Layers. Our maps are mostly small-scale, *Universal Transverse Mercator* (UTM) maps of areas a hundred to several thousand miles in extent. The scales range from about 1:8,000 in a few of the detailed (large-scale) maps of parts of towns, to about 1:70,000,000 in the small-scale maps of the U.S. used in the climate diagrams. A typical full-page map of the Adirondack Park has a scale of 1:1,260,000.

We provide scale bars on all large-scale maps, but omit them from the small-scale maps with conventional boundaries. In these maps the boundaries themselves can easily be used as a scale. New York State, excluding Long Island, is about 325 miles wide; New York north of Albany is about 146 miles wide; and the Blue Line defining the Adirondack Park, the boundary used most often in the *Atlas*, is 101 miles wide and 127 miles high.

THREE COMMON MAP PROJECTIONS

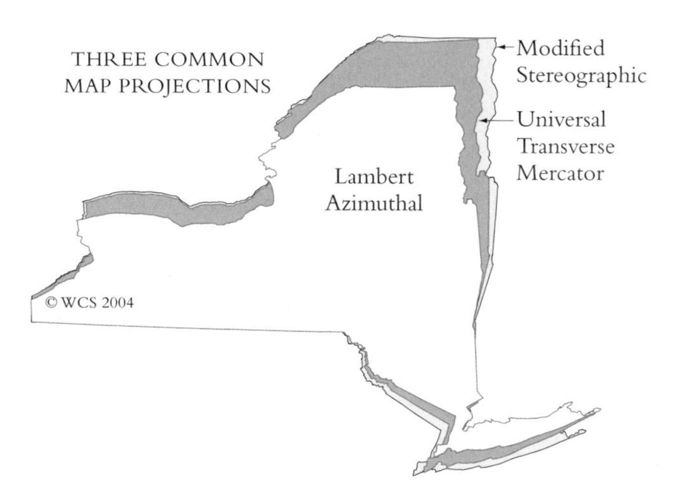

All flat maps have distortions created by the projection process. At the scale of a town or a county the differences between map projections are very slight; at the scale of a state they are easily

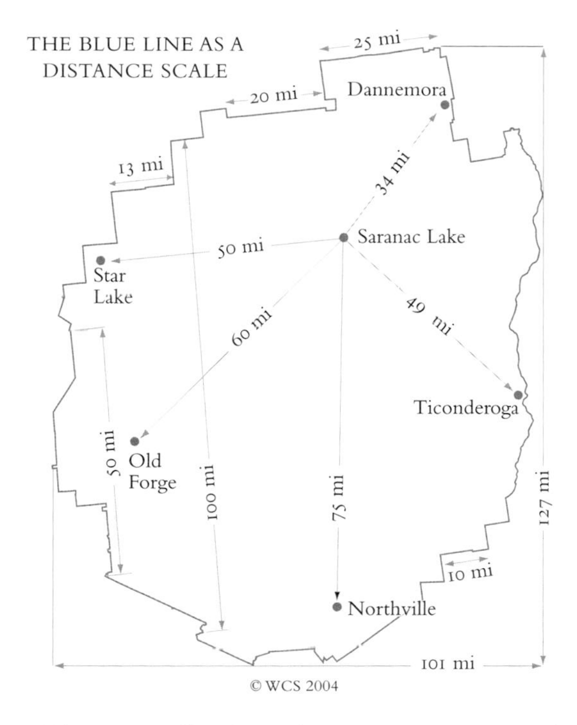

visible, especially if you have an accurate eye for proportions or are trying to combine maps in different projections. The UTM projection that we use represents angles and north-south distances quite accurately but exaggerates areas and east-west distances, especially near the east edge of the state.

In many of the maps, an unlabeled layer of geographic information underlies the main theme. In biological maps we often show elevations or waterbodies; in social and economic maps we often show roads and settlements or, because they are of economic importance, the major lakes. We call these layers *reference layers* because, as common elements that reoccur on many maps, they function as grids that locate the thematic material, and allow the accurate comparison of maps. The major lakes, for example, have characteristic shapes that are easily remembered and will identify many of the settlements in the park. A collection of labeled maps at the end of the *Atlas* serves as a reference to the reference layers and also as a gazetteer for the *Atlas* as a whole.

Errors & Lies. Trees have leaves and maps have errors. We have caught some of them. We can assure you that, regardless of what other sources claim, the 1980 population of Glens Falls was not zero and there is not even one two-hundred-year-old mobile home in Mayfield. But doubtless there are other errors that we have not caught and some that we have added ourselves. We warn you, sternly, that only your own good sense will protect you against the hegemony of reliable sources. If a number smells wrong it may well *be* wrong, and if right and wrong numbers matter to you, you should go back to the original sources and not take anyone else's word for it. We give the sources at the end of the *Atlas*. The rest is up to you.

Lies—the distortions maps introduce—are more subtle and pervasive than errors. Because maps are filters, even an accurate map may deceive. The map of public roads that opens this chapter is deceptive because two-thirds of the roads in the Adirondacks are not public. The maps of population density in chapter 7 are deceptive because they are snapshots of a mobile population. They show where the residents sleep at night but not where they go in the day, and tell us nothing about the non-resident population of workers and visitors at all.

Almost every one of our maps has similar distortions and limitations. Please consult the notes and think carefully about what the maps really say, and even more about what they do not say.

Darkness & Hope. A final thought, before we commence. As honest map makers, we have tried to show everything about the Adirondacks that we could, bright and dark, good and bad. Our thought was that, whatever the Adirondacks mean to you and whatever your vision of their future, the more you know about them the more wisely you will be able to act on their behalf.

This has meant including not only the beauties of the park but some of its sad and fearful things as well. The wild rivers, the people boating on them, the small towns and the tapestries of events and games and classes and meetings in them are beautiful. AIDS and poverty and hunger are sad. The gradual loss of places that were once wild to noise and crowds is also sad, but in a different, less violent way. The acids and toxics in the soils, the sick trees and animals, and the predictions of heat and stress to come are, for those of us who try to look ahead, deeply fearsome.

We would not have included these things if we did not think that the landscape will survive them and that we have the power to change them. The Adirondacks, though imperfect and contradictory, are also lovely, inspiring, and strong. All of us who love them and have lived afield in them have been the beneficiaries of that loveliness and strength. It is now time, we are convinced, to use that strength in their service. This book is our way of beginning. We hope you, in whatever way you think best, will do the same. There is much, much, to do, and it will take all of us to do it.

The Childwold Park House, 1878

COLORED RELIEF MAP

Alpine tundra. Two mountains are over 5,000': Algonquin (5,213') and Marcy (5,344').

5,000'

Subalpine krummholz and small areas of tundra. Fifteen mountains are in this range.

4,500'

Low, slowly growing, subalpine spruce-fir. About twenty-nine mountains are in this range.

4,000'

Spruce-fir zone, with some birch and mountain ash; trees small, openings common.

3,500'

Spruce-fir zone, with white and yellow birch; trees larger.

3,000'

Transition zone between birch-beech and spruce-fir zones; the highest logging reached this zone.

2,500'

Upper hardwood zone, mostly yellow birch and beech, with some sugar maple; logging widespread where terrain and laws allow.

2,000'

Sugar-maple and beech zone, with much spruce in valley swamps. Almost completely logged, only occasionally farmed.

1,500'

Maple-oak zone of the Adirondack edge, once extensively cleared for farming.

1,000'

Oak zone of lower valleys. Growing season about 120 days, still a number of active farms.

500'

Bottomlands of the Champlain Basin; growing season over 150 days; the most intensive modern farming is here.

0'

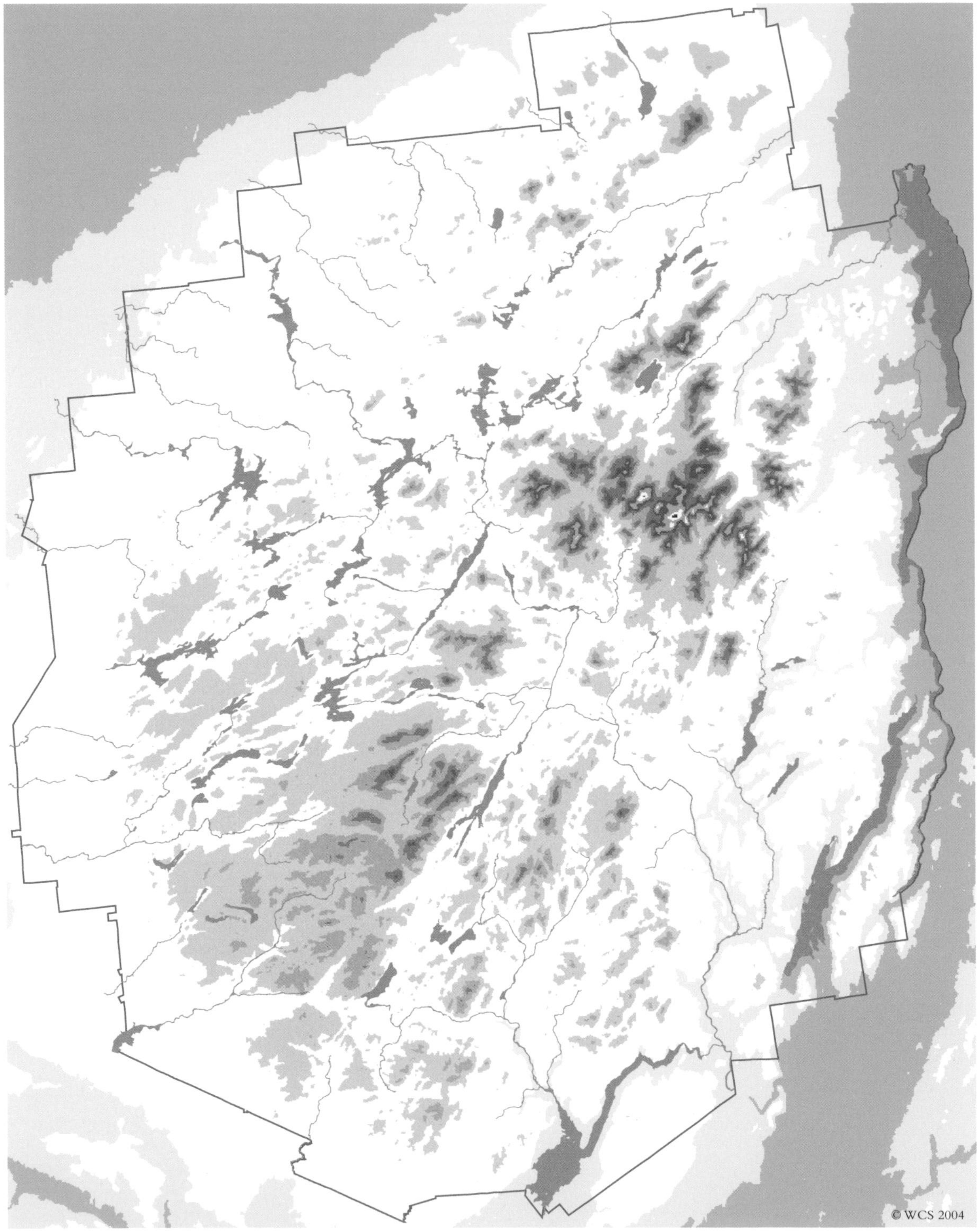

2 *Environments*

THIS chapter is about the physical features of the Adirondacks: their landforms, rocks, soils, minerals, waters, and climate. Wetlands, which are both physical and biological features, are introduced here and described in more detail in Chapter 4.

Geologically, the Adirondacks are a bedrock dome, pushed up from about fifteen miles below the surface and dissected by erosion into valleys and highlands. They are near, but quite separate from, the Taconic and Green Mountains. The Taconics and Greens are old, quiet, sedimentary mountains. The Adirondacks are young, granitic, and actively rising.

Just what is making them rise is not known. Hot spots where the magma beneath the crust is rising can create domes, but such domes usually have hot springs or geysers, or at least a telltale excess of warmth in deep mines or wells. The Adirondacks have none of these. Like the Black Hills of South Dakota, another dome-without-a-reason, all we can really say is that the Adirondacks are either a new hot-spot dome whose heat hasn't reached the surface or something different and unknown.

The most striking physical features of the Adirondacks are that they:

- Are underlain by hard, acidic granitic rocks.
- Are consistently high, with most of the valley floors between 1,000 ft and 2,000 ft elevation.
- Are most mountainous in the east and south, and smoothest and gentlest in the northwest.
- Are dissected by systems of shallow valleys which follow ancient Grenville-age faults.
- Are drained by rivers that run radially outward through these valleys.
- Have glacial sands and gravels in the valleys, stony glacial tills on the middle slopes, and almost no soil on the upper slopes.
- Have many open river corridors maintained by flooding and ice.
- Have large numbers of wetlands, including characteristic northern types like conifer swamps and large open bogs.
- Have a cold-temperate climate, with cool summers, short growing seasons, and a deep snow-pack in winter.

The consequences of these features will be explored in this and other chapters. In brief they had three effects: first, they made the Adirondacks almost unfarmable and so forbidding to settlers interested in agricultural homesteads; second, they made them mineral- and timber-rich and so inviting to wilderness industries; and third, they made them cold and coniferous and so home to many northern plants and animals not otherwise found in New York.

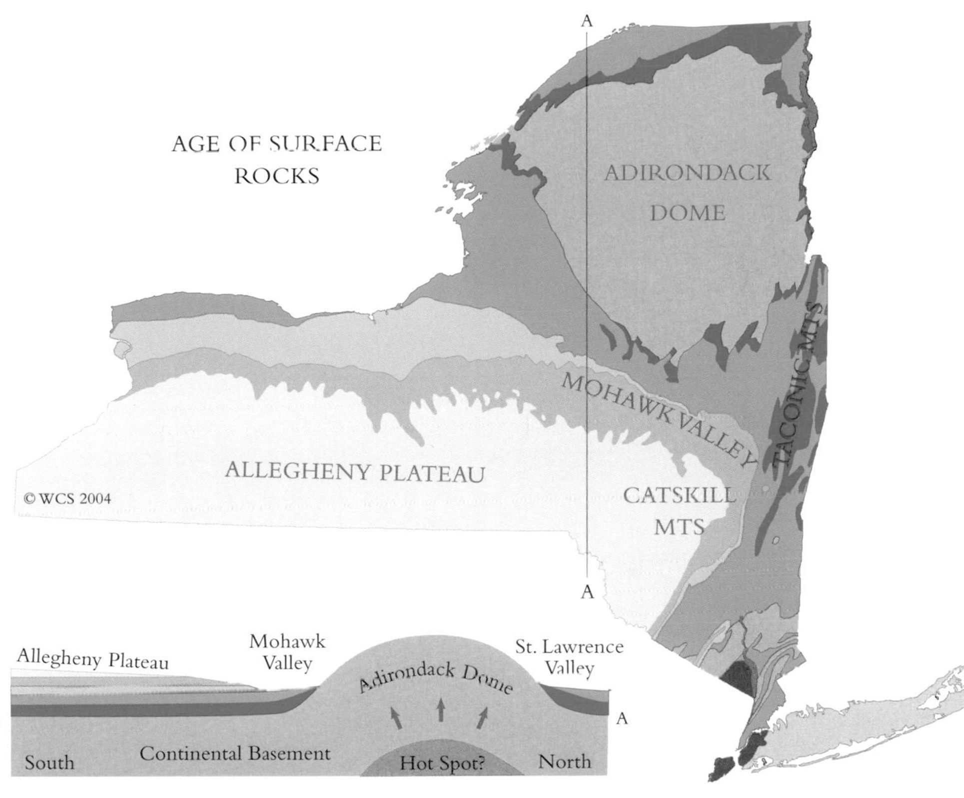

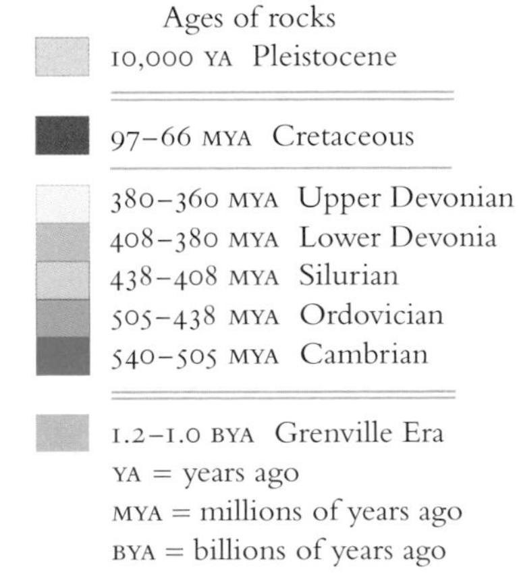

Most of New York is covered by Paleozoic rocks, deposited in a shallow sea between 360 and 540 million years ago. The Adirondacks, in contrast, are basement rocks, once deeply buried, that have recently been thrust to the surface. The lower Hudson Valley was resubmerged in the Cretaceous and contains relatively young rocks. Long Island, a sand and gravel bar deposited by the glaciers, has no bedrock at all.

THE making of even a small mountain range often involves world-scale events. This is especially true when, as with the Adirondacks, the rocks from which the mountains are made are very old.

The Adirondacks are part of the Grenville Province—a group of rocks, extending from Labrador to Texas, that was deposited along the edge of the North American continent over a billion years ago. Most of the Grenville rocks are buried by younger rocks; they occur on the surface only in places where the basement has been pushed up and the overlying rocks worn away.

Getting the Adirondack rocks from the basement to the surface took two continental collisions, two continental breakups, one mysterious uplifting, and a lot of behind-the-scenes work by frost and water. The story goes like this.

An Ocean Closes

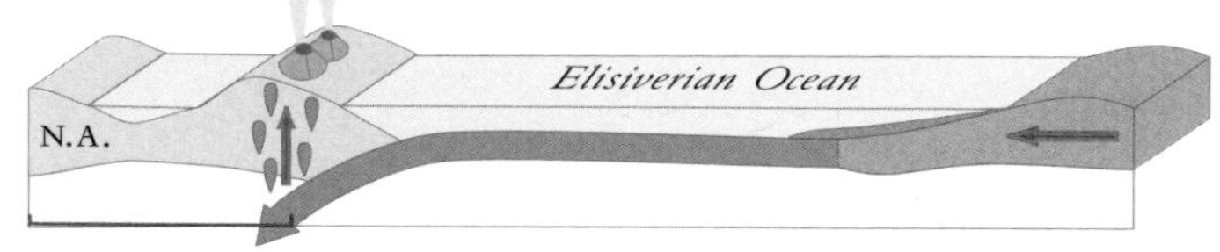

1160 MYA
Middle Proterozoic

An ancient ocean is closing and an ancient continent, perhaps Amazonia, is approaching North America from the east. An arc of volcanic islands has formed, made by rising magma from the subducting plate. Eastern North America has high mountains, like the Andes.

Grenville Orogeny

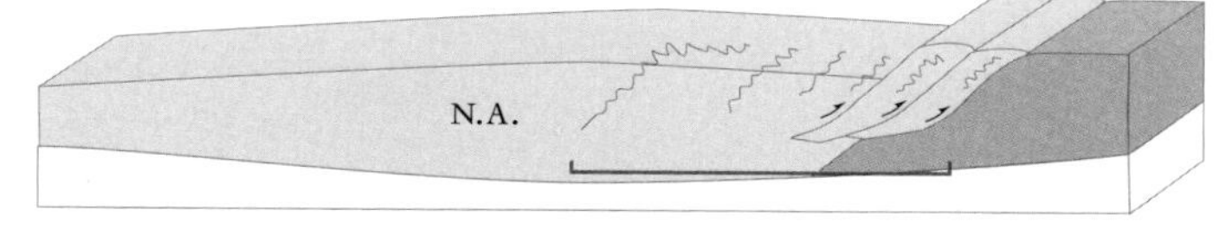

1080 MYA
Middle Proterozoic

The continents have collided, greatly thickening the crust, mushing the island arc into the mainland, metamorphosing the rocks, and making deep folds and faults. These folded rocks, a billion years later, will become the Adirondacks. New York is now a high plateau like the Himalayas. It probably has giant mountains and deep valleys, but we know nothing of them.

Relaxing & Cracking

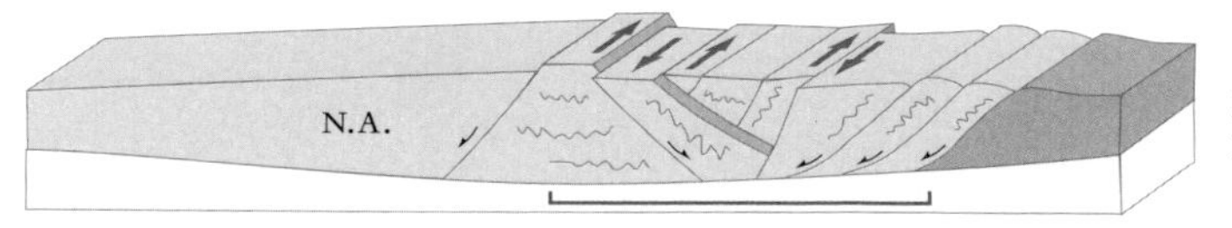

1060 MYA
Middle Proterozoic

When the pressure from the collision is relaxed, the crust expands, cracks, and slides both up and down and sideways, creating a complicated system of faults.

A Quiet Time

Nothing is recorded in the rocks for three hundred million years.

1060–700 MYA
Middle Proterozoic

Doming & Rifting

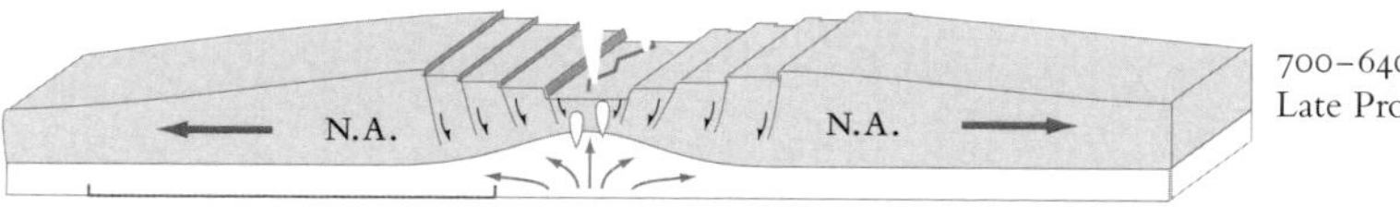

700–640 MYA
Late Proterozoic

A hot spot lifts the crust and forces it apart, creating faulted terraces and a rift valley.

Spreading

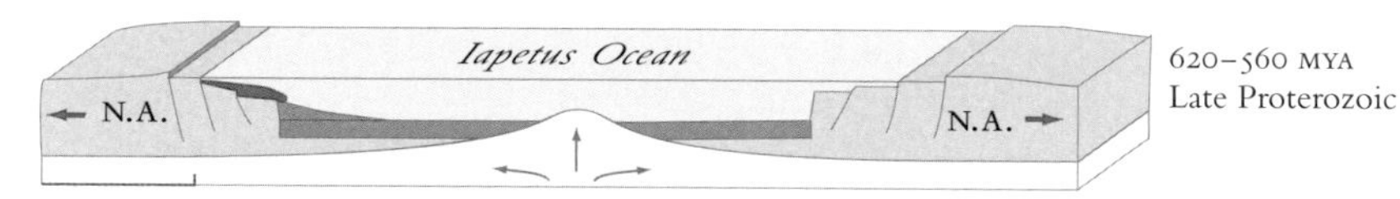

620–560 MYA
Late Proterozoic

A new ocean develops in the rift. The sediments that will become the Taconic Mountains accumulate on its western slopes. Limy rocks form in the shallow waters, muddy ones in the deep.

A New Arc

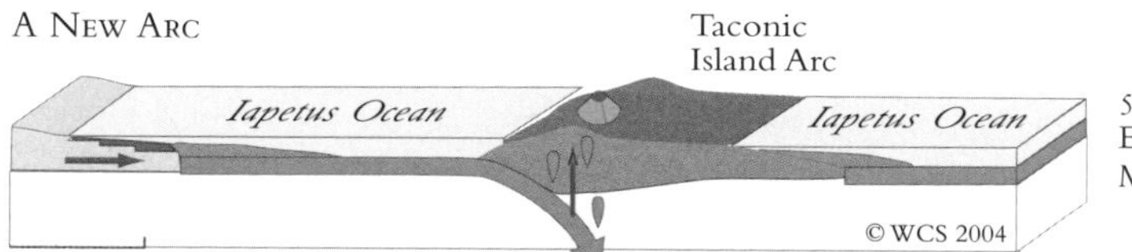

540–480 MYA
Early Cambrian to Middle Ordovician

After a hundred million years, middle age for an ocean, the Iapetus Ocean begins to close. Its crust sinks and a volcanic arc forms. The tectonic cycle has begun to repeat.

Taconic Orogeny

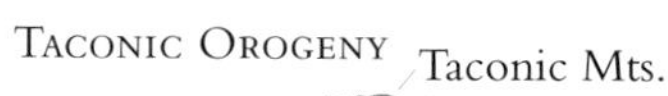

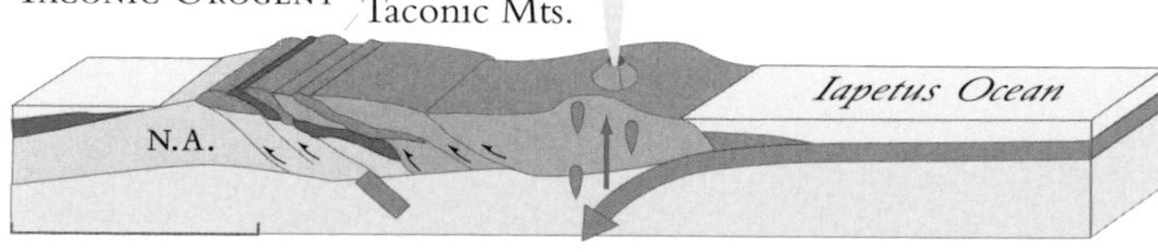

460–440 MYA
Middle & Late Ordovician

The island arc collides with North America, buckling it downwards and creating a shallow sea to the west. The rocks at the edge of the continent break and slide over each other. A messy wedge of sediments is pushed several hundred miles west and becomes the Taconic Mountains. The oceanic crust east of the island arc begins to sink, and the ocean continues to close.

Erosion & Sedimentation

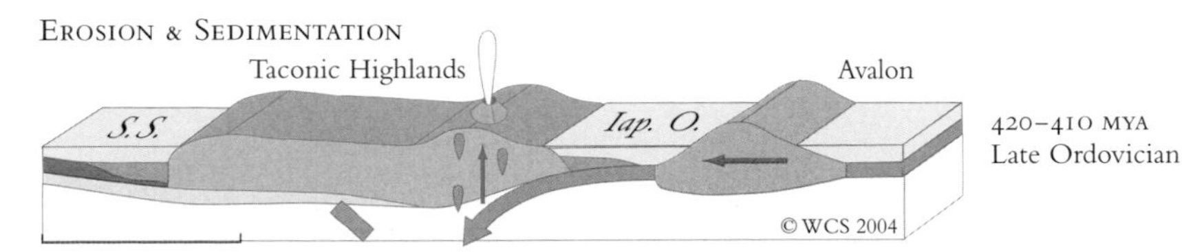

The Taconics have worn down and the sediments from them, which will make the Paleozoic rocks at the edges of the Adirondacks, have been deposited in a shallow sea to the west. Avalon, a noble but carelessly handled continent, is approaching from the east.

Acadian Orogeny

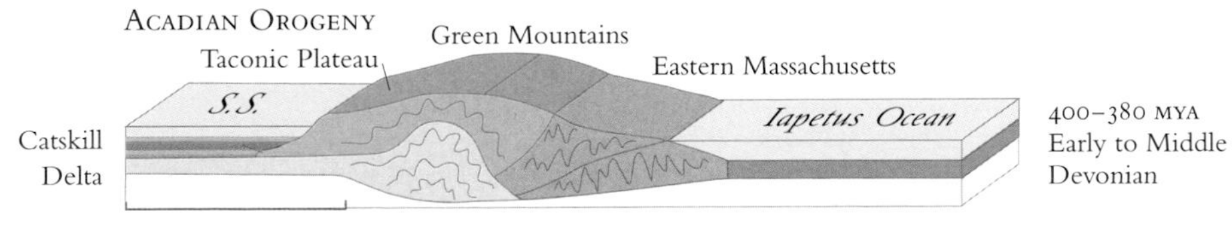

Avalon collides with and sticks to North America, becoming eastern New England. The crust thickens and buckles, and the northern Appalachians become high mountains. Sediments from them accumulate in the interior sea to the west, forming the Catskill Delta. All the Adirondack neighbors now exist, and it is quiet again. But we are still not done with mountain-making.

Appalachian Orogeny, Pangaea Forms

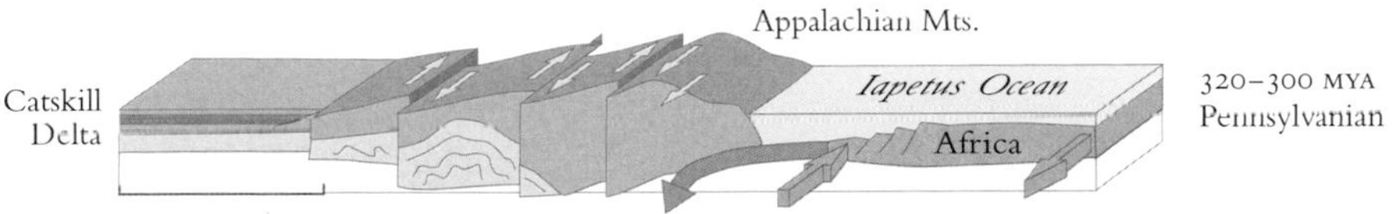

The African plate, sliding southward on a transform fault, rotates enough for its submerged margin to strike North America. The collision creates large north-south faults and builds the Alleghenies and southern Appalachians. The Catskill Delta, containing the youngest rocks in eastern New York, rises above sea level. Pangaea, the last supercontinent, has now formed.

Rifting Again

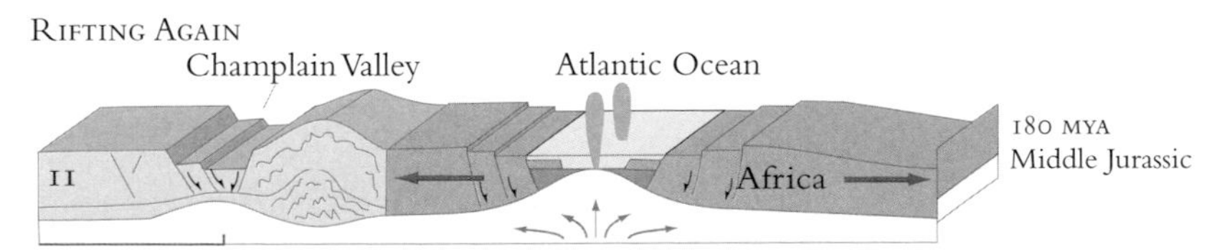

Pangaea stays together for a hundred million years. Then rifting to the east of us separates Africa and North America and creates the Atlantic Ocean. Some of Avalon is left behind and becomes eastern New England. Partial rifting at the edges of the Adirondacks creates the Champlain Valley and some of the interior Adirondack valleys.

The Adirondacks Rise

Something deep in the crust, likely a hot spot, pushes the continental basement and the overlying sediments upwards.

And Are Exposed

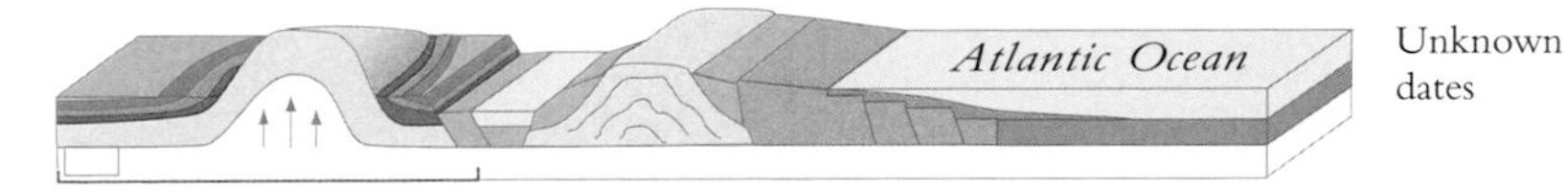

The sediments over both the Green Mountains and Adirondacks wear away, exposing the billion-year-old continental basement at their cores.

And Are Dissected as they Rise

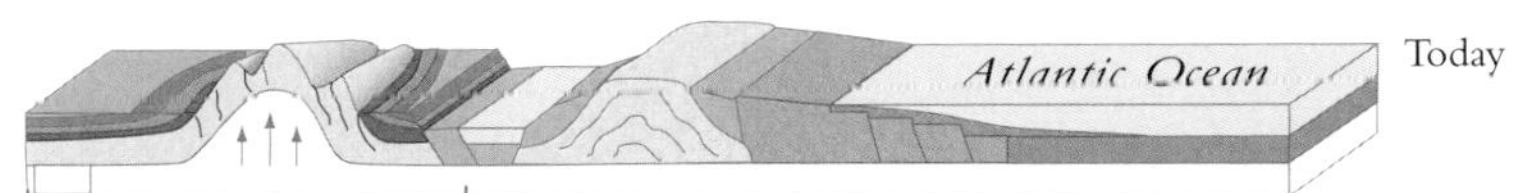

Erosion, following old, long-inactive faults, creates systems of parallel valleys. The Adirondacks continue to rise at a rate of 2-3 mm per year or almost one foot per century. How long they will do this, and to what end, we have no idea.

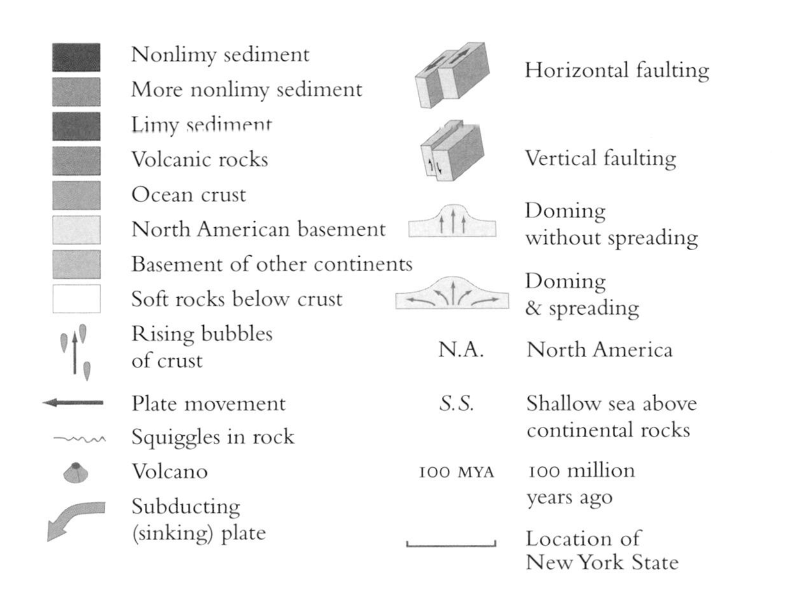

THE surface of the Adirondack dome, as defined by the valley bottoms, rises gently from elevations of one hundred to five hundred feet at the edges of the park to nearly two thousand feet in the center. The mountains are another one thousand to three thousand feet higher.

The dome was apparently lifted smoothly and quickly, with little faulting of the valley walls or downward erosion of the valley floors. In consequence the valleys are high and shallow; unlike the Green Mountains and Taconics, the Adirondacks have no lowland routes into or through them. Were there such routes, and farmable intervales to go with them, Adirondack cultural history might have been very different.

Most Adirondack valleys follow bedrock structures that are much older than they are. The curved east-to-west valleys of the southern Adirondacks lie in soft bands of metasedimentary rocks—the blue swirls in the map on the facing page—that were created in the Grenville Orogeny, a billion years ago. The straighter, northeast-trending valleys, like those that contain Long Lake and Lake George, are fault-bounded rift valleys. Their floors, called *grabens*, moved downwards during the rifting that broke Pangaea apart 180 million years ago (*Map 2-1*).

The red dots are the epicenters of the known earthquakes prior to 1980. Most had an intensity of Mercalli 4 or less and so were barely perceptible at the surface. The Adirondacks are more active seismically than the rest of New York but less active than the Laurentians and St. Lawrence Valley to the north. There are interesting quake-clusters around Warrensburg, Raquette Lake, and Chateaugay Lake but little seismicity in the most actively rising parts of the dome around Lake Placid and the High Peaks. Whatever it is that is making the High Peaks rise is subtle and powerful and, thus far, has left us very few clues about its location or method.

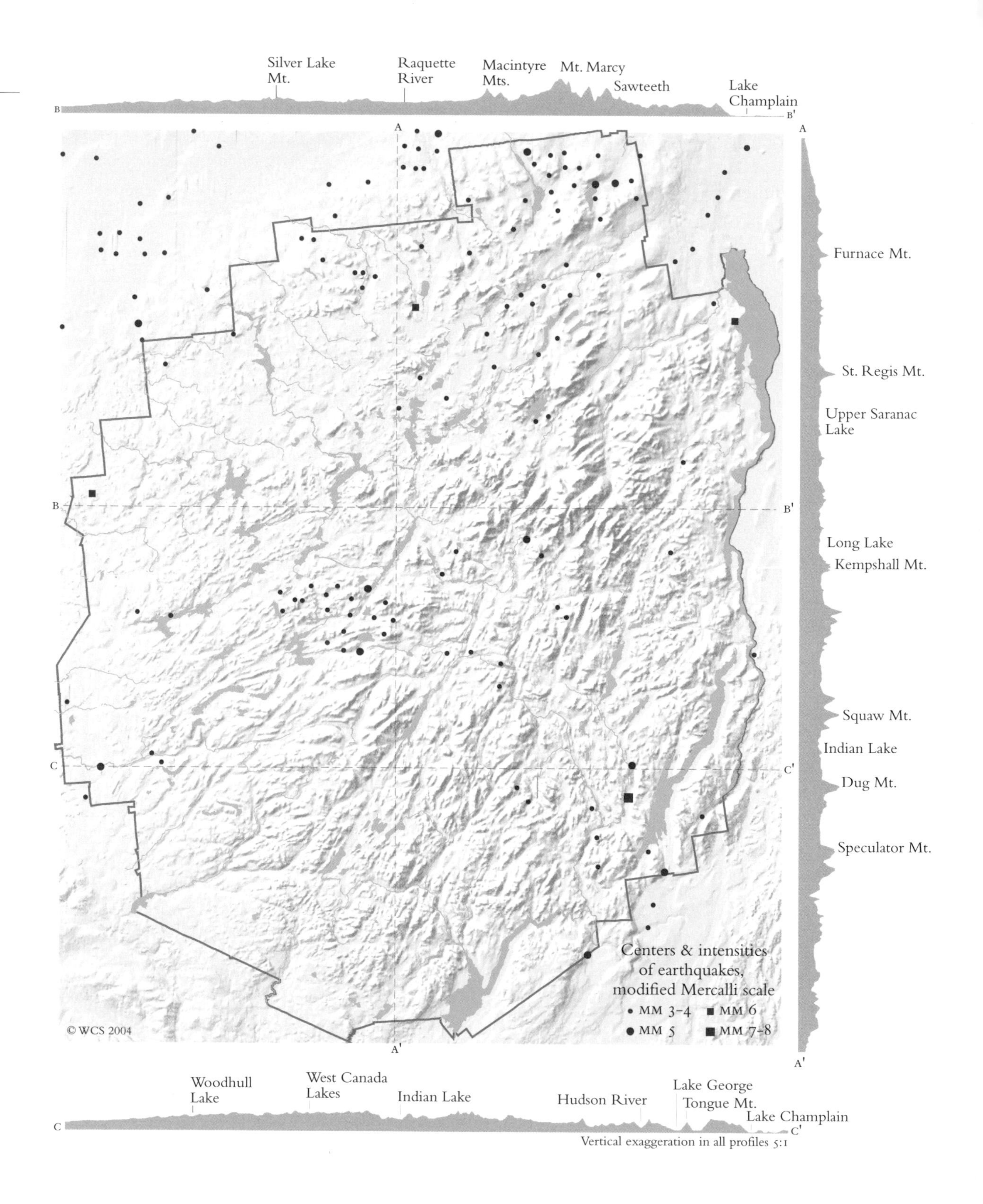

THE Adirondack dome is an island of old deep-earth rocks. Some were originally ocean sediments. Others, perhaps most, were igneous rocks which were intruded from below. All were swirled together and metamorphosed in the Grenville Orogeny, when North America collided with another continent a billion years ago. After a century of study, the meaning and chronology of the swirls are only just being deciphered.

The *granitics*, GR, including granites and granites metamorphosed to gneisses, are the hard, white-gray, acid rocks found throughout the Adirondacks. They produce rounded hills with steep sides, smooth sheer cliffs, acid soils, soft waters, and the low-diversity forests and wetlands characteristic of granitic landscapes everywhere.

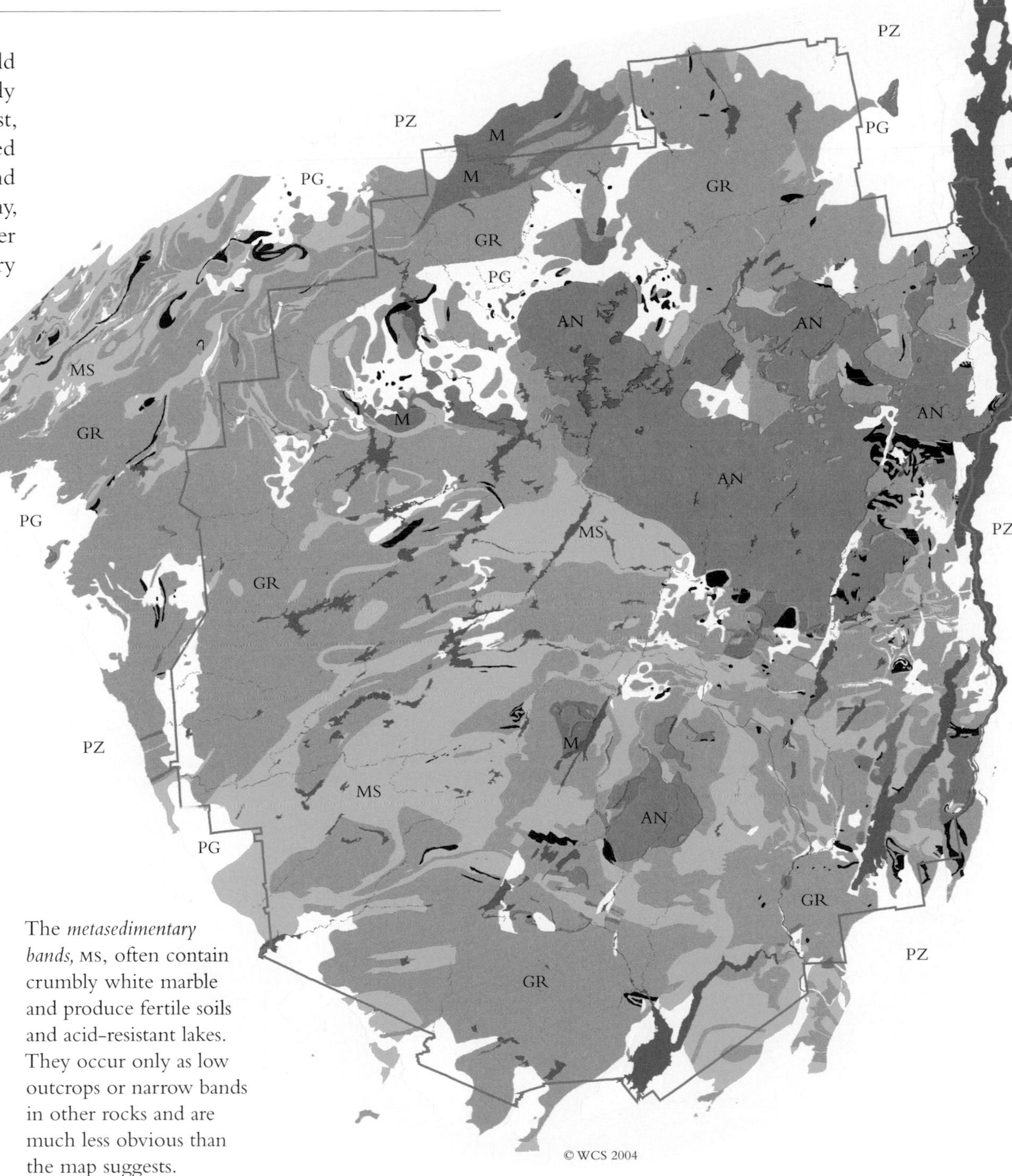

Younger rocks and surface materials

PG	Postglacial sands and gravels
PZ	Various Paleozoic rocks

Proterozoic rocks of the continental basement

AN	Anorthosite
M	Mixed anorthosite, granite, gneiss
GR	Granite and gneiss
	Gabbro and amphibolite
MS	Layered metasediments

The *metasedimentary bands*, MS, often contain crumbly white marble and produce fertile soils and acid-resistant lakes. They occur only as low outcrops or narrow bands in other rocks and are much less obvious than the map suggests.

The *Paleozoic rocks*, PZ, are metamorphosed sediments from the Iapetus Ocean, pushed here from the east in the Taconic Orogeny.

Anorthosite, AN, the basic rock of the High Peaks, is a feldspar-rich igneous rock that makes more fertile soils than the granitics.

Gabbro and *amphibolite*, shown in purple, are chemically specialized metaigneous rocks, rich in plagioclase feldspar and dark minerals.

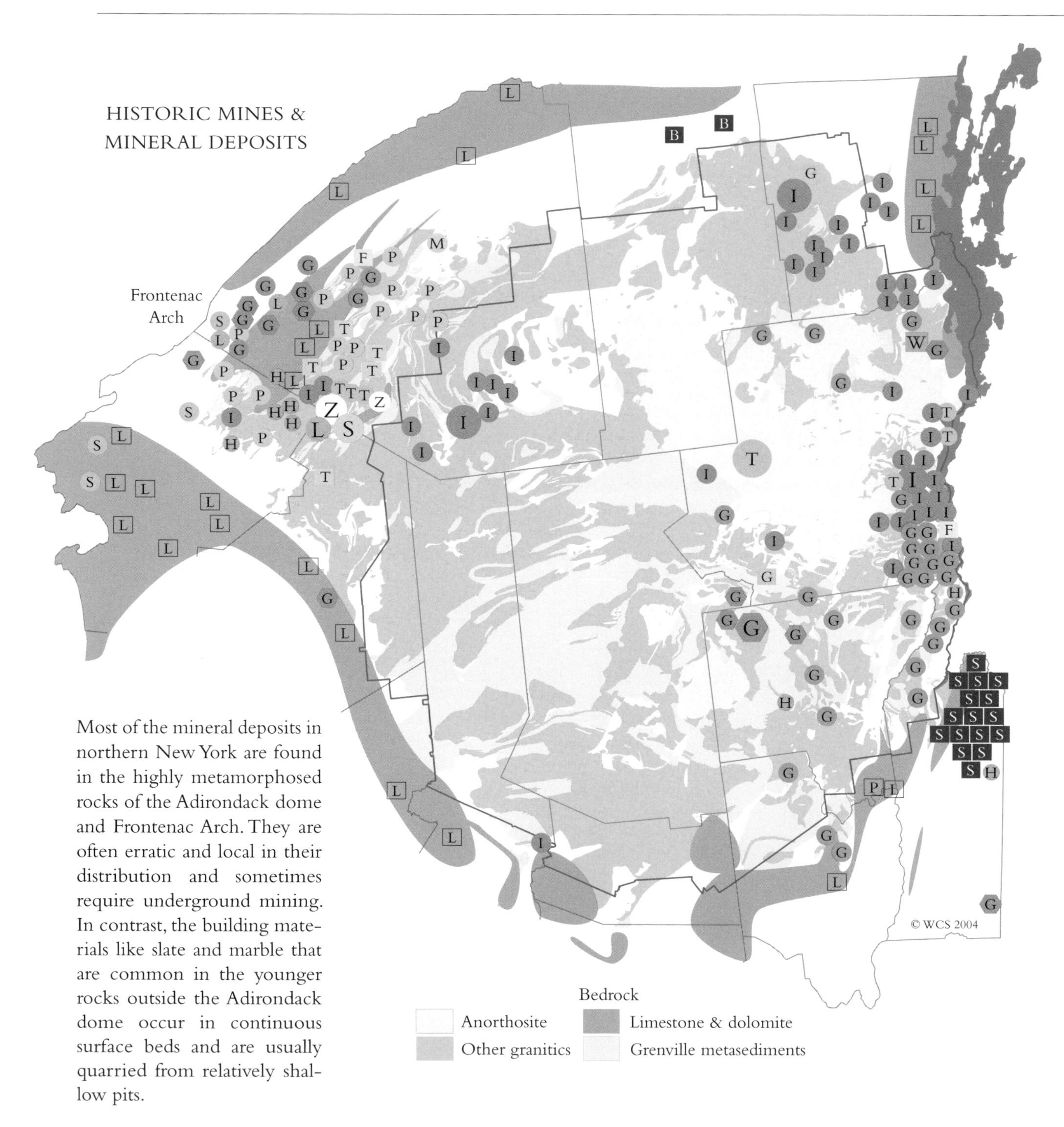

Most of the mineral deposits in northern New York are found in the highly metamorphosed rocks of the Adirondack dome and Frontenac Arch. They are often erratic and local in their distribution and sometimes require underground mining. In contrast, the building materials like slate and marble that are common in the younger rocks outside the Adirondack dome occur in continuous surface beds and are usually quarried from relatively shallow pits.

IN 1850, most Adirondack towns were frontier settlements. Like other frontier towns the world over, many had been created to extract the natural resources needed by settled areas. In the early Adirondacks the key resources were timber, bark, and iron.

Iron was the first of these to be exploited. In 1840, when the big hemlock-bark tanneries were just starting to be built and the big river drives were still a decade away, almost thirty Adirondack towns were already mining or making iron.

Adirondack iron deposits are widespread but very local. While small amounts of iron occur in many rocks, the concentrated deposits required for profitable iron-making are much rarer. One lot in a town may have a million-dollar bed of magnetite while the rest have nothing. Mineral prospecting, then as now, was a gambler's game—months of searching, lots of worthless samples and blind holes, once in a long while a bonanza.

Why the rich deposits are one place rather than another is still a mystery. We know that, because hot fluids dissolve metals from the country rocks and redeposit them in veins of concentrated ore, the old, much-heated rocks of metamorphic terranes like the Adirondacks are usually rich in minerals. But why, exactly, the veins are where they are—why Gore has garnet, Hague has graphite, and Lot 21 in Moriah had, once, the richest single iron deposit in the world—we have no idea.

Metals

G Graphite
I Magnetite
H Hematite
L Lead
M Molybdenum
P Pyrite
S Silver
S Strontium
T Titanium
Z Zinc

Nonmetals

B Bluestone
F Feldspar
G Granite
G Garnet
L Limestone, marble
P Portland cement
S Slate
T Talc
W Wollastonite

ADIRONDACK mining began with the opening of iron mines and forges near Lake Champlain during the American Revolution, and for the next 180 years iron was by far the most important Adirondack mineral. By 1860 the iron industry was the largest employer in the eastern Adirondacks, and the Adirondacks were one of the largest producers of crude iron in the United States. In 1924 Adirondack ore production was averaging a million tons per year. Port Henry, the largest iron town, had blast furnaces and ore docks and looked like a suburb of Pittsburgh.

Besides iron, three other Adirondack minerals were once of national importance. Two are still mined today.

Graphite, a soft, greasy form of carbon used in pencils, crucibles, and lubricants, was mined near Ticonderoga. The Joseph Dixon Company, famous for the yellow Dixon-Ticonderoga pencil (and also for having, for two years, the twenty-year-old Henry Thoreau among its Massachusetts employees), began mining Adirondack graphite in 1830. In the 1880s, their company town Graphite, near Lake George, was producing three million pounds of graphite a year. The mines at Graphite operated until 1921; Dixon continued to make pencils and crucibles in Ticonderoga until 1968.

Garnet, a hard, deep-red gemstone associated with veins of gabbro, is used for sandpaper and as an abrasive for fine polishing. H.H. Barton, the Philadelphia sandpaper-maker who invented garnet paper, opened the Barton garnet mines on Gore Mountain in 1878. The company built a processing plant and a small, self-contained village on the north side of the mountain and developed a large open pit mine there. In 1980 it relocated the main operation to North Creek. The Barton Mines are the Adirondack's oldest working mine and the second-oldest Adirondack industry of any kind; in 2002, under the management of the fifth generation of Adirondack Bartons, they completed their 124th year of operation,

Wollastonite, a form of calcium silicate with long slender crystals, is used in ceramics, paints, plastics, and as a substitute for asbestos in brake pads and clutch plates. Most "sanitary porcelains"—toilets, sinks, bathtub—are made with wollastonite in the clay body or glaze. Wollastonite was first discovered in Willsboro in 1810 but not mined commercially until 1943. The mines, originally run by Godfrey Cabot Inc. and now by NYCO Minerals, currently produce about 100,000 tons a year.

The map below gives a snapshot of Adirondack mining in 1955. This was a time when the total production of Adirondack mines was near its all time high. Most of the small nineteenth-century mines were long gone, but the big mines that remained were very big indeed. Republic Steel and Jones & Laughlin were the two largest. Republic produced 21 million tons of iron ore at Mineville between 1938 and 1955. In 1953, its best year, it produced 2,006,866 tons. Cabot Inc. was the only producer of wollastonite in the world. The Barton Mines on Gore Mt. were the largest garnet mine in the world and, given their processes for heat-treating garnets and producing very fine garnet powders for precision optical work, perhaps the most technically advanced as well.

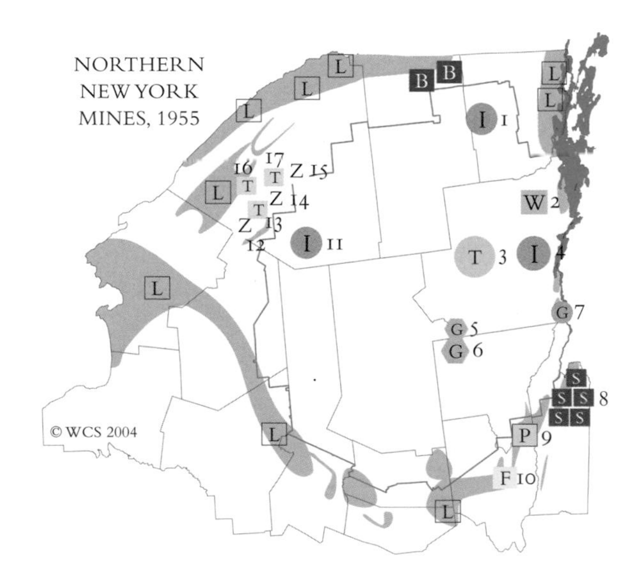

NORTHERN NEW YORK MINES, 1955

1. Lyon Mountain Mine & Eighty-one Mine, Republic Steel, Lyon Mountain
2. Fox Knoll Mine, Cabot Inc., Willsboro
3. Lower Works Mine, National Lead Company, Tahawus
4. Fisher Hill, Barton Hill & Bonanza Mines, Republic Steel, Mineville
5. Ruby Mt. Mine, Barton Mines Corp., North River
6. Gore Mt. Mines, Barton Mines Corp., Johnsburg
7. Dixon Crucible Co., Ticonderoga
8. About 15 slate pits, various owners, Granville, Hampton, & Whitehall
9. Glens Falls Cement Co., Glens Falls
10. Dry Brook Quarry, Corinth
11. Benson Mines, Jones & Laughlin, Star Lake
12. Balmat Mine, St. Joseph Minerals Corp., Balmat
13. Gouverneur Talc Co. Mine, Balmat
14. Hyatt Mine, Talcville
15. Edwards Mine, St. Joseph Minerals Corp., Edwards
16. International Mine, Hailesboro
17. Reynolds Talc Mine, Talcville

To soil scientists, soils are wonderfully complicated things that deserve to be mapped in complicated ways. To the farmers that use them and the animals and plants that live on them, they are much simpler—light or heavy, stony or soft, fertile or barren. The maps, deliberately simple, shows the soils as the animals and plants might see them.

We recognize three main groups of soils, two brought by the glaciers and one home-made.

The *unsorted glacial soils*, including tills and moraines, are made from materials deposited directly by melting ice. They are the typical hillside soils of middle elevations in the Adirondacks. Most are dense, stony, and poorly drained. If the bedrock from which they were derived was rich in nutrients, they can be moderately fertile and grow fine hardwood timber and passable crops. If it was not, they will be infertile and acid, growing softwoods better than hardwoods and crops only poorly if at all.

The *sorted glacial soils* are glacial materials that were subsequently moved by water and sorted into the homogeneous deposits we call clays, silts, sands, or gravels. Geologists separate them further, according to their origins, into outwash plains, lake deposits, stream deposits, and terraces; we leave them together here.

Sorted soils are found in all the valleys below 2,000 feet, either in narrow bands or, especially in the northwest, in broad outwash plains. Commonly, they are soft and stone-free. Where they are fine and silty, as in the Champlain Valley, they can be spectacularly fertile. Where they are coarse and sandy, as in the northwest Adirondacks, they are inherently infertile and usually acid.

The *peats*, treated in more detail in *Map 2-6*, are organic soils formed from undecayed plant material. They are invariably dense, wet, and low in mineral nutrients. They are the typical soils of northern swamps and bogs—infertile and unfarmable, loved by orchids and moose and spruce, and, ultimately, the source of much of the beauty and wildness of the north.

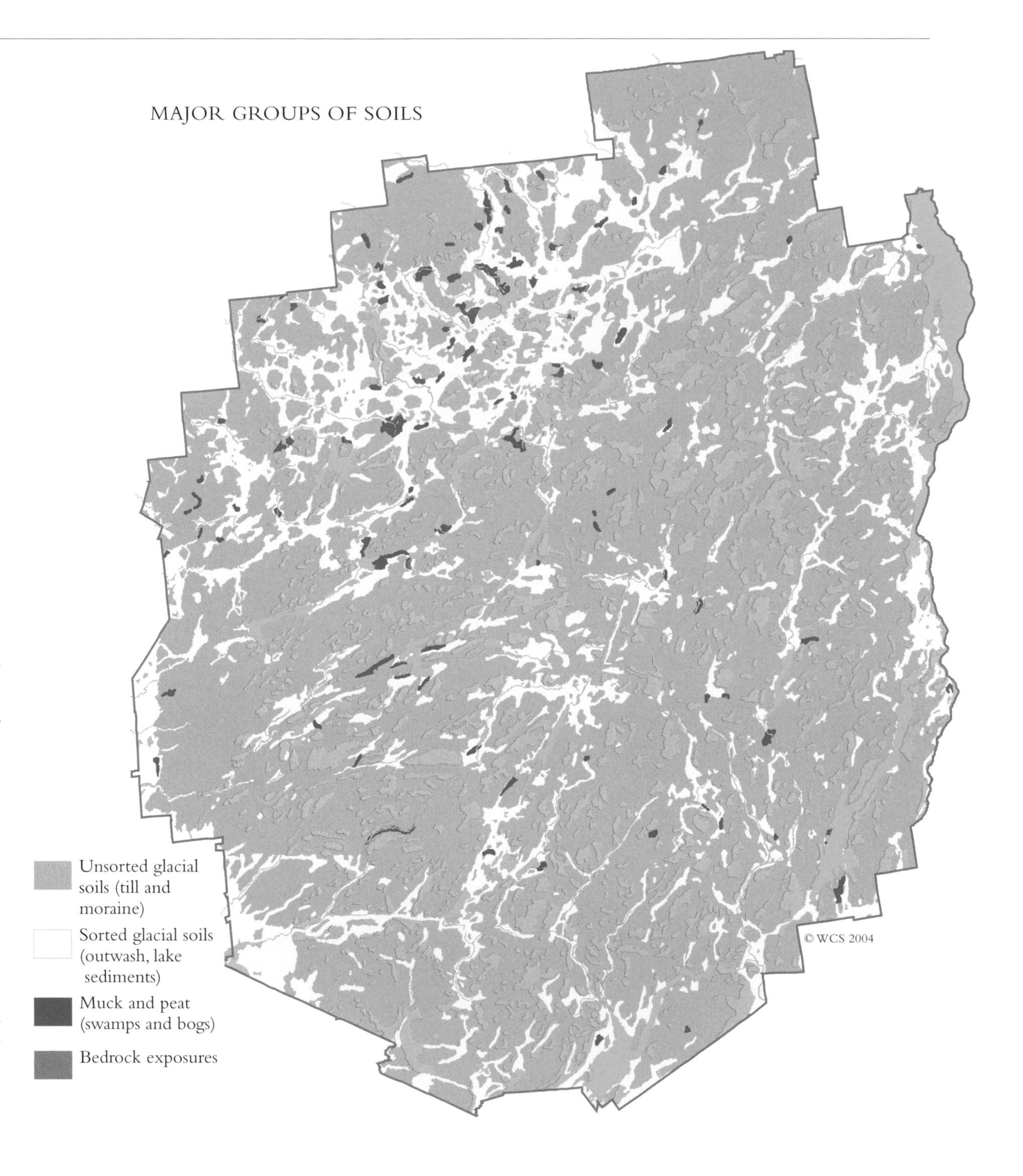

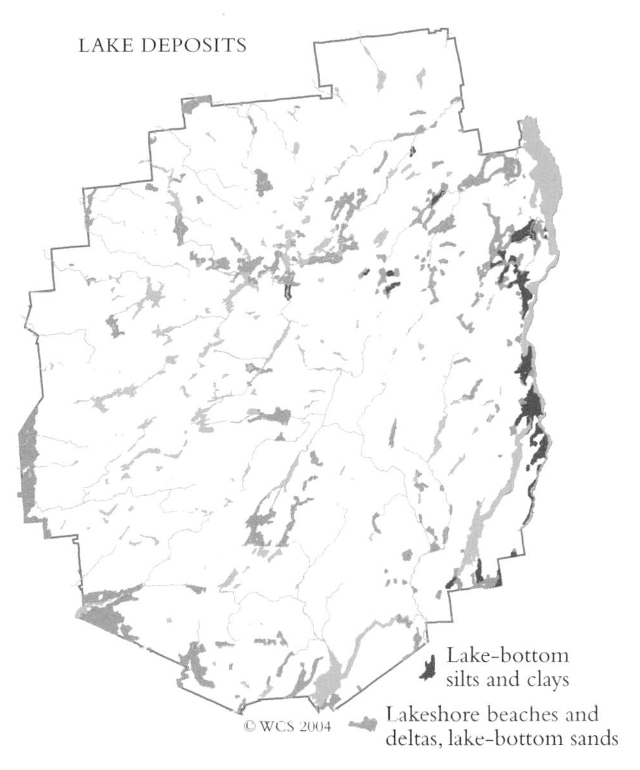

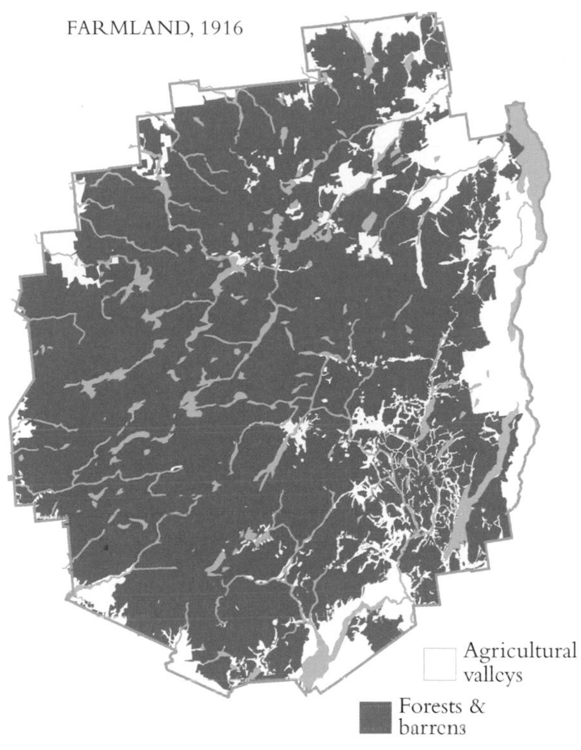

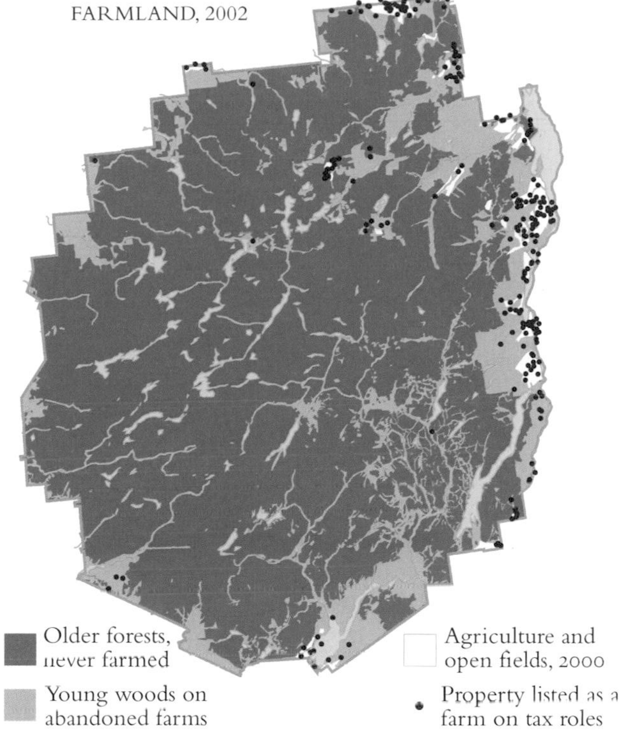

The only Adirondack soils that are both soft enough to plow and fertile enough to raise a decent crop after plowing are the lake deposits. The lake bottoms at the edge of the park were by far the best. Almost all Adirondack *high farming*—the sort of intensive agriculture that could recycle nutrients and maintain soil fertility for many generations—was on lake bottoms. The other lake deposits supported mostly *hill farming,* which is really a kind of nutrient-mining using cows and sheep. Hill farming lasted for nearly two hundred years in New England, producing a little wool and cheese and a lot of maple syrup and poetry. In the Adirondacks, perhaps more practical or less Jeffersonian, it was abandoned as soon as alternatives became available.

Serious Adirondack farming was limited to the Champlain, Ausable, Saranac and Hudson Valleys in the east and to a few valleys at the edge of the park elsewhere. All of these valleys were below seventeen hundred feet and usually had at least some lake-bottom or lake-terrace sediments.

There were also many small farms in the Adirondack interior. They were less visible than the more continuous farmlands in the big valleys but were still of great importance in Adirondack history. In the stagecoach days transportation costs were high and every settlement had to produce its own meat, vegetables, grain, and hay. The towns where there was enough flat farmland to do this grew and became year-round settlements. The ones where there wasn't, didn't.

Farming in the Adirondacks, as elsewhere in the Northeast, reached its greatest extent over a hundred years ago and has declined greatly since then. The reasons for the decline are largely economic—western competition, high costs, low prices. For the last hundred years, each generation of farmers has worked more acres and produced more per person, per animal, and per acre than its predecessors but still has not made a living doing it. Several thousand Adirondack farms have gone out of business in this century. The two hundred that remain are limited to the most productive soils at the edges of the park. Even there, on some of the finest land in the Northeast, all, even the biggest, are fighting to survive.

The fourteen major Adirondack rivers run outwards from the center of the Adirondack dome. The center where their watersheds meet is flat, and the drainage pattern there wandering and complex. The watersheds interlock like puzzle pieces, and without a map or a good memory it is hard to know which watershed you are in.

Most Adirondack rivers have stillwaters and swamps near their headwaters and become steeper and rockier at the edges of the dome. None were navigable by large boats, but all could be negotiated, both upstream and downstream, in canoes and bateaux.

The valleys of the eastern rivers are dry and relatively warm. They were important biological and cultural corridors, used by lowland plants and animals after the glaciers, by Indians for at least the last thousand years, and by settlers and tourists for the last two hundred. The valleys of the northern and western rivers are swampy and boreal and were not used as corridors by either plants or people. In the northwest the trees and railroads followed the ridges instead of the valleys; the farmers and road builders took one look and went somewhere else.

For a hundred and forty years, the twelve largest rivers were the major routes through which logs left the park. The driving of loose logs—logs not assembled into rafts—began on the Schroon and Hudson Rivers about 1810. By 1860 almost every river big enough to hold a thirteen-foot log was being driven. Each watershed had a timber-owners association that conducted drives, and every big sawmill was on a drivable river.

River-driving did not last long. By 1910 most of the virgin timber near the rivers had been cut, and the biggest drives were over. Pulpwood was driven down the big rivers for a few years more, but soon it too was gone. The last large pulpwood drive on the Moose was in 1947; the last one on the Hudson, where river driving began, was in 1950.

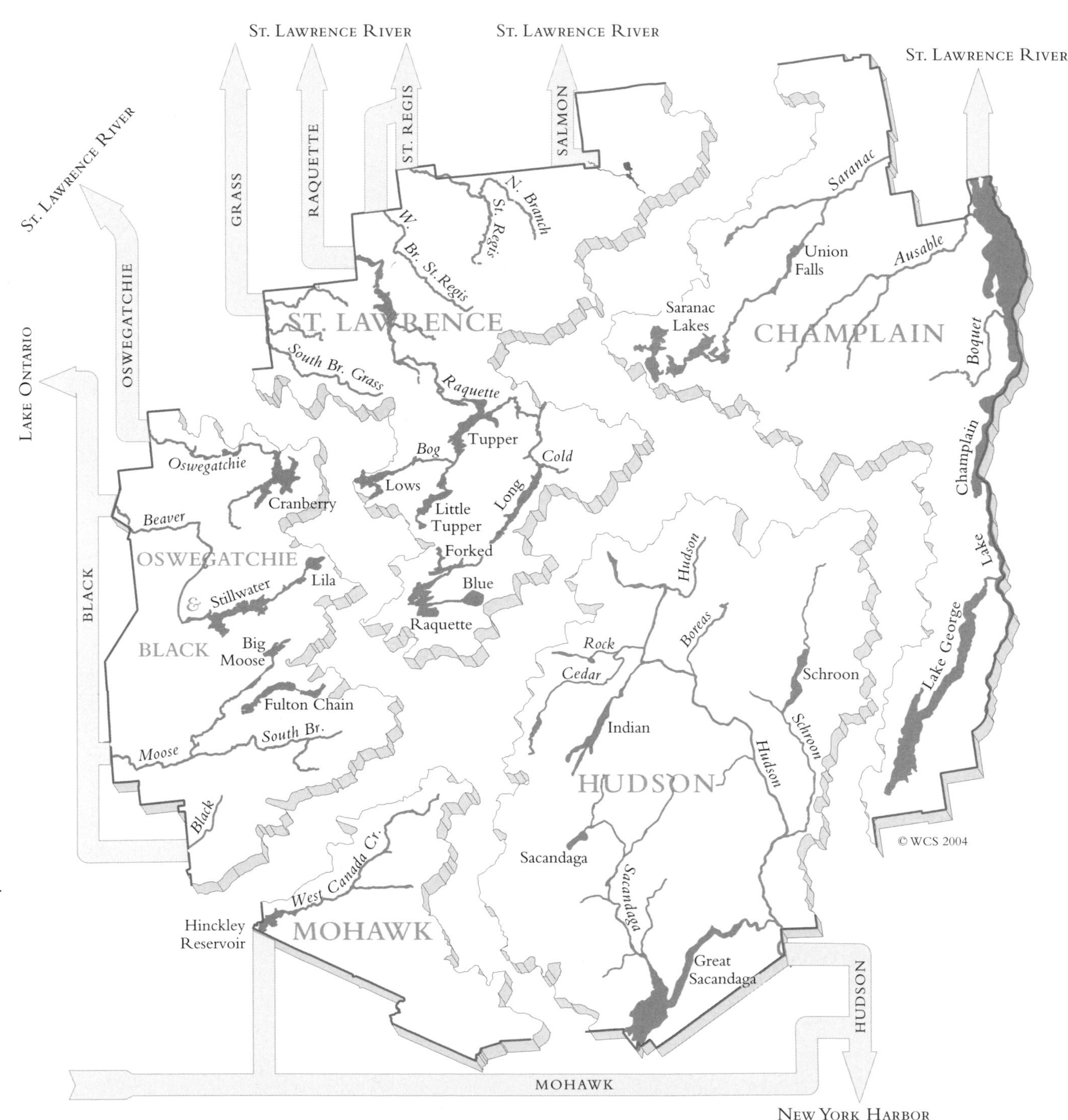

A century ago, impounded water was as essential to transport and manufacturing as fossil fuels are today. New York became a hydraulic economy when the Erie and Champlain Canals opened in the 1820s and remained one for nearly a century and a half. Nineteenth-century New York had twelve hundred miles of canals, thousands of water-powered mills, and a five-hundred mile log-transport system that relied on river drives. Twentieth-century New York moved of ore, oil, and lumber by canal until the 1980s and still generates 16% of its electricity from hydroelectric plants.

The early Adirondacks had much usable water and lost no time in putting it to work. Most Adirondack watersheds were dammed before they were settled. The map shows thirty-three large nineteenth-century dams, most built to supply water for canals and river drives. Hundreds of smaller ones are long gone.

Even with the drives and mills gone, impounded water is still a primary Adirondack commodity. While still used for energy—the map shows twenty-nine extant hydro-electric plants generating 260 megawatts—today it is even more important for powering the tourist economy than for powering light bulbs. Woods and mountains and swamps are nice, but, as every hosteler and storekeeper knows, it is the lakes that drive the tourist economy, and the tourist economy, especially in the small seasonal towns, that drives everything else.

What neither the tourists nor many of their hosts appreciate is how few of these lakes were here a hundred and fifty years ago. Of the fifty-five most-visited and most-settled lakes in the park, only twenty-two are natural or almost natural. The other twenty-three were artificially created or have been substantially enlarged. Dam-building, though often destructive and controversial, is both an honored Adirondack tradition and, judged by popularity and profits, an enormously successful one.

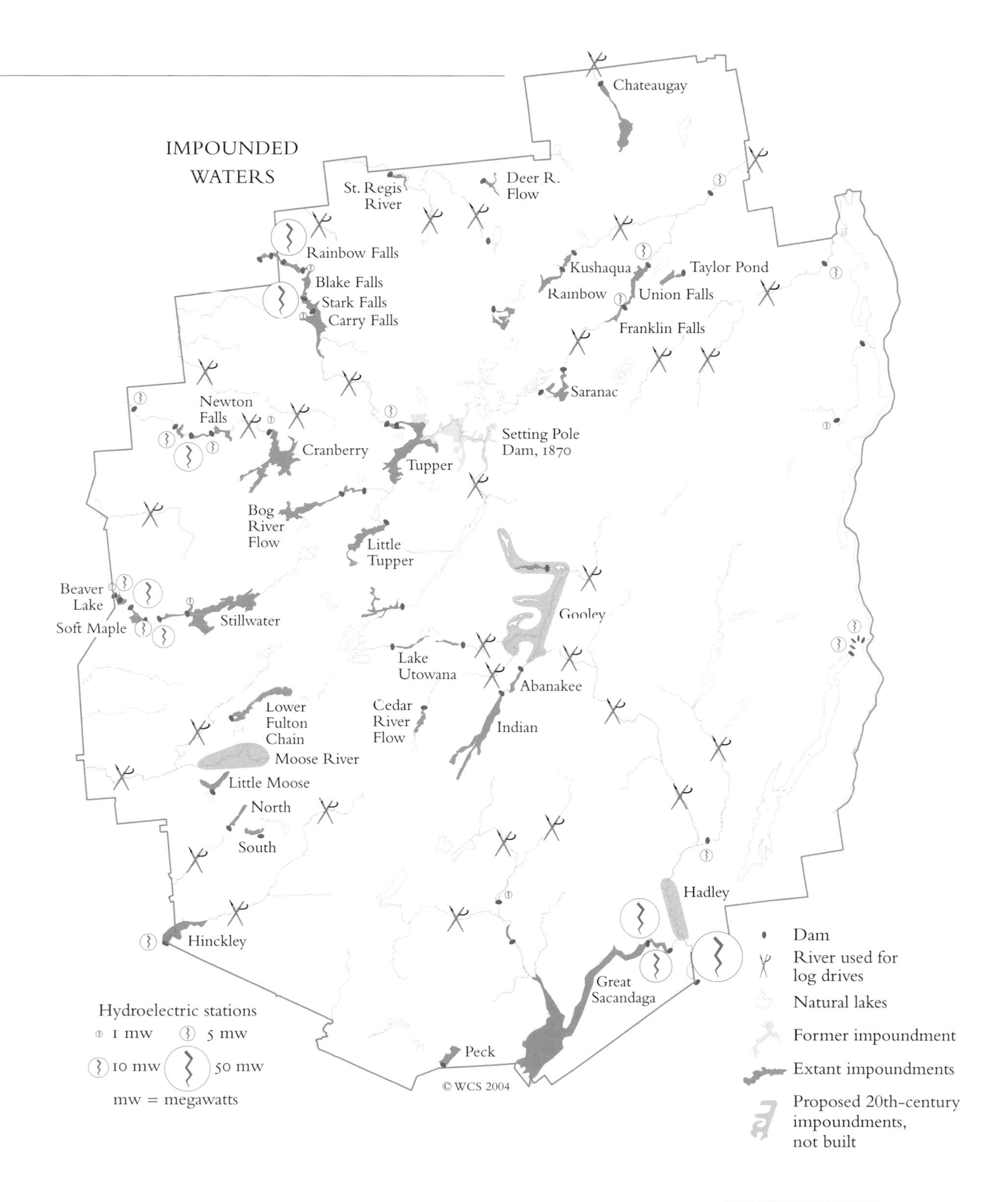

THE Adirondacks are full of wetlands. The largest and in many ways the most unusual are in the flat, outwash-filled valleys of the northern and western rivers. Here many wetlands are more than five miles long and over five hundred acres in size. At least three major types that occur here—*open river corridors, floating bogs,* and *large open bogs*—are thoroughly northern in character, with boreal features like floating mats, nutrient-patterned bog forests, and extensive sphagnum-sedge meadows. These sorts of wetlands are common in Canada and northern Maine but rare this far south.

The boreal big wetlands of the northwestern drainages are, for those who have been fortunate enough to explore them, among the most distinctive and beautiful places in the park. Sadly, because we live in a century of rapidly warming climates, they may prove to be some of the most fragile places as well.

Sphagnum hummocks, dwarf evergreens, delicate sedges

The *sphagnum mosses,* which make hummocks or carpets in many northern wetlands, are keystone species which can raise water tables, tie up nutrients, produce acids and peat, and stunt the growth of other plants. The plant ecology of these wetlands may be seen as a three-way contest between sphagnum, trees, and shrubs and herbs. Where water levels are variable, trees and mosses suffer and shrubs and herbs prevail. Where levels are constant and surface water is fertilized by mineral soil, trees prevail. And where levels are constant but the surface water is separated by a peat layer from mineral soil, sphagnum and its associates prevail.

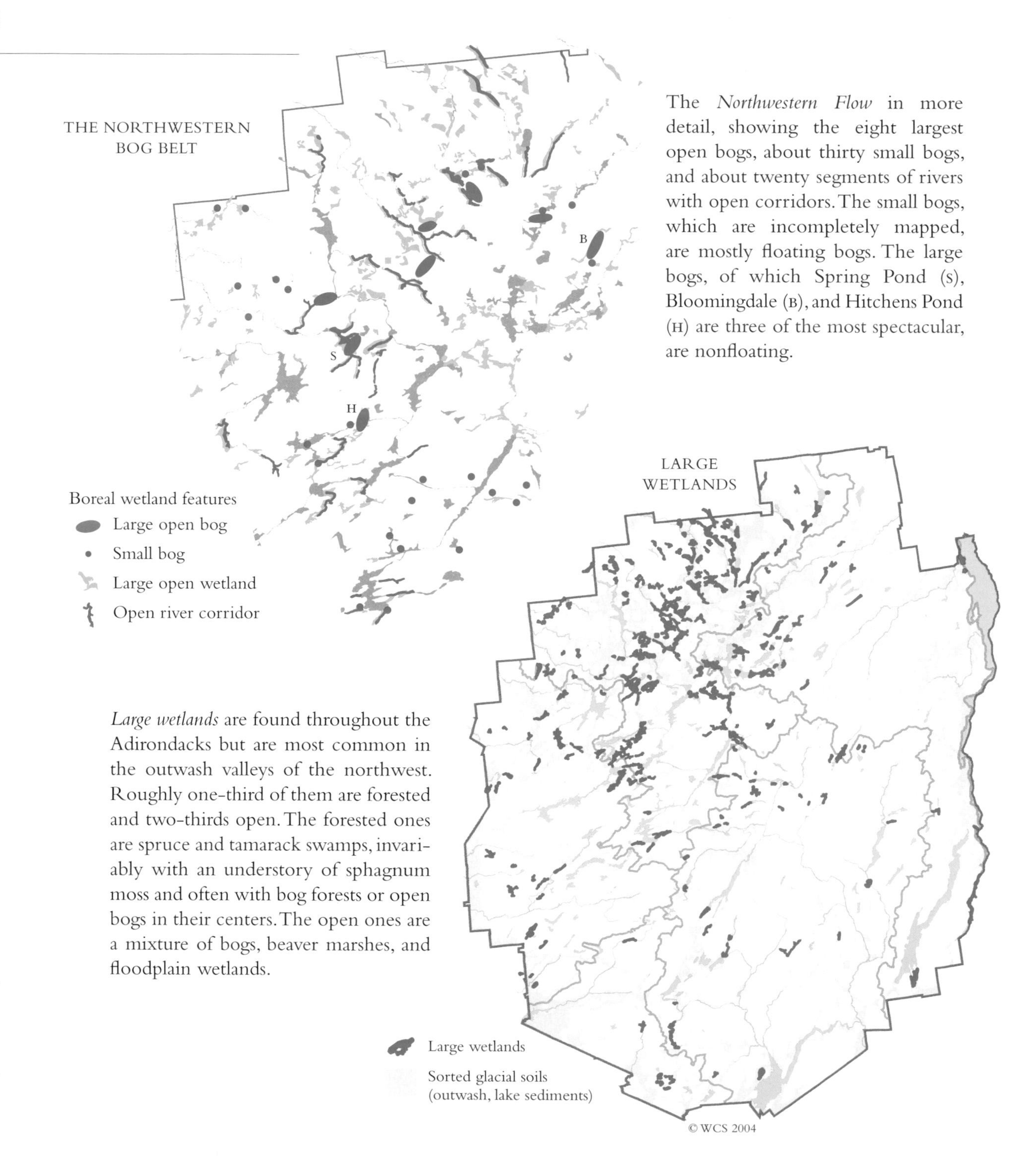

The *Northwestern Flow* in more detail, showing the eight largest open bogs, about thirty small bogs, and about twenty segments of rivers with open corridors. The small bogs, which are incompletely mapped, are mostly floating bogs. The large bogs, of which Spring Pond (S), Bloomingdale (B), and Hitchens Pond (H) are three of the most spectacular, are nonfloating.

Large wetlands are found throughout the Adirondacks but are most common in the outwash valleys of the northwest. Roughly one-third of them are forested and two-thirds open. The forested ones are spruce and tamarack swamps, invariably with an understory of sphagnum moss and often with bog forests or open bogs in their centers. The open ones are a mixture of bogs, beaver marshes, and floodplain wetlands.

Open river floodplains are ubiquitous in the Canadian north and typical of many northern rivers in the park. The Oswegatchie and St. Regis are particularly fine Adirondack examples. The floodplain is kept open by a mixture of channel migration, annual flooding, ice action, beaver activity, and peat formation. Beaver ponds are common, and large floodplain bogs sometimes occur. Some of the best summer canoeing in the Adirondacks is in the open floodplains of the northern and western rivers.

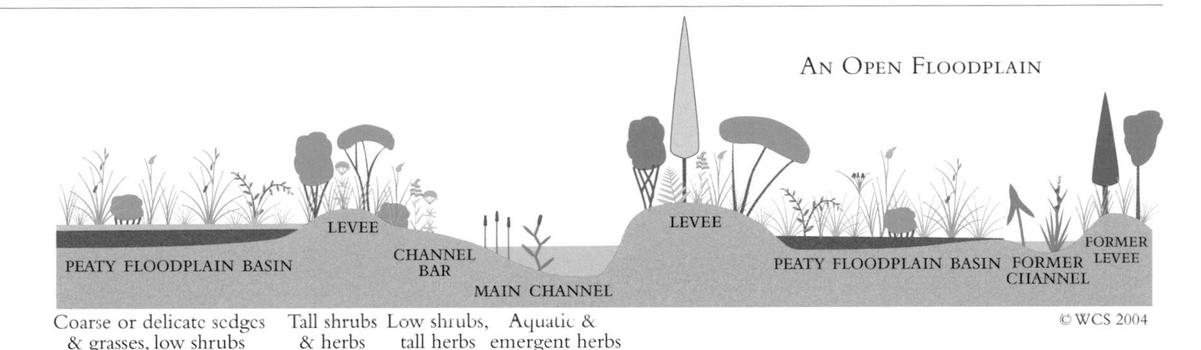

Marginal sphagnum bogs are the one of several kinds of *biogenic shorelines* found in the protected bays of northern lakes and slow swampy streams. The south inlet of Raquette Lake and Quebec Brook below Madewaska Pond have classic Adirondack examples. The mat forming plants, principally leatherleaf and sedges, grow out from the shore. Sphagnum mosses, herbs, and eventually trees colonize the older portions of the mat. Such mats, which can be very large, once occurred on almost every protected shore in the Northwestern Flow. Probably over half of them have been destroyed or altered by dams. Floating mats can accommodate small changes in water level, up to a foot or two, by moving up and down. When the changes are larger, they can be destroyed altogether, or flooded and converted to sedge mats, or torn loose and converted to floating islands.

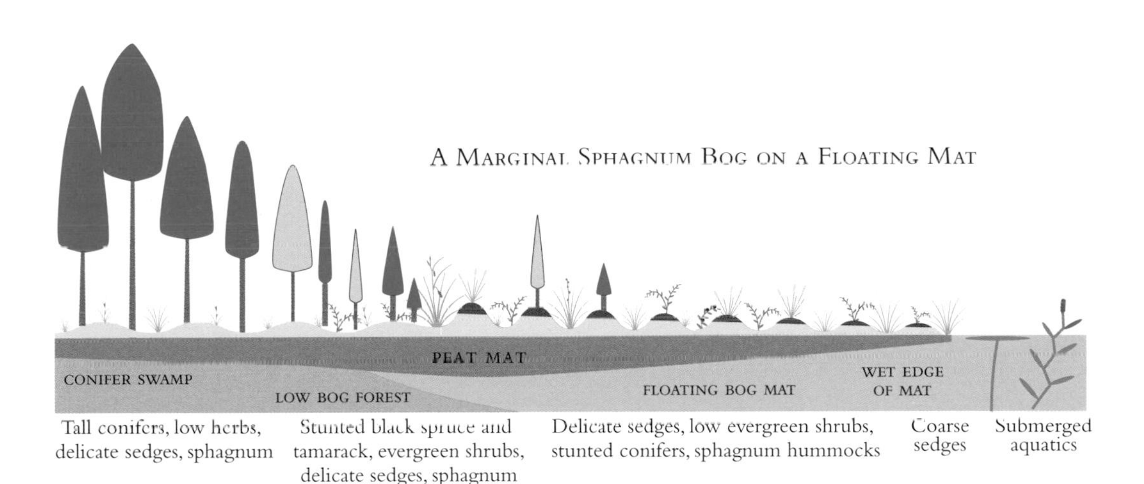

Open, nonfloating shrub-sphagnum bogs, surrounded by bog forest and a conifer swamp are common boreal habitats. From the trees' point of view, the bog is a dysfunctional portion of the forest where sphagnum has locked up so many nutrients that trees can no longer grow. From the sphagnum's point of view it is a place where, with the help of water and peat, something has finally been done about the shade problem. Open bogs are found somewhere in almost every large Adirondack conifer swamp. Most, like the one in the illustration, are small. A few giants in the north and west have taken over a whole wetland basins and are a mile long or more.

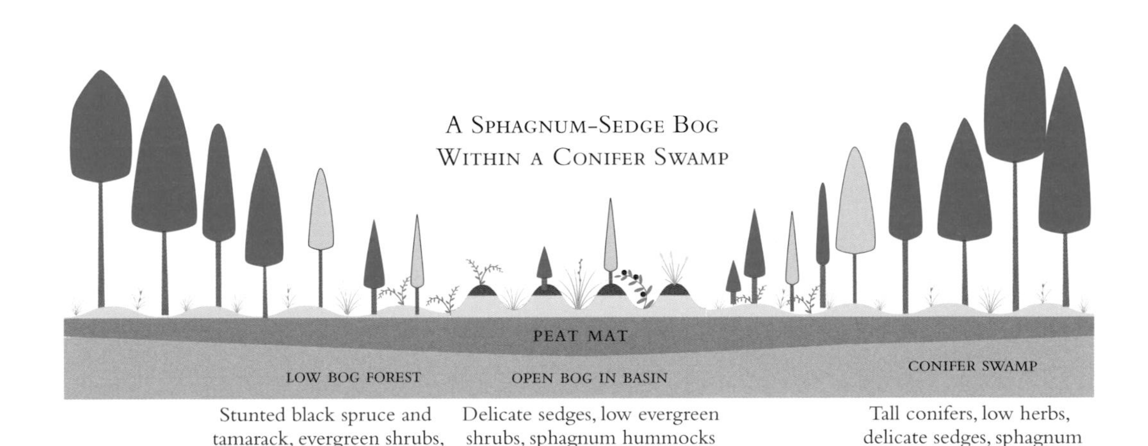

NORTHERN New York has a moist, cold-temperate climate with large daily and seasonal variations in temperature. Compared to the rest of the contiguous U.S., it is wetter than average and as cold or colder than anywhere except the northern Midwest and western mountains.

Average monthly temperatures are between 10 and 20°F in the winter and 70 and 80°F in the summer. The average daily temperature range is about 15°F in the winter and a striking 20–25°F in the summer. Precipitation is relatively constant, between 3 and 4 inches per month, year round.

Within this familiar pattern there are some less familiar details. Summer and fall are wetter then spring. The snowfall curve is asymmetric, rising rapidly in the fall and tapering more gradually in the spring. The temperature curve has a sharp summer peak and flatter winter bottom. As many poets have told us, our winters linger and our summers flee.

The climates of the Adirondack edge and interior are surprisingly similar: people who want to get away from Tupper Lake in the winter don't go to Watertown. The major difference is that the interior is about five degrees cooler in winter or summer and has a one-month to two-month shorter growing season. Corn needs a long growing season and almost cannot grow in the interior valleys; there are big silos in the Champlain and Black River Valleys but none within the Adirondacks.

There is also a strong east-west difference. The west gets moisture from Lake Ontario and is much wetter, especially in fall and early winter. The east is much dryer and receives its peak rainfall, mostly from local thunderstorms, in mid-summer.

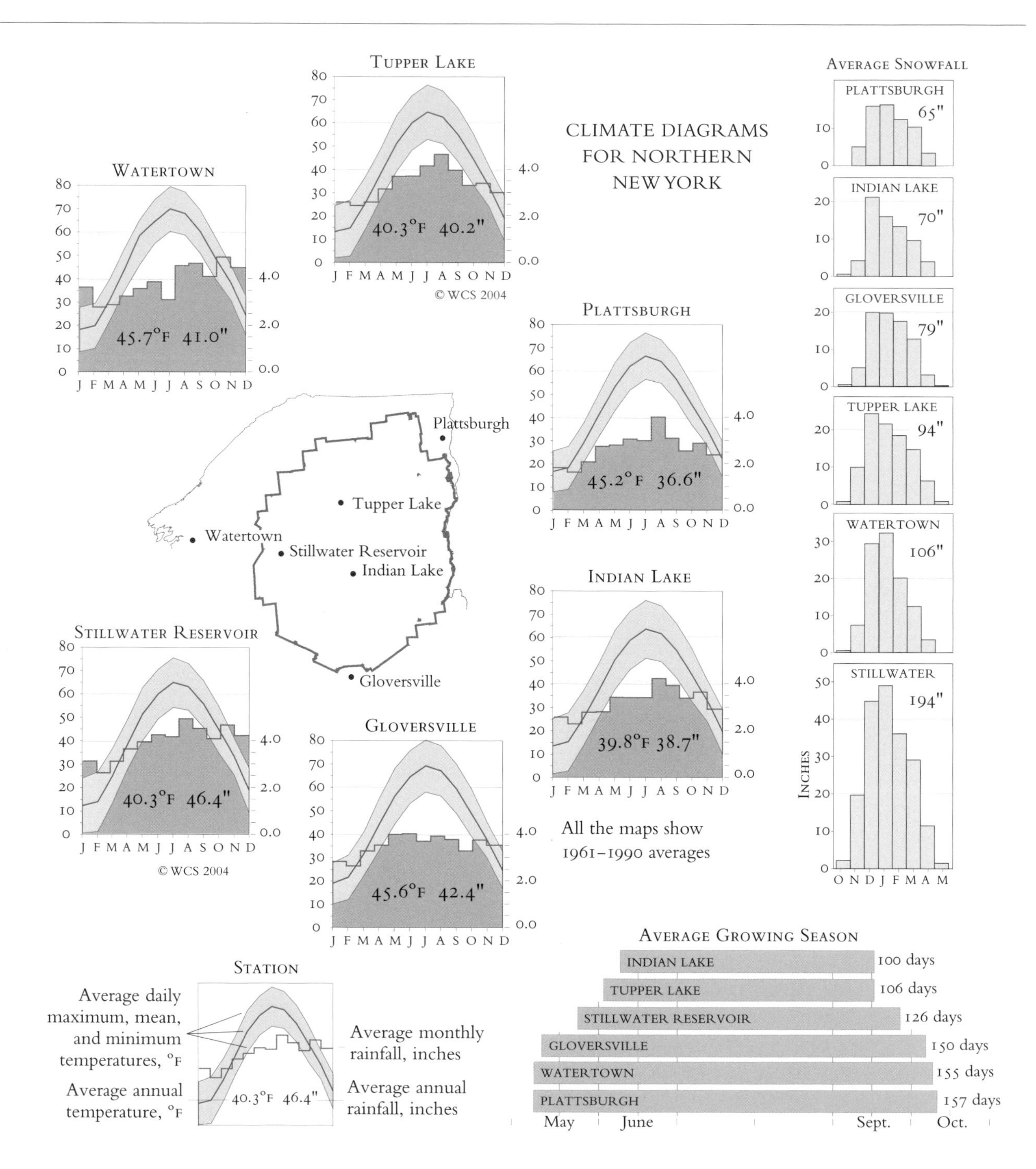

Our understanding of the global climate and what we are doing to it is now detailed and well established. The most important conclusions are that:

- The world climate is very sensitive to atmospheric concentrations of CO_2 and other greenhouse gases.
- The concentration of CO_2, which has been below 300 ppm for the last 160,000 years, is now 370 ppm and increasing about 1% per year.
- Almost all of the recent increase in CO_2 is from the burning of fossil fuels.
- Over the last hundred years the Northern Hemisphere has warmed faster than at any time in the last thousand years. A substantial amount of this warming has been caused by the increases in CO_2.
- Computer models of global climate, while still uncertain, are now good enough to reproduce many details of past climates and so are probably capable of predicting many of the important features of future climates.

The computers' predictions for the Northeast, shown in the graphs, are dramatic. Current CO_2 concentrations are about 30% above their pre-industrial levels. If current fossil fuel use continues, they will reach twice their pre-industrial levels sometime in the next century and not stop there. This will cause a rapidly accelerating warming. By the end of the century temperatures are predicted to be 6–13°F warmer than today, giving the Adirondacks a climate like New Jersey or Virginia. This much warming will not destroy the Adirondacks—Virginia and New Jersey are habitable and have trees and birds—but it will change them and the lives we live in them greatly.

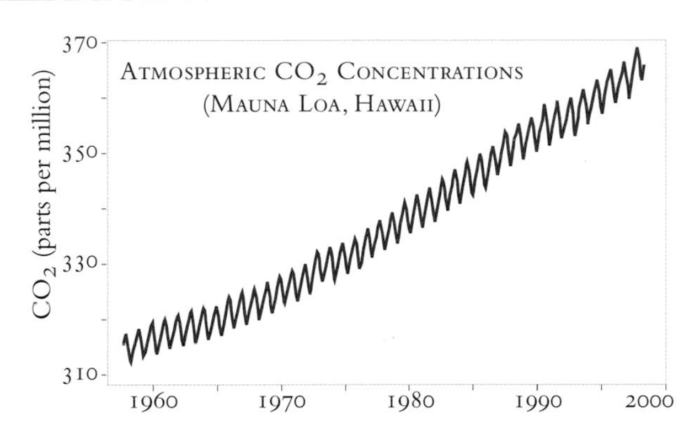

Atmospheric CO_2 concentrations, which control global temperatures, have risen by 90 ppm in the last 150 years and at their present rate will rise another 200 ppm in the next 60 years. The fluctuation in CO_2 in the last ice age was less than 100 ppm.

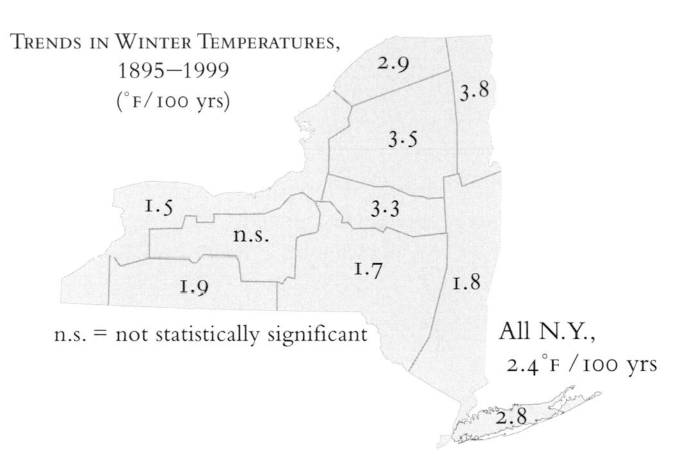

Over the last century, winter temperatures have risen significantly, and snowfall and snow cover have decreased. Summer temperatures have only changed slightly (p. 244).

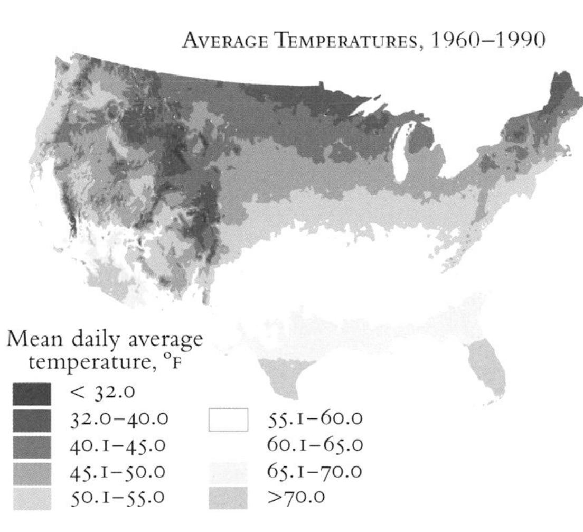

Average temperatures vary with latitude; the temperature gradient is steepest and hence the color bands narrowest in the north. New Jersey is about 10°F warmer than the Adirondacks; the Carolinas about 20°F warmer.

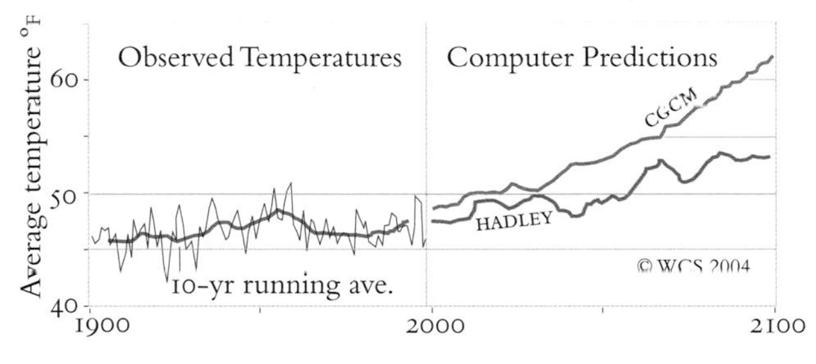

The future temperatures predicted by the Canadian Global Coupled Model (CGCM) and the Hadley Center Model, both assuming a continued 1% per year increase in CO_2, show increases of between 6°F and 13°F in the next century. The sudden increase is a consequence of the rapid increase in CO_2 shown in the upper left diagram. A warming of 10°F would be as much warmer than the present climate as the present climate is warmer than the ice ages.

THE CLASSIFICATION OF FOREST PRESERVE LANDS
Wildernesses are areas, ideally over 10,000 acres in size, where the "imprint of man's work [is] substantially unnoticeable," and which have "outstanding opportunities for solitude or a primitive and unconfined type of recreation." Motorized recreation is forbidden in wilderness areas, and many structures have been removed from them. But wildernesses commonly contain trails, shelters, former roads and railroads, impoundments, and artificially maintained fish populations.
Wild forests are areas that lack "the sense of remoteness" required for a wilderness and where the "resources permit a somewhat higher degree of human use." Almost all wild forests allow some kinds of motorized use—commonly snowmobiles, much less commonly four-wheel-drive vehicles, float planes or ATVS—in restricted areas. None allow unrestricted mechanized use.
Classification of state lands
Wilderness, 1,089,000 ac
Primitive, 45,000 ac
Wild forest, 1,289,000 ac
Intensive use, 19,000 ac
Administrative, 1,500 ac
Historic, 500 ac
Water, 374,000 ac
Unclassified, 34,000 ac
Administrative areas include offices, garages, shops, barracks, labs, and other state facilities.
Primitive forests are similar to and managed like wildernesses but generally have inholdings, structures, or roads that decrease their wilderness character.
Historic areas are "managed for historic objectives" and may have buildings, signage, rest rooms, etc.
Intensive use Areas include forty-three state campgrounds and three state-operated ski areas.
Debar Mt.
Whitehill
Taylor Pond
St. Regis Canoe Area
Whiteface Mt. Ski Area
Raquette Boreal
McKenzie Mt.
Saranac Lakes
Sentinel Range
Jay Mt.
Hurricane Mt.
Split Rock
Mt. Van Hoevenberg Olympic Facility
Cranberry Lake
Horseshoe Lake
High Peaks
Giant Mt.
Aldrich Pond
Five Ponds
Dix Mt.
Crown Point
Lake Lila
Whitney
Hammond Pond
Pepperbox
Blue Mt.
Vanderwhacker Mt.
Hoffman Notch
Sargent Ponds
Pigeon Lake
Hudson Gorge
Pharaoh Lake
Independence River
Blue Ridge
Ha-Da-Ron-Da
Moose River Plains
Gore Mountain Ski Area
Siamese Ponds
Lake
West Canada Lake
George
Jessup River
Black River
Wilcox Lake
Ferris Lake
Silver Lake
Shaker Mt.
© 2004 WCS

3 The Adirondack Park

THE Adirondack Park is a 5.9-million-acre tract, surrounded by a legislatively designated boundary called the Blue Line and regulated by two comprehensive land-use plans, one for private lands and one for public lands. It currently contains about 3 million acres of private lands, 2.5 million acres of public lands, and 0.4 million acres of water, of which roughly 0.3 million acres are public and 0.1 million acres private.

The state lands in the park belong to the New York State Forest Preserve. The Forest Preserve was created in 1885 from lands that the state had bought at tax sales. It was made permanent in 1894 by a constitutional amendment that prevented any sale or removal of timber and required that Forest Preserve lands be "forever kept as wild forest lands."

The lands that made up the original Forest Preserve had been acquired haphazardly and in fairly small pieces (*Map 6-3*). The early lumbermen would often buy, log, and abandon a property without paying any taxes on it. The abandoned property reverted to the town, which then, when it could find a buyer, sold it to recover the unpaid taxes. Often however, when the best timber was gone, there were no buyers. In that case the state stepped in, bought the land that no one wanted, and added it to the Forest Preserve.

At first the state only held these lands until somebody else offered to buy them. In effect the original Forest Preserve was a kind of revolving land bank into which the timber owners placed lands of no immediate value and from which they bought lands or stole timber when the trees regrew.

The result was a destructive cycle of cutting and abandonment that rapidly depleted the resource base (*Map 6-2*). The creation of the Forest Preserve in 1884, the creation of the Adirondack Park in 1892, and the constitutional protection of the Forest Preserve in 1894 were all attempts to stop this cycle. The first two were meritorious failures; the third did the trick.

The creators of the park had differing visions of its future. Some felt that the Forest Preserve was only a temporary necessity and that once the forests had recovered they could be returned to commercial use. Others felt that it should be a permanent wilderness and that the state should acquire, by purchase if possible or condemnation if necessary, all the private land in the park.

Neither vision prevailed. The state never created a systematic acquisition program or took large amounts of land by condemnation. But neither did any of the numerous attempts to harvest or dispose of the Forest Preserve ever prevail. Both the park and Forest Preserve have continued to grow, though neither to the extent that their proponents once imagined nor their critics once feared.

The result of this growth is an unusual park, created piecemeal over a hundred and thirty by purchases from willing sellers and containing towns and industries as well as wildernesses. In both the mixture of public and private lands and the gradual and relatively benign way that it was created it is almost unique in the U.S. In the level of protection afforded to it by the "forever wild" clause of the state constitution, it may be unique in the world.

CLASSIFICATION OF ADIRONDACK PARK LANDS

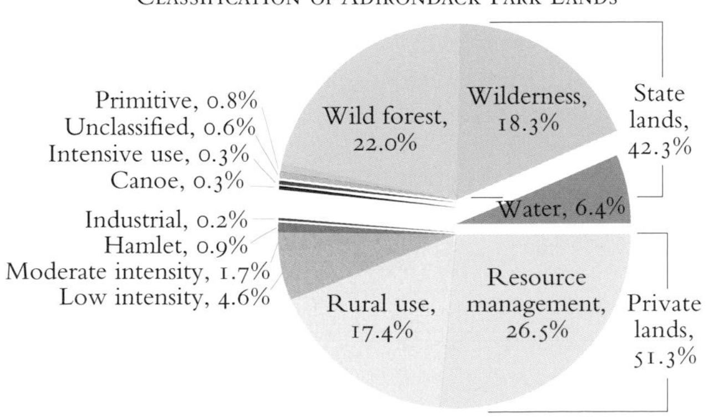

GROWTH OF THE ADIRONDACK PARK

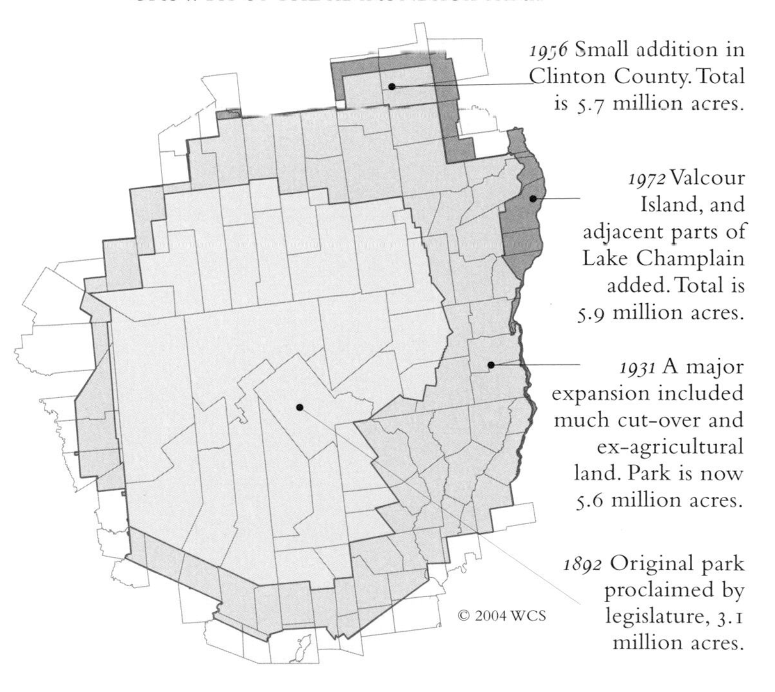

THE Adirondack Park and the Adirondack Forest Preserve overlap but are not identical.

The Forest Preserve is older and smaller than the park and includes some land outside it. It was originally created as an economic resource—a land bank that would assure continued supplies of water and timber. Only later, and much to the surprise of the interests that created it, did it become a dedicated wilderness in which no timber could be cut.

The park, created seven years after the Forest Preserve, was originally conceived as an area of public virgin or near-virgin forests in which there would be little economic activity or private land. Like the Forest Preserve, it rapidly left its original conception behind. Today it is twice its original size and still includes, as it did when it was created, a mixture of forest, agricultural, industrial, and developed land.

There was considerable ecological logic in the growth of the park. The original Blue Line of 1892 included most of the remaining virgin forest but excluded the heavily used industrial and agricultural lands to the east. By the mid-twentieth century many of the industrial lands were reforested and were included in the 1912 and 1931 expansions of the park.

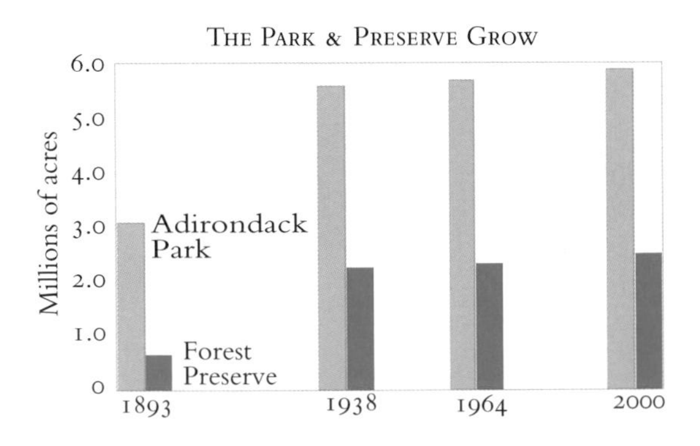

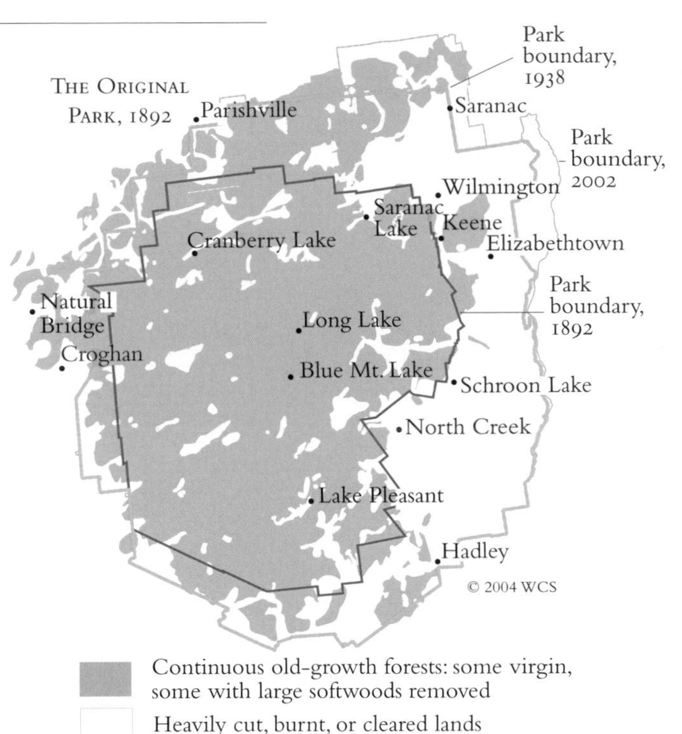

For the first seventy years, the lands within the park were managed no differently than any other state lands. The Forest Commission and its successor, the Department of Environmental Conservation, cut trails on the public lands, fought fires, rescued hikers, licensed sportsmen, and enforced the game laws. The towns managed development on private lands through zoning ordinances when they had them and let it go largely unmanaged when they didn't. No one dealt with large-scale planning, or endangered species, or wetland protection, or the question of what uses were compatible or incompatible with wilderness.

In 1971 everything changed. The state created the Adirondack Park Agency and charged it with creating and enforcing two park-wide master plans, one for private lands and one for the Forest Preserve. The Agency worked quickly, creating a State Land Master Plan in 1972 and a Land Use and Development Plan in 1973. The State Land Classification (p. 24) was the first real attempt, in the eighty years since the Forest Preserve had been declared to be "forever wild," to ensure that some parts of it stayed that way. It placed nearly a million acres into fifteen wilderness areas and one canoe area, prohibited motorized recreation in these areas, and required the closing of existing roads and the removal of structures which conflicted with the wildness of these areas. The Land Use Plan (*Map 3-2*) divided the private lands into six classes, largely based on existing uses, and specified the density of development that would be allowed in each class and the kinds of projects that would require regulatory reviews.

The two plans were, for better or worse, far ahead of their time. In 1971 there were only a few designated wildernesses of any kind in the U.S. and no areas comparable to the Adirondack Park in which development was regulated by a master plan. Needless to say they were also highly controversial—praised as visionary by some, condemned as elitist and undemocratic by others.

Thirty years later they remain controversial, but the controversy has shifted considerably. Few people today doubt that some sort of master planning is necessary, or that the plans have in fact accomplished many of their original goals. The present debate is over the adequacy of the goals and the details of the plans: how new lands should be classified; how much, if at all, the Forest Preserve should expand; and, most important of all, whether the master plans, which are now thirty years old, provide the protection that the park will need in a new century.

1868 The state begins buying lands that towns have confiscated for unpaid taxes at tax sales. In the next twenty years it acquires over half a million acres this way.

1873 The Commission on State Parks says that the protection of Adirondack forests from wanton destruction is "absolutely and immediately required" and recommends the creation of a park of 1.7 million acres.

1883 The legislature forbids the resale of the lands that have been acquired in tax sales.

1884–85 A state commission recommends and the legislature then establishes a Forest Preserve and a Forest Commission to oversee it. The Forest Preserve lands are to be "forever kept as wild forest lands," but, in a conflict of interest as old as government, the timber on them may be sold to support the Forest Commission.

1892 The legislature creates an Adirondack Park of 2.8 million acres, defined by a blue line on a map within which the state was to acquire lands and create a large, continuous park.

1893 The islands in Lake George are added to the park.

1894 The legislature, alarmed by heavy cutting and continuing timber thefts from Forest Preserve lands, amends the state constitution to prohibit all timber sales on the Forest Preserve. This is a surprise and threat to the Forest Commission, which depends on timber sales for its funding. The Commission lobbies for many years to repeal the amendment.

1900 The Town of Webb is added to the park.

1912 About a million acres of land are added to the east and north edges of the park.

1931 1.5 million acres are added, including the Champlain shore north to Crown Point, bringing the park to 5.6 million acres.

1956 100,000 acres, including twenty-four mountains in Clinton and Franklin Counties, are added.

1973 234,000 acres are added, including the Champlain shore from Crown Point to Valcour, bringing the park to 5,927,600 acres.

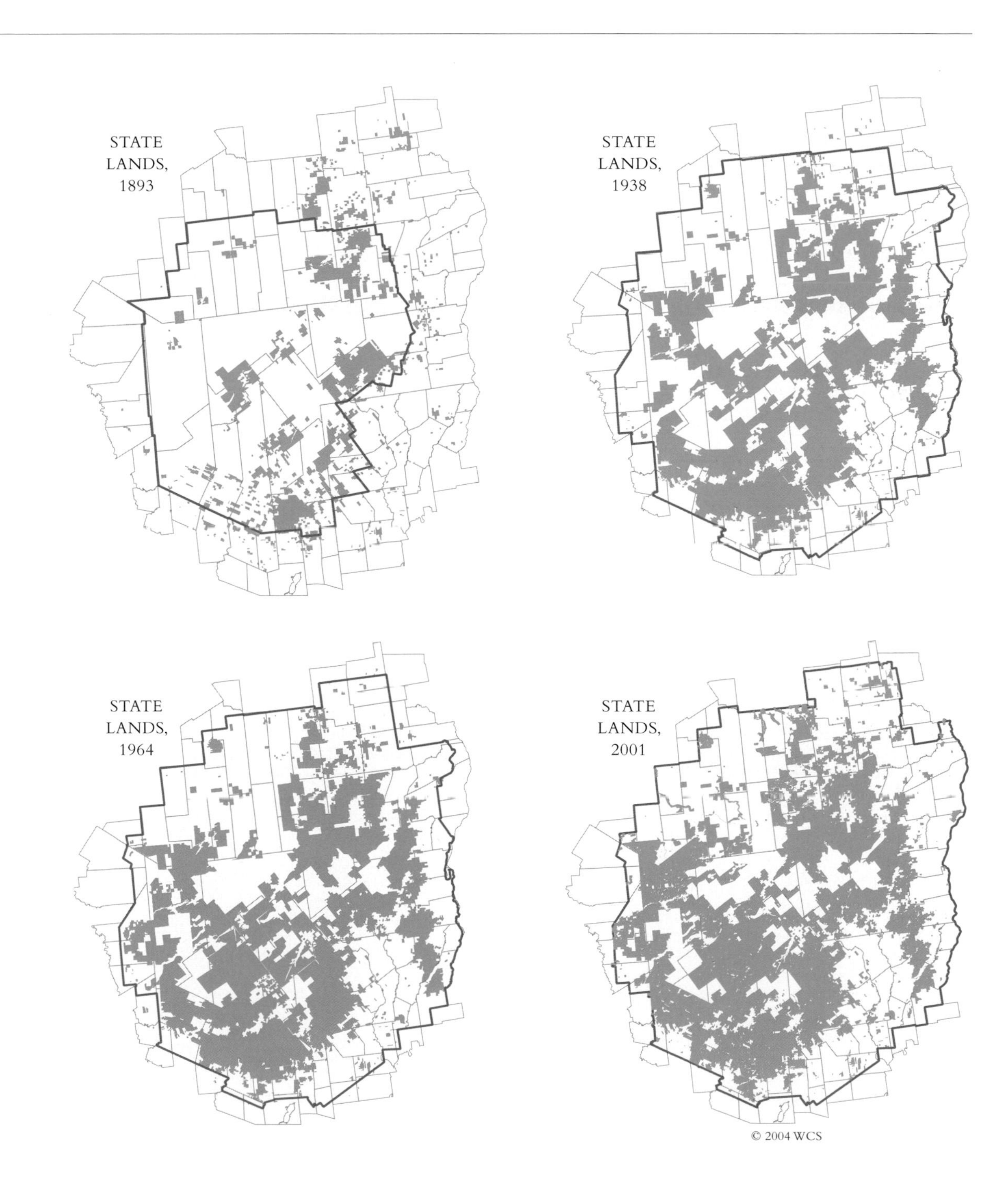

THE goal of the *Land Use and Development Plan* that the New York State legislature approved in 1973 was radical then and remains so today: it was intended to channel new development into already developed areas and thus preserve the wildness of the remaining private lands. It did this through two tools: a *land classification* that specified allowable density limits for new construction, and a series of guidelines that determined when a development or other project was sufficiently big to require an individual review.

The 1973 plan, which has withstood a number of legal challenges, in effect creates a park-wide zoning code administered by the Park Agency. It was intended to regulate and slow but not to prevent development. Under its provisions almost any piece of private land in the park can be developed, as long as the density standards are met and the Park Agency determines that the project is appropriate to the site and does not violate environmental standards or the wetland-protection or river-protection laws.

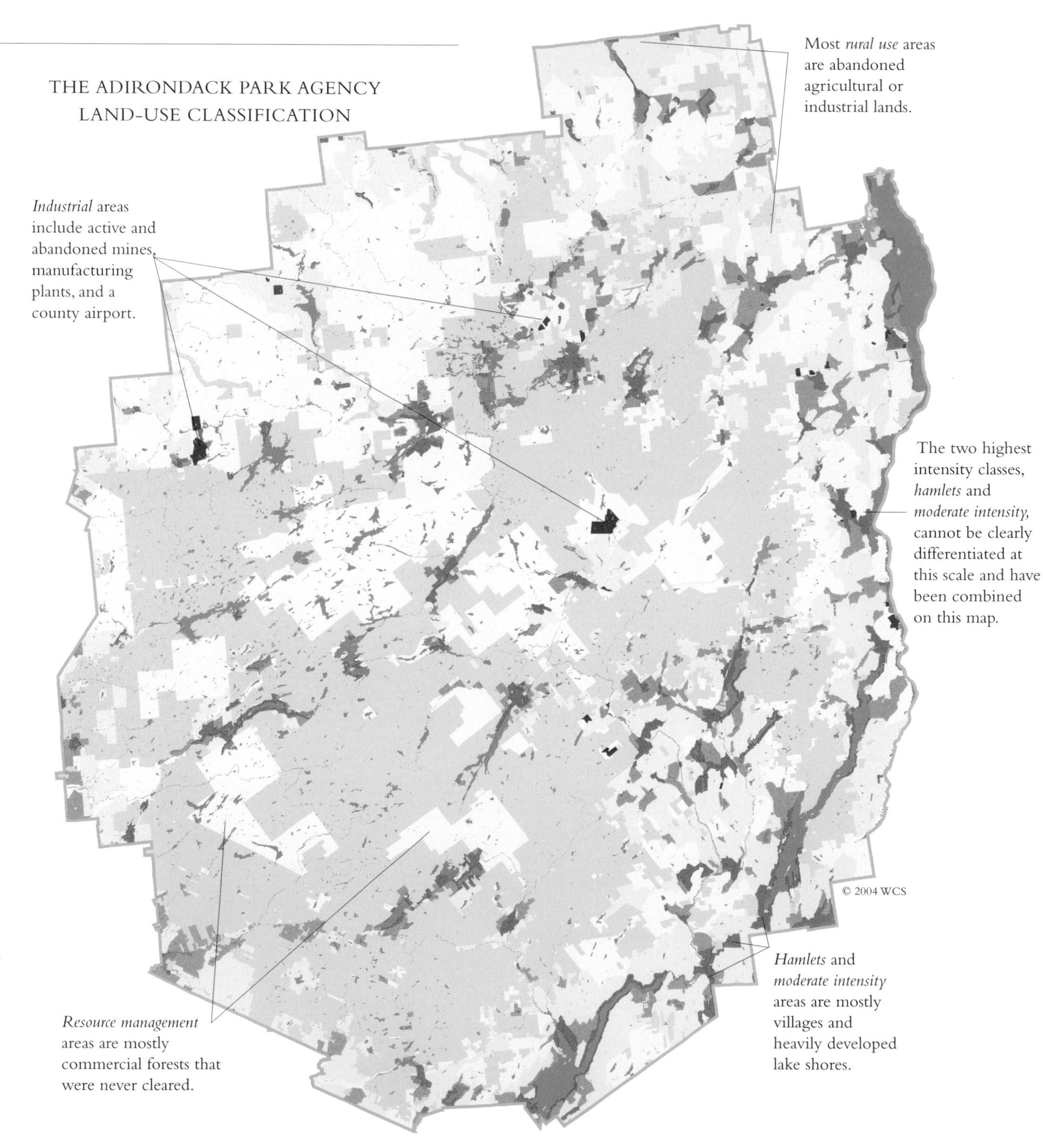

Classification of private lands

- *Industrial* sites, no limit on density
- *Hamlets*, no limit on density, and *Moderate intensity*, 500 buildings per square mile
- *Low intensity*, 200 buildings per square mile
- *Rural use*, 75 buildings per square mile
- *Resource management*, 15 buildings per square mile

Public lands

- New York State Forest Preserve and other state lands

The *Hamlet* and *Moderate intensity* zones are too close together to distinguish at this scale and are mapped as one color.

THE map on the opposite page shows a slightly simplified version of the current *Adirondack Land Use and Development Plan*. The plan concentrates development on private lake shores, in the town and village centers, and in the old farm and industrial lands of the Adirondack periphery. The remaining lands are mostly woodlands. The public and private woodlands interlock in a complicated way, an inevitable but not necessarily undesirable consequence of the decision to obtain public lands only from willing sellers.

The land-use classification distinguishes sharply between the "resource-management" lands which may be logged or developed at a low density, and the state lands which cannot. But in fact these lands are often quite similar. While the resource-management lands are currently used for timber, most of the state lands were used for timber in the past. While the state lands will have no development, the resource management lands will have at most restricted development.

The map to the right has been made by combining the state lands with the resource-management lands to show which forests will remain undeveloped or sparsely developed. This is not a map of wilderness because it does not distinguish between logged and unlogged lands. But it is a picture of how an animal who avoids open or densely developed lands might see the park. Note how little development there is (or will be) in the whole west half of the park, and how a truly shy animal, or for that matter a shy woodswoman or woodsman, could reach almost every portion of the park without crossing developed land. Northern Maine is the only other place in the lower forty-eight where the woods are this large and this continuous.

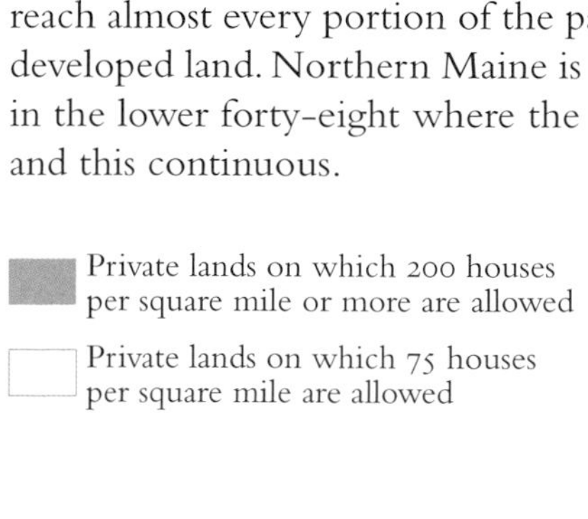

Private lands on which 200 houses per square mile or more are allowed

Private lands on which 75 houses per square mile are allowed

State adminstrative and intensive use land, mostly developed

State Forest Preserve lands, which may not be developed or logged, and private timberlands on which logging and development of 15 houses per square mile are allowed

A hundred years ago the Adirondacks contained just two types of land: constitutionally protected public lands and private lands that could be used or abused as their owners desired. Forestry was the most common use of private lands and was, with some justice, often seen as more an abuse than a use.

Much has changed. Many industrial ownerships are gone. Timber investment groups, almost unknown twenty years ago, now own more than 250,000 acres of land. Housing development has replaced forestry as the conservationist's worst nightmare, and conservation groups now routinely partner with timber investors and paper companies to prevent the development of commercial forests. It is indeed a different century.

Further, roughly a third of the 1.3 million acres held by the thirty largest Adirondack landowners is protected by easements—legal agreements which limit or prevent development. Almost none of these easements existed twenty years ago. Now they are ubiquitous. In the last ten years almost every sale of a large block of land has included a development easement.

The result is a new sort of map, unexpected twenty years ago and barely visible ten years ago, in which many large tracts of private land are undevelopable or quasi-public or both.

There remain some large tracts of land without easements, especially in the center of the park. But if present trends continue, as it seems likely they may, there may eventually be easements on these lands as well.

If this happens, the founders' original goal of a broadly protected Adirondack forest may yet be achieved. But if so it will be achieved in very different way than the founders imagined and will include a mix of ownerships and activities that they could not have imagined and might well not have approved.

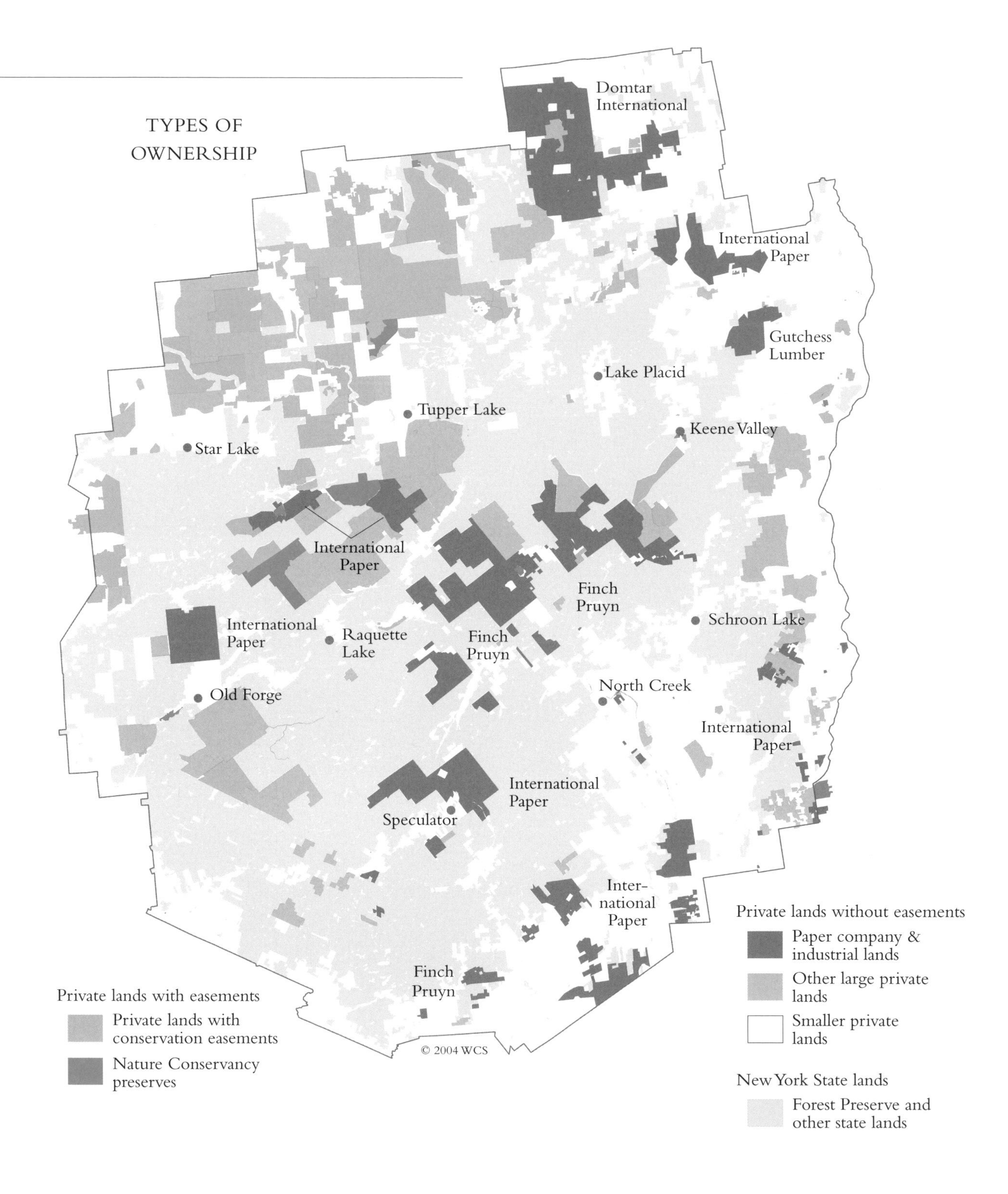

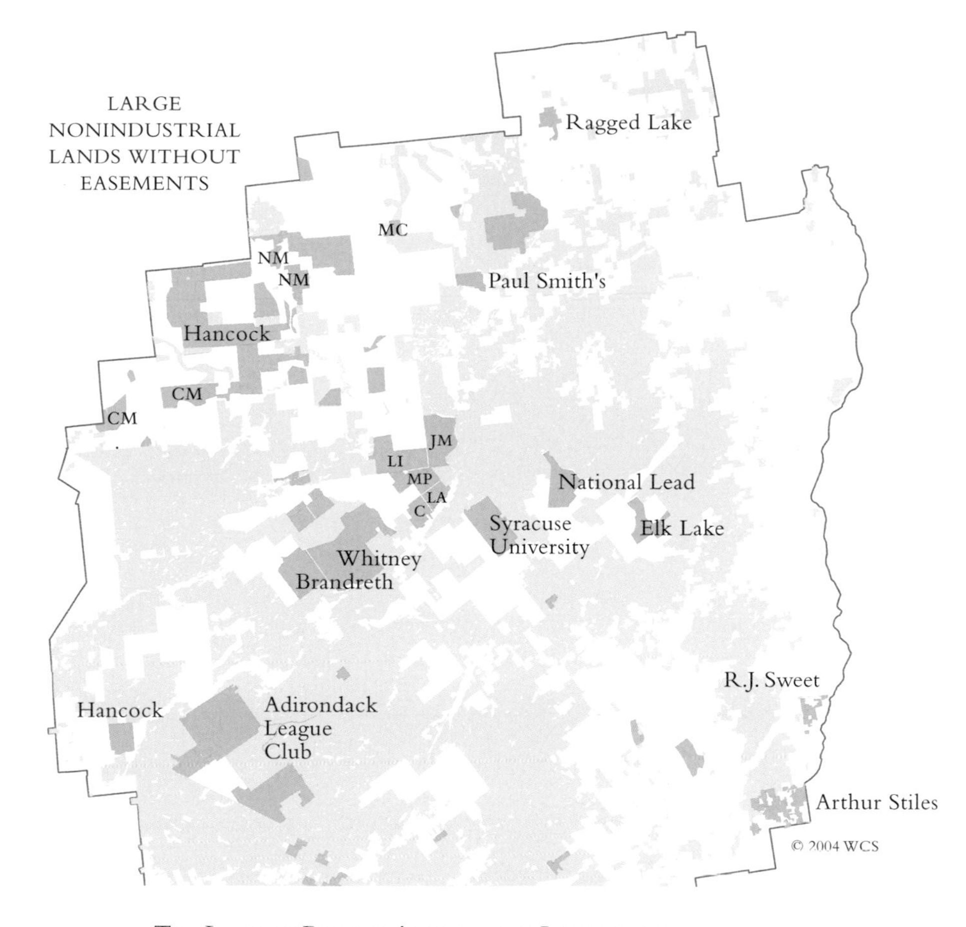

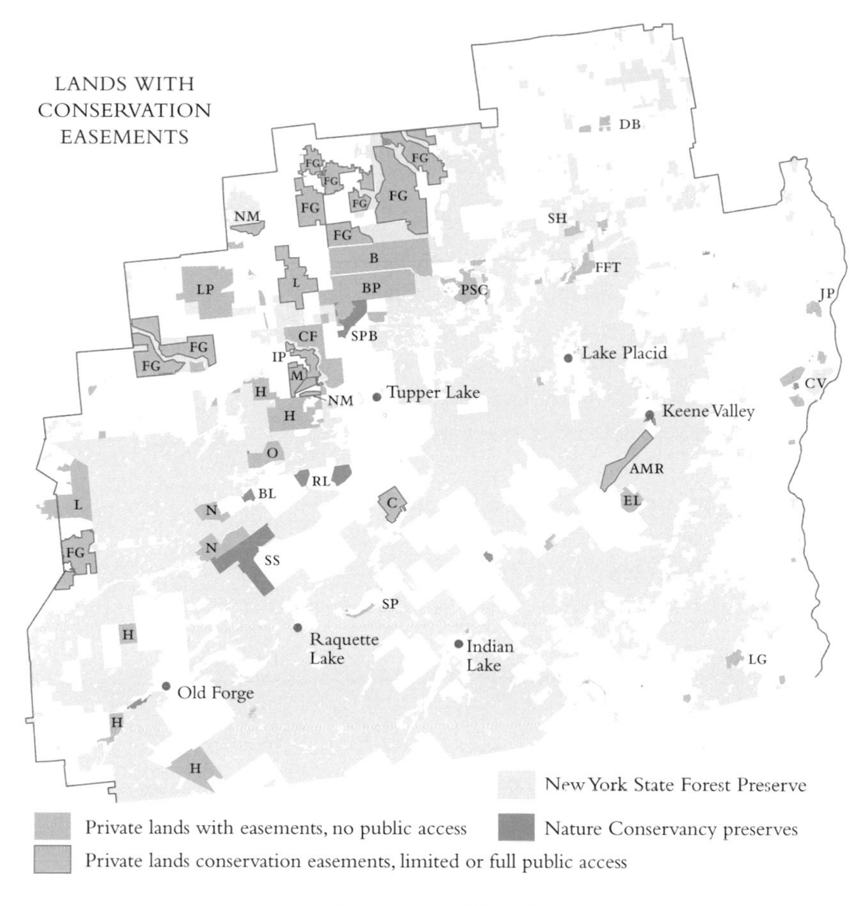

The Largest Private Adirondack Landholders

International Paper (1%)	327	Paul Smith's College (PSC, 56%)	14
Finch Pruyn (0%)	165	William Rockefeller WR, 100%)	13
Forestland Group (FG, 100%)	110	Conservation Fund (CF, 100%)	13
John Hancock Timber (H, 37%)	103	Syracuse University (0%)	12
Domtar (0%)	102	Litchfield Park (LI, 0%)	11
Lassiter Industries (L, 56%)	86	National Lead (0%)	11
Adirondack League Club (ALC 0%)	47	Gutchess Lumber (0%)	11
Whitney Industries (0%)	36	R.J. Sweet Inc. (RJS, 0%)	10
Clerical Medical Forestry (CM, 0%)	36	McCavanaugh Pond Club (MC, 0%)	10
Wilhelmina Ross Trust (RO, 100%)	27	Nature Conservancy (100%)	9
Brandreth Associates (0%)	21	Adirondack Mt. Reserve (AMR, 100%)	9
Long Pond (LP, 100%)	19	Bay Pond Condominiums (BP, 100%)	8
Elk Lake Land (EL, 20%)	17	Upstate Colony (C, 0%)	8
Niagara Mohawk (NM, 0%)	16	Langley Park (LA, 0%)	7
John McCormick (JM, 0%)	15	Moose Pond Investors (MP, 0%)	7

The columns give the size of the holding, in thousands of acres; the figure in parentheses, the percentage of the land on which development is restricted by conservation easements.

Other Properties With Easements

B	Brandon	FG	Forestland Group	O	Otter Brook
BL	Bog Lake (TNC)	H	Hancock Timber Invest.	PSC	Paul Smith's College
BP	Bay Pond	JP	Jones Point	RL	Round Lake (TNC)
C	Cedarlands	L	Lassiter	SH	Stone, Hames, & Donald
CF	Conservation Fund	LG	Lake George	SP	Sargent Pond
CV	Champlain Valley	LP	Long Pond	SPB	Spring Pond Bog (TNC)
DB	Debar Mountain	M	Massawepie	SS	Shingle Shanty (TNC)
EL	Elk Lake Land	N	Nehasane	WP	Windfall Pond
FFT	Franklin Falls Timber	NM	Niagara Mohawk		

The large Adirondack landholdings include *industrial lands*, held by paper and lumber companies; *commercial timberlands*, held mostly by timber investment groups; *recreational lands*, held by clubs and families; and *institutional lands,* held by a variety of nonprofit organizations. The ownership of the commercial and industrial lands has been changing rapidly. In 2001 Champion International sold all its Adirondack lands to the state and the Forestlands Group. International Paper and Whitney Industries have both sold parts of their lands, and in 2003 Domtar offered all its Adirondack lands for sale.

AMERICANS, though historically distrustful of governments, have created over 80,000 of them. At least 180 of these have jurisdiction over some part of the Adirondacks. Besides the state and federal governments, there are 12 counties, 92 towns, 13 incorporated villages, and 61 school boards. In addition, several state agencies have broad rule-making and permitting powers and are very much part of the government of the park.

The federal government, while barely visible within the park, is extremely important here. It sets air quality standards, regulates hydroelectric plants and transmission lines, protects wetlands, rivers, and endangered species, and funds highways, schools, municipalities, and much environmental research. It maintains a large army base at Fort Drum and a small federal prison at Ray Brook. During the Second World War it built the railroad to the titanium mines at Tahawus, and during the Cold War it operated Air Force bases at Rome and Plattsburgh and Atlas missile silos within the park (*Maps 5-13, 14*).

New York State is the most important government in the park as a whole. The elected branches make grants to schools and towns, pay property taxes to the towns with Forest Preserve lands, build and maintain the state highways, fund new acquisitions of state land, and approve the Land Use and Development Plans. It is their decisions that ultimately control what the park contains and, through highway construction, how people move about the park and what they see.

The state agencies (which are, effectively, unelected branches of government) have less decision-making power but are more intimately involved with the park. The Department of Environmental Conservation (DEC) and the Adirondack Park Agency (APA) are the most directly involved with the day-to-day management of the park; the Departments of State, Corrections, and Transportation work at a larger scale but also have important responsibilities within the park.

DEC ADMINISTRATIVE REGIONS

© 2004 WCS

The DEC is large, statewide, and versatile: it builds trails and shelters, runs campgrounds, issues hunting and fishing licenses, manages game, inventories endangered species, issues discharge permits, regulates waste disposal and recycling, monitors acid rain, fights fires, conducts search-and-rescue operations, and maintains public safety and order through two uniformed services, the Forest Rangers and the Environmental Conservation officers.

The APA, in contrast, is a specialized Adirondack organization. It maintains the maps that implement the Land Use and Development Plans; determines, by investigations and hearings, whether proposed Adirondack projects are consistent with the plans; maps wetlands, rivers, and shores and enforces the laws protecting them; and provides technical environmental and planning assistance to towns and villages.

While the state government is broadly responsible for the park as a whole, it is the counties and towns that take care of the people in it.

The counties handle a series of activities too local for the state and too regional for the towns. They record deeds and diseases, keep vital statistics, maintain highways, operate courts and jails, certify weights and elections, run airports and landfills, and provide a wide range of social and emergency services. They are some of the largest regional employers, hiring judges, bailiffs, coroners, attorneys, probation officers, inspectors, social workers, sheriffs, and many clerks and administrators.

Like those of the state agencies, almost all county employees are appointed rather than elected. You can work for, deal with, or depend on state and county governments for many years without ever meeting an elected official.

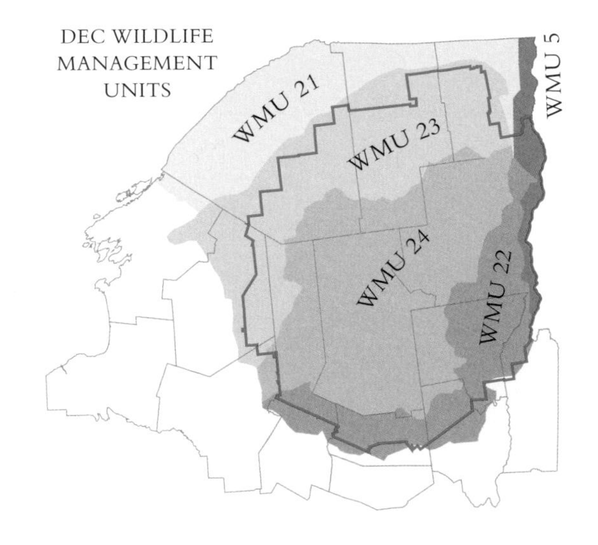

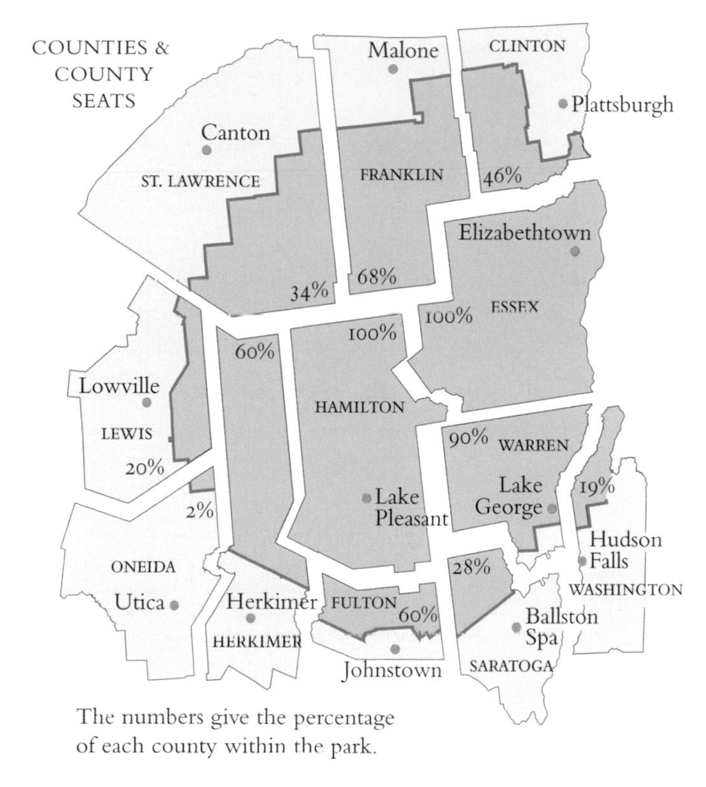

The numbers give the percentage of each county within the park.

The local governments of the park include the 105 Adirondack towns and villages, the 61 school boards, and at least another hundred "special districts" that administer local utilities. These local governments maintain roads, water, and sewer systems; provide police and fire protection; run schools, beaches, transfer stations, ski areas, and information centers; and have plans, ordinances, and codes and the boards necessary to administer them. They also issue building permits for about 800 new structures and 3,500 modifications of existing structures each year. Since less that half of the structures and only a fifth of the modifications are ever reviewed by the Park Agency, this means that the towns are the *de facto* regulators of most new construction in the park.

How well do the different levels of government work together? Jurisdictionally, most of the lines are clear, though not always welcome. The problems arise, as always, where state or national authority preempts local choice.

If, for example, the legislature is considering a new land acquisition or the DEC a proposal to close a lake to float planes to make a new wilderness wilder, the eventual decision will be made at the state level. There will be public notices and hearings, opportunities for public comment, reviews by county officials and the Local Government Review Board, and, of course, lobbying by individuals and organizations. But eventually the state will make the decision. If you are, say, a hiker or conservationist or nonvoting landowner who believes that the Adirondacks belong to the whole state, this may seem appropriate. But if you are a local resident or town official who believes that what is done in the Adirondacks is primarily the business of the people who live in the Adirondacks, your lack of direct political power may be very annoying.

The same is true in the federal domain but with the important difference that the decision will be made farther away and by people who know less about the park, with very little input from the residents or the managers of the park itself. This has been important in the past and could easily be important again. Should wolves arrive from the north or should a developer or a federal agency propose a new dam or pipeline, it will be the federal government that decides whether the wolves will be listed as an endangered species and whether the dam or pipeline will be built. New Yorkers will be consulted but they may or may not be listened to; the final decision, by law, will be Washington's and not theirs.

The spatial overlap of the governments is more confusing than their jurisdictional overlap. State, county, and town borders are nested in a geographically pleasing way. But there the orderliness stops. Village boundaries can cross town lines: the Village of Saranac Lake, for example, is contained within and receives taxes from, but still is legally separate from, the towns of Harrietstown, St. Armand, and North Elba. The park boundary divides towns and counties, making the compilation of park-wide statistics effectively impossible. And worse, a single agency, the DEC, divides its responsibilities for the park into some 17 administrative subregions, 32 ranger districts, and over 120 land management units, few coinciding with town and county boundaries, and none coinciding with each other.

A WILDLIFE SCORECARD

Nearly Extirpated, 1650–1800

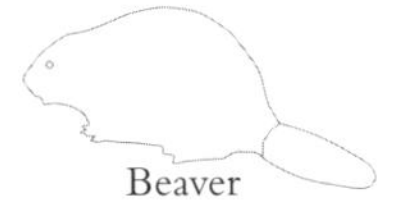
Beaver

Extirpated, 1850–1900

Moose

Cougar

Lynx

Wolf

Greatly Decreased, 1800–1900 Recovered, 1900–2000

Black Bear

White-tailed Deer

[illegible]cat

Otter

Marten

Fisher

Reintroduced Successfully, 1900–1915

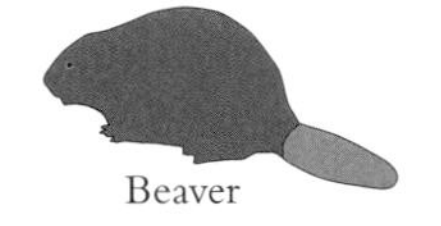
Beaver

Introduced, Didn't Survive, 1910–1990

Moose

Elk

Lynx

Wild Boar

New Arrivals, 1930–2000

Opossum

Coyote

Turkey Vulture

Turkey

Extirpated, Then Reintroduced, 1950–1990

Bald Eagle

Peregrine

Greatly Decreased 1970, Natural Recovery 2000

Osprey

Near Extinction, 2002

Spruce Grouse

Ceased Breeding in Adirondacks, 1985

Golden Eagle

Reintroduced Themselves, 1960–1990

Moose

Cougar

Currently Threatened by Toxic Pollution

Loon

Wildlife vs. People, 1700–2000		
Extinctions 7	New Arrivals 4	Successful Introductions 3
Severe Declines 9	Natural Recoveries 7	Failed Introductions 4
Currently Missing 3	Natural Returns 2	Centuries to Play ?

4 *Animals & Plants*

THE biological geography of the Adirondacks is a large subject, which deserves a book of its own. Here we treat four central themes, illustrating a few of the many possible examples of each.

The first is northerliness. The Adirondacks, which in many ways are like a detached piece of the Canadian Shield, contain more boreal evergreen forest than any of the mountains around them. But the evergreen forests of the Adirondacks, unlike the solid-conifer forests of the Canadian Shield, are surrounded by and interspersed with deciduous ones. What they really are is a temperate-boreal mosaic, equally different from the Catskills and from Canada, with features all their own.

This chapter looks at a number of these features: the forest types and their component trees (*Map 4-1*); the local distribution of these trees, which turns out to be a miniature version of their continental distribution (*Map 4-2*); the way beaver create an almost unclassifiable continuum of vegetation types (*Map 4-6*); the boreal birds associated with these wetlands (*Map 4-3*); and rare habitats and rare plants (*Map 4-13*). If there is an overall message here, it is that while the Adirondacks are clearly northern, they are northern in their own way.

The second theme is decline and recovery. In the past two hundred years, the animals in the Adirondacks have been hunted, trapped, and exposed to toxic pollution. For the last fifty years they have had to cope with habitat changes caused by acid rain and, for at least the last twenty, with ecological changes resulting from a rapidly warming climate. At least sixteen species have either suffered major declines or gone extinct; nine of these have recovered or returned naturally and three were artificially reintroduced. Three are currently missing; another of these is near extinction.

Maps 4-4, 5, 7, and *8* summarize the history of the extinctions and returns. A complex and very interesting picture emerges, which refuses to illustrate the conventional truths of conservation biology. We have animals going extinct before their habitats changed; animals returning to altered habitats and being themselves altered in the process; and uncooperative animals ignoring the corridors and wildernesses we have established for their benefit and making instead for the agricultural edges where the food, the excitement, and the danger is.

The third theme (*Maps 4-10, 11, 14,* and *15*) is the increasing overlap between where people live and where animals live. In the last century, as the center of the park has become older and wilder, both the hunters and many of the animals they hunt seem to have shifted to the edges of the park and the lands just outside of it. This has greatly changed the amount of hunting within the park. Currently black bear and marten are the only animals that are taken more frequently in the interior of the park than at the edges; more otter, deer, and coyotes are now taken outside the park than in.

While the animals and the hunters have been moving outward, settlement has been moving inward. Currently there are some 70,000 residences in the Adirondacks and a year-round human population of 150,000 or more. As a result, Adirondack animals, while less hunted and trapped than in the past, are encountering more humans and roads than they used to and spending more time near human settlements.

All this leads to some important questions. Are some animals like deer, which in the past were controlled by predators and hunters, now controlled by nothing except the automobile and their food supply? If so, how will their populations change now that the mountain lion and coyote, have arrived? What will the arrival of the moose, our biggest herbivore and currently neither hunted not subject to predation, do to the food supplies of other herbivores? Or to the forests? How well will people tolerate suburban moose and bear, not to mention mountain lion? Or, perhaps sometime in the future, wolf? To what extent will the expansion of human animal contacts promote a group of generalist animals—red foxes and crows and coyotes—at the expense of specialists like the bobcat and the marten. And finally, given that accidents, anger, and disease are ubiquitous, to what extent is the increasing human-animal contact dangerous for us both?

The fourth theme, treated nowhere explicitly but underlying many of the maps, is the extent to which the park has changed biologically in the last two centuries, and the surety that there is as much change ahead as behind. The woods of 2002, with moose and beaver and beech disease, are very different from the woods of 1850 and even from those of 1950. The woods of 2050 and 2100 will be equally different from those of today.

ADIRONDACK forests are made from about twenty-five trees. We show the twenty commonest, along with a few of their typical mixtures. The mixture of trees varies gradually from place to place and so forms a forest *continuum* rather than a collection of separate forest types.

The most deciduous and most diverse forests are on fertile soils in the valleys at the edge of the park. As you go higher and onto thinner soils, the forests become less diverse and more coniferous. Middle elevations may be dominated by either conifers or hardwoods, depending how wet and rocky the soils are and where the beaver, which prefer hardwoods, and the moose, which prefer conifers, choose to feed. High elevations are uniformly spruce and fir.

A unique feature of the Adirondacks, shown clearly in the map of forest cover, is the extent to which conifer forests, often in tiny patches or thin lines, are woven into the ubiquitous background of hardwoods. Except in the high mountains, we do not have a distinct, climatically determined, boreal *zone*. Instead we have something quite different, a boreal *mosaic* or *web*, determined by soils and herbivory. In a climate-change century when many species are going to change their boundaries, the difference between these two—the zone which has one boundary and the web which has many—may come to be very important.

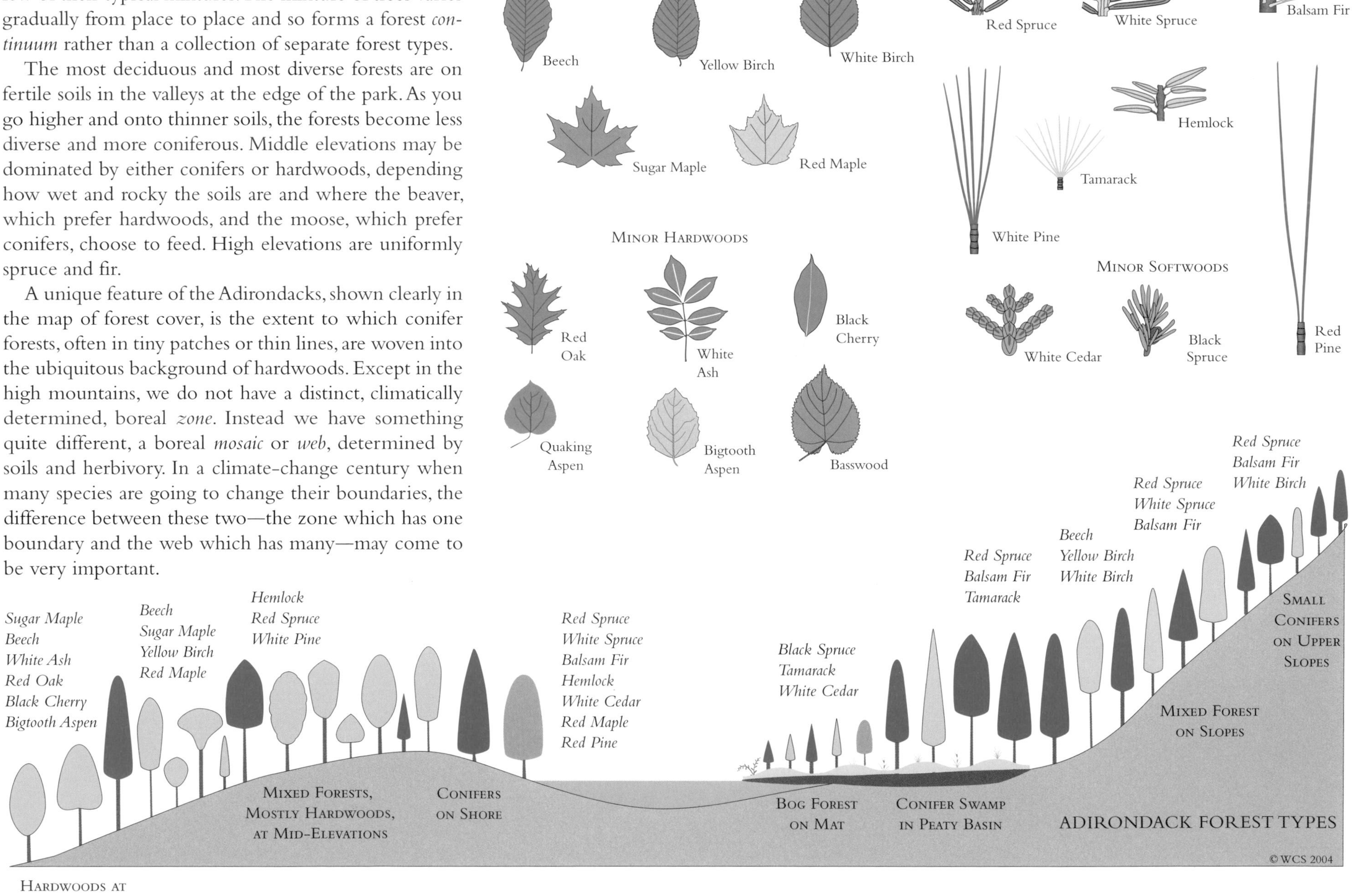

THE FOREST COVER

HERE is an approximate map, made by assigning arbitrary colors to the pixels of a satellite image, showing open lands and four types of forest. Wetlands, which cannot be mapped reliably from satellite images, are not included. Some are mapped as open lands, others as conifer forests.

The park is mostly hardwood forest. Early explorers, perhaps because they traveled along the waterways or perhaps because the heavy logging hadn't begun, reported more conifers. At present the largest conifer areas are on mountain slopes; elsewhere conifers occur in slender lines along drainages and in small patches among hardwoods. Cleared land is mostly around towns or in agricultural and ex-agricultural areas. The curved band of open land northwest of Upper Saranac Lake is the result of the Bay Pond fire of 1934 (F), the last large fire in the park. Small red patches are rocky, paved, or barren areas. Those marked A are rock outcrops in alpine areas, those marked U are urban areas, and those marked M are strip mines.

- Hardwood forests: principally beech, red and sugar maple, white and yellow birch, with some white ash and red oak at lower elevations.
- Hardwood forests mixed with small amounts of white pine, red spruce or hemlock.
- Conifer forests with some maple, birch, or beech.
- Conifer forests: principally hemlock, red spruce, and balsam fir, with scattered white spruce and white pine in uplands and tamarack and black spruce in lowland basins.
- Open vegetated areas: wetlands, agricultural land, highways, developments, and suburbs.
- Barren areas: rock outcrops, strip mines, urban areas.

ADIRONDACK trees come from three main geographic groups. The conifers, birches, and aspens are *transcontinental,* cold-tolerant species, near their southern range limits here. The sugar maple, hemlock, and a number of others are *northeastern* species, tolerant of cold but not of drought, whose distributions center on New England and the Great Lakes. The red oak, black cherry, and several others are *southern hardwoods*, tolerant of drought but not of deep cold, and have their main distributions south of the Great Lakes.

Because the park has a cold center wrapped in a warmer shell, the distributions of trees within it mirror their continental distributions. The transcontinental species are most abundant on the boreal core. The northeastern species can live almost everywhere in the park but are less abundant on the warmest and coldest sites. The southern hardwoods, unable to tolerate the cold and wet in the interior, are only found at the park edge and in their lower valleys.

As a result the park is layered like an onion, southern trees on the outside, northern toward the center. This results in some interesting juxtapositions. By moving five miles up the Hudson River from Blue Ledge you pass from very Appalachian-looking forests of oak and cherry and big-tooth aspen to very northern-looking forests of spruce and birch. By descending a few hundred feet in elevation in the hills along the Oswegatchie River you can leave a typical New England maple-beech forest and find yourself in a Canadian spruce-tamarack swamp. The Adirondacks, an ecologist would say, are a *tension zone* where dissimilar forests exist, side by side. Such zones are diverse and fascinating but, biological history suggests, rarely enduring.

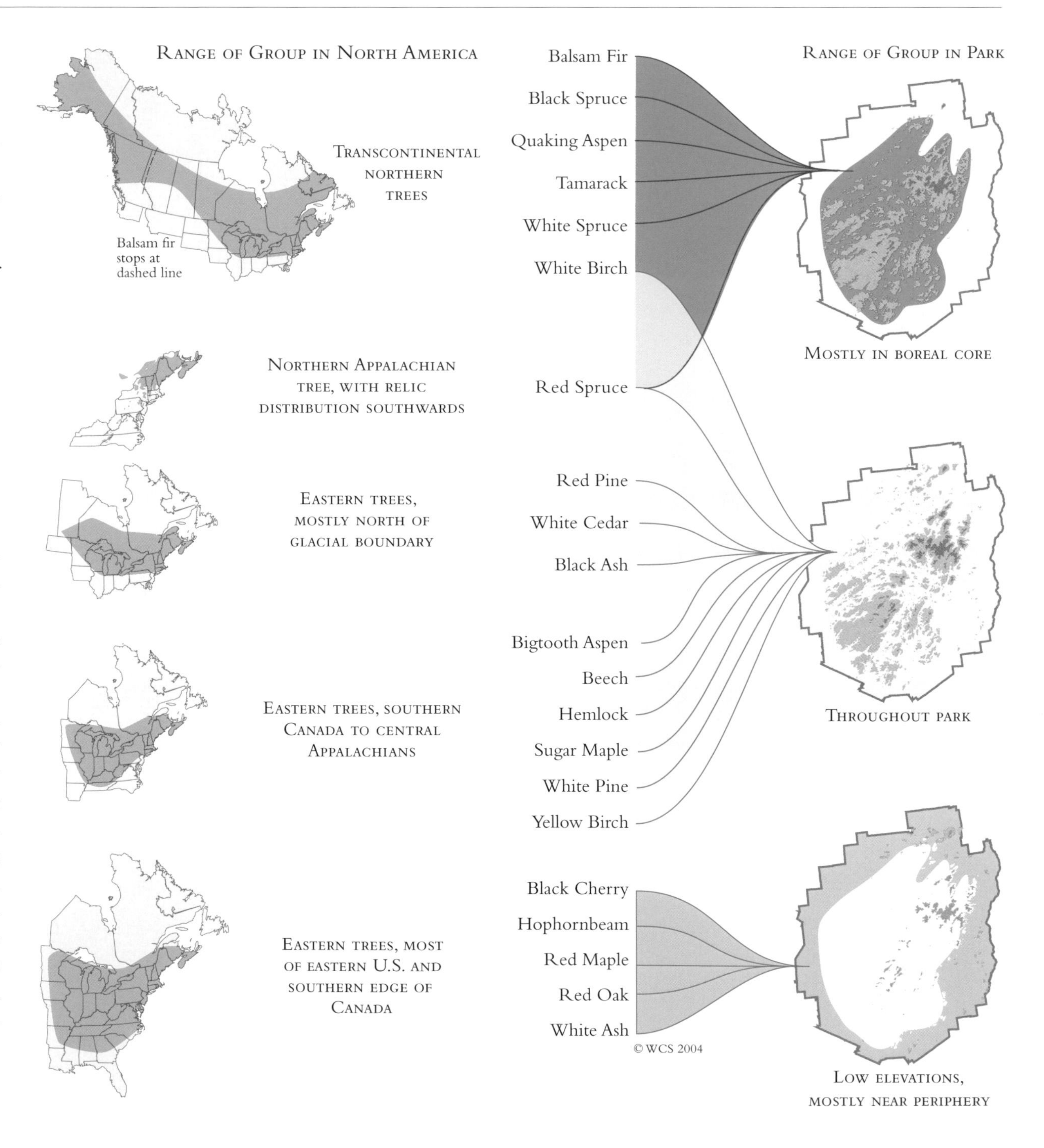

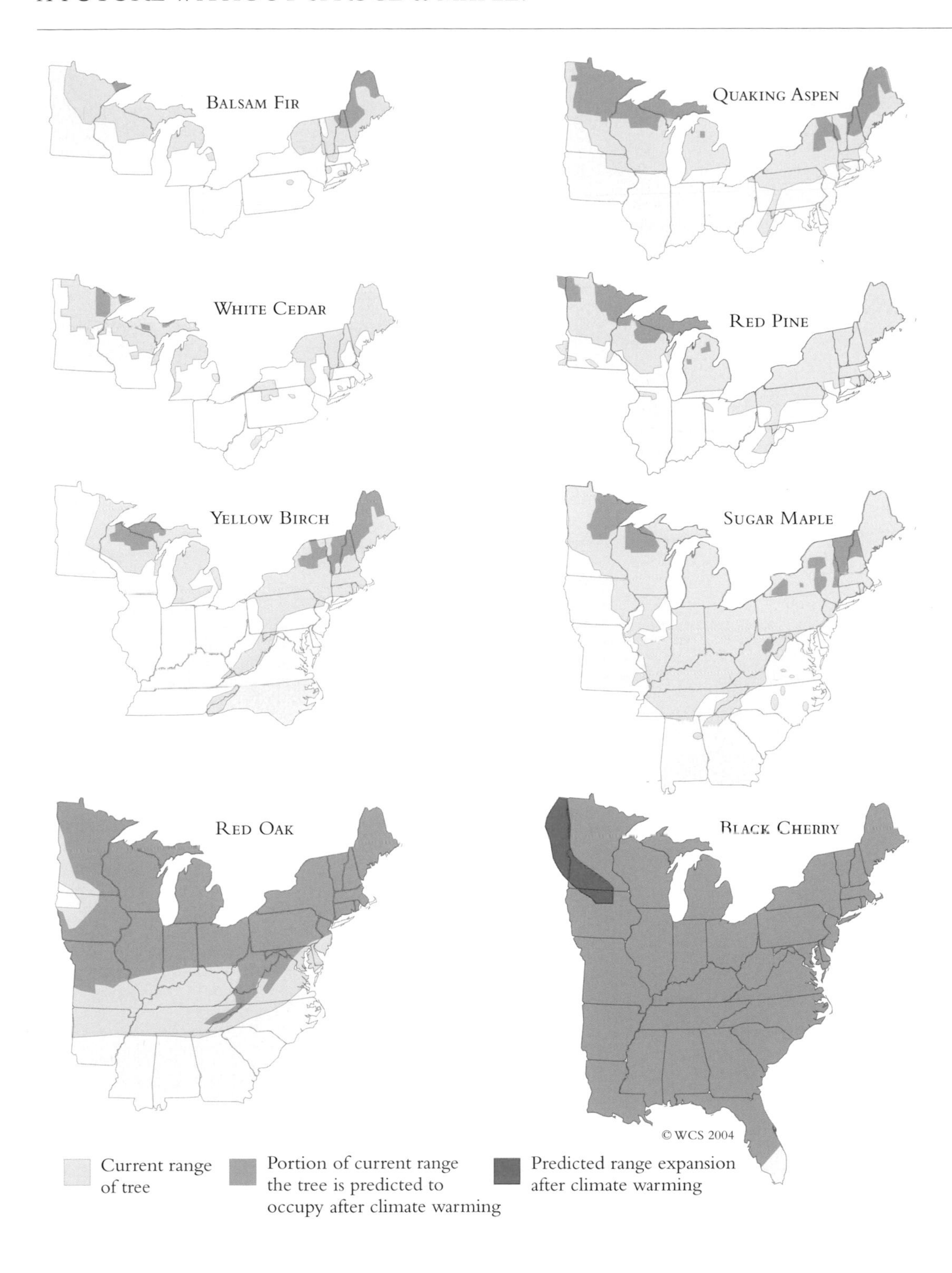

Both history and ecology suggest that the ranges of trees are determined by climate and that when the climate changes the trees will change. But neither predicts how fast or how thorough the change will be.

Currently, world climate is warming faster than at any time in the last thousand years, and computer models predict even greater warmings to come (*Map 2-7*). If these predictions are correct, the climate a century from now will be too warm for any of the characteristic Adirondack trees to prosper. The maps are from a Forest Service research project which used the current relations between trees and climate to predict what may happen to several common Adirondack species if the average temperature warms about 5°F.

The Forest Service models predict that climate change will eliminate the northern trees of the boreal core of the Adirondacks or reduce them to relic populations. We may be the last generation to see spruce-covered mountains or walk through the great spruce-tamarack bogs. The northeastern trees like sugar maple and yellow birch will be limited to the cold core; the southern species like oak and hickory, which are currently restricted to the Adirondack edge, will prosper and spread through the park.

If both the climate and the trees do as predicted, we will see great changes, though not necessarily sudden ones. It seems likely that mature trees will survive for at least a forest generation, and then, as they fail to reproduce, will gradually be replaced by more southern species. And because the landscape is a mosaic, the changes will not be uniform. Some of the conifers may survive longer in cold swamps where none of the southern trees will want to live. And some of the northern hardwoods may migrate to north slopes and ravines where the conifers live at present and, if they can cope with the acidified soils there, may survive indefinitely.

THE boreal forests of the Adirondacks and their associated wetlands contain nineteen species of birds that are near their southern range limits here and do not breed in the adjacent deciduous forests. Five also occur in the Catskills but none in lowland New York. All are birds of the far north, associated with conifers and open wetlands; their presence here is strong evidence of the northerliness of the Adirondack core.

Two, Bicknell's thrush and the blackpoll warbler, are characteristic birds of the *krummholz*—the zone of stunted conifers below the alpine tundra in the high mountains. The others are more commonly found in wetlands, particularly in the transition zone where a band of conifers adjoins an open bog or sedge meadow. They do not require large continuous stands of conifers and do well in the narrow bands and patches of wetland conifers that we have called the boreal mosaic. And indeed, in many ways the Adirondack mixture of open wetlands and thin bands of conifers is very like the open, stunted, subboreal forests and muskegs in central Canada where these birds reach their greatest abundance.

Like the bog plants (*Map 4-13*), the boreal birds are most common in the spruce-dominated wetlands of the northwestern drainages. Every boreal species except Bicknell's thrush occurs in the northwestern drainages. One, the gray jay, occurs almost nowhere else. Almost every five-kilometer survey block in the northwest had one or more boreal species in 1985, and many had six or more.

How long we will have them here is uncertain. Average temperatures are increasing rapidly, and range-limit species like the boreal birds are often particularly sensitive to climate change. Breeding bird surveys suggest that changes are already occurring. Some species, like the rusty blackbird, olive-sided flycatcher, boreal chickadee and Lincoln's sparrow, have recently shown widespread declines on survey routes in both the U.S. and Canada. Others, like the yellow-bellied flycatcher, seem to be increasing. And yet others are increasing in some places and decreasing in others.

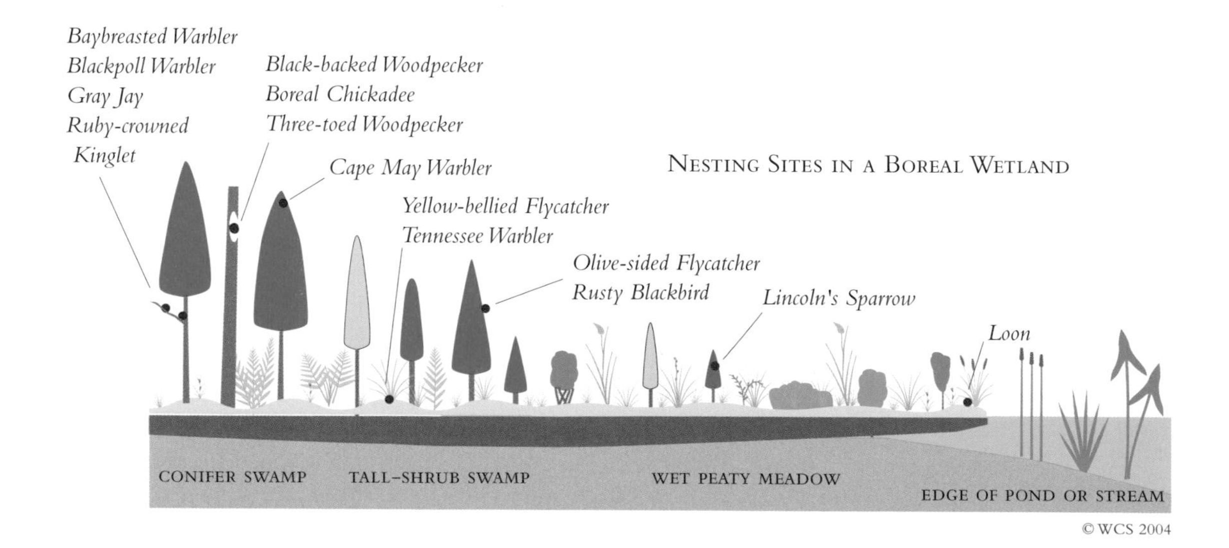

NESTING SITES IN A BOREAL WETLAND

A small, shy, reasonably common flycatcher that nests in mossy hummocks in conifer swamps or mountain woods. Eastern Canadian and U.S. populations are apparently increasing.

A hard-to-see sparrow with a pretty song. Nests in wet sedge-shrub meadows, often where there is a sphagnum moss carpet. Widely distributed in the western Adirondacks in 1985 but decreasing on many eastern breeding bird survey routes since then.

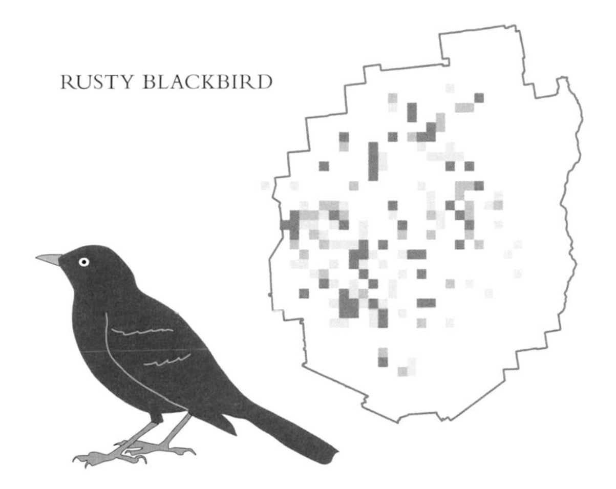

Thirty years ago a characteristic bird of open Adirondack wetlands with a border of conifers. Now decreasing rapidly all across eastern Canada and the adjacent U.S. and becoming much rarer in the Adirondacks.

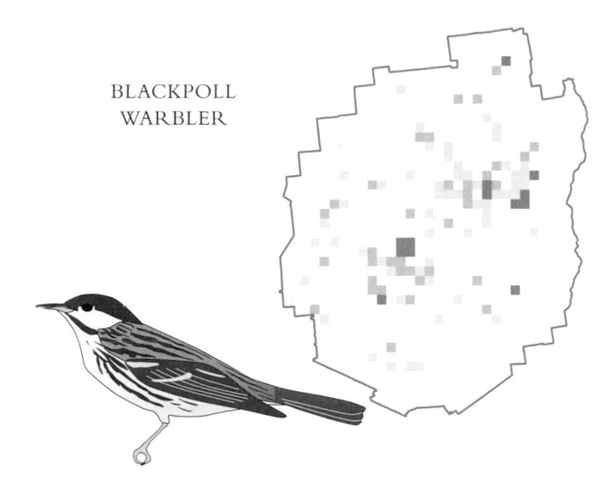

A characteristic species of high-elevation conifers, typically nesting in a crotch of a stunted fir or spruce. Almost ubiquitous above 3,000 feet elevation; also found in a few conifer swamps at lower elevations.

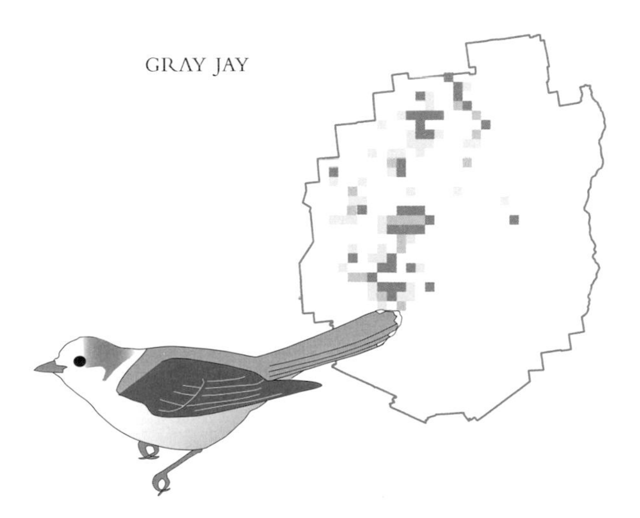

The whisky-jack, a familiar scavenger and camp-robber everywhere in the far north. Rare in the Adirondacks and limited to the largest swamps of the northwest drainages.

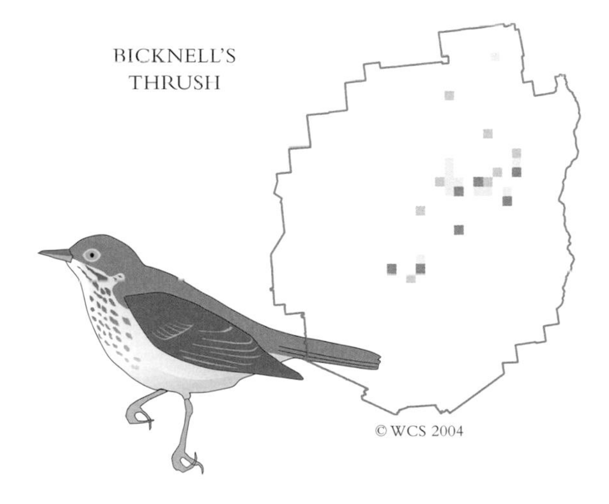

A high-elevation specialist, found in stunted conifers near mountain summits, mostly above 3,500 ft elevation. Occurs in much the same habitat as the blackpoll, but much rarer and harder to see. It has a quiet, very sweet song at dusk and dawn.

NUMBER OF RESIDENT BOREAL BIRD SPECIES, 1985

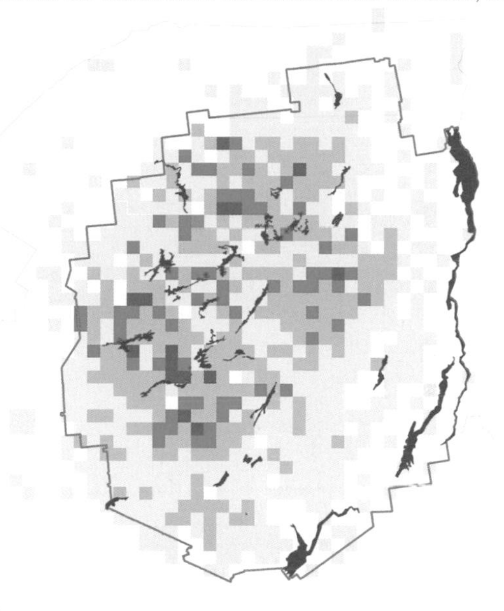

Boreal species recorded per 5-km survey block

Boreal birds are found throughout the Adirondacks but are most continuously distributed in the northwest and spotty and infrequent in the agricultural and industrial valleys of the east and south. They are almost absent outside the Blue Line. They are most diverse, averaging about five species per survey block, in the High Peaks and the bog belt of the Northwestern Flow. The maps use data from the 1985 Breeding Bird Atlas project. We suspect that the current Atlas 2002 project, when complete, will show decreases in diversity.

Possible breeding: bird seen in suitable habitat in breeding season.

Probable breeding: singing male on several dates, territorial defense, courtship, apparent visits to a hidden nest.

Confirmed breeding: coition, nest, young birds, or adults carrying food or nest material.

In the nineteenth-century Adirondacks, animals were either resources to be captured or nuisances to be eliminated. This was, of course, hard on the animals. But the Adirondacks were a wilderness, and for its settlers, as for pioneers everywhere, the harvesting of meat and fur was an economic necessity.

Necessary or not, it was not a sustainable harvest. By the end of the century every large Adirondack mammal had been affected by harvesting: five were gone and another six greatly reduced in numbers.

The twentieth century, with its remarkable series of arrivals and recoveries, was much better for animals. But it did not see a return of the original wilderness and was not without losses of its own. The forests to which the animals returned were younger and less healthy than the original ones. The genetic stocks of the returned animals were different from those of the animals they replaced. The mid-century brought toxic pollutants and a new round of extinctions. The end of the century has brought acid rain and changing climates, whose consequences we are only beginning to discover. All in all, for almost every animal larger than a mouse, it has been a turbulent two-hundred years. Here are some landmarks:

1811 Hamilton County institutes a bounty on wolves.

1825 Nat Foster kills twenty-five wolves, making $1,250 in bounties, about $20,000 in current dollars.

1844 John Cheney, who in his first thirteen years in the Adirondacks has killed 600 deer, 400 martens, 48 bears, 19 moose, and 30 otters, says there is one beaver colony left in the Adirondacks.

1850 Thomas Meacham dies, after killing 2,550 deer, 219 bears, 214 wolves, and 77 mountain lions.

1855 Wolves and mountain lions are scarce but not protected: every town in Hamilton County has a bounty on both.

1860 Though the official New York deer season begins August 1 and ends January 1, in the Adirondacks deer are hunted year-round: by tourists in the summer, by residents in the fall, and by professional hunters supplying logging camps and hotels in all seasons.

1861 The last three moose killed in the Adirondacks are shot near Raquette Lake.

1867 The legislature moves the start of the deer season to October 1, beginning a sixty-year tug of war between the downstate summer hunters and the upstate fall hunters. The professionals who hunt illegally year-round pay no attention.

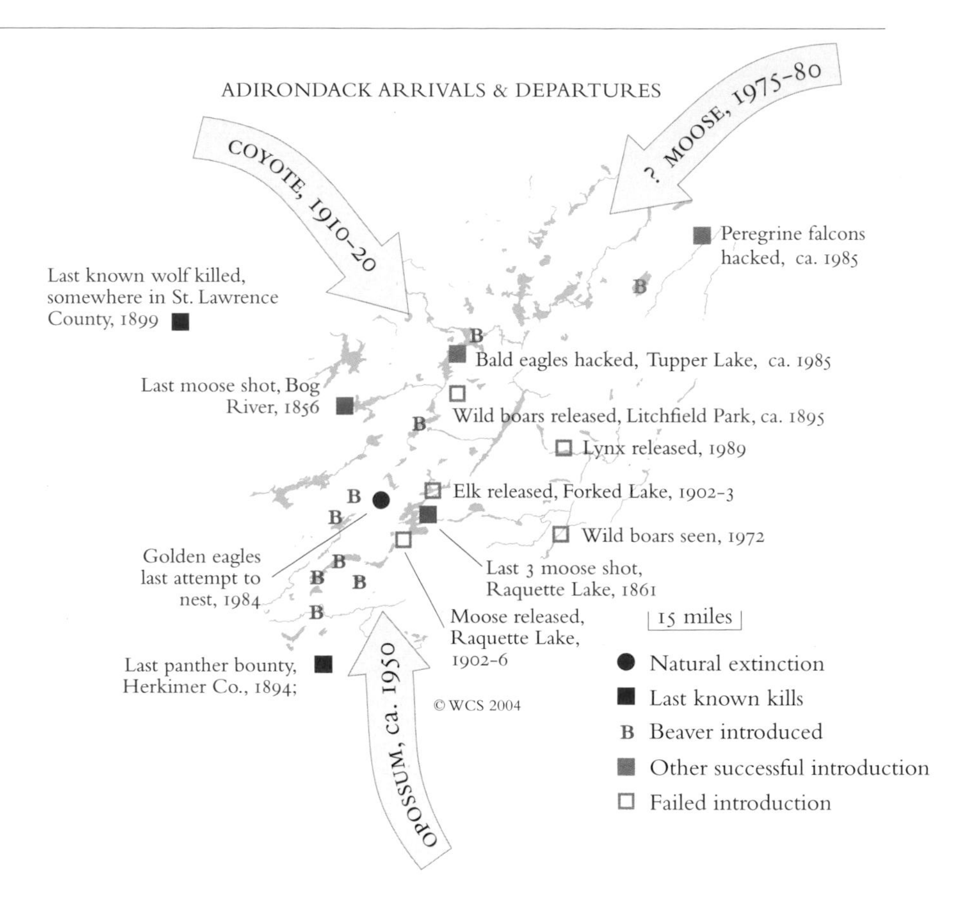

1871 Even though mountain lions and wolves are nearly extinct, New York State places bounties on them. The bounty on an adult wolf is $30, equivalent to $400–$500 today. In the next eleven years, hunters receive bounties on 46 mountain lions and 45 wolves.

1880 Sylvester Palmer of Indian Lake and John Liberty of Elizabethtown are appointed as game protectors by the legislature and become the first game wardens in the Adirondacks. They have their work cut out for them. Game animals have been largely extirpated from the rest of New York, and everyone who want to hunt something larger than a rabbit must come to the Adirondacks.

1894 New York pays its last mountain lion bounty, making a total of approximately 153 mountain lion bounties paid between 1860 and 1894.

1899 Ruben Cary kills a wolf in St. Lawrence County, the last documented wolf kill in New York State. David Mix, a trapper, reports mountain lion tracks near Newcomb; occasional reports of mountain lions continue for two decades.

1901–1906 The owners of large private preserves begin stocking them with large animals. William Whitney and others release at least 250 elk in seven Adirondack counties. Most die quickly. After six tries and much controversy, the legislature eliminates hounding, the hunting of deer with dogs.

1902 Fifteen moose are released at Camp Uncas. Most die or are killed within a year or two.

1902–1909 At least 35 beaver are released in the central Adirondacks. Essex County pays bounties on 40–50 bears per year.

1907 A bill appropriating $20,000 to establish bison herds in Essex, Hamilton, and Warren Counties is introduced in the New York Legislature but does not pass.

1915 The elk herd has been declining steadily. The Benevolent and Protective Order of Elks donates a last boxcar load of western elk, which go the way of the others. The new beavers, on the other hand, are thriving. They have already raised the level of the Fulton Lakes and the Conservation Commission begins issuing permits to remove them.

1919 The Conservation Commission requests that the recipients of hunting licenses destroy the "enemies of useful wildlife." These include most shootable animals smaller than a deer: lynx, bobcat, red fox, gray fox, weasel, porcupine, otter, fisher, and red squirrel.

1920–1930 The eastern coyote, first called the brush wolf and apparently a coyote-wolf hybrid, enters the Adirondacks from the northwest.

1937 The Conservation Commission reports that the setting of open seasons and bag limits for beaver is now "guided largely by the mounting number of complaints regarding damage by beaver."

1950–60 Beaver are now established throughout New York State and trapping is allowed everywhere except in New York City. Adirondack trappers are taking about 5,000 beaver per year. Agricultural spraying with DDT begins; by 1960 DDT has made the peregrine falcon extinct in New York, reduced the bald eagle population to a single active nest, and reduced the osprey population to a few nests on Gardiner's Island.

1970–80 Seven wild boar are discovered near Indian Lake but do not survive. Small numbers of moose begin to arrive in the Adirondacks from Ontario and Vermont. Mountain lions are reported in Massachusetts, Vermont, and New York. Fisher populations have recovered, and fisher are now common in many parts of the Adirondacks. Beech-bark disease arrives from the northeast.

1983 The last pair of golden eagles in the Adirondacks raises a single chick. Mink near polluted rivers outside the park are found to be carrying high levels of PCBs.

1980–90 Bald eagles and peregrine falcons are released from artificial nests and survive to breed.

1989 Eighty-three lynx, trapped in the Yukon, are released near Newcomb. A number are killed by cars; others disperse widely and are found in Ontario, New England, and Maryland.

1990 Raccoon rabies enters New York from the south, beginning the largest documented epidemic of animal rabies anywhere. In the next ten years the state health department examines over 100,000 animals and confirms rabies in more than 14,000 of them.

1998 A survey finds no evidence that lynx have survived in the release areas in Newcomb and the High Peaks. Essex County supervisors, worried by talk about wolf reintroduction, declare that the release of dangerous animals in Essex County is illegal.

2000 Mercury pollution of Adirondack lakes, originating from midwestern power plants, is found to be widespread. The state health department advises against eating more than one meal a month of mercury-contaminated fish from many Adirondack lakes.

2001 Although dangerous animals are prohibited in Essex County, at least 129 people there are forced to seek medical attention for bites from domestic dogs and cats.

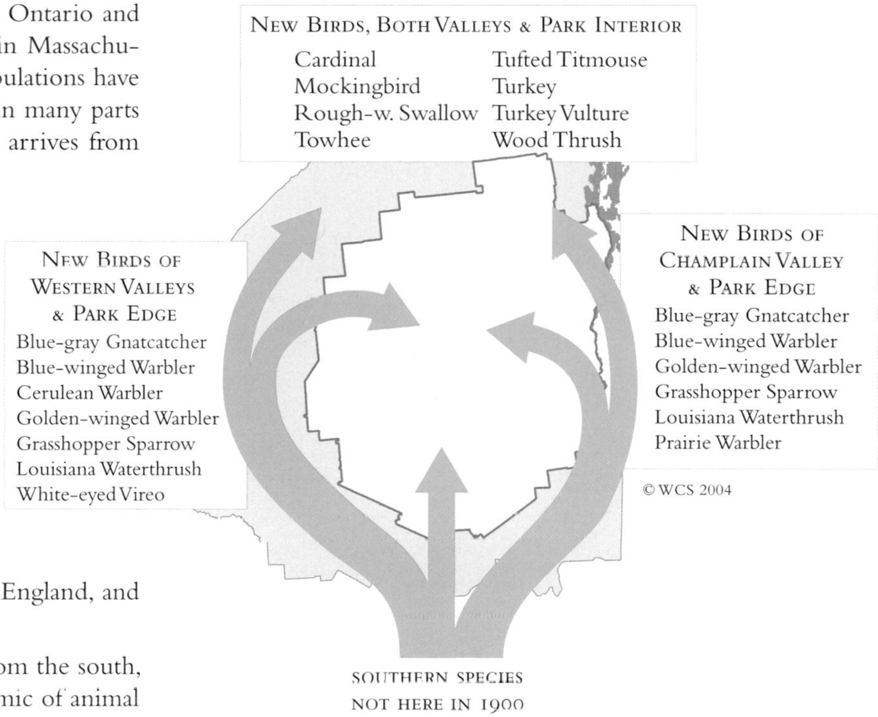

SOUTHERN BIRDS ARRIVE

In the last 120 years, New York State has gained 25 new species of breeding birds from the south. Northern New York has gained at least 17 new species of breeding birds and the Adirondack interior at least 8. Some of the new Adirondack birds, like the wood thrush and towhee, previously occurred in southern New York. At least six others, including turkey vulture, rough-winged swallow, and tufted titmouse, were not even in New York in 1880. The timing and ecology of the northward expansion strongly suggest that climate change is responsible.

By 1700, seventy-five years after the start of the European fur trade, there were very few beaver anywhere east of the Great Lakes. Continued trapping kept them rare for the next two centuries. In 1895, when trapping was first prohibited in New York, the entire New York beaver population was believed to be one or two families near the north end of Saranac Lake.

About 35 beaver were introduced to the central Adirondacks between 1902 and 1909. The surviving natural population, now protected, may have begun to expand as well. Habitat was abundant, and predators big enough to kill beaver were all but absent. The new beavers throve. By 1915 they were flooding summer cottages, and the state was issuing permits to remove them; by 1920 the state was trying to assess the amount of timber they had flooded.

Trapping resumed in the Adirondacks in 1924. The first few harvests were quite heavy and were followed by closed seasons; later harvests, though taking more animals, probably took a much smaller fraction of the population.

Despite harvesting, the beavers spread rapidly. By 1934 they were being trapped in eight counties. By 1940 they were being trapped in nineteen counties and by 1955 in every rural county in the state. By 1970 the harvests were double those of 1950; by 2000 they doubled again.

While beaver harvests were increasing elsewhere, the harvests in the central Adirondacks, where the beaver renaissance began, stayed constant or decreased. Partly this was because the Adirondacks had reached their carrying capacity. And partly it was because both the beavers and the trappers had discovered their promised land in the big marshes and abandoned farmlands to the north and west.

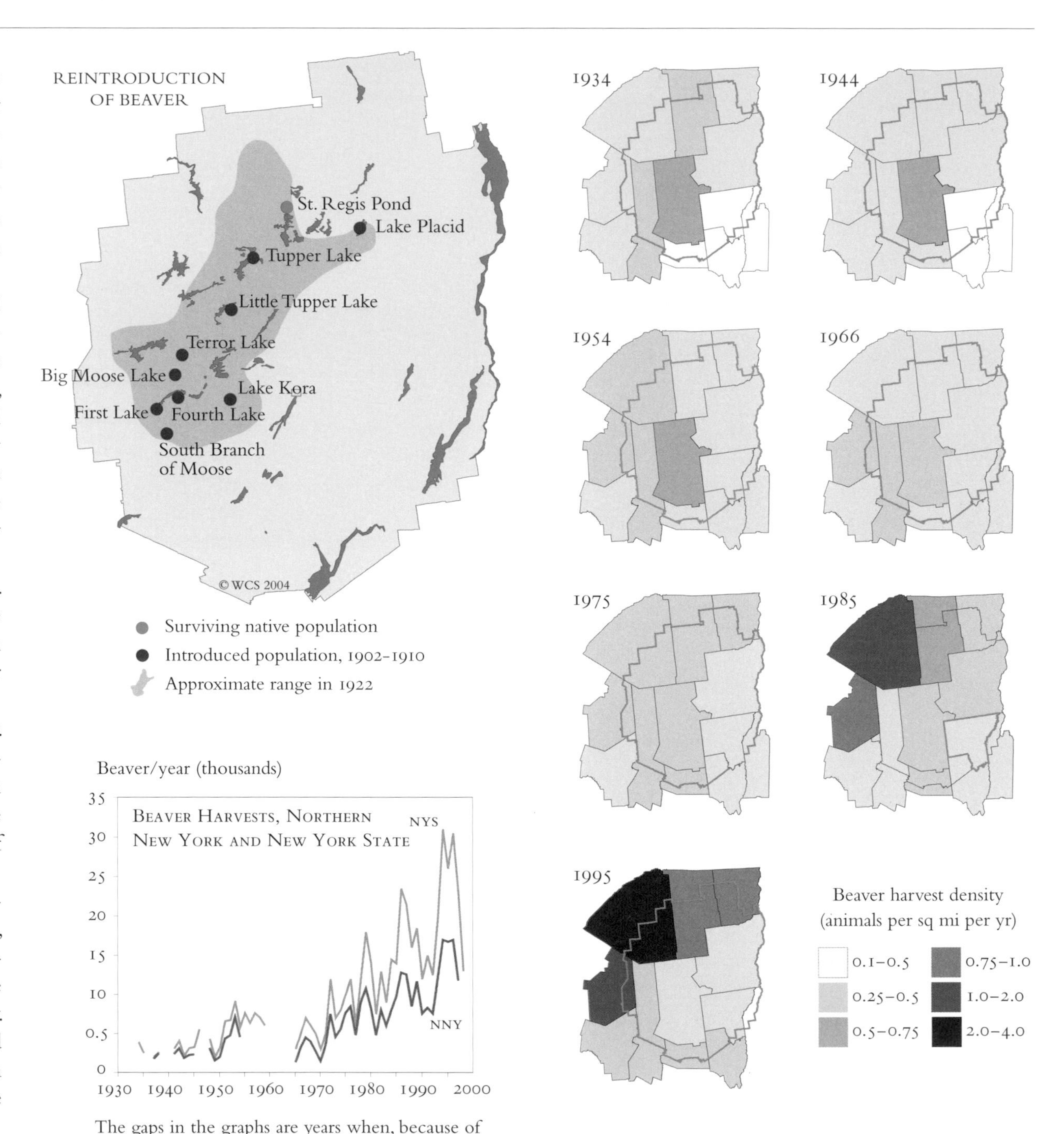

The gaps in the graphs are years when, because of previous heavy harvests, no trapping was allowed.

For much of the past century, the wholesale *price* of a beaver pelt has been between fifteen and twenty-five dollars, only rarely falling to around ten dollars or rising to near thirty. Since the prices of almost everything else have increased steadily, this means the *value* of a pelt has decreased greatly, and a beaver pelt now buys the trapper less than a tenth of what it did eighty years ago.

One consequence of this is that there are now very few full-time trappers who make a living from furs and who travel long distances and run long trap-lines in remote areas. Most trappers today only trap part-time, and so trap in easily accessible areas near where they live.

The maps reflect this. Over the last twenty years, most of the beaver harvested in northern New York have been taken from private lands in the St. Lawrence Valley and along the northern and western edges of the park. In part this is because the young, thin-barked aspen and maple forests on abandoned agricultural lands are a banquet for bark eating animals like beaver and deer. But in equal part it is because these areas, unlike the Adirondack interior, are overlap zones, where people and animals are both plentiful and have easy access to each other.

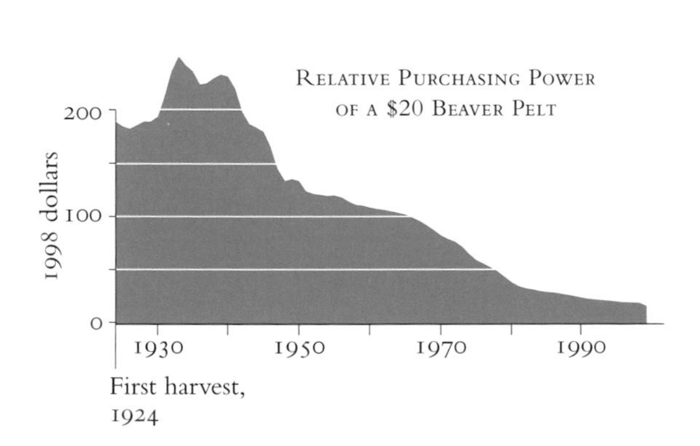

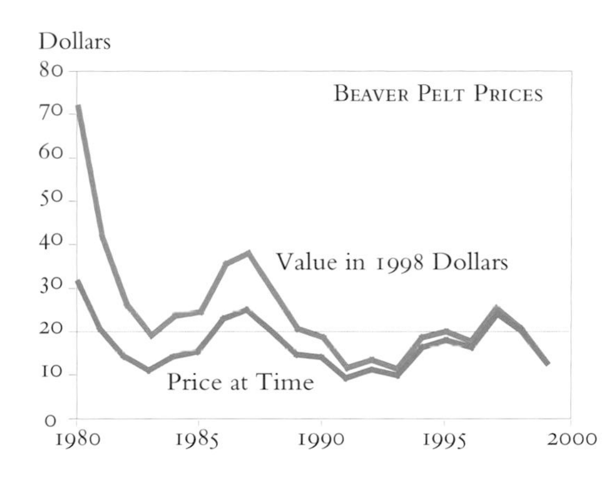

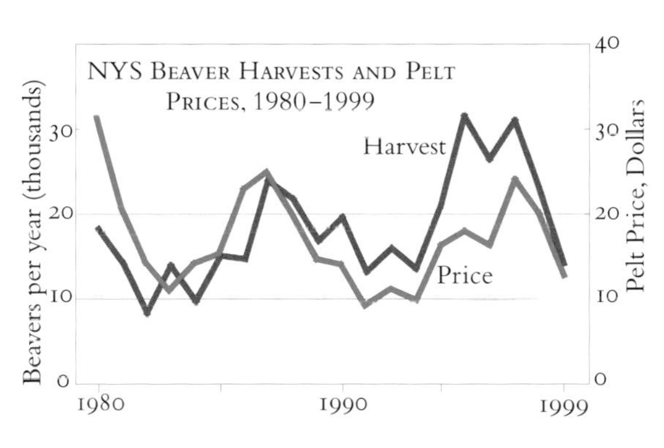

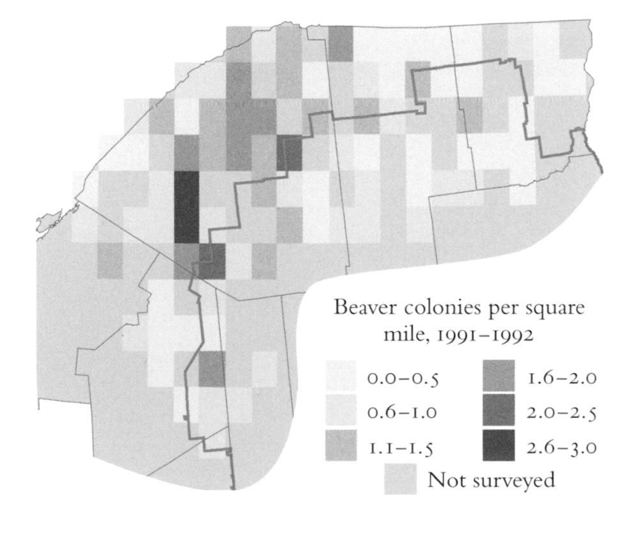

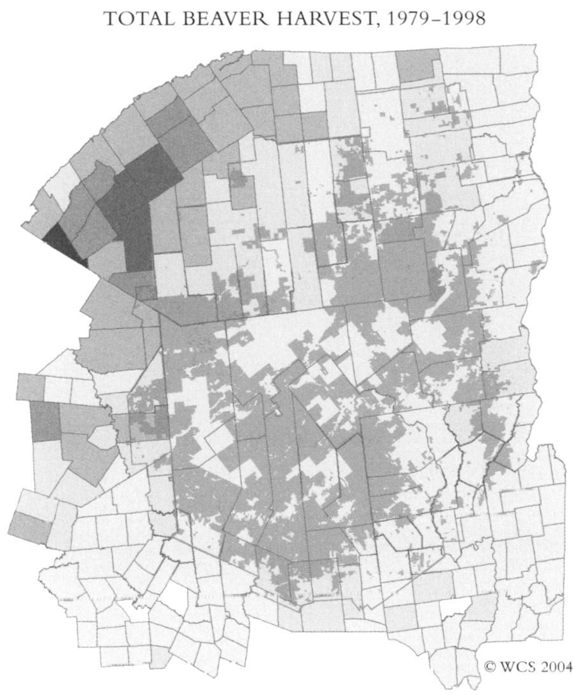

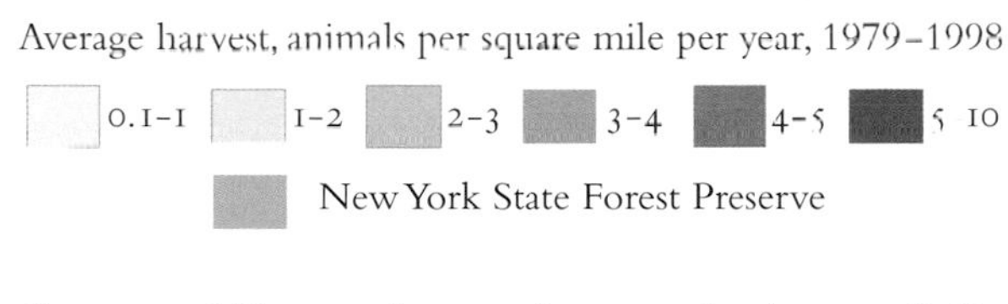

Surveys and Harvests. A 1992 air survey by the DEC (left) shows a concentration of beaver colonies just outside the park in St. Lawrence County, in much the same area where records (above) show the highest average harvests. This is one of the very few instances in which we can show that harvest data do reflect population densities. No air surveys were done on state lands, and so we have no idea whether the low harvests in the park interior reflect low populations of beaver or low effort by trappers.

THE Adirondacks contain enormous number of open wetlands. In this respect they are a typical northern landscape, like Maine and Canada and unlike the rest of New York. Depending on the watershed, ten to twenty-five percent of the land surface is permanently wet, and well over half of this is open.

Because many plants like open wet places, and because animal diversity often depends on plant diversity, the open wetlands are biodiversity hot spots. Virtually every Adirondack animal uses them at some time or other. Over half of the Adirondack birds breed in them or next to them, and between half and two-thirds of the vascular plants in the Adirondacks either prefer wetlands or are wetland specialists.

The reason that open wetlands are diverse is that they change frequently, and the reason that they change frequently is the beaver. The diagram shows the main pathways. Undisturbed wetlands move, quite slowly, along the green paths toward the late-successional communities (boxes with black borders) in the diagram at right. Beavers reverse these processes, moving the wetlands rapidly back along the orange paths toward early-successional communities.

Because of the frequent changes, most Adirondack watersheds contain a mixture of open and forested wetlands and open- and forested-wetland animals. The open wetlands typically accommodate ecological generalists that specialize in transient habitats. The later successional wetlands often contain ecological specialists, which prefer more permanent habitats.

The most interesting feature of the wetland cycle is that it is a *diversity-accumulating* cycle. As wetlands move through the cycle they usually retain some of the species from previous phases and so are more diverse than they would be if they simply remained in one phase. The result is a complex mixture of types—bogs within meadows, meadows within ponds—that is not only interesting to us but seemingly very consistent with the beaver's long-term interests.

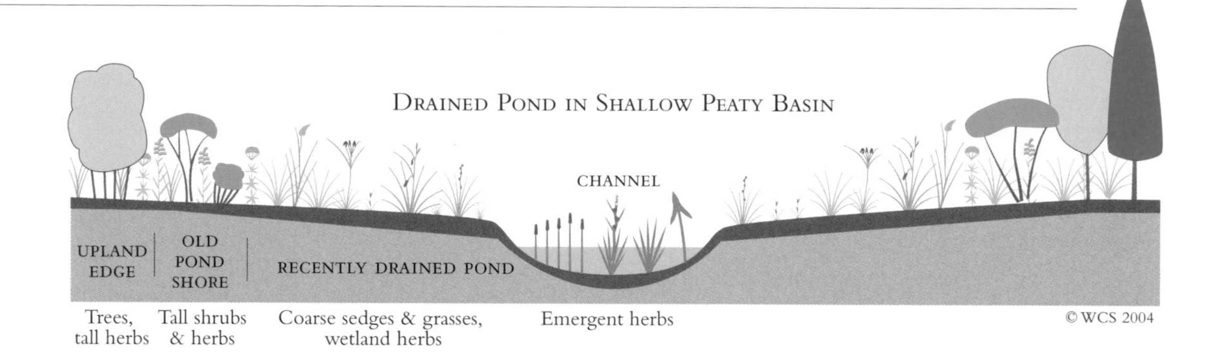

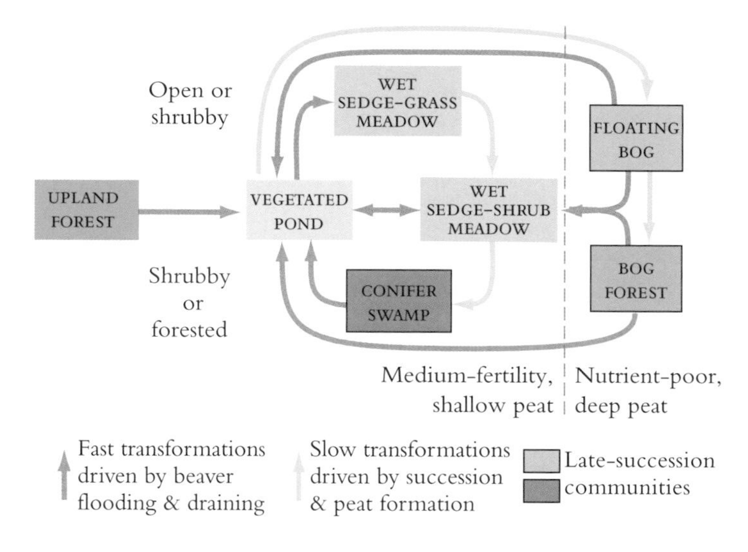

The six main types of Adirondack wetlands. Given the right combination of time and water level, any type can be converted into any other type. Flooding by beaver, a fast process, kills woody vegetation and increases fertility, creating early successional communities like the flooded bog shown below. Drainage, plant succession, and peat formation, much slower processes, reduce fertility and allow woody plants to establish, creating late successional habitats like the sedge-grass meadow with a shrubby rim shown above.

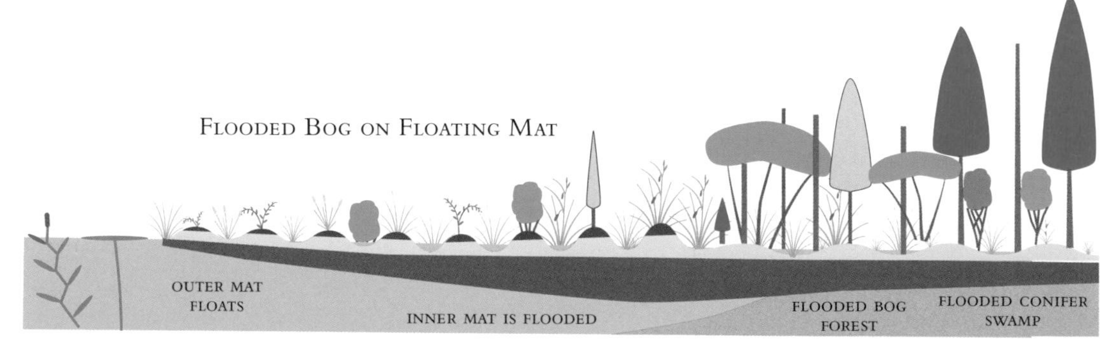

BECAUSE most wetlands contain mixtures of plants from different stages in the wetland cycle, they are hard to classify and map. Like forests, they form a continuum in the field. The best we can do is to map some of the most conspicuous types and hope the reader remembers that there are always others we haven't named.

The map shows about 300,000 acres of the upper Oswegatchie watershed, an area which, even by Adirondack standards, is remarkably wet. About 26% of the land is vegetated wetland and another 4.8% lake or river, for a total of about 95,000 acres of wetland and open water.

The beaver-modified wetlands total only about 9,000 acres, or 9.6% of the wetland and water area. About 2,000 acres of this are the ponds—415 in all—that the beavers are currently occupying or have recently occupied. The other 7,000 acres are drained ponds in other stages of the cycle.

Because the beaver wetlands contain early successional habitats that don't occur elsewhere, their importance for animal and plant diversity is out of proportion to their total area. Fully 55% of the graminoid wetlands, 65% of the ponds, and 95% of the wetlands with standing dead trees are beaver wetlands. These are significant numbers, especially if you are an animal or plant with particular needs. If you simply need a pretty place with water almost any wetland may do. But if you are a moose looking for good grazing, or a merganser looking for a nice private pond, or a woodpecker or bluebird looking for a nest hole, the 9,000 acres of beaver modified wetlands begin to look very important indeed. And if you are a blue gentian looking for a sweet place in the sun, they are heaven itself.

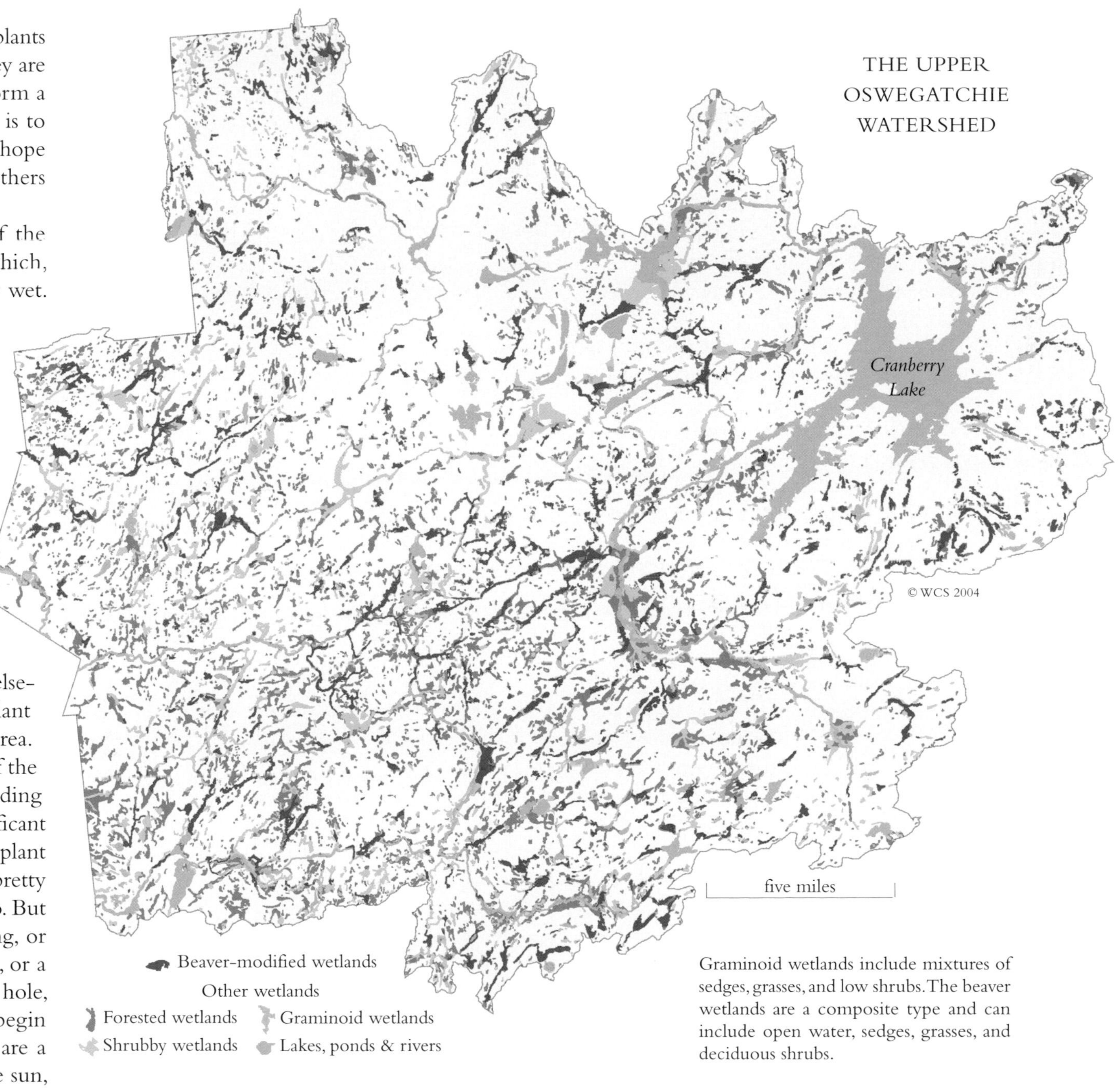

Graminoid wetlands include mixtures of sedges, grasses, and low shrubs. The beaver wetlands are a composite type and can include open water, sedges, grasses, and deciduous shrubs.

SOMETIME between 1920 and 1930 a new animal arrived in New York from the northwest. It was first called a coydog, then a brush wolf, then a coyote. Its ancestors clearly included coyotes from the upper Midwest, but it was larger than a western coyote. In some ways it resembled the eastern Canadian wolves, but it was smaller and did not seem to form permanent packs and defend permanent territories the way they did.

By 1950 it was well established in northern New York. It was not well received there, but predators never were. By 1950 most counties had bounties on coyotes, and the DEC had started trapping adults and destroying dens and pups.

Coyote control, which had not worked in the West, didn't work in the East either. By 1975 coyotes had reached northern Maine and the suburbs of New York City. By 1990 they were found throughout upstate New York and had reached Nova Scotia, Newfoundland, and the northern shore of the Gaspé.

Although coyotes have been living in the central Adirondacks for some time, it is only recently that, by the efforts of a number of researchers, we have begun to learn what they are doing there. The map of their expansion is from research by Heather Fener; the chart of their diet from a carnivore study by Justina Ray, Matt Gompper, and Roland Kays (*Map 4-11*).

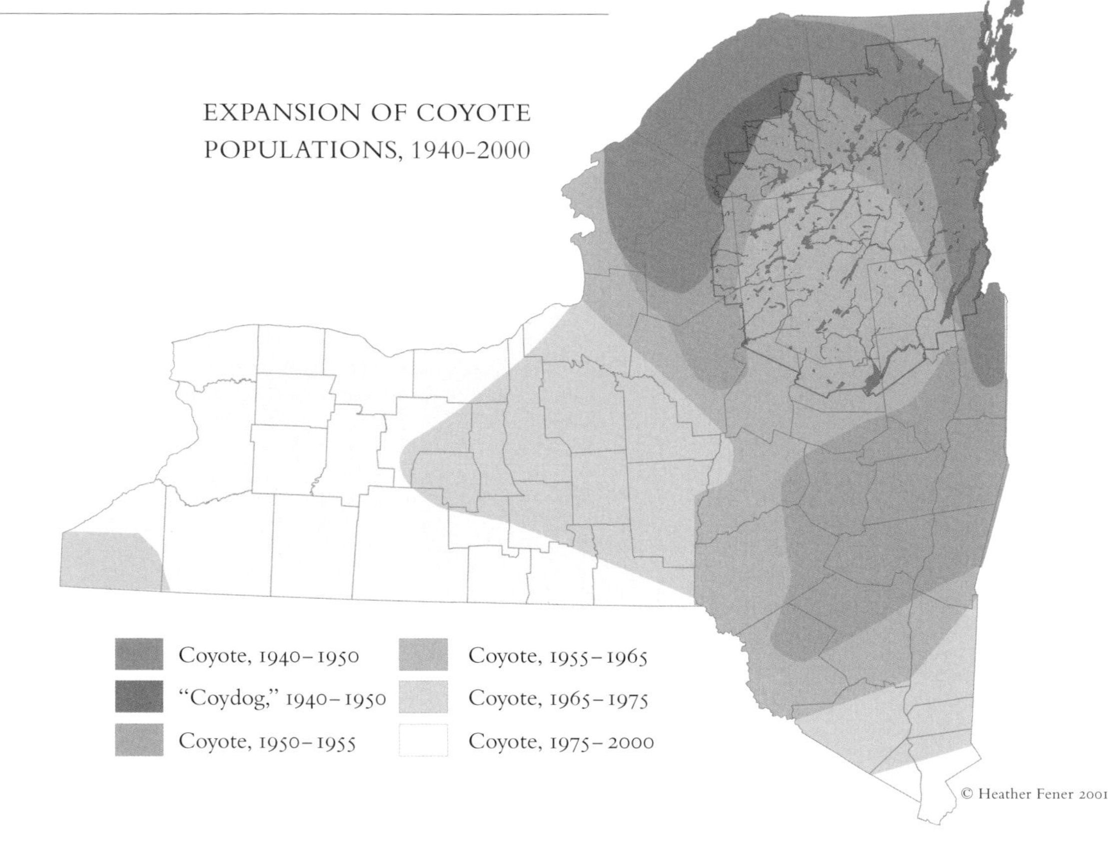

SUMMER DIET OF ADIRONDACK COYOTES, 2000–2001

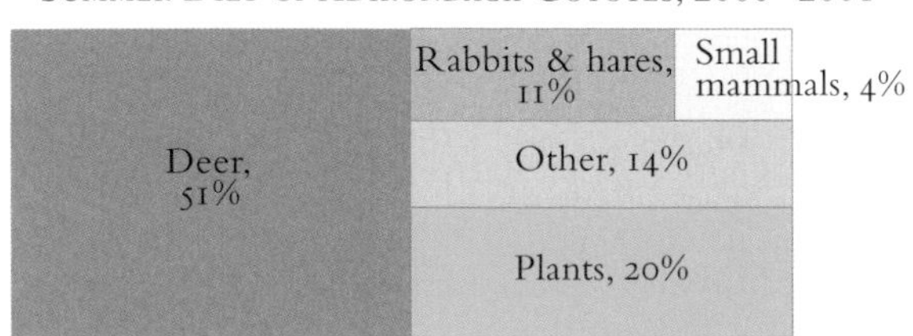

Exactly who the eastern coyotes are is still a mystery. The standard theory is that they are western coyotes with an admixture of wolf traits that they picked up while migrating eastward. But this may be too simple: while eastern coyotes are believed to have at least a few wolf genes, no one knows whether these genes are responsible for their "wolflike" traits.

A further complexity is that some eastern wolves may not be pure wolf. According to one group of geneticists, both the extant wolves of Algonquin park and the extirpated wolves of the Adirondacks are "red" wolves, genetically intermediate between the gray wolf and the western coyote. If this is so, it raises very interesting questions about how close our present coyote-with-wolf-traits is or isn't to its Adirondack predecessor, a wolf-with-coyote-traits.

Still this may be too simple. The more we learn of the relations between extant and extinct canid populations, the more it seems that the names *wolf* and *coyote* may refer to the endpoints of a genetic continuum. If this is so, many contemporary canid populations are neither pure wolf nor pure coyote but somewhere in a very interesting evolutionary landscape between the two. Even more interestingly, they are apparently able to move quickly around this landscape, from wolf towards coyote or coyote towards wolf, as geography and circumstances dictate.

THE moose, our largest animal, was gone from the Adirondacks for a hundred and twenty years (*Map 4-4*). It survived to the north and east of us but was not expected to return. Informed opinion, ignoring the overlap of deer and moose in Canada and Maine, said that a parasitic nematode, innocuous in deer but fatal in moose, made it impossible for moose and deer to coexist.

No one told the moose this, and in the 1970s they started moving south and west, using the new wetlands created by the return of the beaver. By 1990 there was a moose explosion in Vermont and a smaller but very promising moose presence in the Adirondacks.

The maps, based on sightings reported to the DEC, give the best picture we have of the return of any native animal. The first records suggest a few wandering animals. Then temporary clusters develop but do not persist; the patterns suggest a few animals moving, perhaps meeting, then moving again.

About 1990 the patterns change. The density increases and the clusters become larger and more persistent. Soon afterwards, animals begin to disperse out of the park to the south. All this suggests that a critical threshold had been reached, and the park now had areas with permanent breeding populations and was now producing and distributing, rather than just attracting, moose of its own.

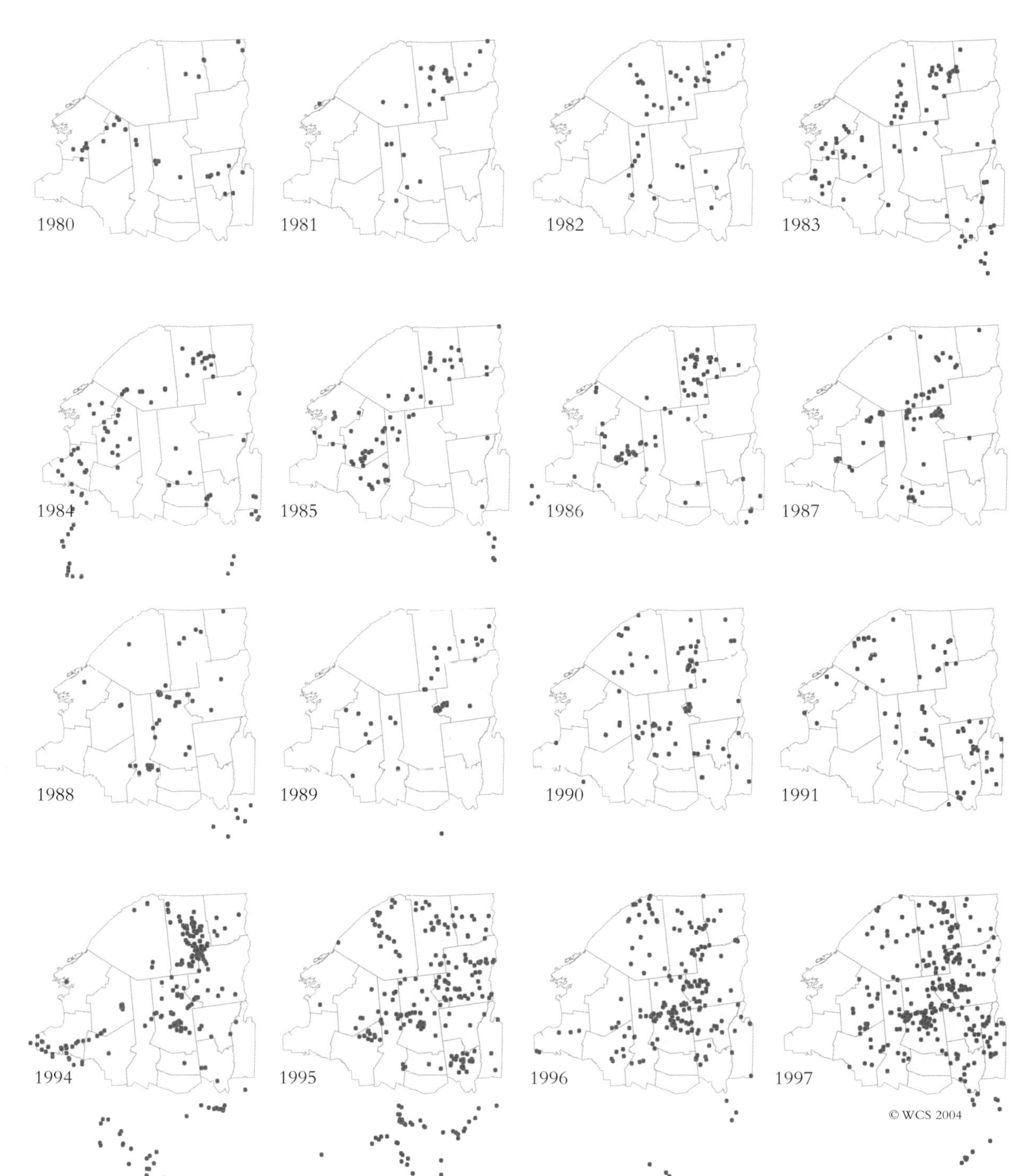

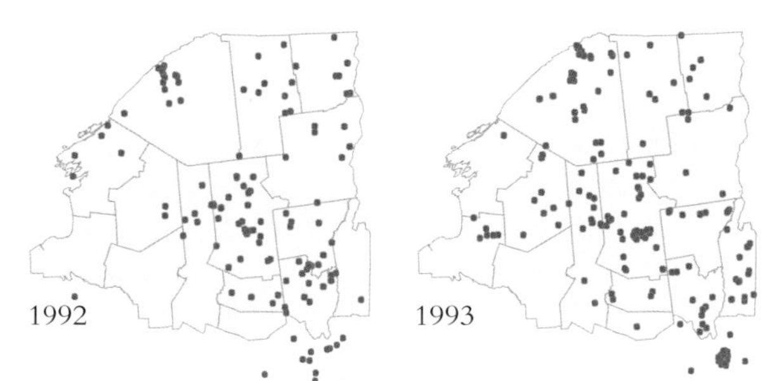

THE lynx is a short-tailed cat with long sideburns and ear tufts. It is closely related to the bobcat but larger, less spotted, and more of an ecological specialist. If you see a tawny cat with spots and several bars on the tail, it is a bobcat and you are lucky or skilled because bobcat aren't easy to see. If you see a grayer, more ghostly cat with long ear-tufts and just a little black on the tips of ears and tail, it is a lynx, and you are very lucky or very skilled indeed.

The lynx is a deep-snow cat, specialized for hunting the snowshoe hare and highly dependent on it. It is most at home in the arctic and subarctic north where hares are abundant and there are no other predators its size. Southwards, where hares are less abundant and it has to share them with able (and aggressive) generalists like the fisher and coyote, it does less well.

The Adirondacks were probably borderline lynx country at best. Almost certainly there were lynx here, but we have no idea how many there were, or if they were permanent residents or wanderers, or who shot or trapped the last one. Today, with the area of conifers decreased by logging and with coyotes and bobcats reducing the hare population, the Adirondacks may not be lynx country at all. Lynx are still reported from time to time, but because they look so much like bobcat, there is no way to be certain of the reports.

In the late 1980s the reintroduction of lynx was attempted. Eighty-three lynx, most yearlings, were trapped in the Yukon and brought to Newcomb. Some were released quickly, others acclimatized in portable cages before being released. Lynx urine was spread around one release site to encourage them to stay in the area.

Young lynx, however, are not staying animals. Within the next few months, all moved away. About forty were eventually found dead, most commonly as road kills. Some went as far as Maryland and New Brunswick. The other forty were not found. Lynx are prodigious dispersers, and our guess is that when they found neither good rabbit populations nor any signs of other lynx, they just kept on going.

In 1998 and 1999 researchers set out rubbing pads scented with a cat attractant to see if any lynx remained in the area. No lynx hair was found, but, disturbingly, almost no bobcat hair either. Perhaps the attractant, which has worked well in the West, doesn't work in the wetter eastern forests. Perhaps Adirondack cats don't rub, or perhaps there are almost no Adirondack cats to rub. At this point we just don't know.

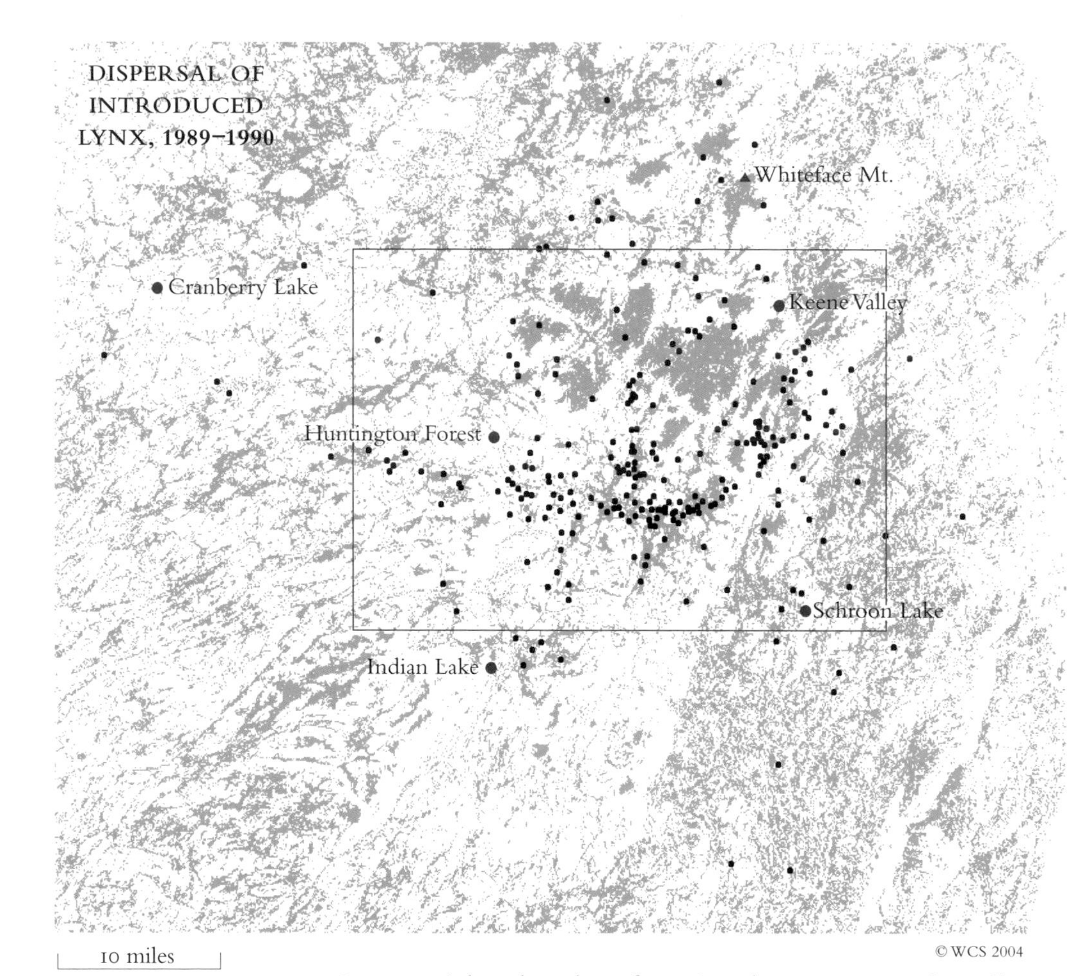

In 1989, eighty-three lynx from Canada were given radio collars, and released in Newcomb. The map shows the radiolocations obtained in the next eighteen months. The lynx mostly dispersed to the east and west, following bands of wetlands, and then scattered to the north and south where they found opportunities. Some went down the Hudson River towards Indian Lake, others up the Hudson to Tahawus. Most apparently avoided the High Peaks, but a few found their way north through Avalanche Pass and the Cascade Lakes. A number went east toward the Schroon River and then north towards Keene Valley. The batteries in the radio collars lasted till 1990; after that the lynx disappeared in the forest.

MOUNTAIN lions, alias panthers, pumas, cougars, and catamounts, were once the most widespread mammal in the western hemisphere, occurring from the Yukon and Nova Scotia to Patagonia. In 1800 they were found throughout the Adirondacks, though always scarce here. By 1890, after over two hundred had been shot or trapped, they were almost gone from the Adirondacks and indeed from most of eastern North America.

Occasional reports of Adirondack lions continued for the next fifty years. No one knows whether these were true. The cougar is a very distinctive animal, but cougar tracks and sign are less distinctive. A good cat tracker can recognize them easily, but when there are very few cats there are very few good cat trackers.

About 1980 things began to change. There were many more sightings, including a number by experienced observers. Furthermore, the observations started in a cluster north of the High Peaks and seemed to be spreading from there. Young cats were seen, and DNA tests confirmed the presence of cats in Vermont and elsewhere in the east.

Clearly, whether or not the mountain lion was ever really extinct in the east, it is not now. But just what sort of nonextinctness it has—whether it is a resident or a wanderer, a native or an alien, very rare or becoming widespread—and, even more, what its future will be, remains completely unknown.

1 Gabriels, July 26, 1982. A cougar seen running directly across a road. "I never expected to see such a beautiful creature."

2 Black Brook, Aug. 8, 1989, 6:30 p.m. A mother and two cubs, watched by two observers for 15 minutes.

3 North Elba, Oct. 16, 1990, 12 p.m. Cougar treed by a rabbit dog, watched for 2-3 minutes.

4 Belmont, Aug. 29, 1992, 9:45 p.m. A cougar, "larger than a German shepherd," observed for thirty minutes while it appeared to be stalking a dog.

5 Lewis, Nov. 10, 1993. A large cat, "the size of a golden retriever," watched for fifteen minutes.

6 Crown Point, Apr. 24, 1997. A large cat jumped across a dirt road; two smaller cats, about two thirds the size of the large, followed more slowly.

7 Santa Clara, July 7, 1998, 9:55 a.m. A bicyclist who came within thirty feet of a large cat with long tail estimates the weight at ninety pounds.

8 Greenwich, July 12, 1999. A cougar, seen standing and walking from 100 feet and less, "close enough to see whiskers," watched for five minutes by an observer familiar with them from the west.

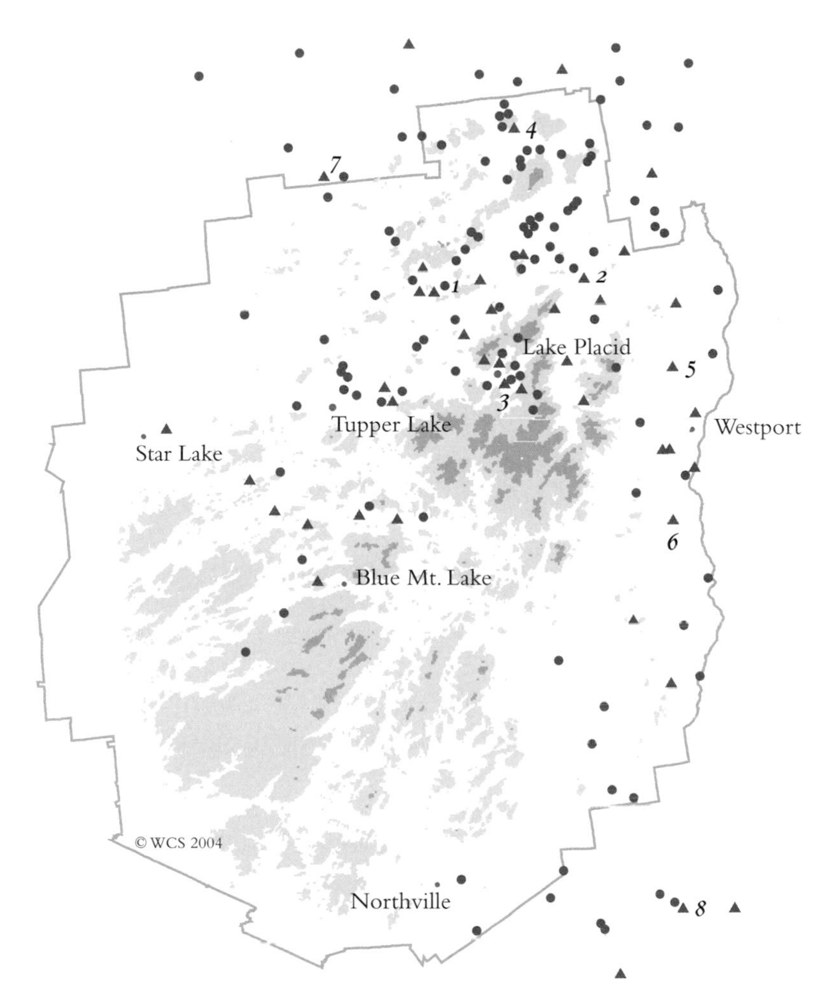

COUGAR SIGHTINGS REPORTED TO THE DEPARTMENT OF ENVIRONMENTAL CONSERVATION, 1950–2000

▲ Highly credible observations: either an experienced observer, a good description of the animal, or a long observation period

● Briefer or less complete observations

Over 3,000' | 1,000–3,000' | Under 1,000'

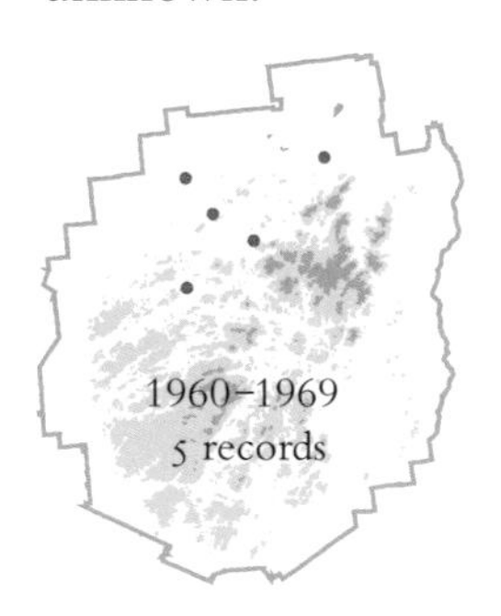

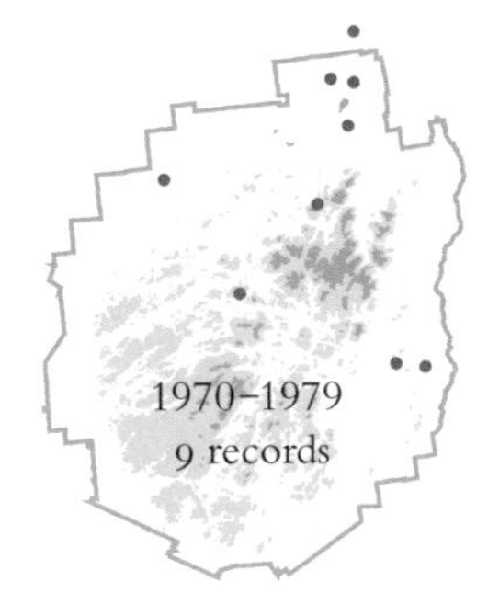

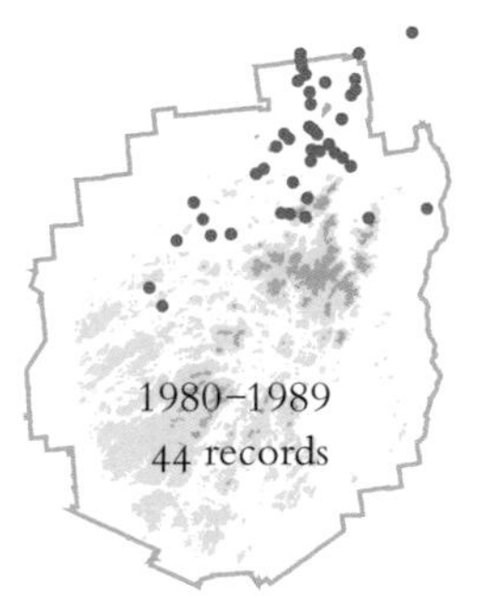

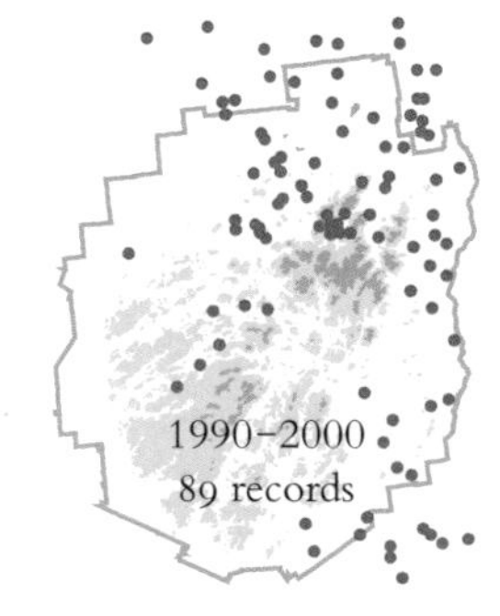

THE deer harvest has changed greatly in the last forty years. Fifty years ago deer were common in the Adirondacks and scarce outside. Everyone came here to hunt, and hunting clubs had all the members that they could use. Now the map has almost reversed. There are more deer outside the Adirondacks than in, Adirondackers go outside to hunt, and hunt clubs are hurting for members.

These changes reflect changes in the hunters, the forests, and the climate. Before wilderness areas were established in the 1970s, many state lands were open to jeeps, motorboats, and float planes. Now motorized access has been greatly restricted. Deer hunters tend to hunt within a few miles of an access point, and some of the change in Adirondack harvests may be because it is now difficult for them to get to the places they used to go.

The woods were changing as well. In the 1950s, perhaps two-thirds of the Adirondacks were covered with young woods and openings (*Map 6-4*) that had excellent winter browse and allowed deer to maintain moderately high populations. Gradually the forests became older and more continuous. The deer population fell and, because of the decreased food supply, became increasingly vulnerable to hard winters. The deep snow in 1969 and 1970 reduced deer populations all across northern New York. The deer in the St. Lawrence Valley, where snow depths have been decreasing and there is abundant food, gradually recovered and went on to a population explosion. Those in Essex and Hamilton, where there is deep snow cover and limited food, still haven't returned to their pre-1970 levels.

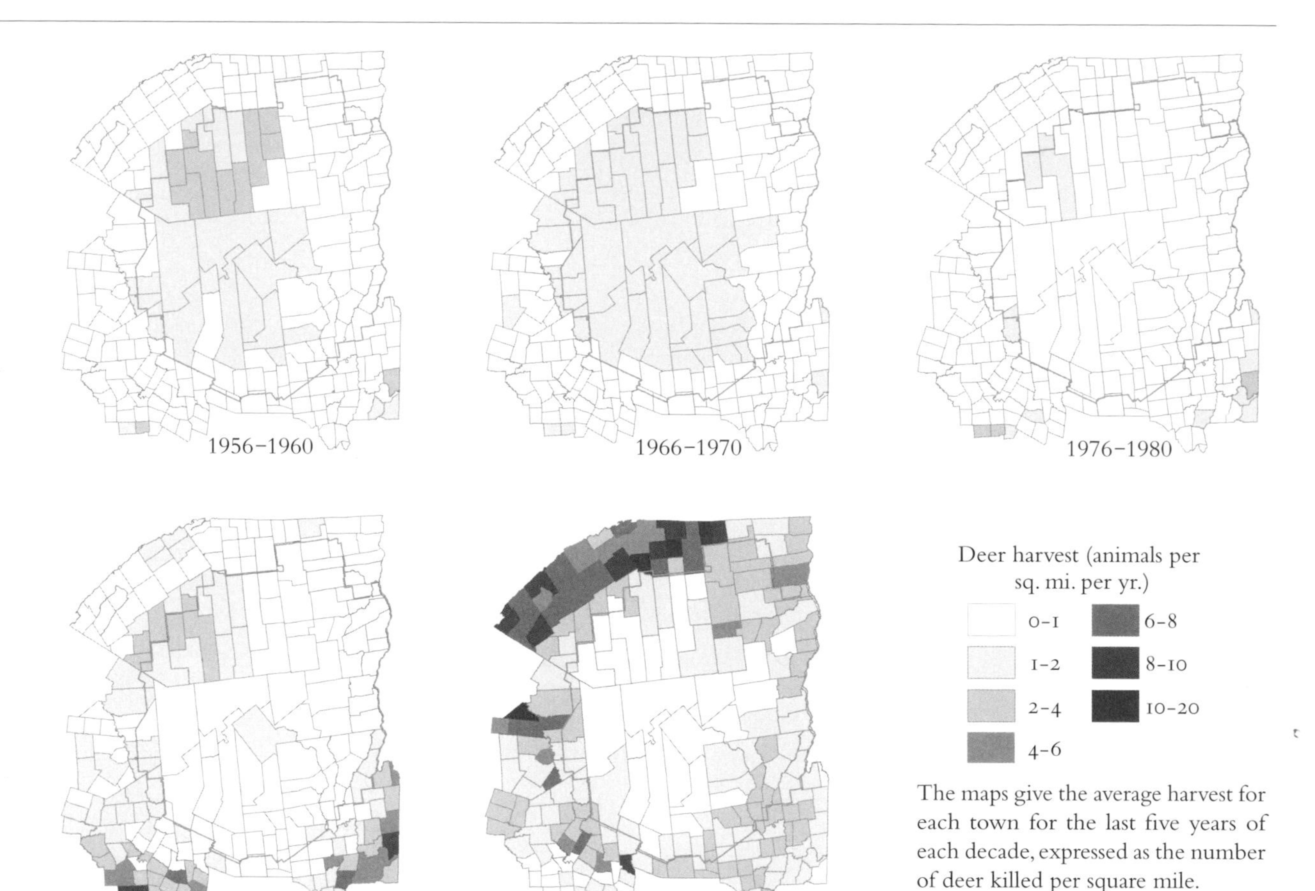

The maps give the average harvest for each town for the last five years of each decade, expressed as the number of deer killed per square mile.

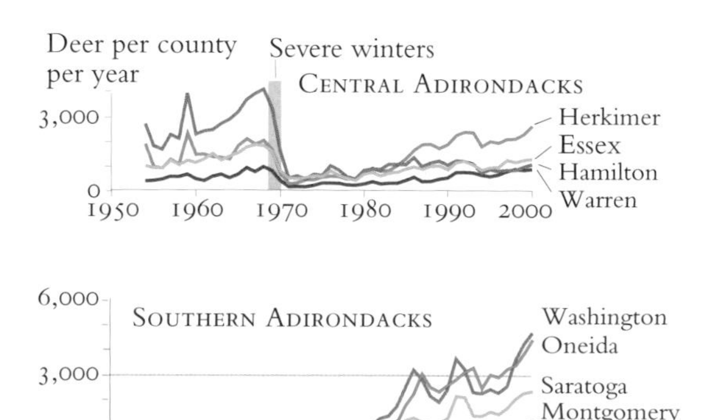

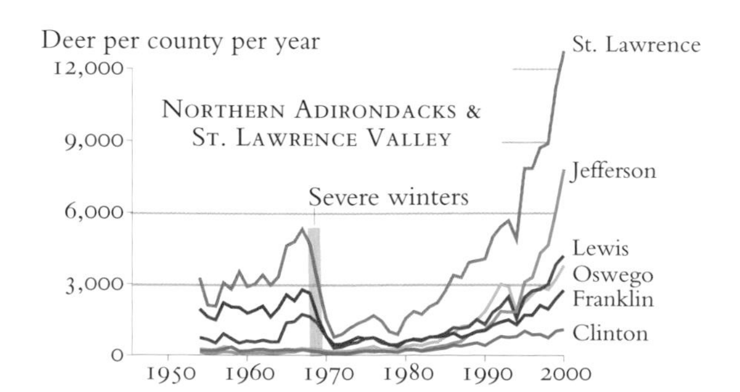

OUTSIDE the park the story was different. In 1950, deer harvests around the Adirondacks were low. Northern New York was too extensively farmed and had too many hunters and too few woods to support very many deer. Farm abandonment and a decreasing population of hunters changed this, and by 1985 deer populations in the valleys around the park were increasing rapidly.

The maps show that this was the northern edge of a larger pattern. A band of high harvests, which had been moving north for the last thirty years, reached the St. Lawrence Valley around 1990. By 2000, deer harvests in St. Lawrence County had reached 4–6 animals per square mile. This was five times higher than their 1950 levels but still lower than the unprecedented harvests of 10–20 animals per square mile across southern New York.

The northwards progression of the harvests suggest that climate may be involved. My guess is that farm abandonment, which was happening all over, provided the fuel for the population increase, and improved winter survival, progressing from south to north as the climate warmed and snow cover decreased, controlled the pattern.

Several details in the graphs on the opposite page suggest climate control. In the past, severe winters, like those of the late 1960s, caused heavy mortality in the northern Adirondack counties but not in the south. Since the 1980s, snow depths outside the park have been decreasing, and the fluctuations in deer populations, both in the north and in the south, have been relatively small.

If this interpretation is correct then, to a deer the winters of the St. Lawrence Valley today are about the same as a Pennsylvania winter of fifty years ago, and the Adirondacks are the only place in New York where there are still, for a while at least, old-fashioned winters.

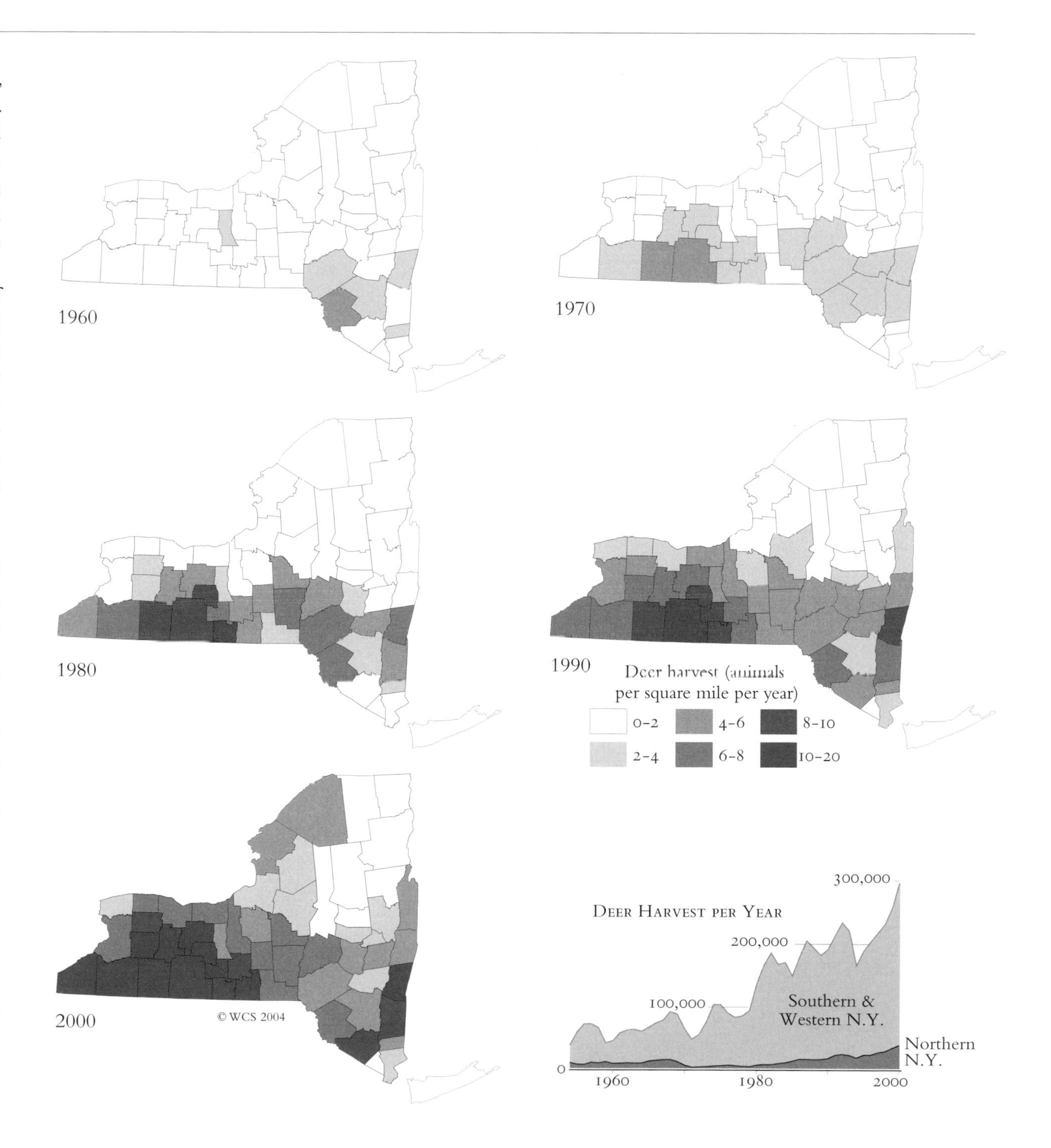

THOUGH the furbearers are all wide-ranging and adaptable animals, each species appears to have its own local geography.

Bears are taken throughout the park but only occasionally outside it. They are most common, or most huntable, in the mid-elevation towns, and less often taken in either the high mountains or the swampy northwestern watersheds. The locally high harvest in Inlet may reflect the annual Bagging of the Dump Bears, a popular if not very sporting event back in the 1980s.

Coyotes, by contrast, are taken mostly outside of the park. They do very well in farm country, and because of their wariness are hunted best in the open. Whether the spotty pattern of harvests reflects the distribution of coyotes, the areas of most active hunting, or simply erratic reporting, we don't know.

Otter are taken in low numbers throughout the park and in much higher numbers in the St. Lawrence Valley. They often use beaver ponds and seem to be most common in the northwestern watersheds where beaver are now spectacularly abundant. It is quite possible that the dramatic increase in otter harvests since 1980 is the result of a similar increase in beaver (*Map 4-5*).

Fisher are harvested over the widest area of any of the animals considered here. Harvests are low in the park interior, where motorized access is limited, and highest in towns that were once agricultural and now have had a century or more to return to woods.

Bobcats are the rarest and most erratically harvested of the furbearers. They seem to be most often taken in rocky hill country near the park border—places with cottontail rabbits and oaks and gray squirrels, where the hills are lower, warmer, and less evergreen than the central Adirondacks.

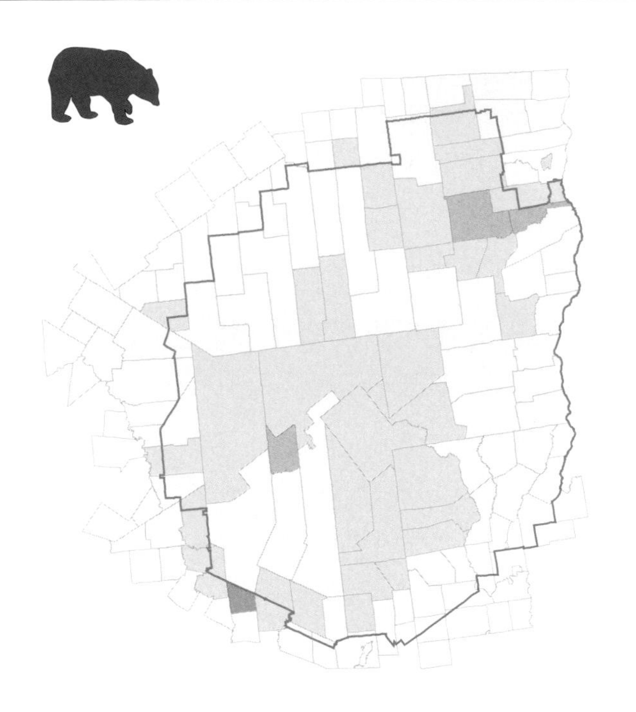

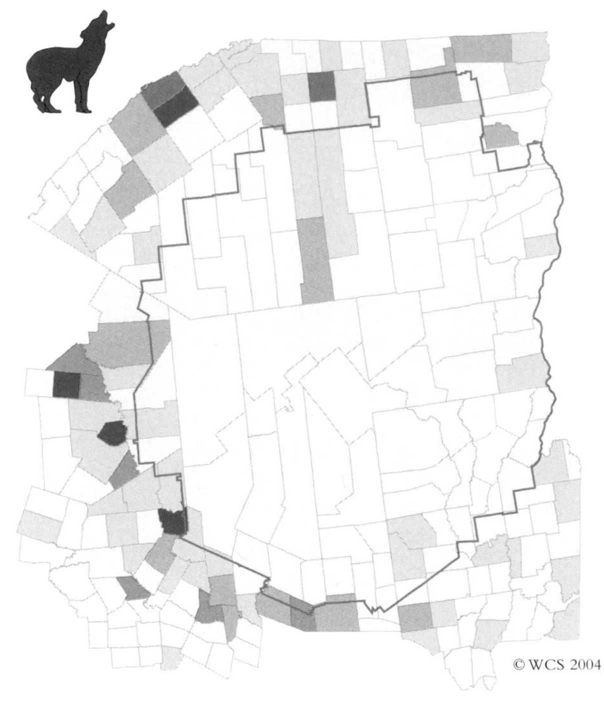

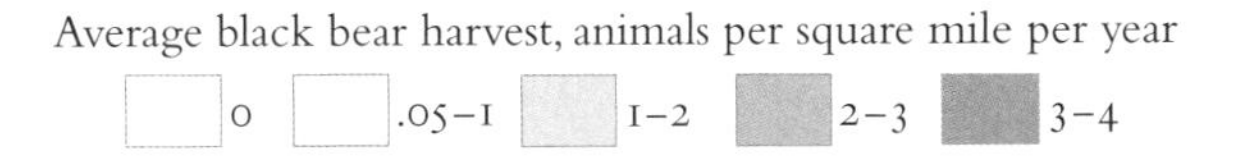

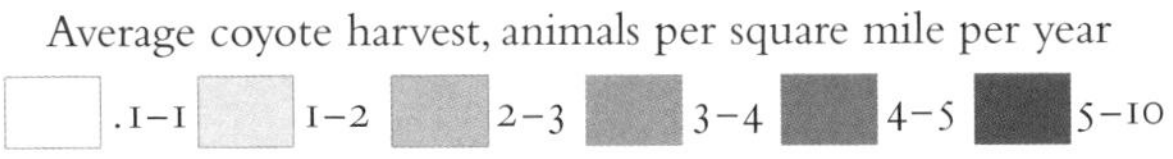

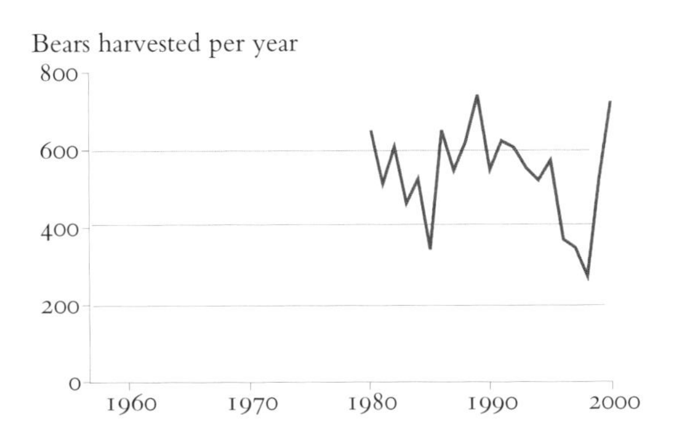

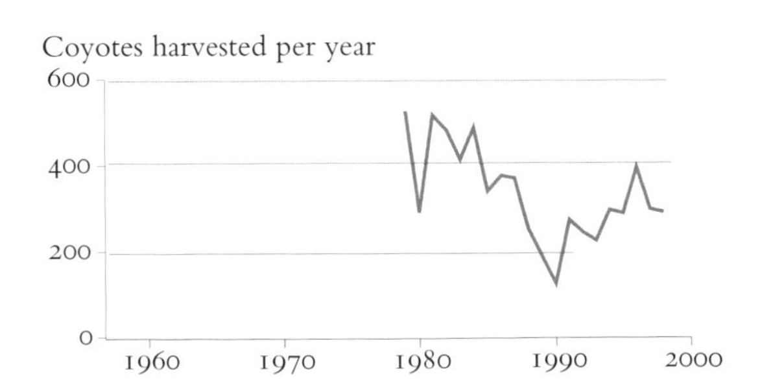

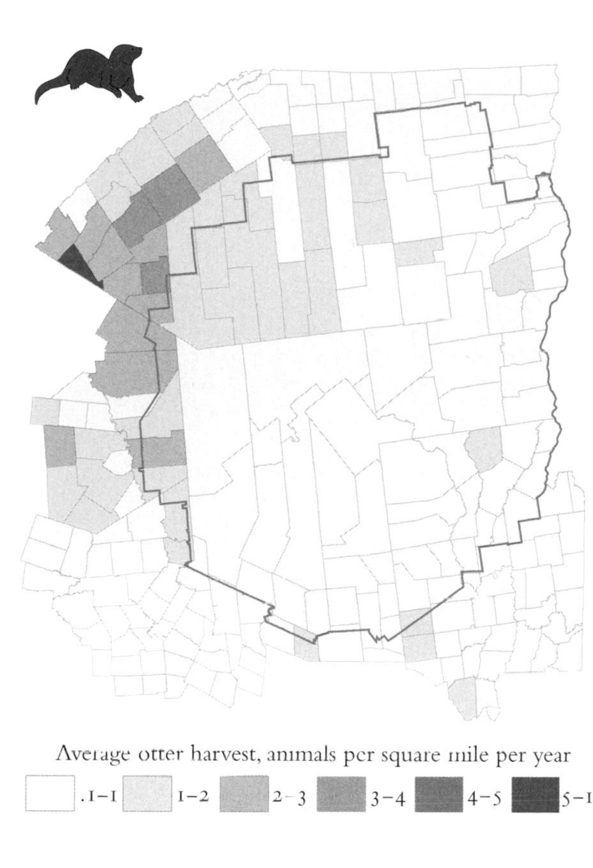

Average otter harvest, animals per square mile per year

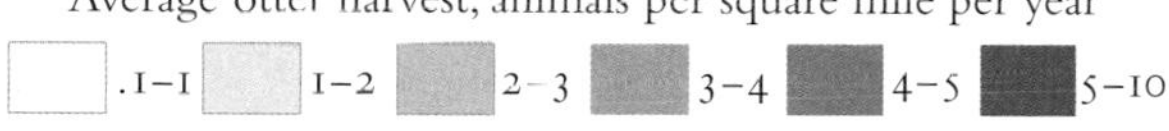

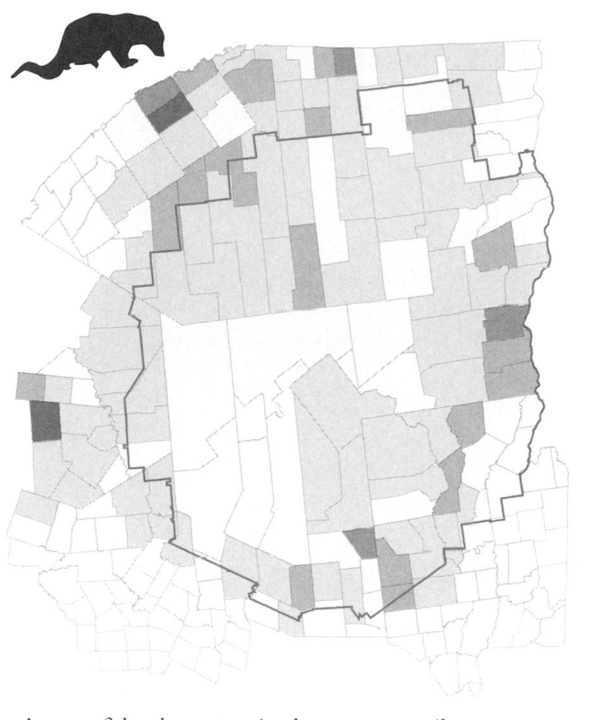

Average fisher harvest, animals per square mile per year

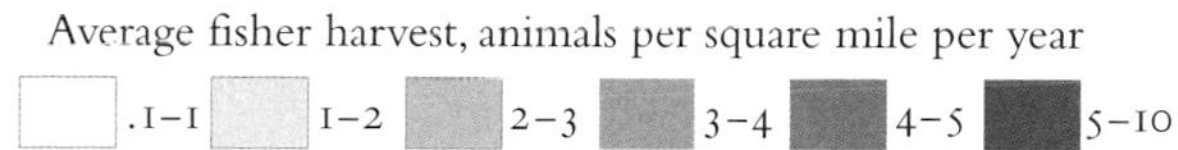

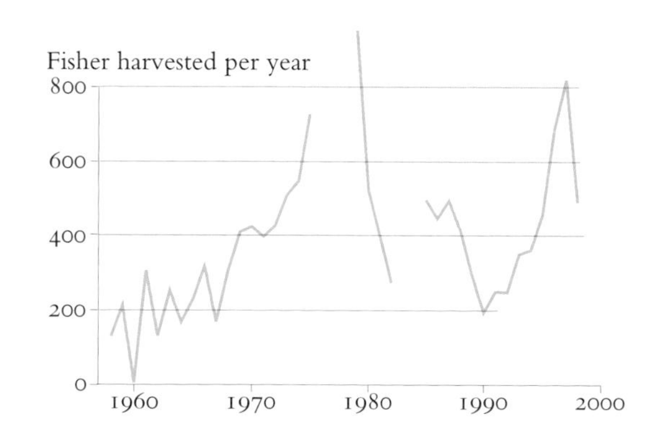

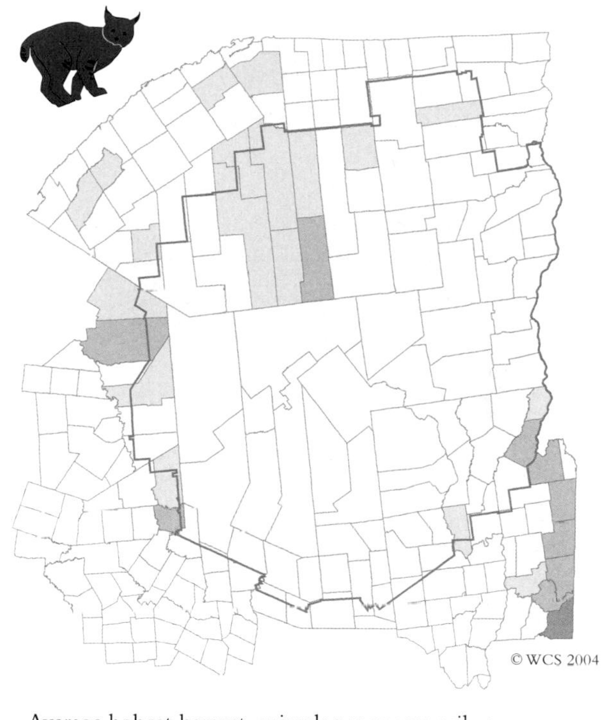

Average bobcat harvest, animals per square mile per year

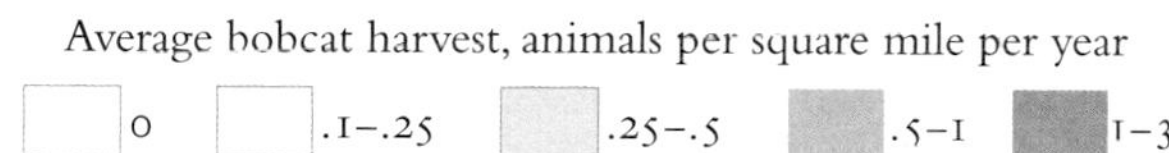

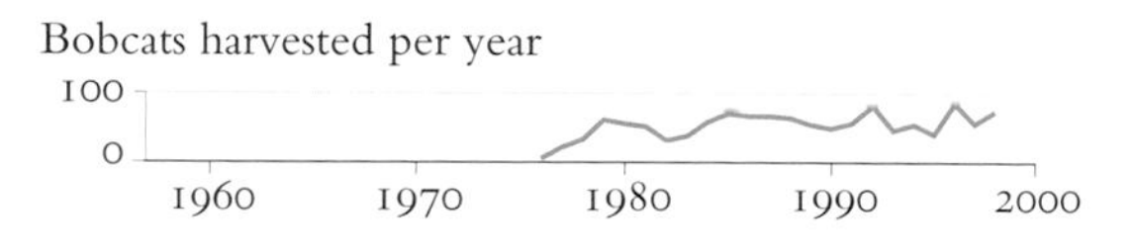

Total Furbearer Harvests

The maps give the average harvest per square mile per year from 1980 to 2000. The bear, coyote, fisher, and otter maps have the same coloring. Bobcats are rarer, and their map uses a different color scale.

HARVEST data like those in *Map 4-10* are interesting but limited. They tell us how many animals are being taken *from* a population but not how many are *in* the population. And worse, because the emphasize the most intensely hunted areas, they tell us almost nothing about the undisturbed and extremely interesting animal communities of the wilderness Adirondacks where few hunters and trappers go.

To study these communities, more systematic and less invasive methods are needed. Three researchers, Justina Ray, Matthew Gompper, and Roland Kays, completed such a study in 2002. They selected fifty-four three-mile study transects, put cameras and track plates along them, and walked them once a month to collect scats. In three years and about 8,000 trap-nights of effort, they collected 400 scats, 266 photographs, and 121 tracks from eleven species of carnivores. The results give us our first picture of carnivore distribution and carnivore communities in the Adirondacks as a whole.

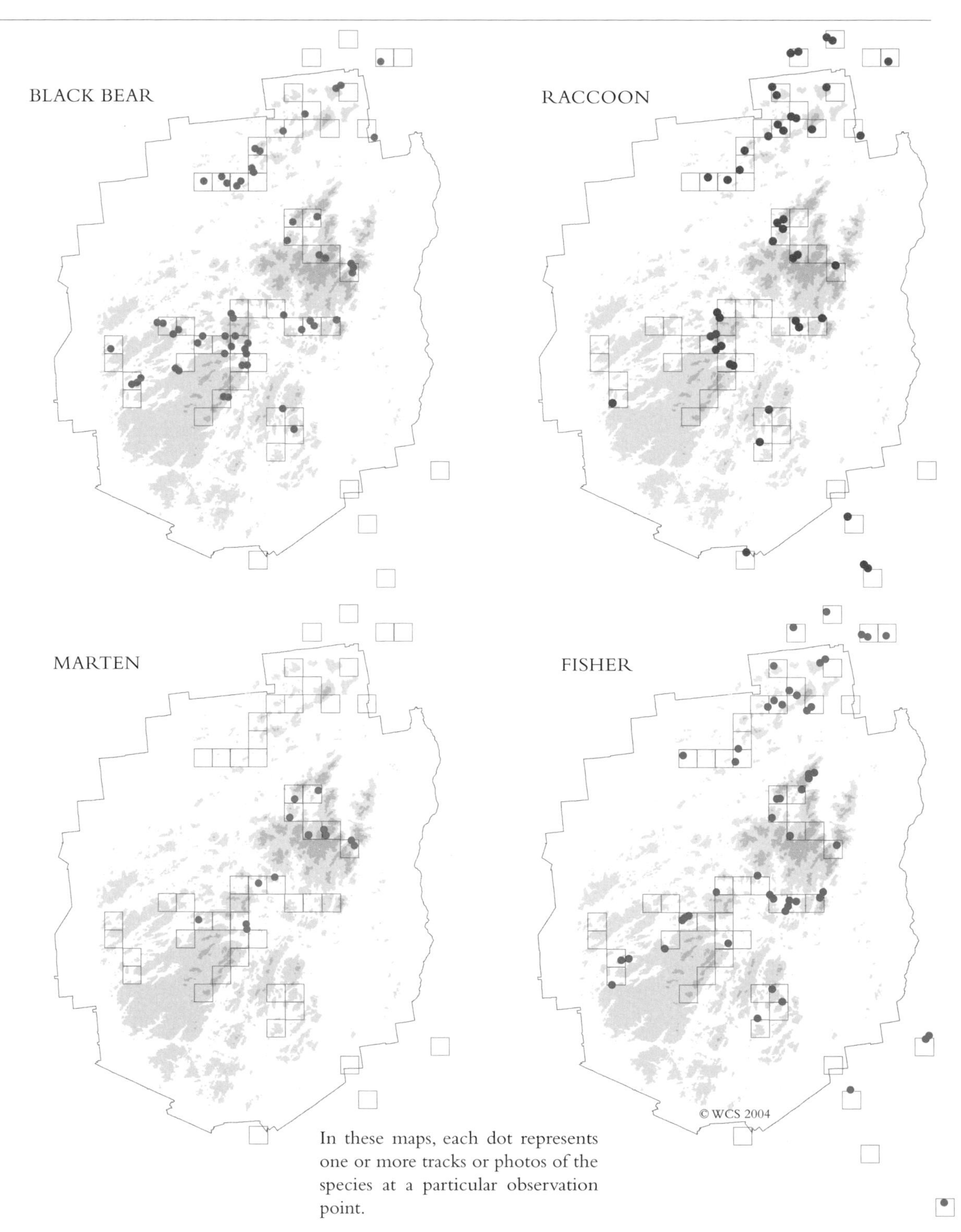

In these maps, each dot represents one or more tracks or photos of the species at a particular observation point.

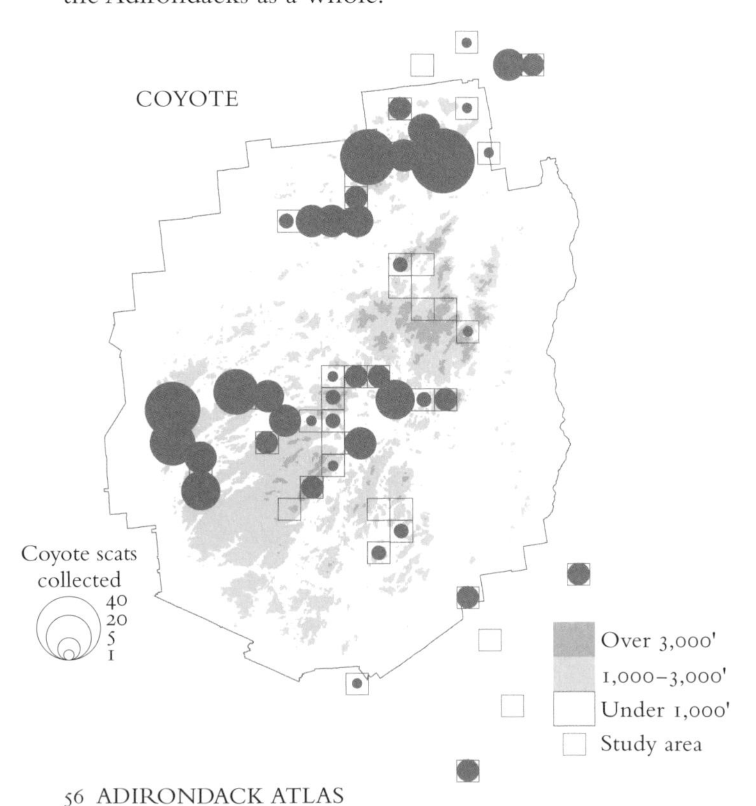

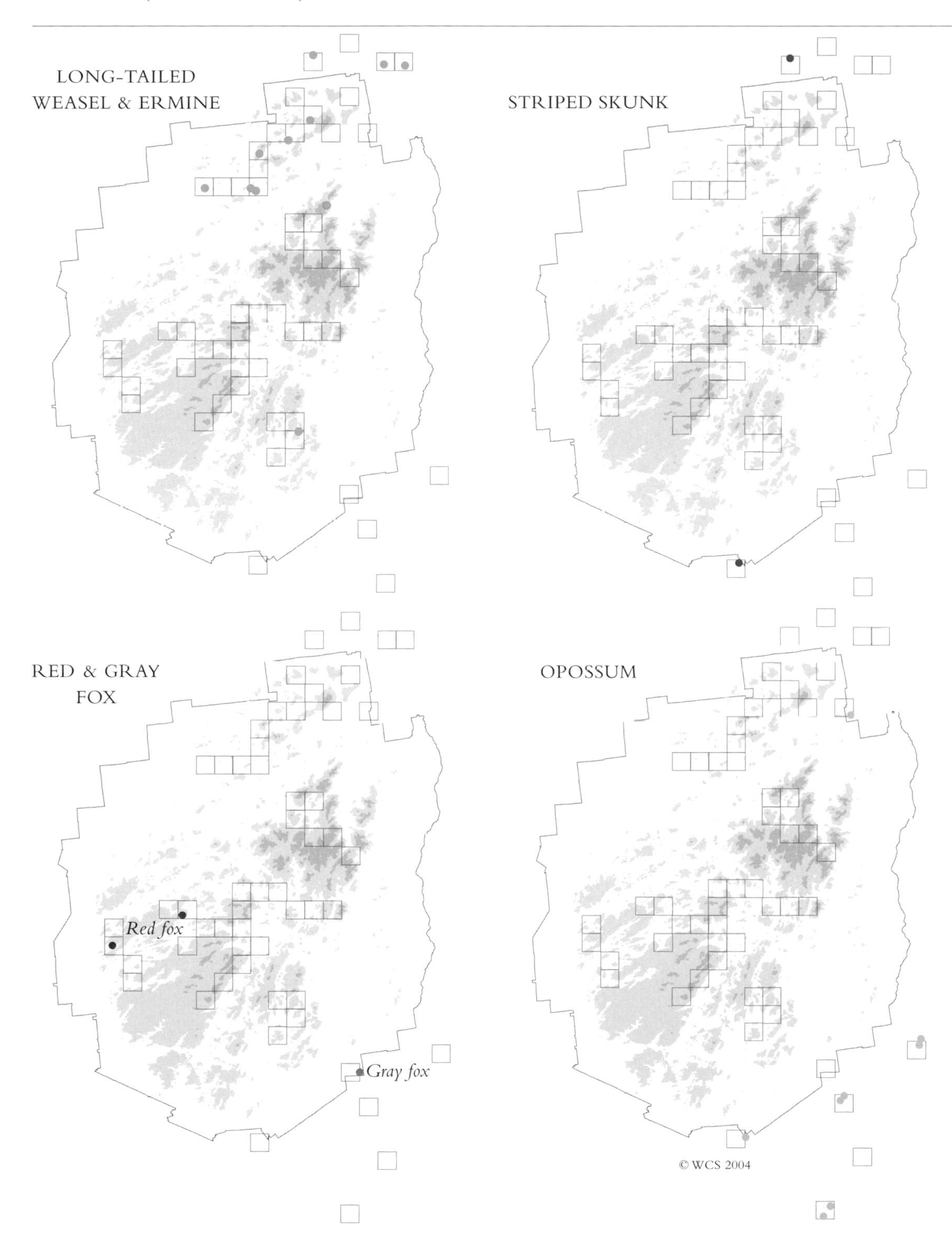

Only five of the ten carnivores documented in this study seem to occur regularly in the park interior. Coyotes and fishers, true ecological generalists, are the most widespread; they are found, though in varying densities, in both wild and settled areas and at almost all elevations. Raccoons are also widespread, though most common at low elevations and near water.

Bears and martens are more restricted. Bears are more common in the interior of the park than on the edges. Martens, which are mostly animals of the boreal forest, are limited to the central mountains. The Adirondack martens are the southernmost marten population known and are separated by over a hundred miles from other martens to the west and north.

The other five carnivores are all commoner outside the park than in. Weasels occur in the park but are rare. Foxes were even rarer and were all found near roads or houses. Skunks and opossum—the latter not a carnivore but in many ways ecologically similar to them—seem to have a very strong association with settled areas. Skunks are common in the Champlain Valley and have been reported from some of the larger Adirondack towns but were not found at any of the park interior sites examined in this study. Possums, which get frostbite easily, were only found outside the park. They are mostly restricted to low elevations and, possibly because good den trees are rare, often shelter in buildings.

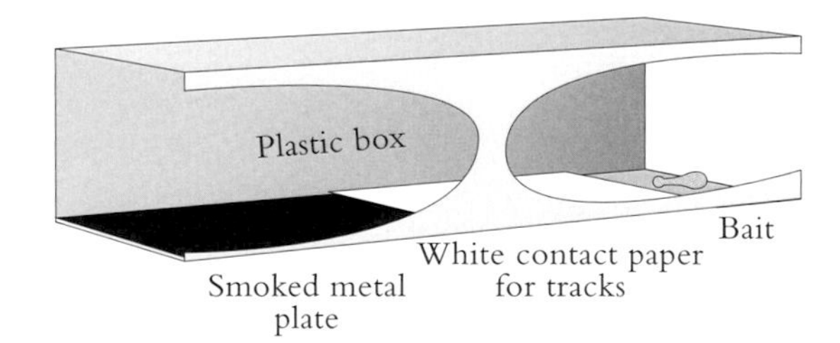

Track boxes are pantries with a memory. The animal enters from the end opposite the bait, picks up soot from the smoked plate, and leaves tracks on the contact paper. The tracks are extremely detailed and sometimes allow individual animals to be identified.

WOLVES, the only formerly widespread predator still missing from the Adirondacks, have recently made a dramatic comeback in the Great Lake States and mountain West.

The comeback was largely unexpected. In 1960 wolves were almost extinct in the contiguous United States. Their populations had been greatly reduced by hunting and poisoning, the ungulates that were their main prey species were relatively scarce, and hunters and farmers were for the most part intolerant of their continued existence and adamantly opposed to their recovery.

In the next forty years much changed. Hunting and farming both declined, and a new and significant pro-wolf constituency developed. Wolves were protected under the federal Endangered Species Act, and federal wolf-restoration programs were undertaken. The Mexican and red wolves, two of America's most endangered mammals, were bred in captivity and returned to the wild. Programs that compensated farmers for livestock losses decreased local opposition to wolf recovery. And, perhaps most important of all, greatly increased ungulate populations in many parts of the wolf's range provided abundant biological fuel for natural population growth.

The results have been dramatic. The current lower-forty-eight wolf population is about 3,000, perhaps five times larger than the population in 1960 and almost half the Alaskan population. The current growth rate will double both the population and the area it occupies in the next fifteen years. Interestingly, most of the increase has been the result of natural recovery and not reintroduction. The federal breeding programs, though essential for preserving endangered genotypes, have provided only a few hundred animals. The wolves' own breeding program, though less scientific, has provided a few thousand.

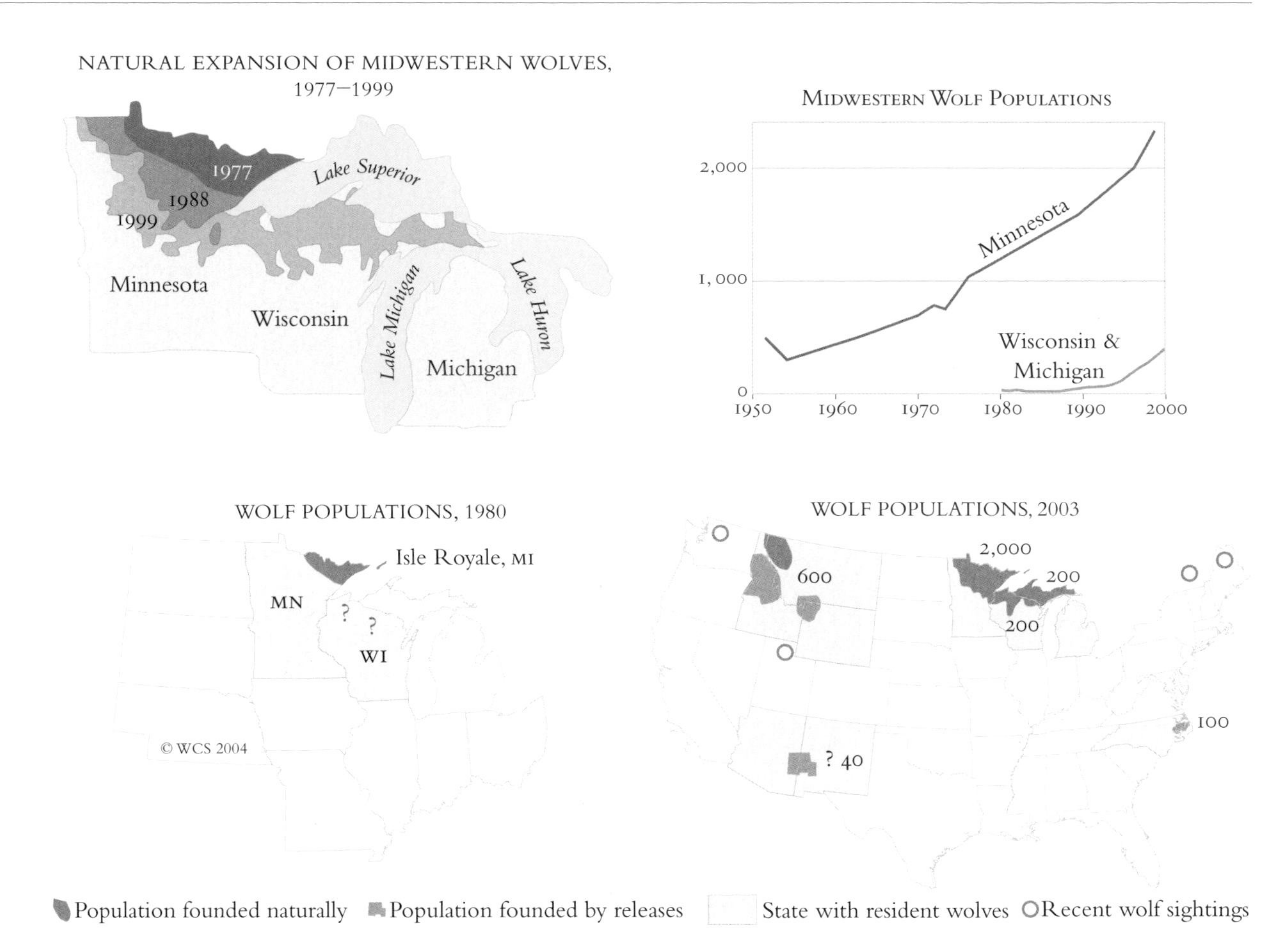

In 1980, the only wolves in the contiguous U.S. were a population of about a thousand gray wolves in northeastern Minnesota and related but much smaller populations on Isle Royale, Michigan, and probably in northern Wisconsin. Canada and Alaska had large populations, estimated at 60,000 and 7,000 animals, probably totaling over half the world wolf population. Gray wolves had been extinct in the East since about 1900, in the mountain West since about 1920, in the other midwestern states since about 1960, and in Mexico since around 1980. The last wild red wolves in Texas and Louisiana were captured for a captive breeding program in 1980.

Since the 1960s, when wolves were declared an endangered species in the lower U.S., wolf populations have been growing. The Minnesota population has quadrupled its size and range and founded new populations in Michigan and Wisconsin. Canadian wolves expanded naturally into the northern Rockies in the mid 1980s and were used for successful reintroductions in Idaho and Yellowstone in 1996 and 1997. Red wolves from captive stock were released in the Alligator River wildlife refuge in North Carolina in 1987, and captive Mexican gray wolves were released in the Apache National Forest in Arizona in 1998.

THE obvious question is whether a similar natural wolf recovery can occur in the Northeast.

Biologically, there seems to be no question that it can. Deer populations are at an all-time high, and there is as much wolf habitat in the Adirondacks as in Yellowstone, and more in northern Maine. Eastern coyotes, which are in some ways similar to eastern wolves, could hybridize with the returning wolves and alter their bloodlines. But wolves and coyotes have coexisted for a long time without losing their identities, and, though wolves returning to coyote country may change in the process, it seems unlikely that they will cease to be wolves.

Geographically, a return is more difficult. The wolves that arrived in Glacier National Park in 1986 came, in gradual stages, down a wilderness corridor where prey was abundant. A wolf that leaves Algonquin Park for the hundred and fifty mile trip to the Adirondacks must cross highly developed corridor where prey is scarce and hunters are abundant (see the *Human Footprint* map on p. ix). A wolf traveling from Quebec to Maine has a shorter distance to go and wilder country to travel in but still faces the formidable task of crossing the St. Lawrence River.

Culturally, there are significant barriers as well. Farming is still important in New York and Vermont, and farmers will not welcome a new predator. But feelings are changing, and I see increasing evidence that the farmers in my area have given up on eliminating predators and regard them as just another farm expense. Fishers and coyotes are a nuisance and mean that dogs sleep out and stock is brought in at night and barn cats disappear more frequently than they used to. But deer are in many ways a worse nuisance and the Lyme disease ticks they carry an unwelcome predator too. So there are trade-offs, and in any event it is easier to tolerate coyotes than to get rid of them. A new predator that was not much larger than a big coyote and kept to itself might not be noticed at first. After a few years, because rural culture is adaptable too, if it kept the deer down and howled prettily on winter nights, it might even be welcomed.

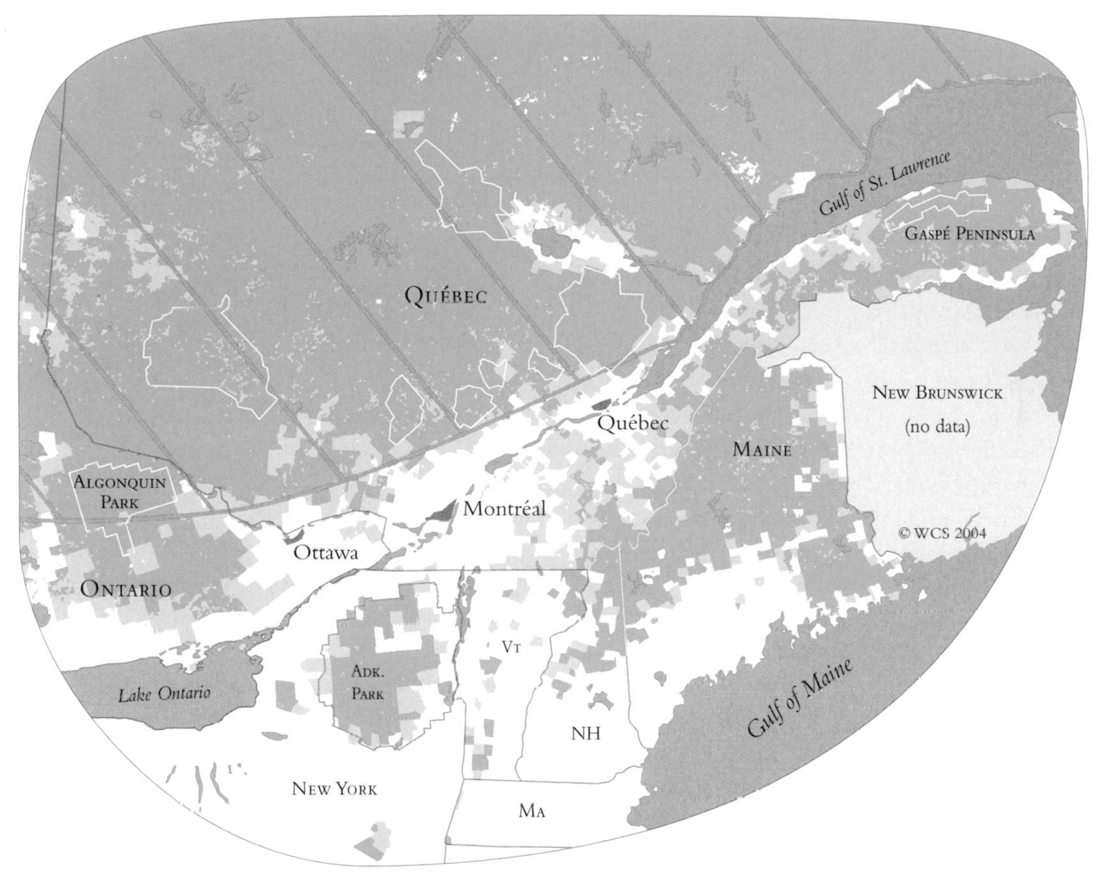

The map shows the human population and road density in areas where wolves do and do not occur. It does not take into account the density of prey, which is not well enough known to be mapped. Large amounts of potential habitat exist in Maine and the Gaspé, and a smaller but still substantial amount in the Adirondacks. To reach this habitat wolves will have to leave wild areas where they have some protection and pass through settled areas where they don't. When they do so they will be at risk: currently humans cause about 60% of all the fatalities in eastern wolf populations. But dispersing animals may be less vulnerable than resident ones, and wolves are prodigious dispersers. Movements of fifty or a hundred miles are routine for young wolves, and movements of several hundred miles are rare but possible. Several recent sightings suggest that young wolves, against the odds, have crossed the St. Lawrence and that one or two may have entered the U.S. One or two wolves is certainly not a population. But it is certainly the raw material from which a population may someday be made.

Current wolf habitat

Potential core wolf habitat, less than 10 persons and 1 mile of road per square mile.

Potential dispersal habitat, less than 26 persons and 1 mile of road per square mile.

Unsuitable for wolves

Protected areas in Canada

Urban areas

LIKE all northern landscapes, the Adirondacks are strongly patterned, and so have a predictable botany. Their flora consists of several hundred common plants, woven by climate and topography into a dozen or so basic communities—woods, wetlands, shores, outcrops, tundra, and so on. To the botanist's eye these communities are subtly and tellingly varied. But, because climate and landscape weave by fixed rules, they are not surprising.

For surprise, which is to say for the places where nature goes beyond pattern and rule, the naturalist seeks out rare places and rare species. Because rarities appear and disappear and because a single botanist (or single generation of botanists) will only see a few of them, this is always an unfinished business.

The map shows what the author and a few other botanists have learned in the last twenty years. Compared to the other maps in this book, it is doubtless very incomplete. But it is different in interesting ways from the map that the previous generation of botanists would have made. And it is probably very different from the one that the next botanical generation, documenting climate change and the loss of boreal species, will have to make in a few years.

The map emphasizes two sorts of special sites. The *open bogs* are low diversity habitats with northern specialties. The *limy outcrops*, are highly diverse but less geographically specialized. Where water and lime overlap, as in limy peatlands, things start to get seriously interesting. And when in addition, as at some of the cliffs on the upper Hudson, the microclimate is peculiar, then we see some exotic things indeed.

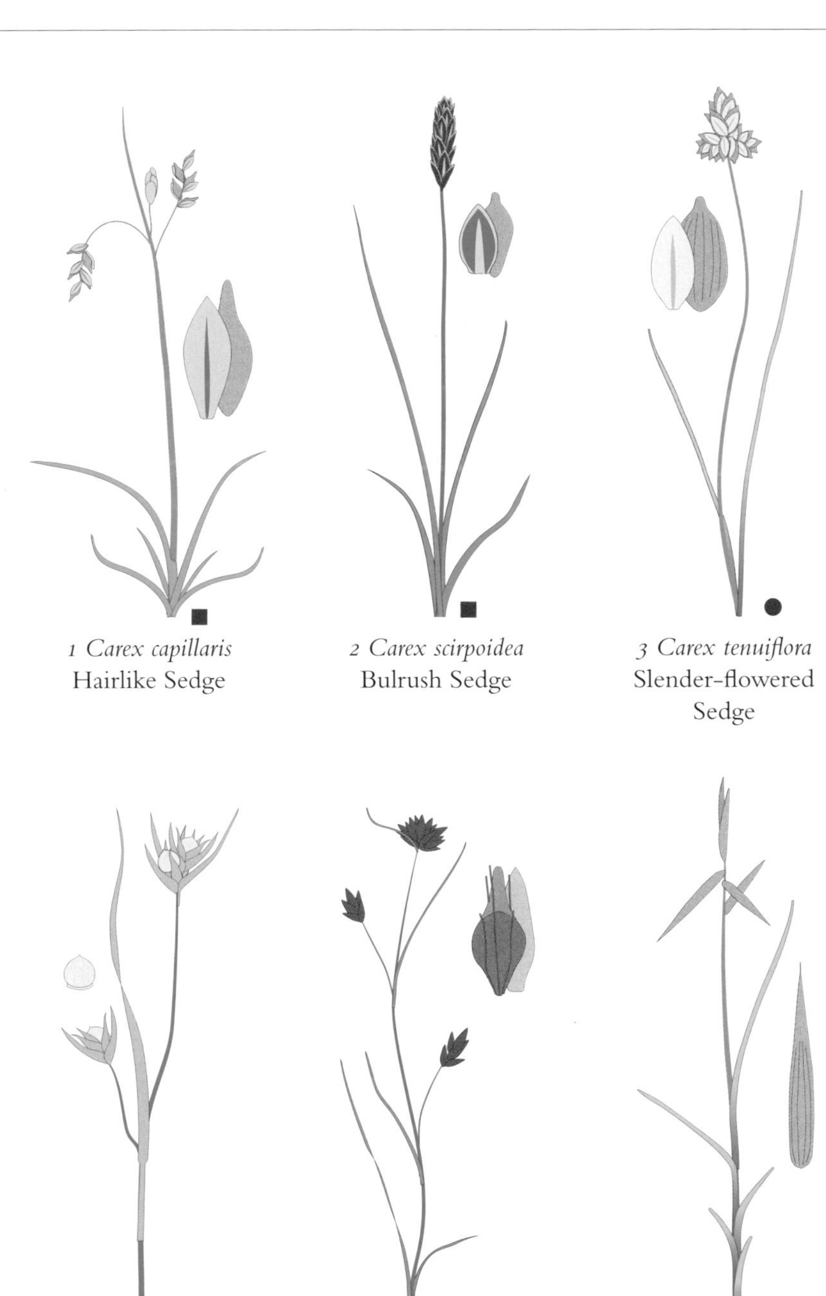

1 *Carex capillaris* Hairlike Sedge

2 *Carex scirpoidea* Bulrush Sedge

3 *Carex tenuiflora* Slender-flowered Sedge

4 *Scleria triglomerata* Three-fruited Nut-rush

5 *Rhynchospora fusca* Black Beak-rush

6 *Carex pauciflora* Few-flowered Sedge

Rare & Uncommon Sedges. The sedges are our largest family of flowering plants and have many unusual species. The six shown here have strong climate and habitat preferences and are limited to, and often good indicators of, fairly special places.

Carex capillaris and *Carex scirpoidea* are arctic species that like wet, calcium-rich rocks. *C. capillaris* lives in cold, spring-fed wetlands and at cliff-bases where it is wet year-round; currently only one plant is known in New York. *C. scirpoidea* lives on exposed cliffs that can get very dry in summer. It is locally common, but only at a very few sites.

Carex tenuiflora is a subarctic species that grows in wet mossy woods where the bedrock or seepage is at least somewhat limy. Limy bogs are the rarest Adirondack wetlands, and it is one of our rarest wetland species. *Carex pauciflora,* though also subarctic and also liking moss, is restricted to dry, very acid bogs and bog woods but can be locally common there.

Scleria triglomerata and *Rhynchospora fusca* are plants of wet sandy shores. The only Adirondack site for *Scleria* is an odd and ecologically violent meadow by the Hudson River which is covered by thick ice-flows till late in the spring. *Rhynchospora fusca* is found in more serene but also unusual places, often at the edges of bogs but not in them. It needs an open, peaty place, perhaps even a slightly unstable or dangerous one, where something excludes the big, conservative, everyday plants and lets the small, delicate, adventurous ones flourish.

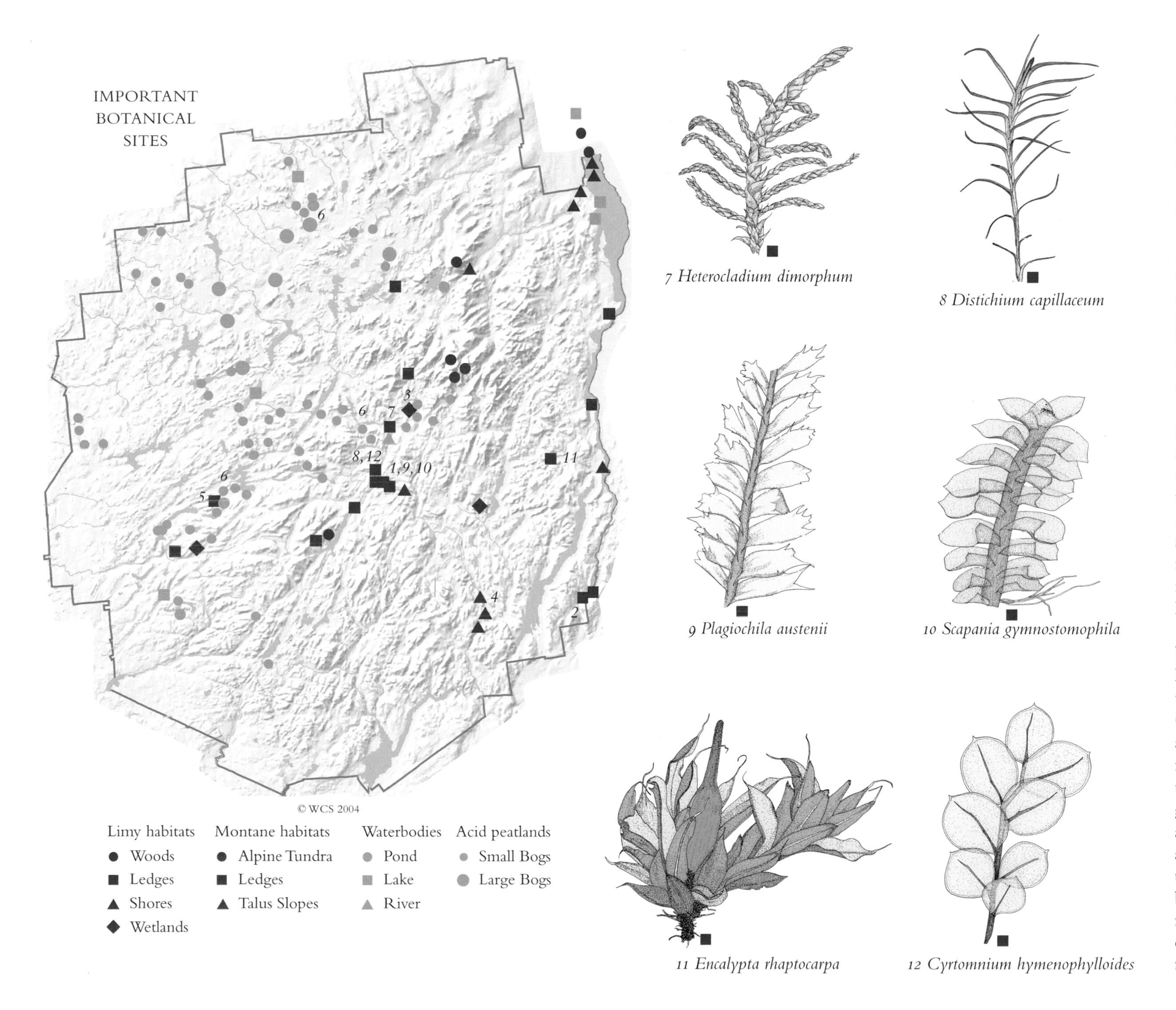

7 *Heterocladium dimorphum*

8 *Distichium capillaceum*

9 *Plagiochila austenii*

10 *Scapania gymnostomophila*

11 *Encalypta rhaptocarpa*

12 *Cyrtomnium hymenophylloides*

Rare Mosses & Liverworts of Limy Cliffs. Here are six of our rarest species, each known from only one or two places in the Adirondacks. All like cool, sheltered cliffs, and are most often seen on river-shore ledges in the Hudson and Champlain drainages. They all like the same sort of places, but to be rare is to be choosy, and we usually find just one or two at any spot.

Most are northerners, and, because they are probably older than the modern continents and oceans, very widely traveled. *Scapania gymnostomophila,* a minute plant found in rock crevices, reaches the northern tips of Greenland and Ellesmere Island. *Cyrtomnium,* a translucent, blue-green moss that also likes cracks, is found across the Arctic from Alaska to Greenland and Asia. It has almost never been seen in fruit. *Encalypta*, which likes to be close to water, also grows in Iceland, Hawaii, and Japan. And *Distichium,* a slender, featherlike moss that is perhaps the most beautiful of these six, is found, and doubtless appreciated, in the mountains of five continents.

RABIES, an ancient disease of unknown origin, is caused by a virus and transmitted by the saliva of infected animals. It can be prevented by immunization with a vaccine, either before or after exposure, but is almost always fatal after symptoms develop. It was probably brought to North America by domestic animals and, until the twentieth century, was largely a disease of dogs and cats.

By the mid-twentieth century, vaccination had largely eliminated rabies in domestic animals. But microbes evolve quickly, and sometime in this century the rabies virus diversified into three strains infecting wild animals. The raccoon and fox strains are epidemic, alternating between outbreaks and latent periods. Because these strains are relatively new, their outbreaks can be sudden and severe. The bat strain is endemic—continuously present at a low level in most bat populations but not subject to outbreaks.

In 1999, New York was near what it hoped was the end of the largest animal rabies epidemic ever to occur in North America. Raccoon rabies had entered New York about 1990 and spread north at a rate of about thirty miles per year. By 1999 it had nearly reached the Canada border, but failed to spread to either the central Adirondacks or eastern Long Island. During the epidemic, over 14,000 cases of rabies were confirmed in twelve species of wild animals and five of domestic animals, including everything from bobcats to cows. Almost none of the confirmed cases involved dogs: rabies is now a disease of wild animals, and domestic dogs are mostly well-vaccinated and not very wild.

Remarkably, though thousands of people were bitten or otherwise exposed to raccoon rabies, there were only two cases of human rabies in New York in this period, and neither involved the raccoon strain of the virus.

Managing the epidemic has required a major public health effort involving hundreds of thousands of tests

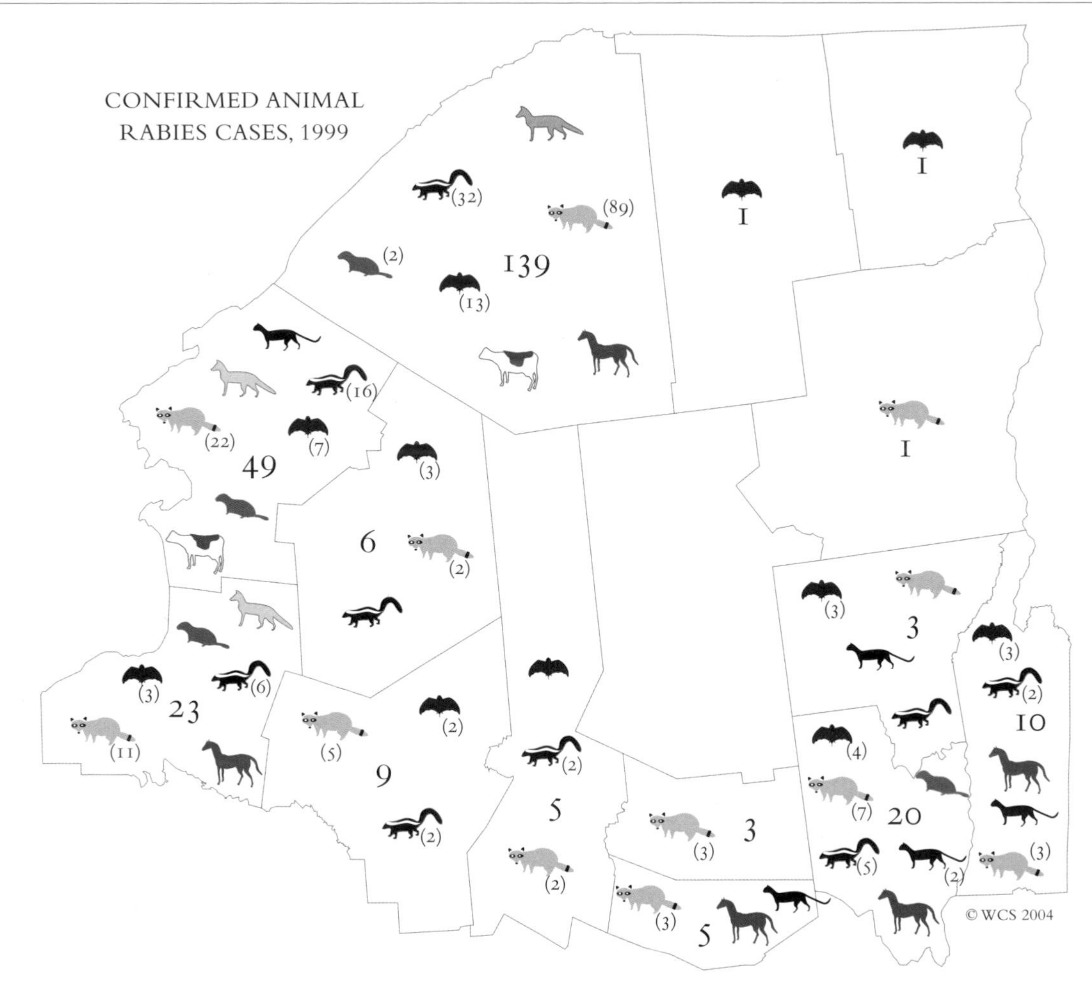

CONFIRMED ANIMAL RABIES CASES, NEW YORK STATE, 1999

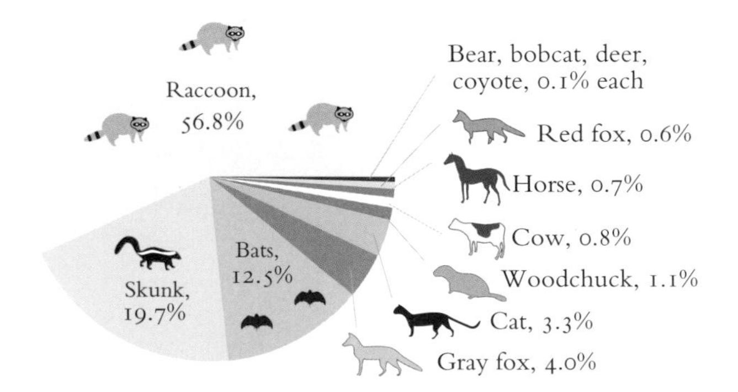

In 1999 the rabies epidemic was at its peak in far northern New York. There were 234 confirmed cases of raccoon rabies in nine species of wild animals and three species of domestic animal, plus 41 cases of bat rabies, all in bats. There were no cases in dogs at all, suggesting that the current vaccines and vaccination programs are working extremely well.

The larger numbers are the total confirmed cases in each county; the numbers in parentheses the number of cases confirmed in that species, when more than one.

and vaccinations, at a cost of over a hundred million dollars. It is an example of how well an excellent public health system, given the resources it needs, can respond to a major epidemic.

There is much to be learned from this. First, and most important, it is a reminder of how biologically active the human-animal interface is and how rapidly diseases can arise and spread there. Rabies is an ancient disease of this interface; AIDS, mad cow disease, West Nile virus, and hantavirus are new or newly emerged ones. There will be many others.

Second, the suddenness and severity of the epidemic warns us how lucky we were that this was a well-known disease for which we had diagnostic tests and good vaccines and a well-funded health system to administer them. For many other human-animal diseases there are neither tests nor vaccines. In many places, including some where the human-animal interface is particularly turbulent, there is almost no public health system at all. With a less-known disease in a less-prepared places, thousands of human fatalities might have resulted.

Finally, the New York distribution of the disease suggests that rabies not only originated from domestic animals but prefers to remain within the world of human-sponsored animals. Like the cats and dogs that carried rabies in the past, the majority of the raccoons and skunks that are carrying it today seem to be human commensals, common near settlements and rarer or less encountered in continuous forests. Although there are raccoons throughout the Adirondacks (*Map 4-11*) and many confirmed animal rabies cases in the rural towns of the Adirondack edge, there were few in the Adirondack interior. Perhaps the virus never got there, or perhaps it did and the animals that it infected died without meeting people. Either way, the edge of the park interior, marked roughly by the long-obsolete Blue Line of 1892, seems surprisingly relevant as a biological boundary today.

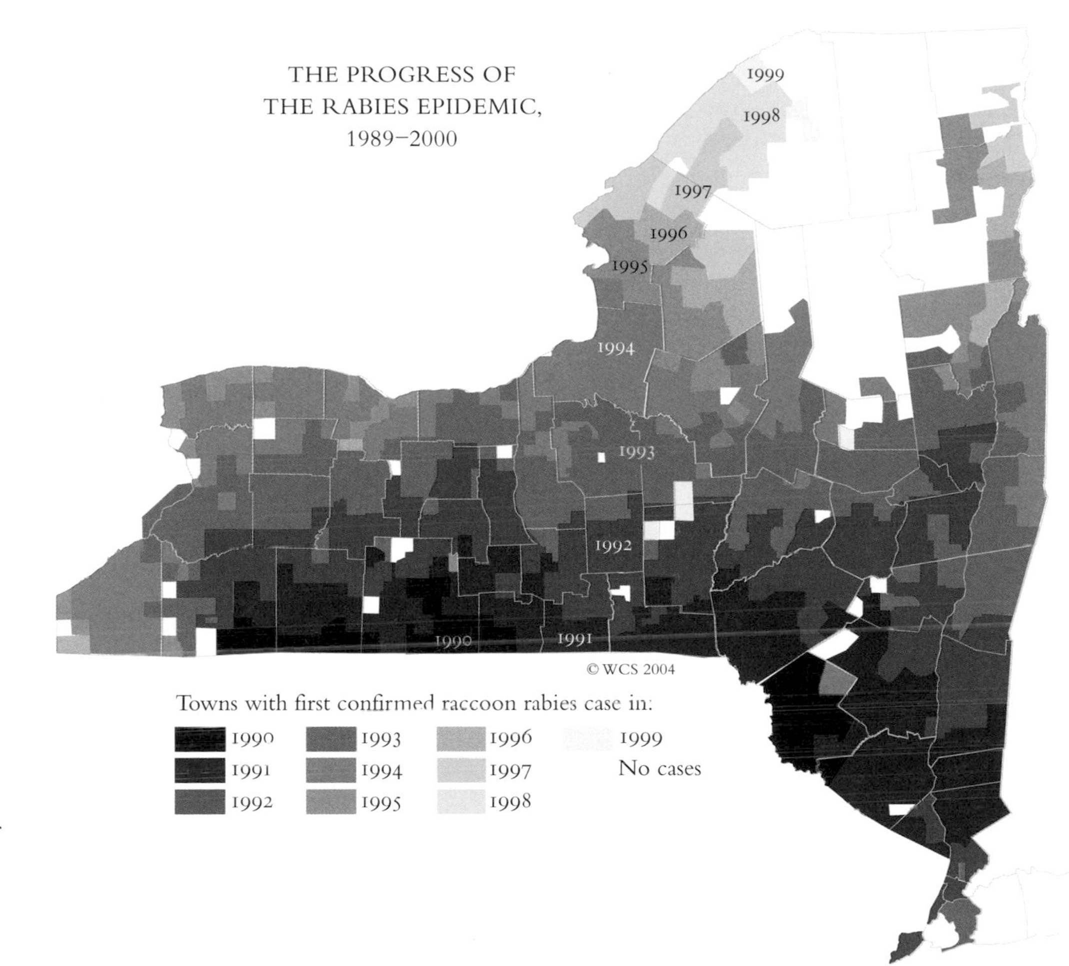

Rabies reached the Champlain Valley and Adirondack periphery by 1995 but did not seem to spread further northwards or inwards.

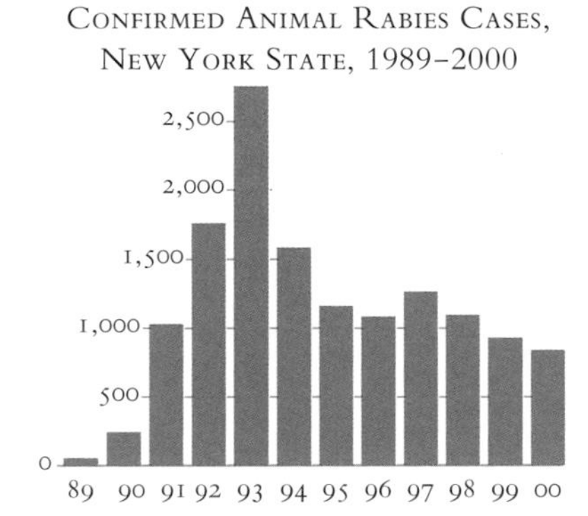

Though rabies cases occurred in Westchester, New York City, and Staten Island, rabies never spread to eastern Long Island, even though raccoons are abundant there. It is possible that, like the California freeways, the north-south parkways on western Long Island are such effective animal killers that the eastern parts of the island are now biologically isolated from the west.

THOUGH not much discussed by the tourist bureaus and agencies, there is a serious bear problem in the Adirondacks. Backpackers in the High Peaks are most at risk, followed by visitors at a number of popular campgrounds. Many residents are affected as well.

In 2000, an average year, at least three hundred people in the Adirondacks were sufficiently scared, annoyed, inconvenienced, or enraged by bears to report the incident. Ordinary bad bears stole animal and human food, broke into freezers, cabins, and vehicles, ate corn, smashed beehives, entered a house without invitation and took a pot roast from the table, scratched at a tent where a six-year old was sleeping, followed two ladies out walking, killed a pig, licked a camper's arm, and cuffed a woman. An exceptionally bad one charged a uniformed ranger who stopped it with pepper spray.

None of these incidents was serious, except of course to the bees and the pig. But many of the 135 hikers who reported bear problems in the High Peaks lost all their food and had to abort their trips. And any one of these incidents could have turned nasty. Sooner or later one probably will. Bears, under any circumstances, are big dangerous animals that are used to having their own way. Dangerous animals in remote areas, where travelers know the rules and the risks, are one thing. Dangerous animals, now thoroughly habituated to people, in popular camping areas are quite another.

For some years the official position was that if campers did the right thing—hung their food, left coolers in their cars—there would be no problems. That turned out, as such advice often does, to be wishful thinking. Portable bear-proof containers, which are now being used extensively in some western parks, may be a better answer. They are

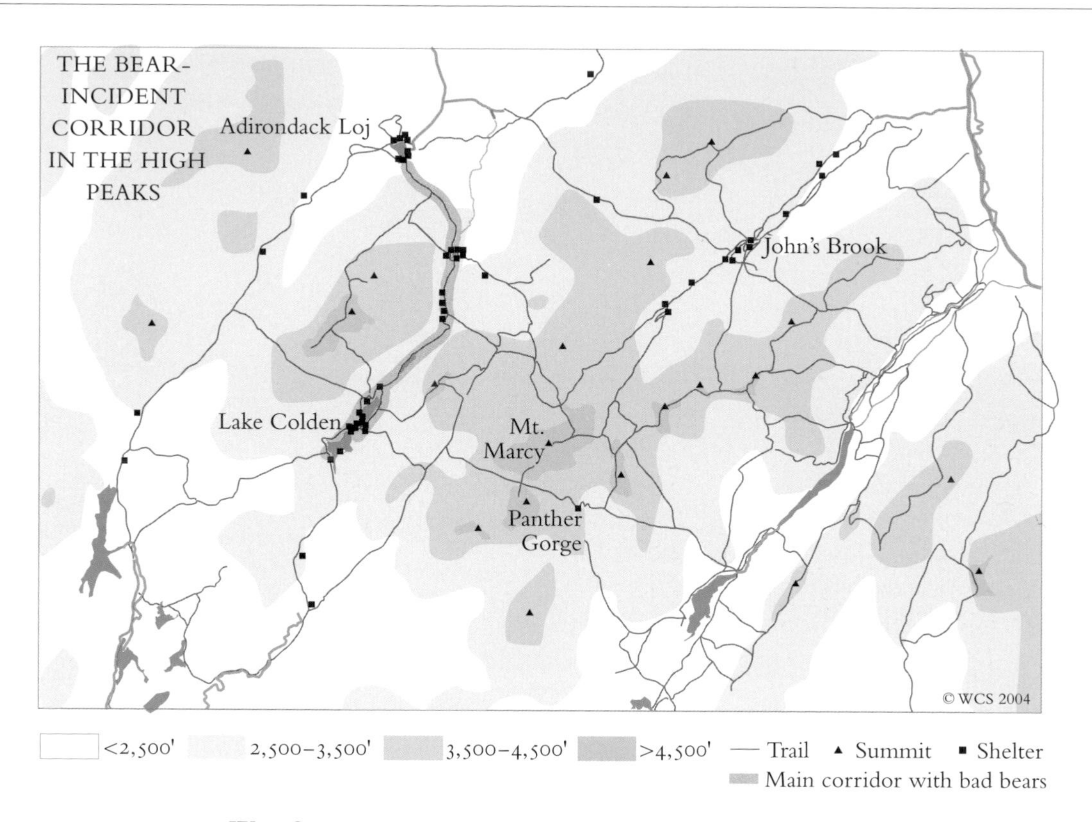

WHY SUSPENDED FOOD SACKS ARE CALLED BEAR BAGS

July 22, 2000, Lake Colden. Bear came to twelve feet of us while having dinner. Ran after our neighbor . . . stole their food, went in neighbor's tent. Roamed in our neighborhood all night. Stole all our food (but cheese) in morning. Very frightening.

July 26, 2000, Panther Gorge Lean-to, 6 p.m. Bear showed up just as we arrived to set up camp . . . He would not be scared away and after an hour he got our food.

Aug. 2, 2000, 5:55 a.m. We had a very well hung bag—between two trees twenty feet in the air. The bear worked on it from 7 p.m. until midnight and finally broke one of the tops of the trees to get the bag. Very large bear.

Aug. 6, 2000, trail-head to Phelps Mt. Food hung very high both nights . . . [Bear] slashed rope and took bags into the woods to chow down on the little that remained. Growled at Dad.

Aug. 6-8, 2000, Colden Dam. Bears got food every night. Yellow-tagged bear watched us hang food. Cable lines don't work without the pulleys and cable—bear climb and chew through the ropes!

Sept. 11, 2000, Beaver Pond Lean-to. Bear broke a bear bag pulley system and ate four days of food at one sitting. Needless to say our trip was cut short . . .

Sept. 17-19, 2000, Colden. Bears got food on all three nights from campers at the shelters; all parties had to leave as food was stolen; two bags were hung on a DEC cable.

moderately expensive but much less so than a ruined backpack and an aborted trip. A possible drawback is that when bears learn that they can't get into food containers at night they may get more daring about daytime and dinner-time raids.

Shooting the problem bears is of course an option, though probably not a good one. Ecologically it wouldn't matter very much. About five hundred bears are shot in the park every year already; taking another dozen would likely make no difference to the population. But it would likely have no effect on the bear problem either. Bears are mobile and well-informed, and experiments with shooting them in the western parks suggest that bear gangs have no trouble finding recruits to fill vacancies. And in any event it would likely upset people: oddly, killing animals for public safety is more controversial than killing them for sport.

Though big wild animals like bears and moose are potentially very dangerous to people, it is the smaller, tamer animals that do the most harm. In 2000, the year when bears charged, cuffed, scratched or licked about six people, 2,700 people in eleven Adirondack counties were treated for injuries from other animals. About a hundred of these were injured by wild animals, mostly rabid raccoons and skunks (*Map 4-14*). The remaining 2,600 were injured by pets. Dogs were responsible for 2,100 injuries, including many of the most serious ones. Cats, though lower and lacking as many opportunities, still managed to get their teeth or claws in at least 474 of their owners and neighbors.

It thus appears, at least at present, that the actively dangerous animals are as likely to be in houses as in forests. The danger of the wild is not a myth—cougars and coyotes and wildcats and bears are certainly capable of causing serious harm. But thus far they are not doing their share.

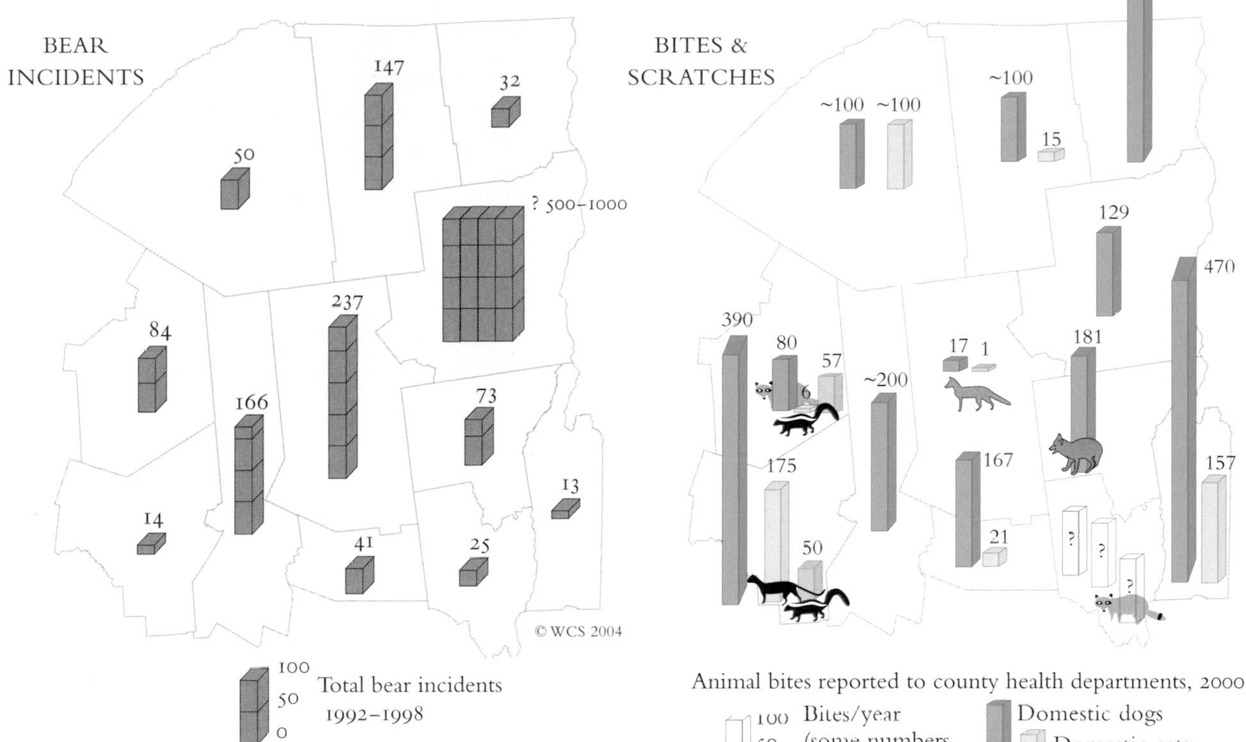

Animal bites reported to county health departments, 2000

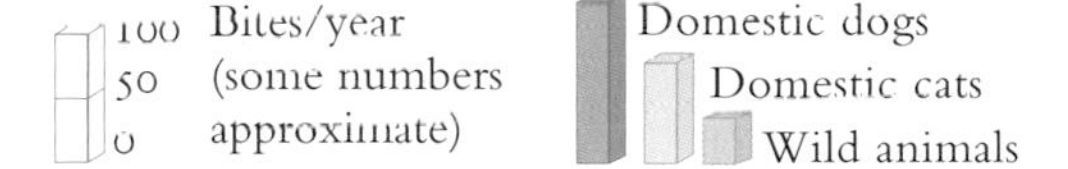

Total Bear Incidents For 12 Counties, 1992-1999, Excluding High Peaks

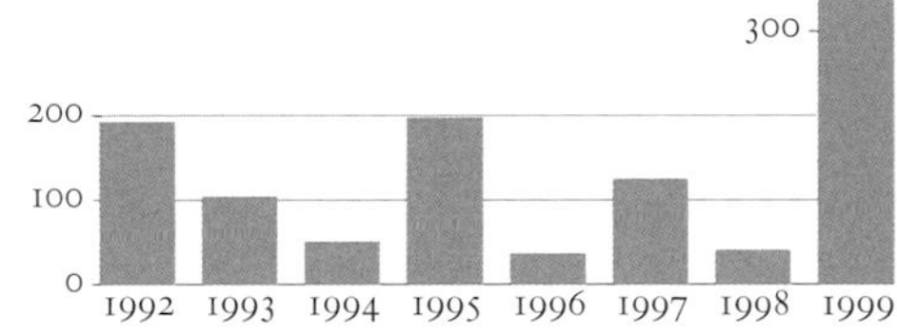

All the counties except Essex are based on DEC reports. The number for Essex County, though surely exceeding that of the rest of the park put together, is not known. Our estimate comes from Adirondack Mountain Club records. The DEC, either as a bookkeeping convenience or a comforting pretense, records the hundred plus bear encounters in the High Peaks each year as a single incident.

Injuries From Animals, 2000

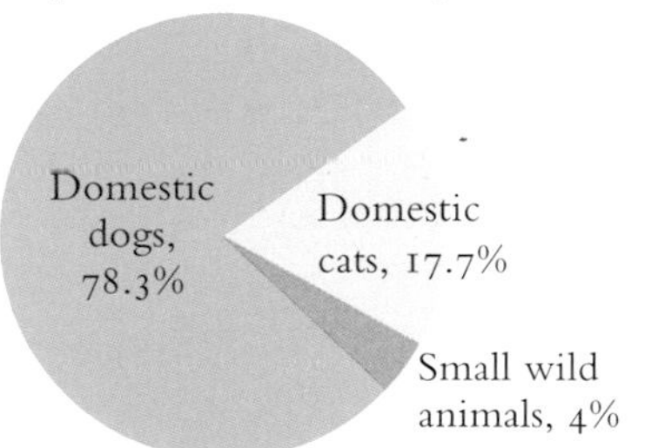

All but one of the 108 bites from wild animals occurred in counties with confirmed cases of animal rabies; many were likely rabid animals.

ALTHOUGH the constitution requires that the state lands in the Adirondacks be "forever kept as wild forest lands," over the hundred and eight years that that provision has been in place a remarkable variety of nonwild activities have taken place on them. The state has constructed campgrounds, cabins, toilets, garages, headquarters, docks, roads, dams, bridges, and fire towers; authorized the use of motorized vehicles of all kinds; cut trees and salvaged timber; and introduced some animals and plants and applied herbicides and pesticides to eliminate others.

If the primary job of the state is to facilitate the comfortable use of the public lands, then much of this is just ordinary management. But if the primary job of the state is to enforce the constitutional requirement that the park be kept wild, then some of these activities are much more questionable. The maps show two biological examples, fish and flies.

The problem with fish is simple: they like to be in the water and human beings like to take them out. Because there are almost as many humans as fish, this creates an imbalance; the millions of people who visit the park can take more fish from the water than any natural fishery could stand.

The obvious solution is to put millions of extra fish in the water. In 1999 approximately 1,500,000 hatchery fish were released in the Adirondacks. About a million of these were native species, the remainder introductions. More than half were of legal size and could be caught the same year they were released.

For the state, the fishermen, and certainly for many otters, mink, and herons, this is a wonderful thing: it is cheap, does no obvious harm, and makes many people happy. The only people it distresses are a few grumpy ecologists, who find it incongruous that the largest wilderness in the east has almost no natural fisheries. I am one of them but conscious of being in a very minor minority.

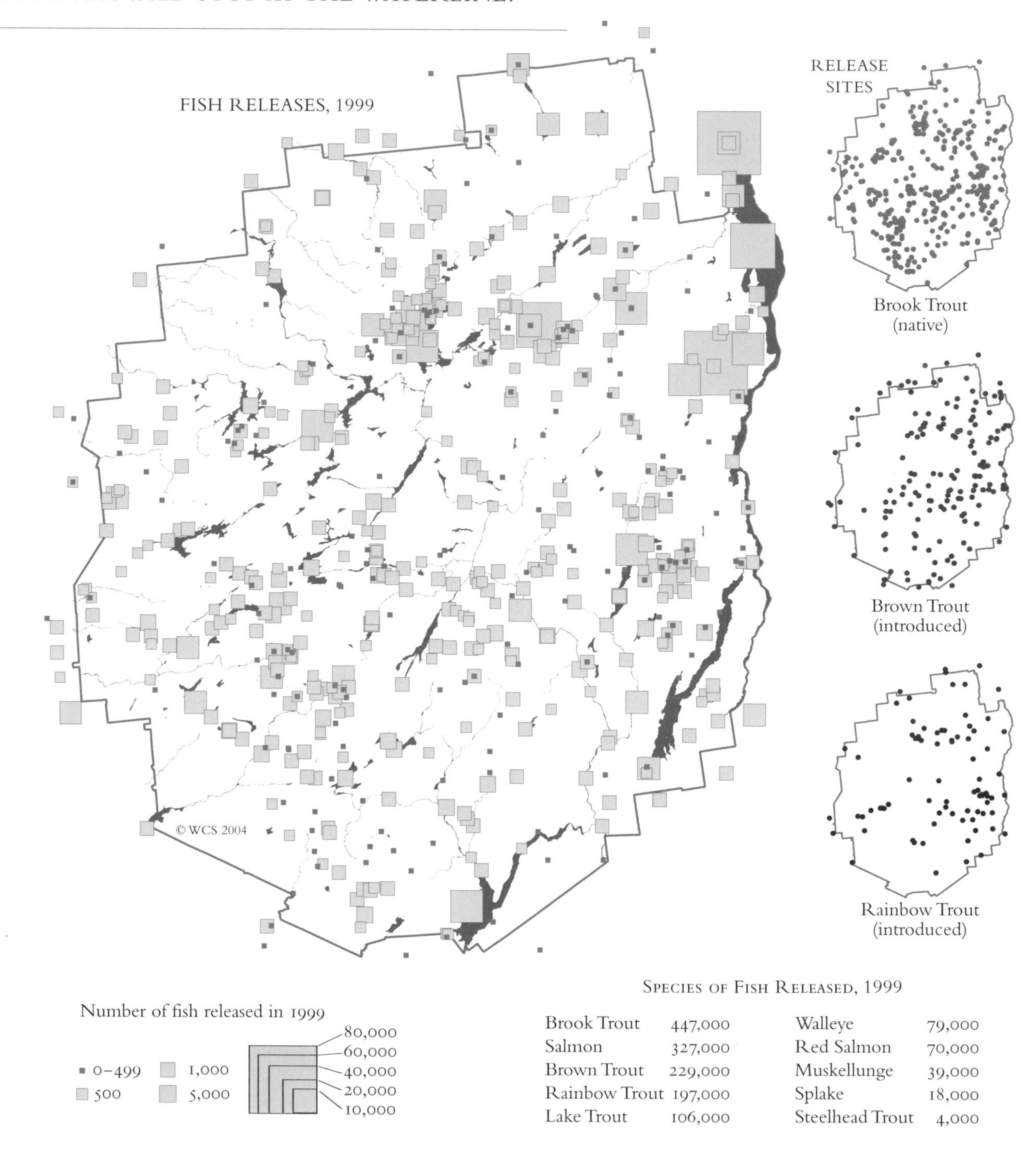

SPECIES OF FISH RELEASED, 1999

Species	Number	Species	Number
Brook Trout	447,000	Walleye	79,000
Salmon	327,000	Red Salmon	70,000
Brown Trout	229,000	Muskellunge	39,000
Rainbow Trout	197,000	Splake	18,000
Lake Trout	106,000	Steelhead Trout	4,000

BTI is a crystalline bacterial toxin that kills flies and their relatives. It is especially good at killing black flies. Because it doesn't bioaccumulate it is believed to be less harmful ecologically than DDT and the organo-phosphorus pesticides. And because it is a natural product it doesn't receive much regulatory scrutiny: towns that want to use it just fill in a form and say there will be no bad effects and they are in business.

But even if safe and natural, it is still a pesticide, and its use raises the standard pesticide questions. As managers we need to know if we are killing enough flies to make a difference or just breeding resistant flies. As ecologists we need to know what else we are killing besides flies and how we are affecting other animals like fish that may depend on flies. And as residents or visitors asking that nature be modified on our behalf, we need to ask ourselves if we really need to kill flies at all.

Once a town decides to kill flies, it has to decide which streams to treat and how heavily to treat them and how much trouble it can expect from the (enormous) numbers of flies produced in nearby streams it can't or doesn't treat.

The ninety-two Adirondack towns answer these questions in very different ways. Sixty-eight towns, the BTI-skeptics, don't treat at all. The remaining twenty-four, which spend a total of about half a million dollars a year to kill flies, use a variety of approaches. Some, like Diana, treat lightly but very thoroughly. Some, like Colton, treat only a few miles of stream but treat them quite heavily. And some, like Wilmington, spend a lot and treat a lot of streams, but put hardly anything in them. If BTI application is an art, its practitioners have many different styles.

Just what good comes from their efforts is unknown. No town monitors flies or, for that matter, any other part of the stream community that the BTI may be affecting. Some towns say BTI is essential, others think it is useless. The rest of us have no idea at all.

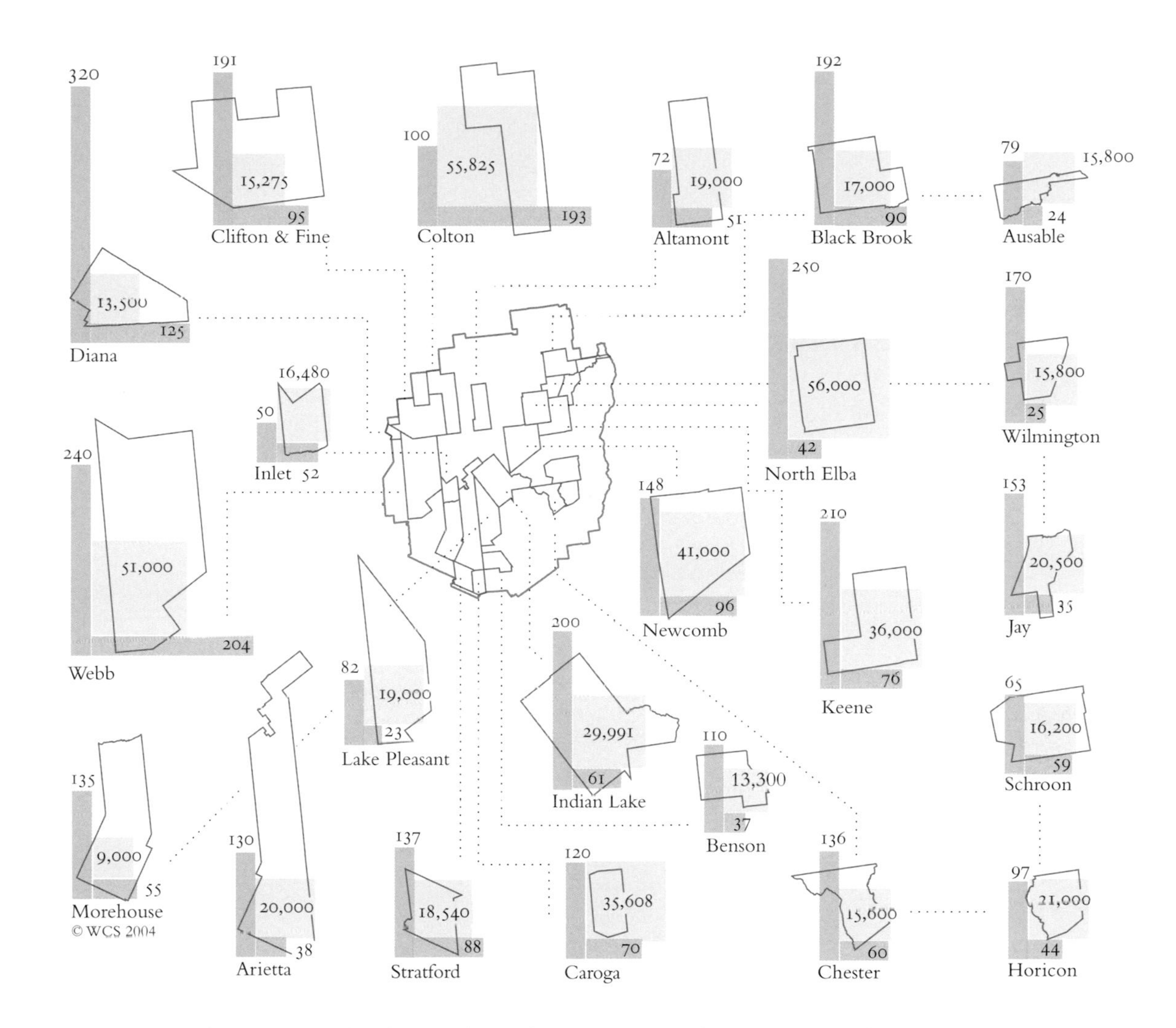

BTI Applications, 2000

200
Miles of stream treated
Cost
$50,000
10,000
Gallons applied 200

The figures show three measures of BTI application: the miles of stream treated, the gallons of concentrate applied, and the total cost. The ratio of orange to blue bars gives the intensity of treatment. Colton treats heavily, North Elba very lightly. The size of the blue bar, relative to the size of the town, gives the thoroughness of the treatment. Arietta treats a small fraction of its streams, Jay a much larger fraction. And the size of the green square, relative to the outline of the town, gives the cost per square mile. It is low in Morehouse but very high in Caroga.

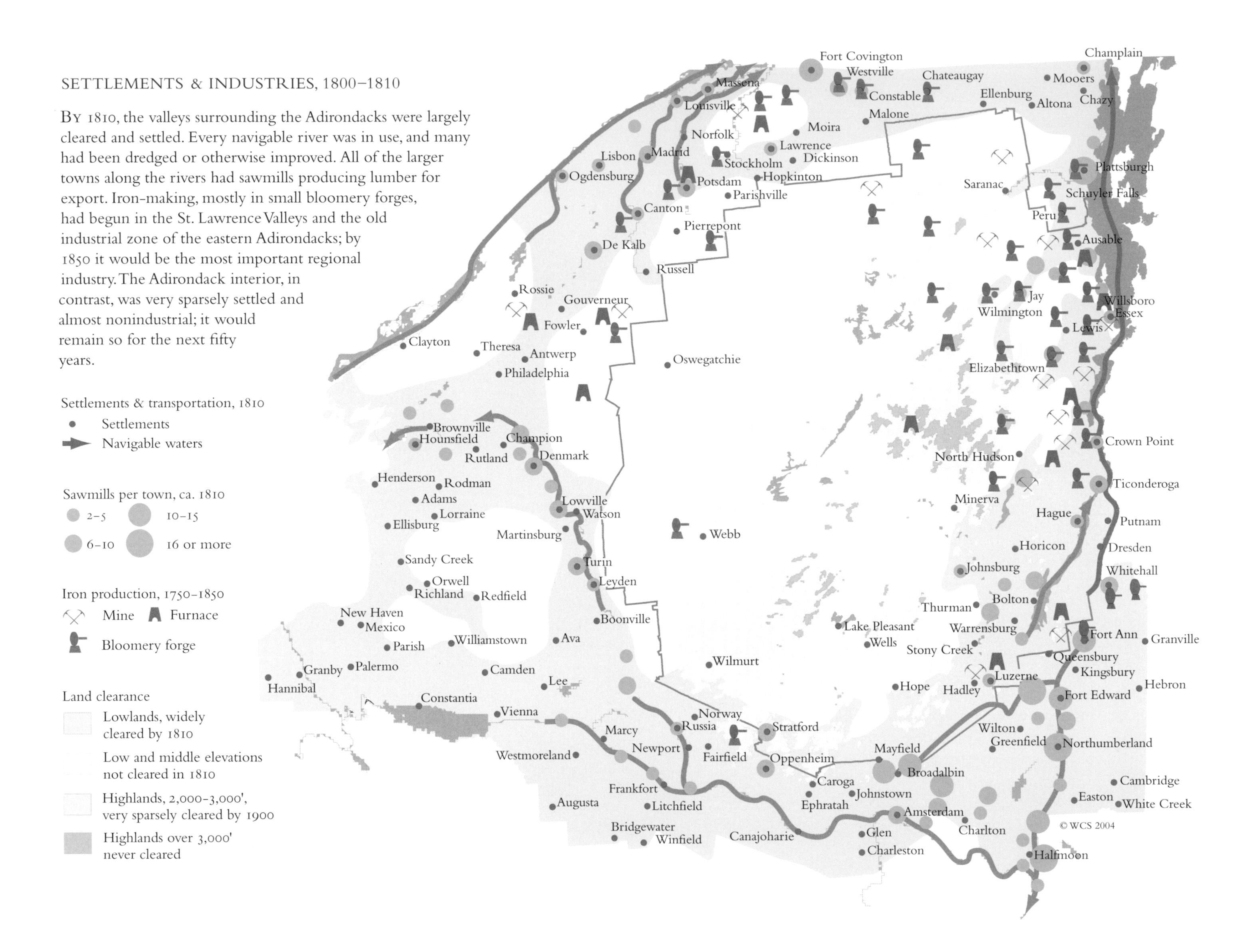

SETTLEMENTS & INDUSTRIES, 1800–1810
By 1810, the valleys surrounding the Adirondacks were largely cleared and settled. Every navigable river was in use, and many had been dredged or otherwise improved. All of the larger towns along the rivers had sawmills producing lumber for export. Iron-making, mostly in small bloomery forges, had begun in the St. Lawrence Valleys and the old industrial zone of the eastern Adirondacks; by 1850 it would be the most important regional industry. The Adirondack interior, in contrast, was very sparsely settled and almost nonindustrial; it would remain so for the next fifty years.
Settlements & transportation, 1810
Settlements
Navigable waters
Sawmills per town, ca. 1810
2–5
10–15
6–10
16 or more
Iron production, 1750–1850
Mine
Furnace
Bloomery forge
Land clearance
Lowlands, widely cleared by 1810
Low and middle elevations not cleared in 1810
Highlands, 2,000–3,000', very sparsely cleared by 1900
Highlands over 3,000' never cleared
Fort Covington
Westville
Chateaugay
Champlain
Mooers
Massena
Constable
Ellenburg
Altona
Chazy
Louisville
Malone
Moira
Norfolk
Lawrence
Dickinson
Lisbon
Madrid
Stockholm
Plattsburgh
Ogdensburg
Potsdam
Hopkinton
Saranac
Parishville
Schuyler Falls
Canton
Peru
Pierrepont
Ausable
De Kalb
Russell
Rossie
Jay
Willsboro
Gouverneur
Wilmington
Essex
Fowler
Lewis
Clayton
Theresa
Antwerp
Oswegatchie
Elizabethtown
Philadelphia
Brownville
Hounsfield
Champion
Crown Point
Rutland
Denmark
North Hudson
Henderson
Rodman
Ticonderoga
Adams
Minerva
Lowville
Lorraine
Watson
Hague
Putnam
Ellisburg
Martinsburg
Webb
Horicon
Dresden
Sandy Creek
Turin
Johnsburg
Whitehall
Orwell
Leyden
Richland
Redfield
Bolton
Thurman
New Haven
Boonville
Mexico
Lake Pleasant
Warrensburg
Fort Ann
Granville
Williamstown
Wells
Parish
Ava
Stony Creek
Queensbury
Wilmurt
Granby
Palermo
Kingsbury
Hannibal
Camden
Lee
Hope
Hadley
Luzerne
Hebron
Constantia
Fort Edward
Vienna
Norway
Russia
Stratford
Wilton
Marcy
Greenfield
Newport
Mayfield
Northumberland
Westmoreland
Fairfield
Oppenheim
Broadalbin
Caroga
Cambridge
Frankfort
Johnstown
Ephratah
Easton
Augusta
Litchfield
White Creek
Amsterdam
Bridgewater
Charlton
Winfield
Canajoharie
Glen
Charleston
Halfmoon
© WCS 2004

5 *War, Settlement, & Industry*

THIS chapter tells the story of the conquest, settlement, and development of the Adirondacks. We begin in the sixteen-hundreds, a violent and tragic century, when the European and Indian nations first encountered each other.

In 1610 the Iroquois were a formidable military-agricultural people, skilled in diplomacy, politically sophisticated, and relatively democratic. The Europeans were recent immigrants, few in numbers and militarily weak. They were less democratic than the Iroquois and, by Iroquois standards, poor diplomats but had the great political advantages conferred by iron-and-gunpowder technology and a monetized economy.

Both peoples valued war, the Indians as a cultural tradition and the Europeans as an economic and political instrument, and both felt it natural for strong states to dominate weak ones. Because even weak states don't like to be dominated, both were at war with some of their neighbors most of the time.

On the New England coast, where the Europeans feared Indians and needed Indian land, the expansion of European agricultural settlements lead quickly to a succession of bloody colonial wars. In Canada and New York, where Europeans wanted furs and the Indians needed trade goods, it was quite different. Instead of seeking to drive the Indians away from the developing settlements, the governors of these settlements sought to attract Indians to the settlements and contract with them for furs and military protection.

Fur supplies, however, are easily exhausted, especially when a voracious industrial market is supplied by expert and highly mobile hunters. The fur trade rapidly moved farther and farther from the coast. Entrepots—trading towns—were established at Montréal and Albany, and fur trade routes soon stretched from them to the Great Lakes and beyond. Competition developed between the Five Nations, who traded primarily with Albany, and the Indians of the Great Lakes and St. Lawrence, who traded with Montréal. Within twenty years after the interior trade began, war had broken out. The issue for the Europeans was whether the French or the English would get the furs. The issue for the Indians was who would take the furs to them.

The eventual result of this disagreement was a hundred and fifty years of war. First, Indians fought Indians, then Indians and Europeans fought other Indians and Europeans, and finally the Europeans, having used up the supply of Indians, fought one another. The wars began as raids along trading routes, escalated to attacks on villages, and eventually ended as full-fledged European-style military campaigns with armies and fleets and battles.

For the Indians, disease, ecological impoverishment, and cultural change came with the wars. They were decimated by epidemics, had their fields, granaries, and villages burned, and saw their game supplies dwindle as European settlement expanded. They became refugees and dependents, relinquished their lands, and moved to reservations, or emigrated to the north and west.

By 1790 the wars were over. French Canada had become British, and the British Colonies American. The Iroquois were much reduced, their homelands occupied by settlers, and their former power and hundred-and-fifty-year partnership with Europeans forgotten.

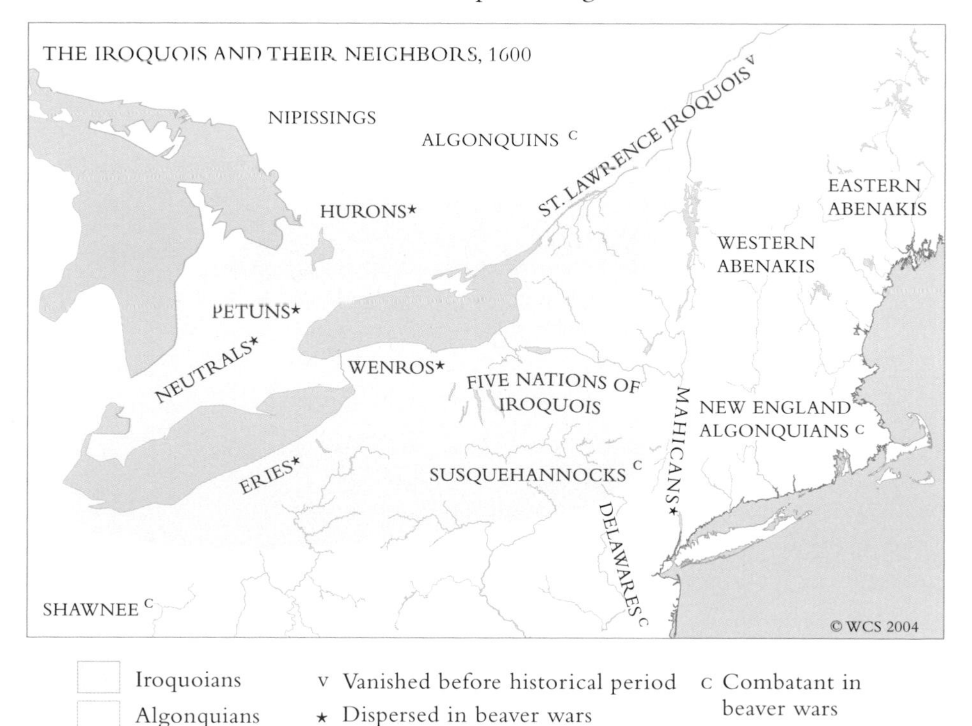

In 1600 the Adirondacks were part of the territory of the Five Nations of the Iroquois, divided between the Mohawk and Oneida Nations. The Iroquois lived in fortified towns south and west of the Adirondacks and controlled the three major valleys—Champlain, Mohawk, and Black—that connected the coastal plain to the west and north. The Adirondacks were neither settled nor part of a trade route but were used as hunting territories.

For much of the seventeenth century the Iroquois were the strongest military force, native or European, in eastern North America. They were deeply involved in the beaver trade and attempted to use their power to divert and monopolize it. For almost fifty years they succeeded but at great cost: the resulting conflicts first exhausted them and then lead to reprisals which they were too weak to resist.

The fur trade began in the early 1600s. Europeans explored the Hudson Valley and Lake Champlain between 1600 and 1610. Shortly afterwards both the French and the Dutch established trading posts. The French initially operated from posts in Québec City and later from Montréal, the Dutch initially from posts in Manhattan and later from Albany.

For the next century the fur trade was an Indian-European partnership. Indians located and trapped the animals, transported them to the trading posts, and defended the trading routes from rival Indians. Europeans ran the trading posts, handled credit and shipping, and, of course, made most of the profits. But for the first century of the trade the Europeans rarely ventured, much less settled, very far from the trading posts. Many probably never saw a live beaver. The geopolitics of the interior trade, the military activities necessary to sustain these politics, and the costs their societies incurred as a result of those activities were entirely Indian affairs.

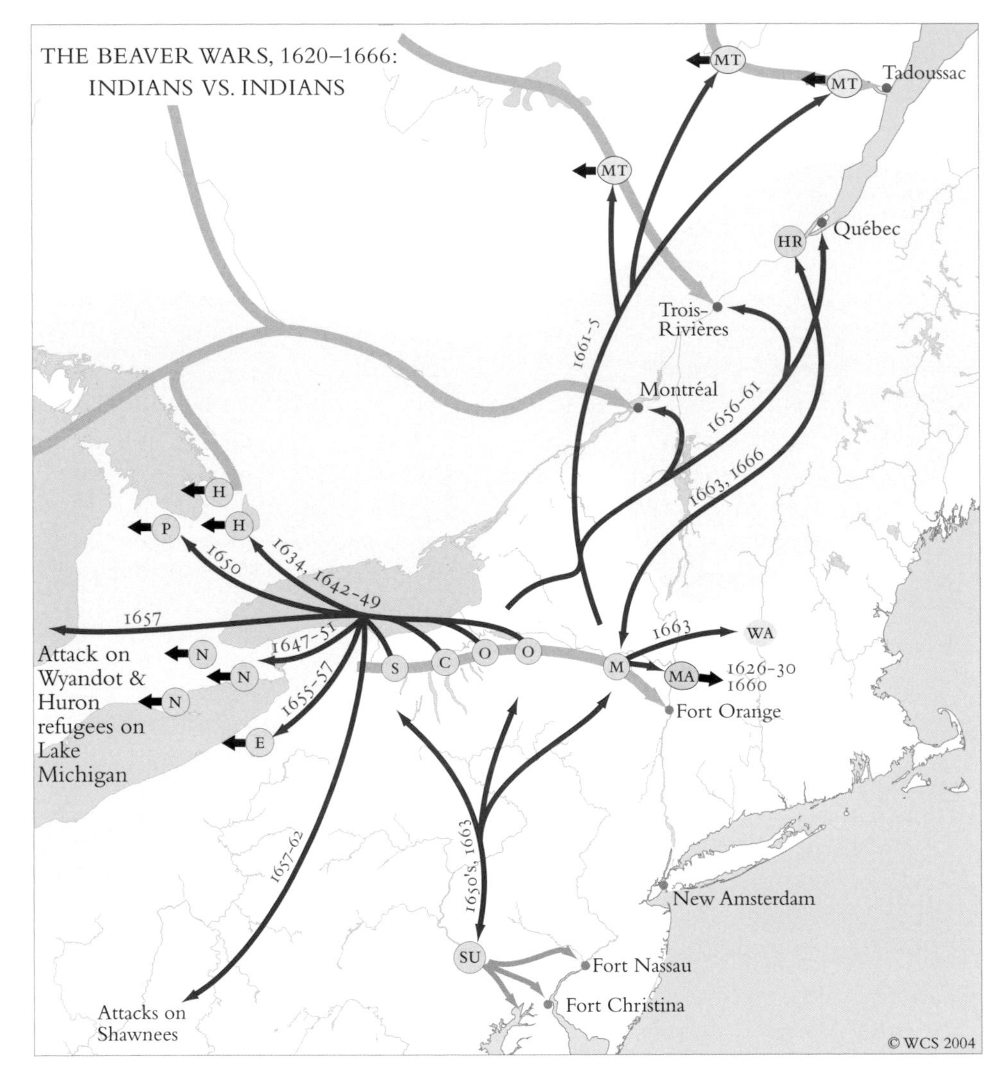

Fur trade route
Iroquois raids
General areas of Iroquois fur raids
French settlements
Dutch or Swedish settlement
Tribe forced to migrate by Iroquois attacks

Five Nations of the Iroquois

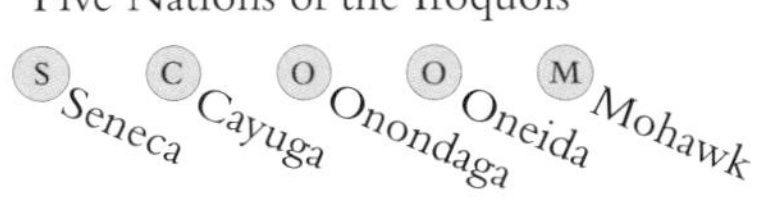

Other Iroquoian peoples
E Eries
H Hurons
HR Huron refugees
N Neutrals
P Petuns
SU Susquehannocks

Algonquins
MA Mahicans
MT Montagnais
WA Western Abenaki

Beaver Wars I. In the first phase of the beaver wars, the Five Nations, lacking adequate fur supplies of their own, tried to dominate or disperse the tribes in the St. Lawrence and thus control the western fur trade. They also sought, by taking captives, to replace losses from war and disease. They began by driving the Mahicans away from Fort Orange, thus making themselves the sole trading partners of the Dutch. They then raided and dispersed the unallied Iroquoian tribes of the eastern Great Lakes, taking control of the western trade routes. Finally, they raided the French settlements of the St. Lawrence, taking furs from the northern trade routes.

THE early fur trade, and hence the politics of fur, focused on beaver. Other furs were mostly luxury items, used whole to make clothing and blankets. Beaver pelts, in contrast, were an industrial raw material and used as a source of hair rather than as furs. The hair was then felted—turned into cloth by a process similar to papermaking—and used in hats, clothing, and many other articles.

Because of the industrial demand, local beaver supplies were quickly depleted, and the beaver-hunting areas moved farther and farther away. In 1624, 7,200 beaver pelts were shipped from the Hudson Valley, most of them trapped locally. By the 1680s some 80,000 were being shipped annually, almost all trapped far to the north and west.

It was this long-distance trade that the Iroquois sought to control. Early in the century, when they could field armies of several thousand warriors and several hundred canoes, they succeeded handsomely. Later, as their strength fell and European strength grew, they suffered military defeats and lost their influence over other tribes.

The first forty years of the beaver wars were a mixture of intertribal raids and piracy. The Iroquois intercepted fur shipments on the St. Lawrence, drove other tribes from prime fur lands, and forced distant tribes with good fur supplies to confederate and trade with them. These were entirely wars of Indians against Indians. The British and the French watched from the trading posts and cheered their teams on but were not strong enough to do much more than that.

About 1660 the Iroquois created, briefly, a beaver-trading empire of subject tribes. It proved, like many empires, easier to build than to sustain; by 1680 it was strained and by 1690 all but gone.

In the next forty years the fur trade moved further to the west, Indian power diminished, and

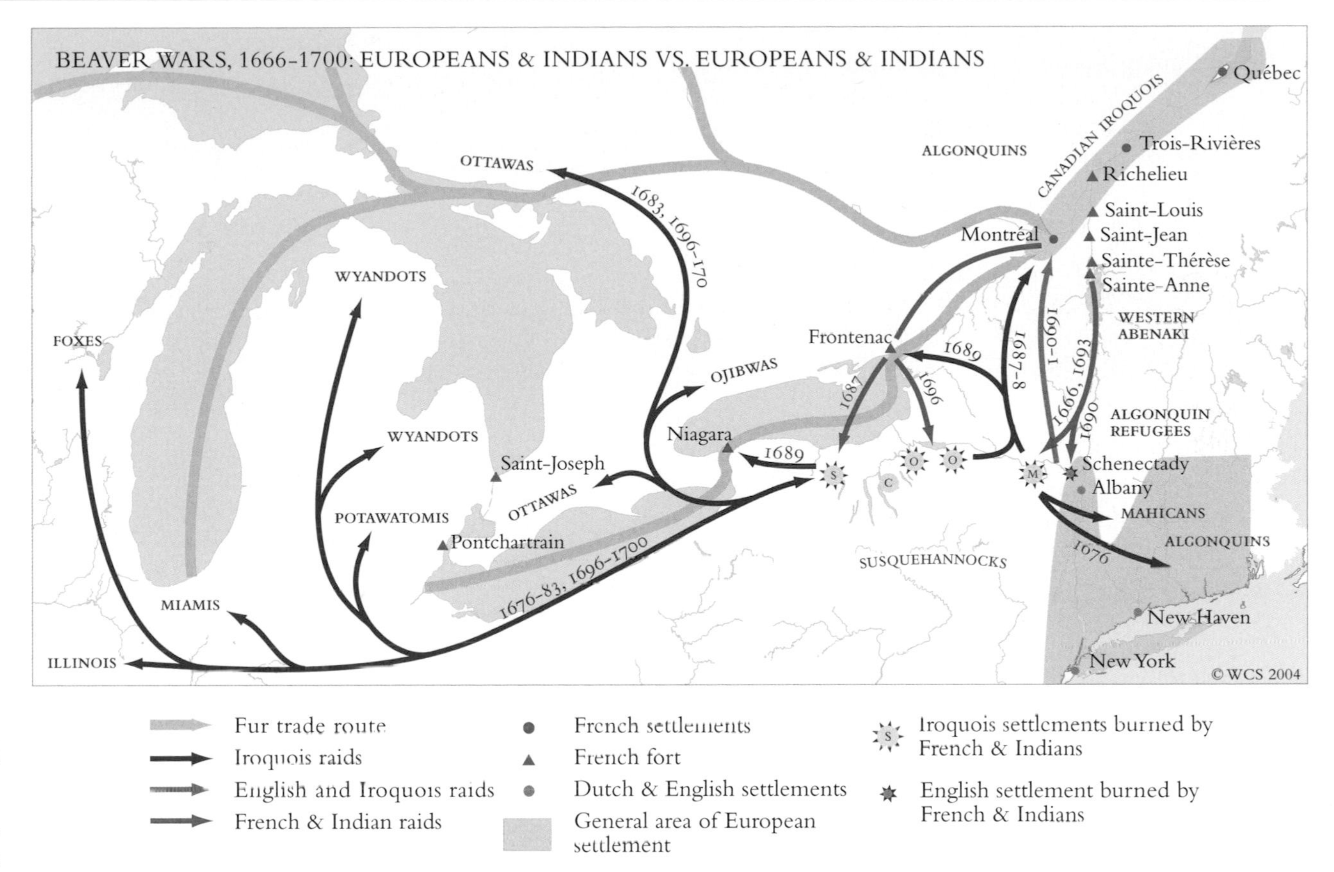

European power grew. The Iroquois, stubborn till the end, fought wars with a number of midwestern tribes and attacked Indian and European villages in French Canada. The French and their Indian allies retaliated, and by 1700 the principal villages of four of the five Iroquois nations had been burned.

The Iroquois remained a military power until the American Revolution, but both the strength they could mobilize and their importance in Euro-American conflicts rapidly decreased. In 1700 they controlled ninety percent of New York. By 1760 they had lost most of their eastern lands; by 1800 they had lost almost all of the rest.

Beaver Wars II. In the second phase of the beaver wars, the Indians were acting partly in their own interest and partly as irregular troops under the direction of the French and English officers. Beginning in the 1670s, the Iroquois raided western Indian settlements but did not disperse or subjugate them. In the late 1680s they attacked and destroyed settlements in the St. Lawrence. The French retaliated by burning the major villages of the Mohawks in 1666, those of the Seneca in 1687, and those of the Oneida and Onondaga in 1696. Only the Cayuga's homeland remained intact. It would be burned by General John Sullivan, on George Washington's orders, a century later.

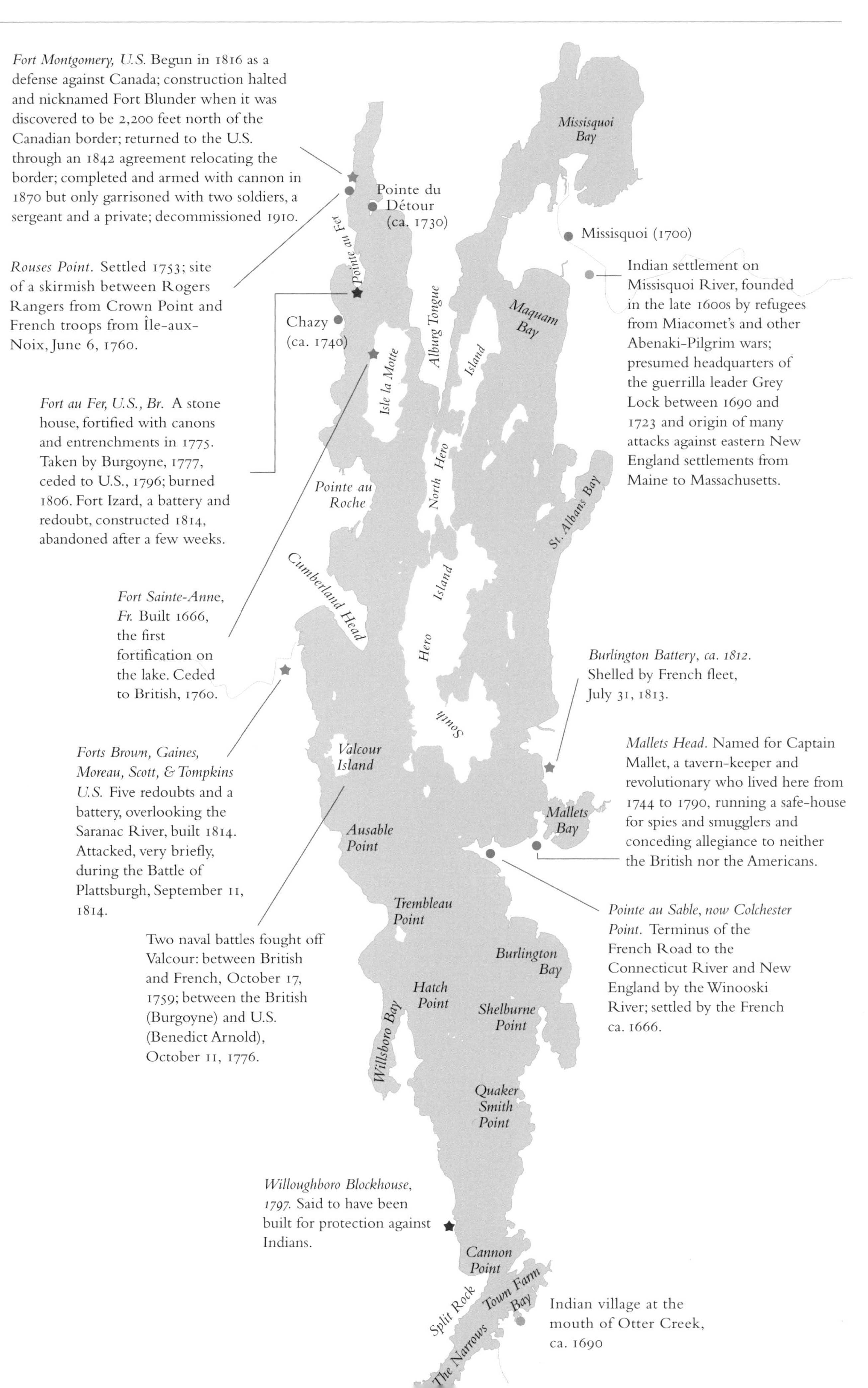
Fort Montgomery, U.S. Begun in 1816 as a defense against Canada; construction halted and nicknamed Fort Blunder when it was discovered to be 2,200 feet north of the Canadian border; returned to the U.S. through an 1842 agreement relocating the border; completed and armed with cannon in 1870 but only garrisoned with two soldiers, a sergeant and a private; decommissioned 1910.
Rouses Point. Settled 1753; site of a skirmish between Rogers Rangers from Crown Point and French troops from Île-aux-Noix, June 6, 1760.
Fort au Fer, U.S., Br. A stone house, fortified with canons and entrenchments in 1775. Taken by Burgoyne, 1777, ceded to U.S., 1796; burned 1806. Fort Izard, a battery and redoubt, constructed 1814, abandoned after a few weeks.
Fort Sainte-Anne, Fr. Built 1666, the first fortification on the lake. Ceded to British, 1760.
Forts Brown, Gaines, Moreau, Scott, & Tompkins U.S. Five redoubts and a battery, overlooking the Saranac River, built 1814. Attacked, very briefly, during the Battle of Plattsburgh, September 11, 1814.
Two naval battles fought off Valcour: between British and French, October 17, 1759; between the British (Burgoyne) and U.S. (Benedict Arnold), October 11, 1776.
Willoughboro Blockhouse, 1797. Said to have been built for protection against Indians.
Missisquoi Bay
Pointe du Détour (ca. 1730)
Missisquoi (1700)
Indian settlement on Missisquoi River, founded in the late 1600s by refugees from Miacomet's and other Abenaki-Pilgrim wars; presumed headquarters of the guerrilla leader Grey Lock between 1690 and 1723 and origin of many attacks against eastern New England settlements from Maine to Massachusetts.
Pointe au Fer
Chazy (ca. 1740)
Isle la Motte
Alburg Tongue
Island
Maquam Bay
North Hero
Pointe au Roche
St. Albans Bay
Cumberland Head
Island
Hero
South
Burlington Battery, ca. 1812. Shelled by French fleet, July 31, 1813.
Valcour Island
Mallets Head. Named for Captain Mallet, a tavern-keeper and revolutionary who lived here from 1744 to 1790, running a safe-house for spies and smugglers and conceding allegiance to neither the British nor the Americans.
Mallets Bay
Ausable Point
Pointe au Sable, now Colchester Point. Terminus of the French Road to the Connecticut River and New England by the Winooski River; settled by the French ca. 1666.
Trembleau Point
Burlington Bay
Hatch Point
Willsboro Bay
Shelburne Point
Quaker Smith Point
Cannon Point
Split Rock
Town Farm Bay
The Narrows
Indian village at the mouth of Otter Creek, ca. 1690

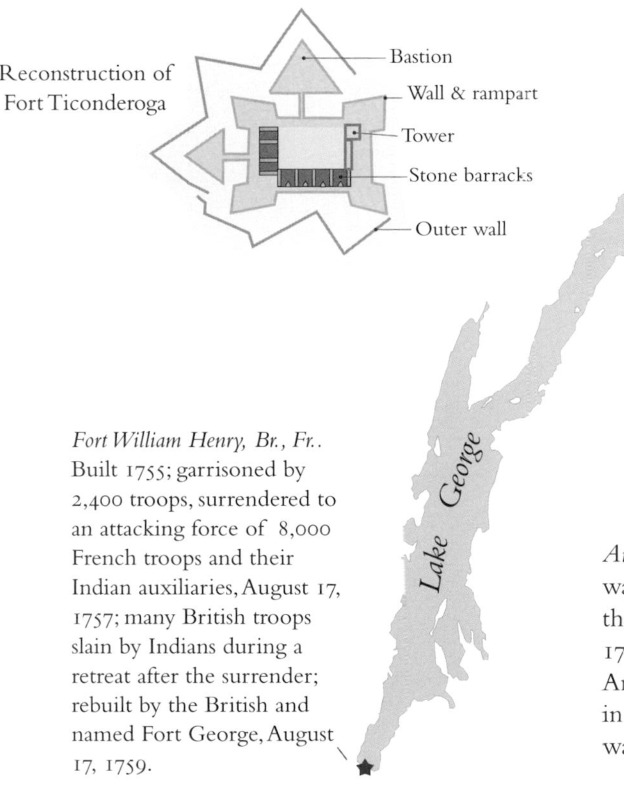

America's Foremost Warpath. For two centuries the Champlain Valley was the most fortified and fought-over corridor in North America. In the 1600s it was the route by which Iroquois fought Algonquins. In the 1700s it was the route by which the British fought the French and the American colonists fought the British. In the early 1800s it was fought in one last time, by the British and the Americans, in a curiously modern war that was about power but not about victory.

War zones, however attractive otherwise, are difficult places to live. The Indians traveled the Champlain constantly but never settled on its shores. The Europeans fortified it but, because of their endless imperial wars, could not hold it or settle it. Only after the 1814 was it gradually demilitarized and settled.

The map is a catalog of the principal forts and engagements of the period of European wars. For narrative maps of individual wars and campaigns see *Maps 5-3, 5-4,* and *5-8.*

EARLY New York was wild and empty but hardly peaceful. At various times about eighteen nations, three European and fifteen Indian, claimed territory within two hundred miles of the Adirondacks. All of them were militarized and most were economically and politically aggressive. When this very uneasy mixture was destabilized by imperialism and the fur trade, the result was a two-hundred-year period in which every generation and almost every person experienced war first hand.

The wars they experienced were largely guerrilla wars in which soldiers attacked trading parties, settlements, and civilians. Only a little of the conflict involved armies and siegecraft and famous generals. Most of it was an intermittent series of raids and ambushes, officially sanctioned but not centrally controlled and often opportunistic rather than planned. Pitched battles were rare; scalpings, kidnappings, torture, and the burning of villages and crops were regrettably common.

In consequence, by 1750 eastern New York had a thoroughly militarized geography and populace. Approximately one hundred forts were built in New York in the first hundred and twenty-five years of settlement. Every settlement of any size, European or Indian, was fortified, and every man was expected to bear arms.

To the Indians this was a source of deep cultural pride. If you were not a warrior, you were not a man. To the Europeans it was less a matter of pride than of necessity. In these violent centuries, there was no way that you could be a farmer or a pioneer without being a soldier as well.

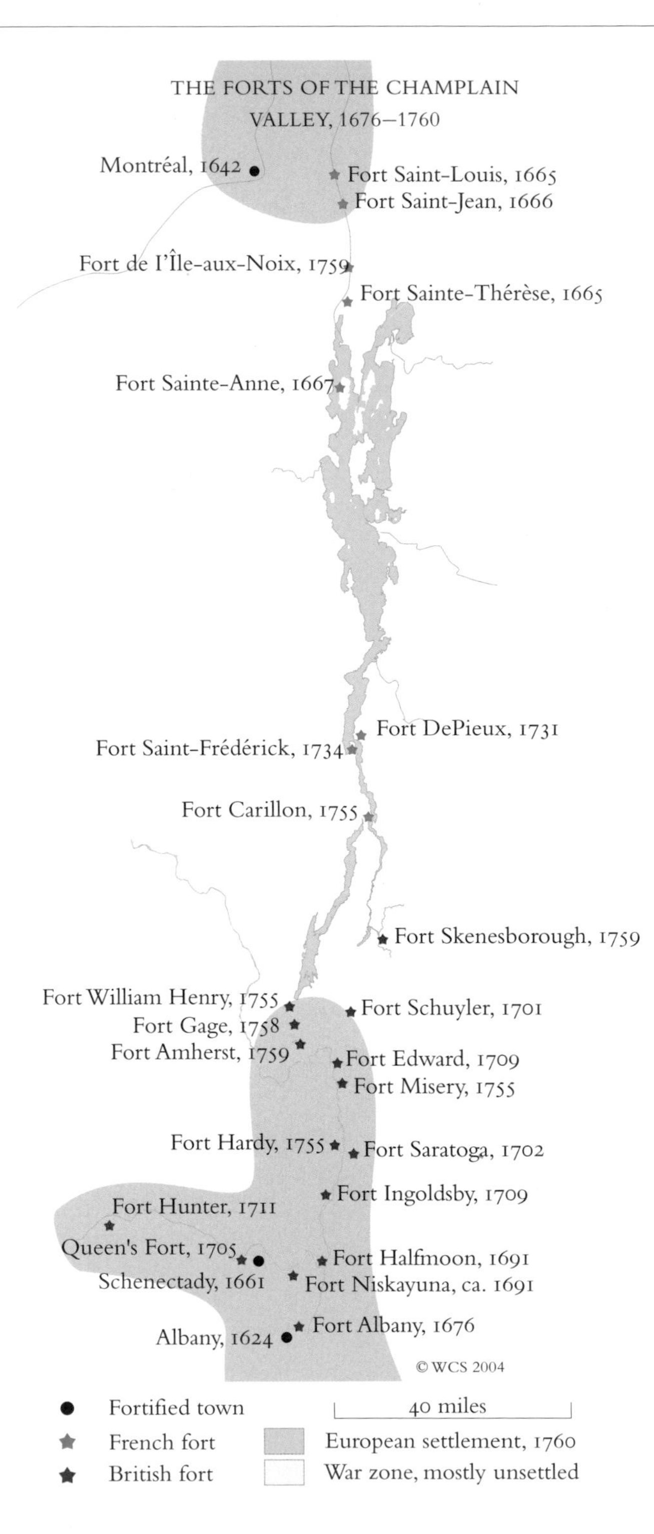

THE first hundred and fifty years of war in the Northeast, bitter as it was, was essentially a series of border conflicts, fur raids, and failed assaults that changed little and settled nothing.

After 1750, things were different. The British by now occupied almost the entire Atlantic coast of America and were crossing the Appalachian Mountains in increasing numbers to settle and trade in the interior. The French had settled, sparsely, the shores of the St. Lawrence and were attempting to control the interior with a far-flung chain of forts extending across the Great Lakes, down the Mississippi, and up the Ohio. The Indians, much reduced but still powerful and defiant, were squeezed between them. All three powers felt that their vital interests were threatened and that the series of wars about to begin would not just be about forts and furs but about who would rule eastern North America. They were right.

The Seven Years' War, the first of the wars for North America, began in the Ohio country, which both France and England claimed. In 1754 the English built a fort at the forks of the Ohio, now downtown Pittsburgh. The French chased them off, the young George Washington was sent to reinforce the English, a French negotiating party got scalped, and a world war started. It was, of course, a case of a "damned silly thing in the Balkans," but when enough troops are mustered and enough cannons loaded, there will always be a damned silly thing somewhere.

Regardless of who started it, the first European-style war in North America had begun. It would be several months before Europe knew it and nearly three years before it was fully underway. By then it would involve the Champlain. American wars always did.

THE FIRST WAR FOR NORTH AMERICA, 1755–1760

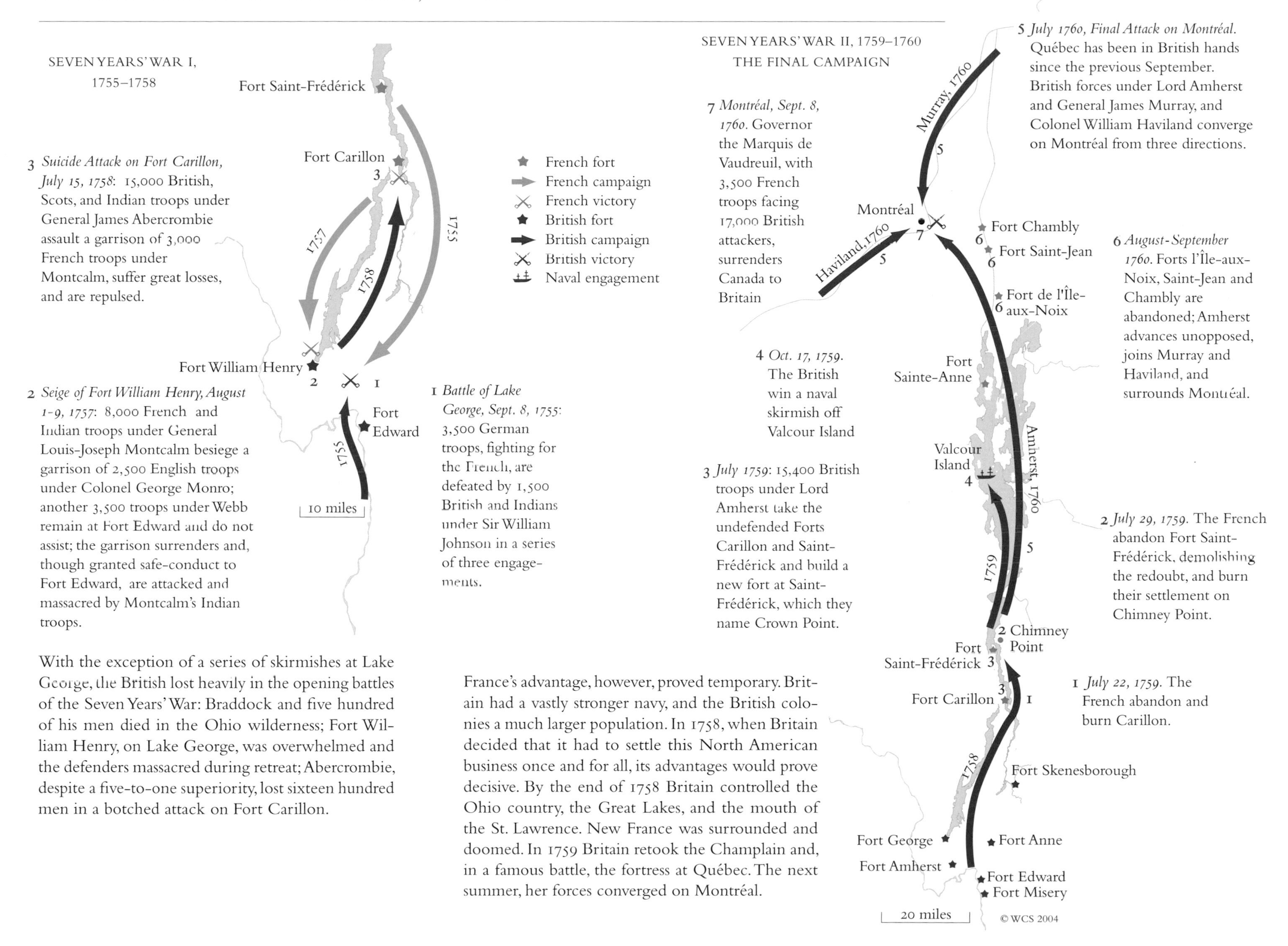

With the exception of a series of skirmishes at Lake George, the British lost heavily in the opening battles of the Seven Years' War: Braddock and five hundred of his men died in the Ohio wilderness; Fort William Henry, on Lake George, was overwhelmed and the defenders massacred during retreat; Abercrombie, despite a five-to-one superiority, lost sixteen hundred men in a botched attack on Fort Carillon.

France's advantage, however, proved temporary. Britain had a vastly stronger navy, and the British colonies a much larger population. In 1758, when Britain decided that it had to settle this North American business once and for all, its advantages would prove decisive. By the end of 1758 Britain controlled the Ohio country, the Great Lakes, and the mouth of the St. Lawrence. New France was surrounded and doomed. In 1759 Britain retook the Champlain and, in a famous battle, the fortress at Québec. The next summer, her forces converged on Montréal.

WHEN the British took control of French Canada in 1760 they had just started their two-hundred-year heyday as the military and economic powerhouse of the world. Surely no one, not even the most radical Bostonians, thought that twenty years later they would have lost half of their North American possessions. But lose them they did, to a small group of their own citizens who at first had neither army nor navy and were not even particularly keen on independence.

The main story of the Revolution, from Concord to Valley Forge to Yorktown, lies to our east and south. The maps show two early northern campaigns, one improvised and ineffectual, one pivotal and grim.

By the twenty-ninth of April, 1775, eleven days after the first fight at Lexington and Concord, the second Continental Congress had declared that the 15,000 militiamen surrounding Boston were the Continental Army and appointed George Washington their commander. His problem was what he could do with a small artilleryless force against a well-armed enemy who controlled the sea and the major port cities. One answer, a natural one for an army of farmers and woodsmen, was to campaign inwards and northwards and try to acquire cannons on the way.

The result was the lightning campaign of 1775: a sneak attack on the forts in the Champlain, a successful but meaningless siege of Montréal, a harrowing advance through the Maine wilderness to Québec, and finally a deadly and futile winter siege of the fortress at Québec. In six months, some rather jaunty soldiering had turned grimly serious. It would stay so for the next six years.

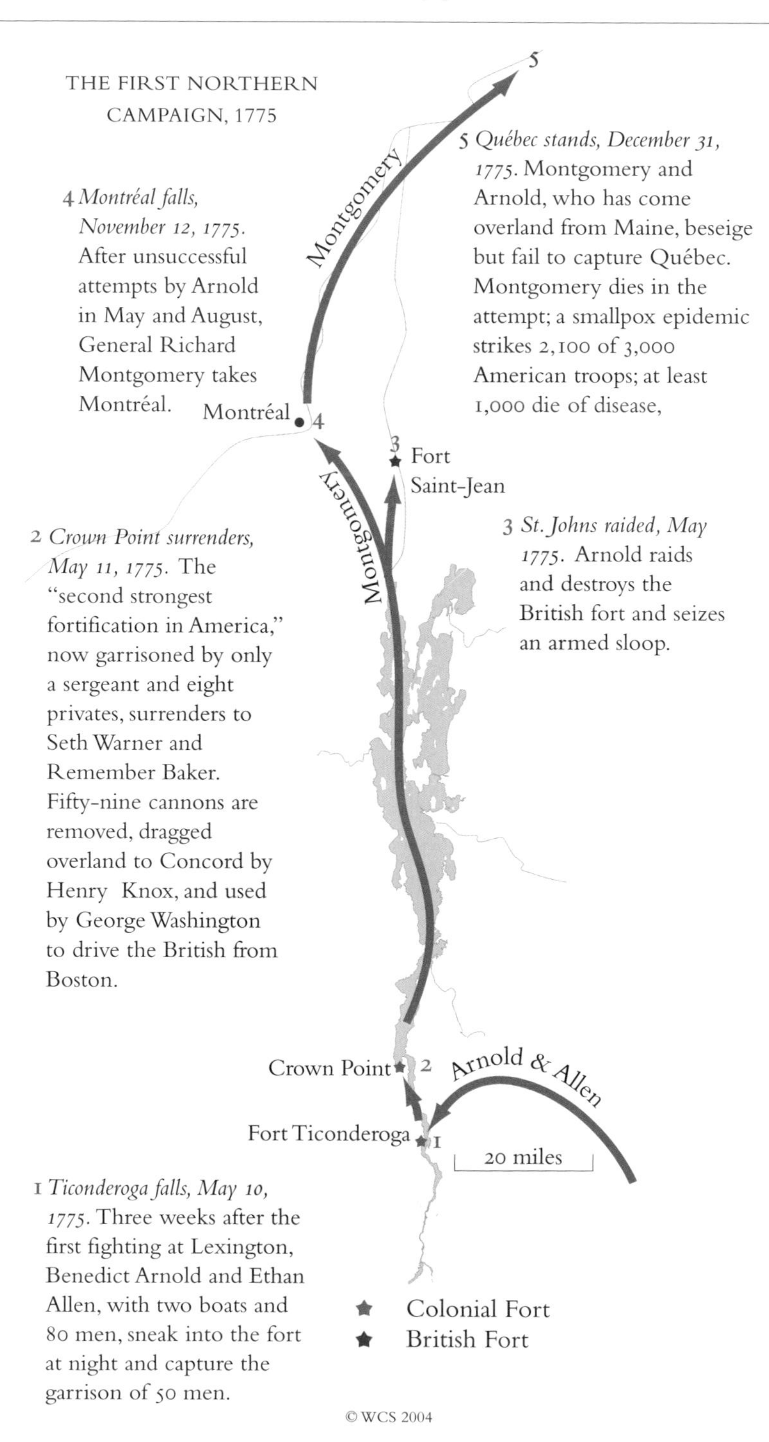

THE war for Lake Champlain, shown in the maps on the facing page, was initiated by the British in 1776. In the spring of 1776 they controlled the St. Lawrence and the Atlantic seaboard and were rapidly reoccupying the French forts in the interior. The Americans held Boston and, briefly, loyalist New York City but nothing much else of strategic importance. The British plan was to drive south from Montréal and north into New York, capturing the New York City and the Hudson Valley and dividing the colonies in two.

Had they had succeeded, the one-year-old Revolution might have ended up as a minor Massachusetts revolt. But they did not succeed. In the wilderness campaign they were undertaking, supply lines and mobility were all-important. The Americans, closer to home and far more comfortable in the woods than the British, were soon able to deny them both.

In 1776, Lake Champlain was defended by American garrisons at Ticonderoga and Mt. Independence. The men were mostly survivors of the winter campaign in Québec; over half were sick or disabled. The British general Guy Carleton had assembled a substantial invading fleet, including the 300-ton *Inflexible*, purpose-built in Québec and hand-carried in pieces up the rapids at Richelieu. Benedict Arnold and a small crew of New England carpenters were cutting trees at Skenesborough and trying to build enough vessels to oppose him. For the British it was still a minor imperial uprising to be put down as cheaply as possible. For the New Englanders, it was a wilderness war of desperation and makeshift.

SUMMER 1776,
CARLETON'S ADVANCE
FAILS

In the fall of 1776, General Guy Carleton invaded Lake Champlain from the north and was opposed by Benedict Arnold. Carleton had all the advantages but chose not to use them.

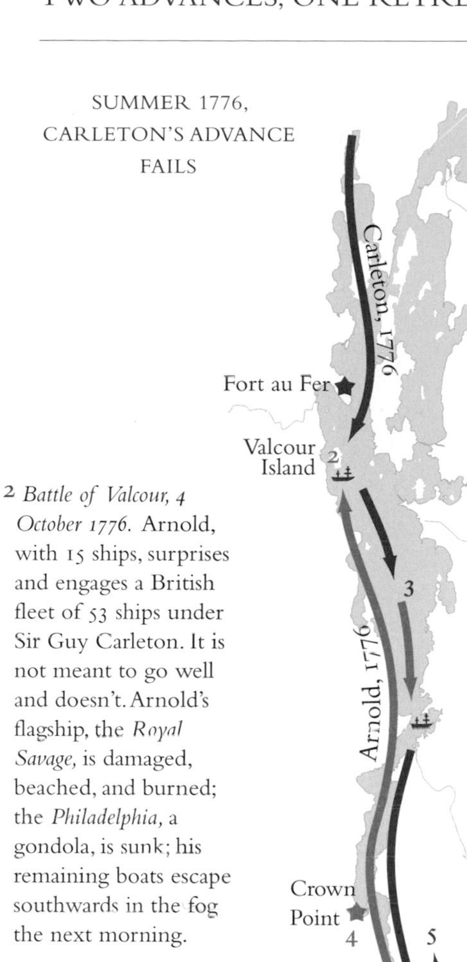

1 *Skenesborough, Summer 1776.* In order to delay the advance of Carleton's fleet, Benedict Arnold assembles a fleet of three schooners, a sloop, seven gondolas, and a galley. In September he sails north with 500 men.

2 *Battle of Valcour, 4 October 1776.* Arnold, with 15 ships, surprises and engages a British fleet of 53 ships under Sir Guy Carleton. It is not meant to go well and doesn't. Arnold's flagship, the *Royal Savage,* is damaged, beached, and burned; the *Philadelphia,* a gondola, is sunk; his remaining boats escape southwards in the fog the next morning.

3 *Pursuit by the British, 5-6 October 1776.* Arnold flees southwards but is becalmed and overtaken; he loses or grounds and burns all but three of his remaining vessels and returns to Crown Point on foot.

4 *Crown Point Lost, October 1776.* The Americans abandon the fort to Carleton and retreat to Ticonderoga.

5 *Carleton Retires, November 1776.* Carleton is delayed for eight days by head winds; he sails to within sight of Ticonderoga and Independence, decides not to challenge them, and retires for the winter.

SUMMER 1777,
BURGOYNE ADVANCES
AND SURRENDERS

In June 1777, General John Burgoyne moved south to attack Albany. He was supposed to be joined by reinforcements from the south and west but wasn't. He reached Fort Edward and then halted two months, suffering increasing difficulties with supplies. Then his bad luck got worse.

1 *June 1777.* General John Burgoyne heads south with 8,000 troops, 138 cannon, 300 officers, and assorted officers' wives and children.

2, 6 *Crown Point.* Abandoned to advancing British, July 6, 1777; Benjamin Lincoln cuts Burgoyne's supply lines, September 1777.

3 *Ticonderoga & Mt. Independence.* Abandoned to the British, July 6, 1777, after Burgoyne places cannons on Mt. Defiance.

4 *Battle of Bennington, August 16, 1777.* A raiding party of 675 men sent by Burgoyne to get food is defeated at Walloomsac, N.Y., by the Green Mountain Militia under General John Stark.

5 *Saratoga September 12 & October 7, 1777.* Burgoyne loses the battles of Freeman's Farm and Bemis Heights; he now has under 6,000 troops.

7 *Stillwater, October 17, 1777.* Burgoyne grounds arms and surrenders at Stillwater; his 6 remaining generals, 300 officers, and 5,500 enlisted men are sent home for the duration.

In the thirty-five years between 1775 and 1810, roughly eighty percent of New York passed out of the control of the Iroquois and into the hands of the European settlers. An obvious question is how it happened and why it happened so fast.

The expansion of Euro-American *settlement*, involving the making of farms and villages, is widely told. The expansion of *sovereignty*, involving the displacement of the Indians and the annexation of their lands, is treated sketchily in most histories.

European sovereignty over northern New York was first asserted in 1713 when the Treaty of Utrecht gave New York sovereignty over all Iroquois lands. Because the Iroquois—who had not been told about this—were still powerful and because even kings can't give away what they do not own, this treaty, though portending much, signified nothing at the time.

Actual sovereignty came rapidly in the last third of the eighteenth century. It was garnered in three ways.

First, substantial amounts of Indian land were purchased by individuals. The Totten & Crossfield Purchase (*Map 5-6*) is the most famous Adirondack example. Some purchases were legal, some illegal; some were free transactions, some forced. Almost all were highly advantageous to the purchasers. The Iroquois were defeated, decimated, and powerless and had to take whatever terms they could get.

Second, much unoccupied land, especially in war zones, was simply taken by the Crown or by the thousands of settlers looking for farms. The Crown had a doctrine of *vacuum domicilium*, meaning roughly that if they didn't see an owner it was theirs. The settlers had an identical doctrine but one framed in the realities of frontier life rather than in the Latin of the lawbooks.

Third, much Indian land was confiscated after Indian-European wars. The Iroquois had fought against the British in Pontiac's Revolt in 1763 and against the New York colonists in a protracted and very violent subconflict within the Revolution from 1778 to 1782. The end of the Revolution was particularly disastrous for the them. Their main villages were burned, their homelands reduced by forced treaties, and a million and a half acres of their upstate lands, including much of the Adirondacks, were seized by the Act of Attainder of 1779. When the Revolution began they were a sovereign people with extensive lands and a significant military presence; when it ended they were refugees and supplicants, ignored in the peace treaty signed in Paris and left to be dealt with as the victors desired.

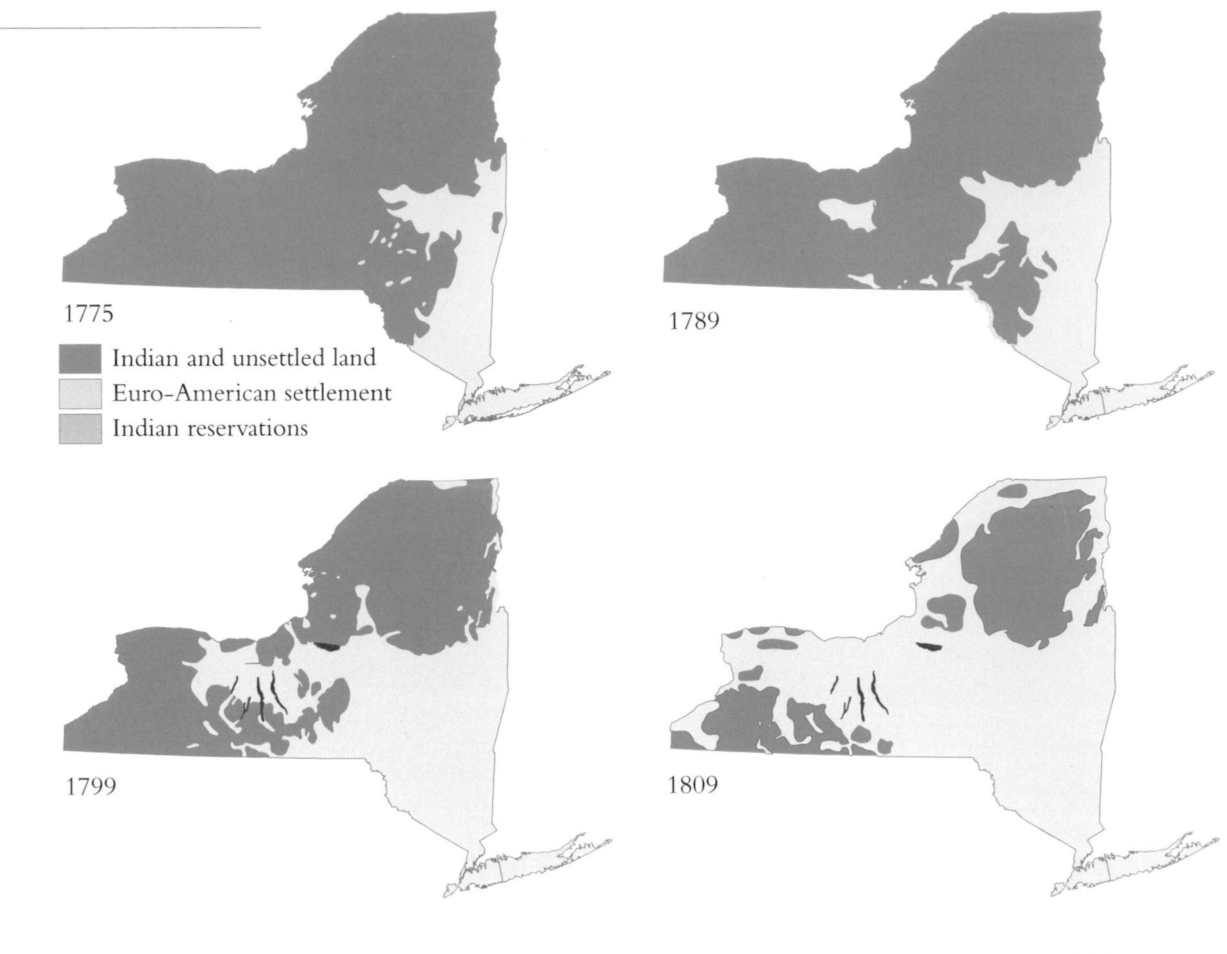

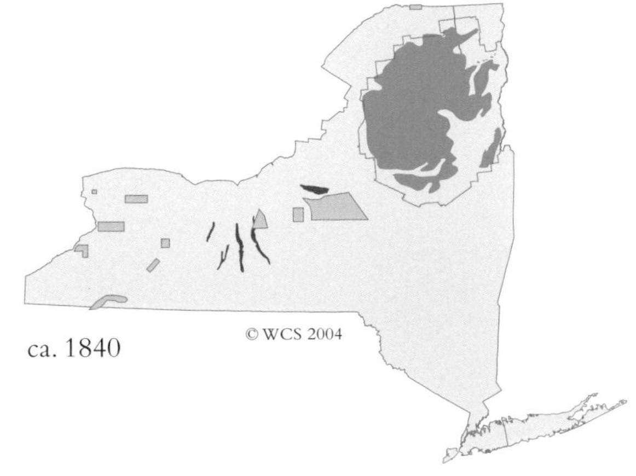

In the middle 1700s, northern New York was a war zone, sparsely settled along the main rivers and hardly settled at all, either by Indians or Europeans, in the war-torn Champlain Valley. Extensive settlement, limited legally to the Hudson and Delaware watersheds but soon extending beyond them, began in 1760 at the end of the Seven Years War and then accelerated at the end of the Revolution when the Iroquois Confederacy was broken and many of the remaining Iroquois left for Canada.

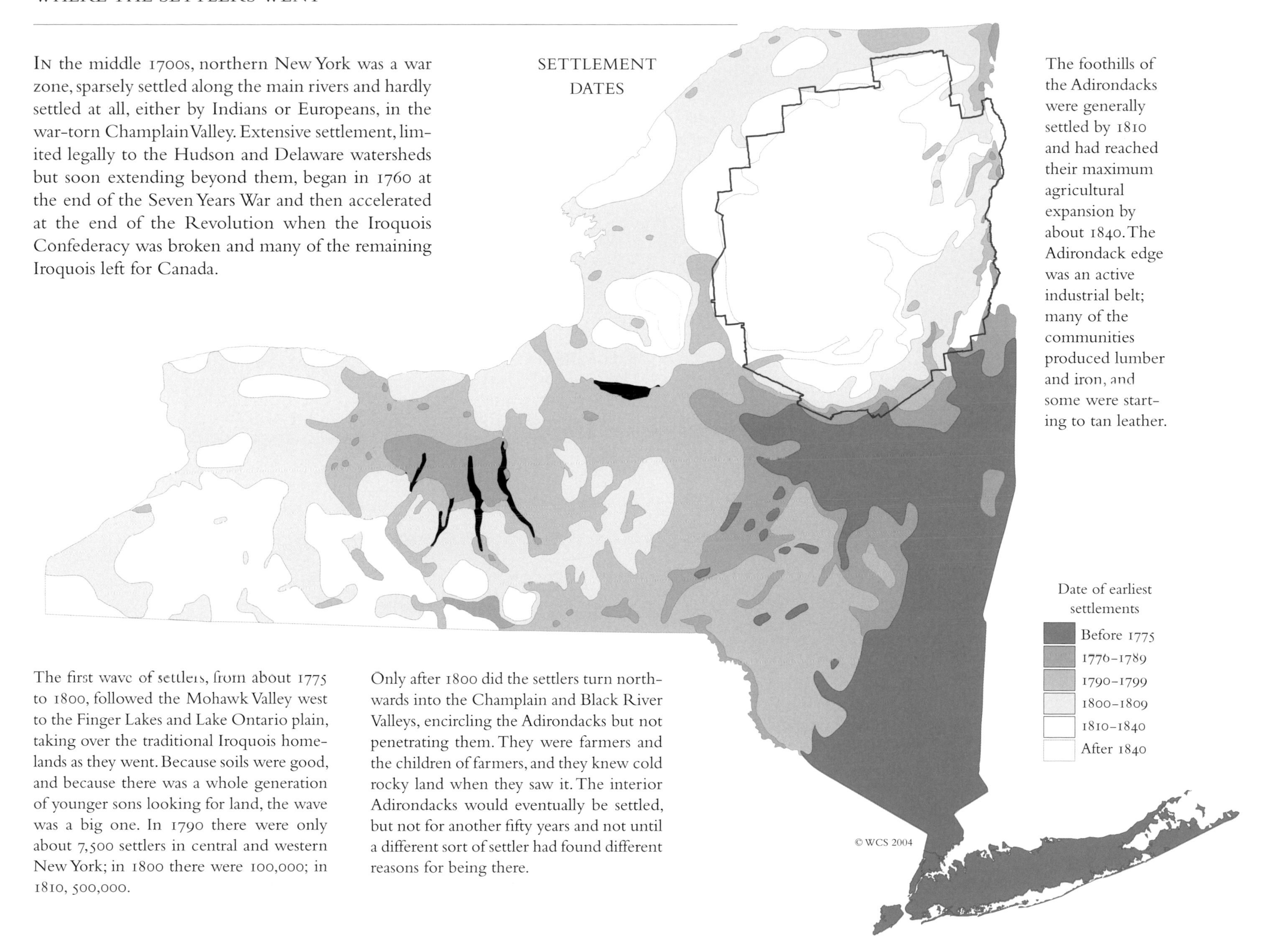

The foothills of the Adirondacks were generally settled by 1810 and had reached their maximum agricultural expansion by about 1840. The Adirondack edge was an active industrial belt; many of the communities produced lumber and iron, and some were starting to tan leather.

The first wave of settlers, from about 1775 to 1800, followed the Mohawk Valley west to the Finger Lakes and Lake Ontario plain, taking over the traditional Iroquois homelands as they went. Because soils were good, and because there was a whole generation of younger sons looking for land, the wave was a big one. In 1790 there were only about 7,500 settlers in central and western New York; in 1800 there were 100,000; in 1810, 500,000.

Only after 1800 did the settlers turn northwards into the Champlain and Black River Valleys, encircling the Adirondacks but not penetrating them. They were farmers and the children of farmers, and they knew cold rocky land when they saw it. The interior Adirondacks would eventually be settled, but not for another fifty years and not until a different sort of settler had found different reasons for being there.

Despite occasional statements to the contrary, only a few pieces of Adirondack land were ever purchased from the Indians. The largest of these, and the one which eventually played the largest role in shaping the future Adirondack Park, was never transferred to the men who purchased it. Instead it was confiscated by New York State, and its purchasers driven into exile.

The story goes like this: By 1770 the English colonists had acquired and settled the Hudson Valley and the eastern end of the Mohawk Valley. Southern New England was completely settled, and a migration to the west and north had begun. Frontier lands, regardless of what they might or might not be good for, were perceived as valuable, and a zone of lands held by speculators extended some distance beyond the settled area.

Purchasing Indian lands in 1770 was complicated by the perquisites of monarchy. Because Indian relations were important and because many purchases from them had in fact been swindles, all Indian purchases now required the approval of the King. And because the King claimed all of North America west to the Mississippi, any lands ceded or sold by the Indians automatically became Crown property. Thus to obtain Indian lands you first bought them from the Indians, then turned them over to the Crown, and then petitioned (read paid) the Crown to grant them back to you.

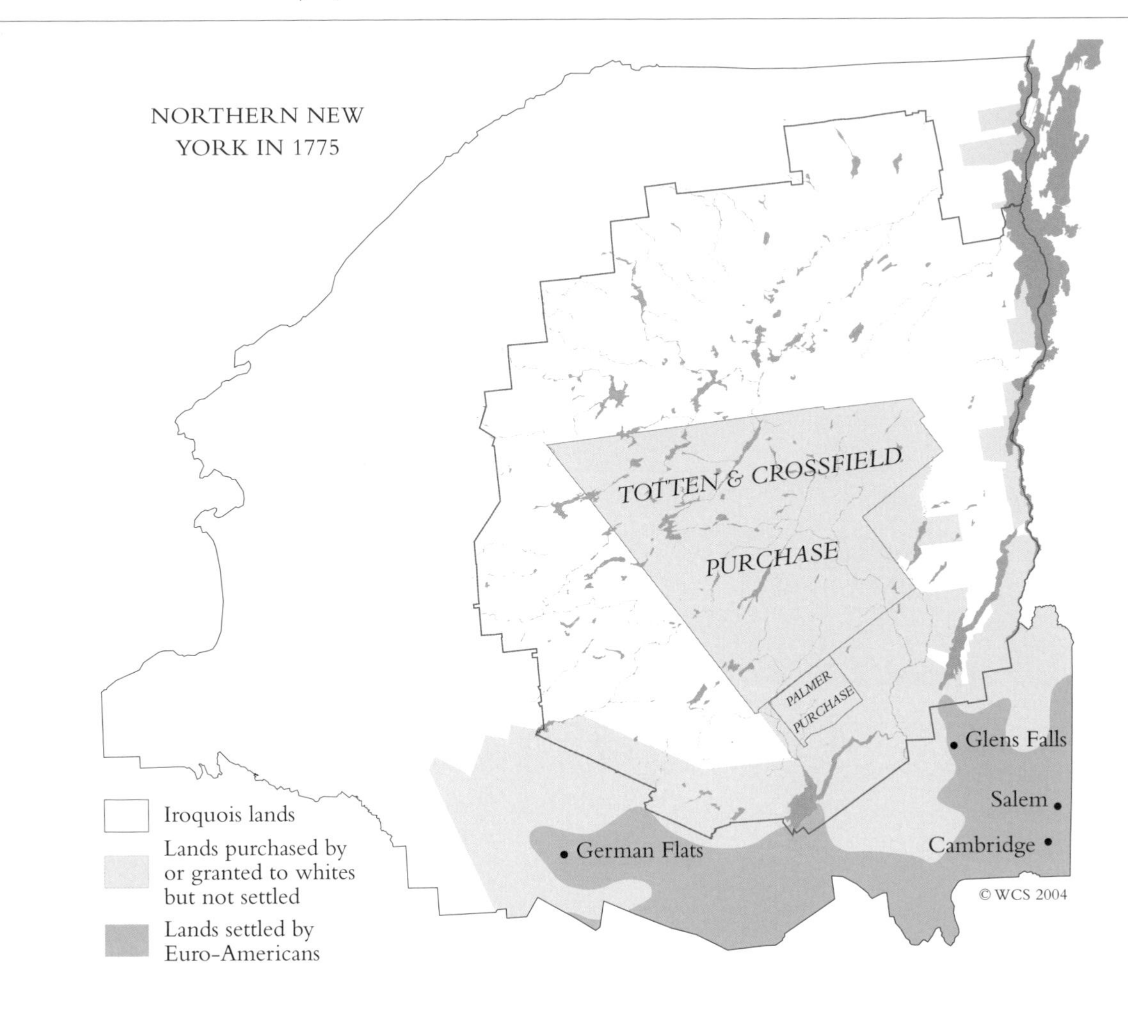

In the case of the Totten & Crossfield Purchase, a war, a poky monarch, and a bill of attainder followed the purchase and prevented the granting. Edward and Ebenezer Jessup were well-connected upstate lumbermen with land holdings scattered across the southern Adirondacks. In 1771 they received permission to buy 1,150,000 acres of Adirondack land and hired Joseph Totten and Stephen Crossfield as their agents. The Jessups first paid the Mohawks about $6,000 for all the Indian land south of a line running west from Port Henry. Then—and this was the real purchase—as loyal British subjects they paid the King $40,000 to grant his new land back to them.

It took King George several years to make up his mind. By then a war was on and his loyal subjects had other things to worry about. The Jessups, along with fifty-three other wealthy Tories, were named in the Act of Attainder of 1779, which required the "forfeiture and sale of the estates of persons who had adhered to the British cause" and, infamously, provided no chance for legal defense or appeal. They fled to Canada, were commissioned as officers in the British army, and have no further roles in our history. By 1786 the Totten and Crossfield lands, had been divided into towns and were being sold to the loyal subjects of the State of New York, in whose hands they came, very quickly, to have a central role in the Adirondack story.

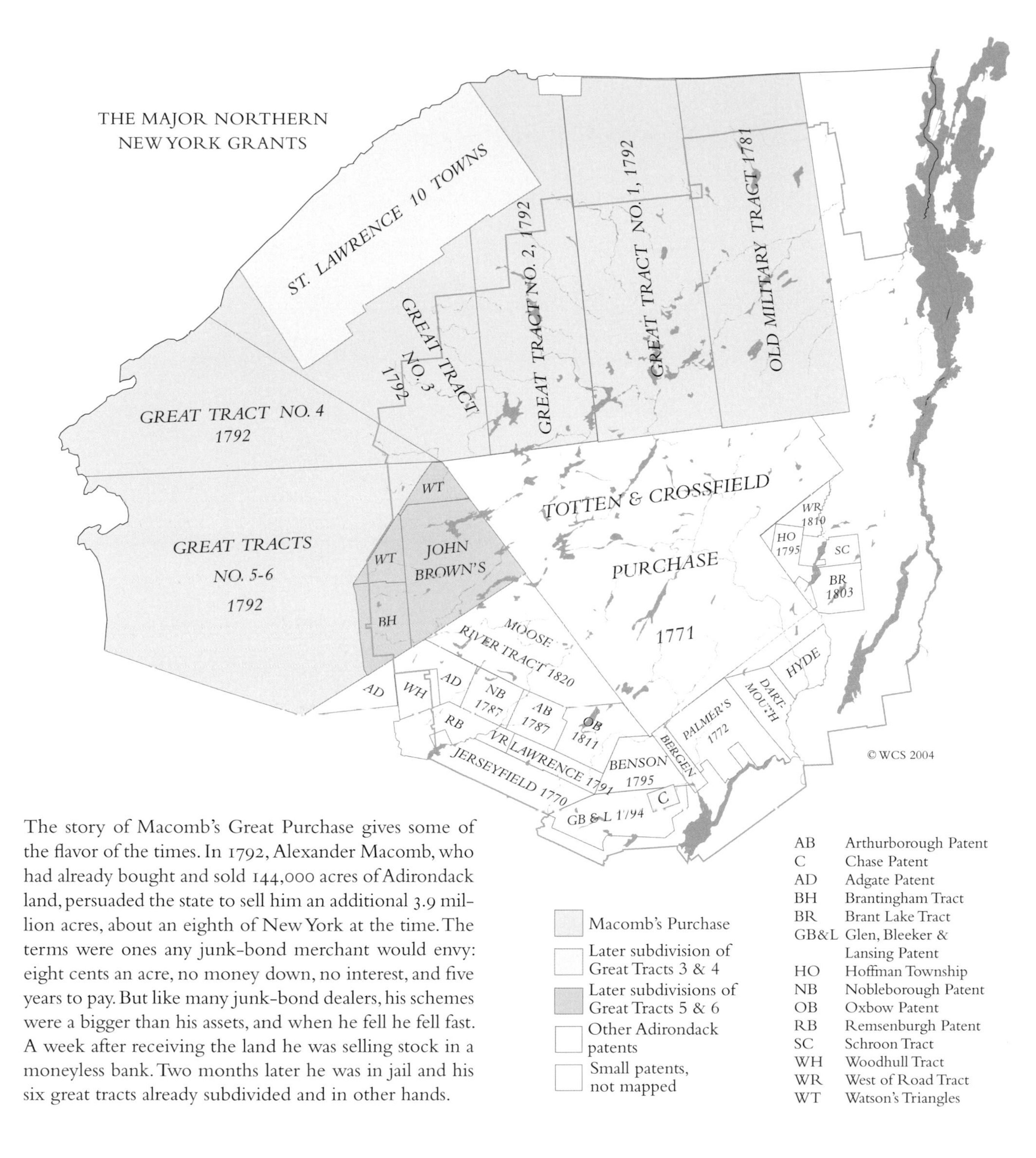

The story of Macomb's Great Purchase gives some of the flavor of the times. In 1792, Alexander Macomb, who had already bought and sold 144,000 acres of Adirondack land, persuaded the state to sell him an additional 3.9 million acres, about an eighth of New York at the time. The terms were ones any junk-bond merchant would envy: eight cents an acre, no money down, no interest, and five years to pay. But like many junk-bond dealers, his schemes were a bigger than his assets, and when he fell he fell fast. A week after receiving the land he was selling stock in a moneyless bank. Two months later he was in jail and his six great tracts already subdivided and in other hands.

At the end of the Revolution New York became the owner of large amounts of Crown and Indian lands, including all of northern New York west of the Champlain Valley. There was no reason for the state to keep these lands—the notion of a forest preserve for public use was nearly a hundred years in the future—and there were many good reasons for it to get rid of them. In any event, the sale of state lands—*alienation* is the technical term—began immediately after the Revolution and proceeded very rapidly. By 1810 about fifteen million acres, half the area of the state, had been sold or given away.

Selling the central and western lands was easy. The river bottoms and Iroquois homelands of central New York were prime agricultural lands, and the growing talk of a western canal kept speculative interest high.

Selling the seven million acres in northern New York was a little harder. The lands were roadless and unexplored and believed, correctly, to be alternately mountainous and swampy. But eventually this hardly mattered. In 1800 land was both wealth and a kind of currency that could be exchanged, on very short notice, for cash or credit. Then as now, venture capital was restless. What the land might someday be good for was unimportant. What mattered was to get the land before someone else did and sell it quickly to somebody else who wanted it. Surveying it, visiting it, or using it could come later, if at all.

ALL the large grants were divided into towns. These first towns had nothing to do with government or settlement or taxes. They were simply square chunks of forest of some 20,000 to 30,000 acres that could be conveniently sold or traded. But some have good stories, and many have shaped current patterns of ownership.

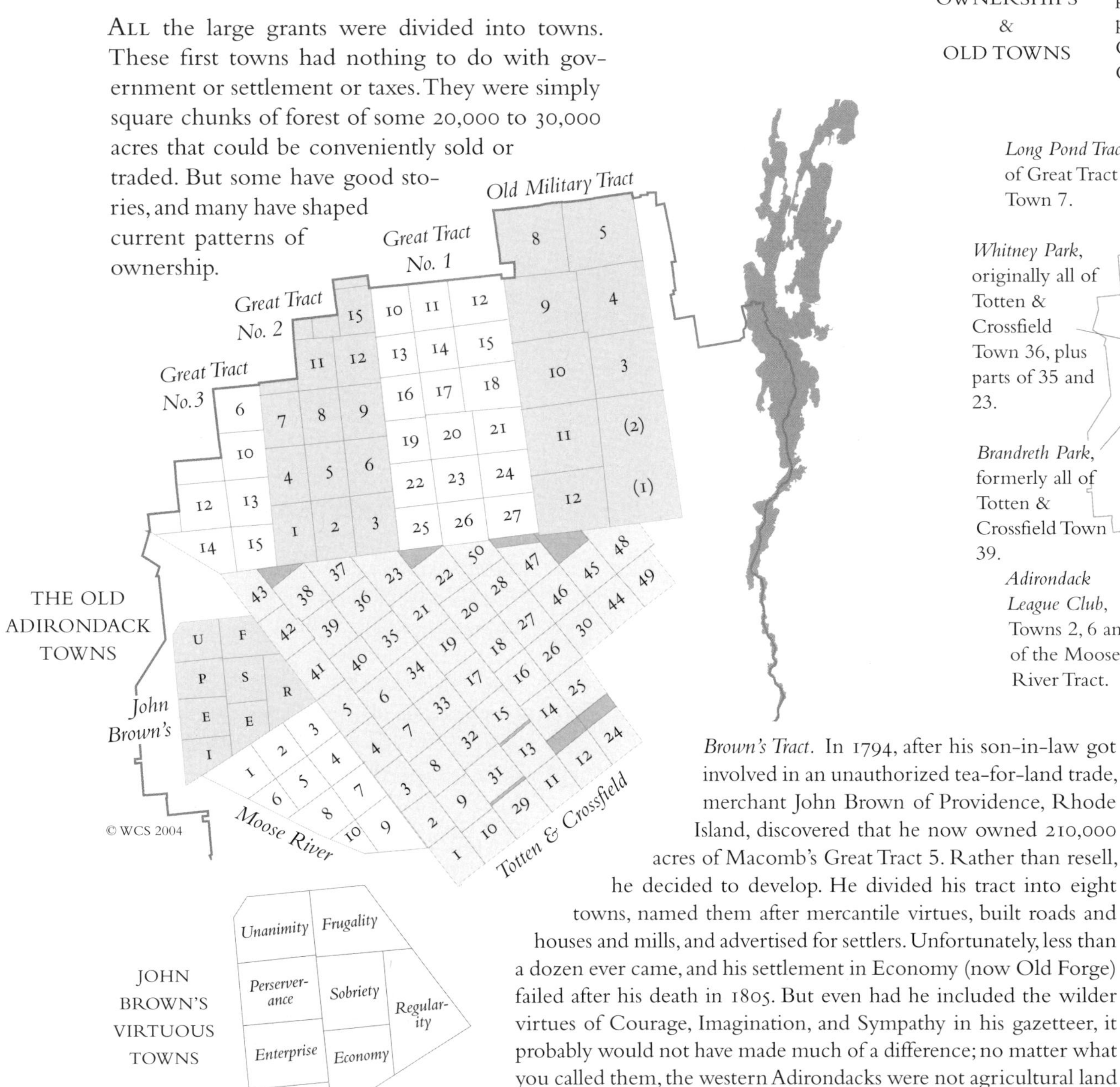

Brown's Tract. In 1794, after his son-in-law got involved in an unauthorized tea-for-land trade, merchant John Brown of Providence, Rhode Island, discovered that he now owned 210,000 acres of Macomb's Great Tract 5. Rather than resell, he decided to develop. He divided his tract into eight towns, named them after mercantile virtues, built roads and houses and mills, and advertised for settlers. Unfortunately, less than a dozen ever came, and his settlement in Economy (now Old Forge) failed after his death in 1805. But even had he included the wilder virtues of Courage, Imagination, and Sympathy in his gazetteer, it probably would not have made much of a difference; no matter what you called them, the western Adirondacks were not agricultural land and were going to have to wait for the loggers and railroads before serious settlement could begin.

MODERN OWNERSHIPS & OLD TOWNS

Former Champion Lands, now partly owned by N.Y.S. and partly by the Forestland Group, Great Tract 1, Towns 11 and 14; Great Tract 2, Town 10.

Domtar Paper, most of Towns 8 and 9 of the old Military Tract.

Fountain Forestry Lands, north half of Towns 16 and 17 of Great Tract 1.

Long Pond Tract, most of Great Tract 2, Town 7.

Bay Pond, originally William Rockefeller, south half of Towns 16 and 17 of Great Tract 1.

Whitney Park, originally all of Totten & Crossfield Town 36, plus parts of 35 and 23.

Finch Pruyn Paper, much of Towns 18, 19, 20, 44, and 46 of the Totten & Crossfield Purchase.

Brandreth Park, formerly all of Totten & Crossfield Town 39.

International Paper Co., Totten & Crossfield Town 9; Town 6 (Sincerity) of John Brown's Tract

Adirondack League Club, Towns 2, 6 and 8 of the Moose River Tract.

Modern Tracts and Old Towns. Many modern Adirondack ownerships were assembled before the modern towns were created, and many Adirondack insiders still refer to properties by their Totten and Crossfield and Great Purchase town numbers. A forester, for example, may say that he has just finished marking trees in Town 20 for a cut this winter. Most of these ownerships were assembled in the late nineteenth century, often from lumbermen who had cut the big spruce and pine and then moved on. All have gained land here and lost it there. Only one of them, Totten and Crossfield Town 39, sold by New York State to Benjamin Brandreth in 1851, is still owned by the descendants of the first purchaser.

THIS map shows the modern town and county lines superimposed on the original grant lines. Early surveyors often laid out north-south lines simply by following the compass needle; as a result the basic Adirondack grid, followed by seven modern county lines and many town lines, lies about eight degrees west of north. (This was the approximate compass declination in the early 1800s.) The more strongly skewed grid lines of the Totten & Crossfield towns are idiosyncratic and not followed by any modern boundaries.

On paper, the old towns were supposed to be straight-sided and rectangular. In practice they were anything but. The early surveyors, working with crude instruments in rough country, invariably ran skewed lines that did not meet where they were supposed to. But since, by both law and custom, ownership is determined by where the surveyor went rather than where he meant to go, the original survey lines, irregularities and all, are now the legal boundaries.

TOTTEN & CROSSFIELD TOWN 25

Bailey's Patent

Thorn's Survey

Minerva

25

Dominick's Patent

Olmstedville

1 mile

Most towns were divided into lots by the original owner. The lots—usually somewhere between a hundred and a thousand acres—were the basic units in which land was subsequently traded and have created the complicated staircase boundaries of many Adirondack ownerships.

Unlike the uniformly lotted towns of the central and western U.S., no two Adirondack towns have exactly the same lotting grid. This can make constructing ownership maps (like *Map 6-3*) exceedingly difficult. Thus Totten & Crossfield Town 25, shown above, was originally sold to three different men, each of whom devised his own lotting grid. Each grid has different east-west lines, and two of the three have different north-south lines. The three patentees—Thorn, Bailey, and Dominick—are long forgotten, but the grids they devised are reflected in every ownership map today.

GOSPEL & LITERATURE LOTS

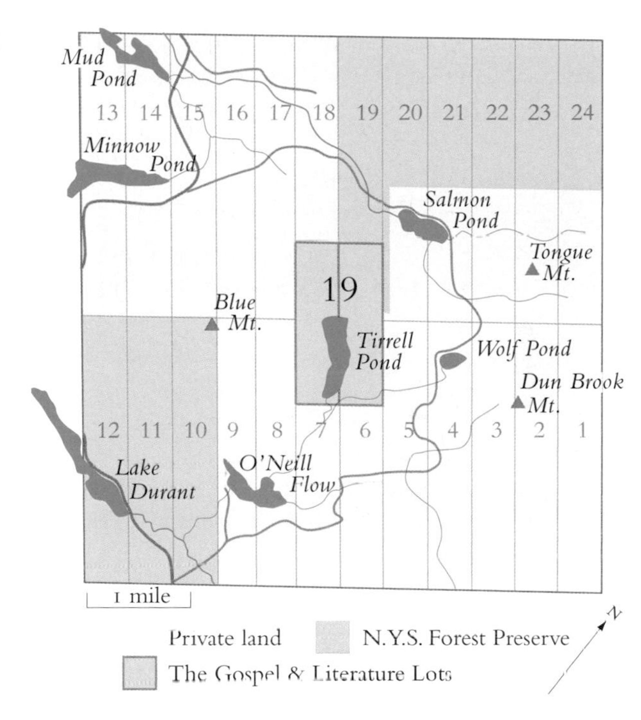

A number of the old towns contain a central block of state land, divided into two lots called the Gospel and Literature Lots. These were mandated by a 1786 law that said that each town shall have one central lot for the support of local schools and the promotion of the gospel and another reserved for the people of the state and applied by the legislature to the promotion of literature. Not all towns had such lots, or, for that matter, schools or churches for the lots to support. And it seems unlikely that the legislature, busy then as now with patronage and power, ever spent much time in the promotion of literature. But these lots still were the first legally protected public lands in the Adirondacks, antedating by over a hundred years the Forest Preserve, Article 14, and the Adirondack Park.

5-8 THE LAST HOT WAR IN THE CHAMPLAIN, 1812–1814 

LIKE the Vietnam War, the War of 1812 was an unpopular war that the United States drifted into gradually and pursued without a clear strategy. It was a real war with real fighting—Burlington, Vermont, was shelled, Washington, D.C., was burned, and many soldiers and sailors died—but, again like Vietnam, the fighting had little to do with the outcome.

The war was linked to the long-standing economic rivalry between Britain and France and started from quarrels about trade, impressment, and neutrality. It was pursued from patriotism and vague hopes that an annexation of some part of Canada might solve the growing Indian conflicts in our own west. It was ended through negotiations, lasting almost as long as the fighting itself, that nullified all new territorial claims and returned the U.S. border to exactly where it had been when the war broke out.

In the Champlain Valley the war was perceived as an unnecessary interruption of cross-border trade and neighborliness. The U.S. commanders in Plattsburgh made a few short overland raids into Canada. The British raided somewhat more successfully, but without lasting effect, along the lake. In the last three months of the war a large British force attacked Plattsburgh, lost their fleet and their combativeness in the first hour of battle, and went home.

Even indecisive wars have aftermaths. The British maintained a garrison at Île-aux-Noix for the next fifty-six years, just in case the braw Lieutenant MacDonough should return. The U.S., determined to prevent another unsuccessful invasion of Plattsburgh, built the giant Fort Montgomery near Rouses Point. The fort took sixty-five years to complete, partly because it turned out to be in Canada and Britain had to be convinced to relocate her border to help us defend ourselves against her. By the time she did the border hostilities were old memories, and the fort was never manned.

WAR IN THE CHAMPLAIN VALLEY, 1812–14

Montréal

Chambly

June 3, 1813. Lieutenant Thomas MacDonough raids the naval base at Chambly and captures two sloops.

October 26, 1813. An attempted attack on Montréal is stopped at Chateaugay.

Chateaugay

Sept. 20, 1813. 4,000 American troops under General Hampton invade Canada and are defeated at Odeltown.

1814. Sir George Prevost builds a small fleet at Île-aux-Noix.

Fort de l'Île-aux-Noix, Fort Lennox

Odeltown

CANADA

U.S.

July 31, 1813. U.S. wins a naval engagement in Missisquoi Bay

November 20, 1812. A detachment from the U.S. army fights the New York State Militia by mistake. The real British appear and the Americans flee in confusion.

American fort
American campaign
American victory
American naval victory
British fort
British campaign
British victory
Multinational blunder

July 30, 1813. British raid Plattsburgh, occupy the town, and burn the arsenal.

Plattsburgh

1812-1814. Extensive but unprosecuted smuggling of lumber, horses, food, and supplies to the British garrisons.

September 11, 1814, Battle of Plattsburgh. The Americans capture the whole British fleet; a large British army under Prevost retires after fleet is captured.

Burlington

July 31, 1813. A British fleet shells Burlington.

Winter, 1813. MacDonough and fifteen shipwrights build three sloops and refit two gunboats in Vergennes.

Vergennes

Winter 1814. MacDonough builds four ships, including the 120-foot frigate *Saratoga* and six gunboats, in eight months.

Since the Champlain and Richelieu valleys were now well settled, the War of 1812 was a rural war of garrisons and militiamen rather than a wilderness war of forts and expeditions. The British and American settlers on either side of the border had close economic and often personal ties with each other. They had no motive to fight one another and often refused to. They had every reason to continue their cross-border trade, though now technically treasonous, and often did.

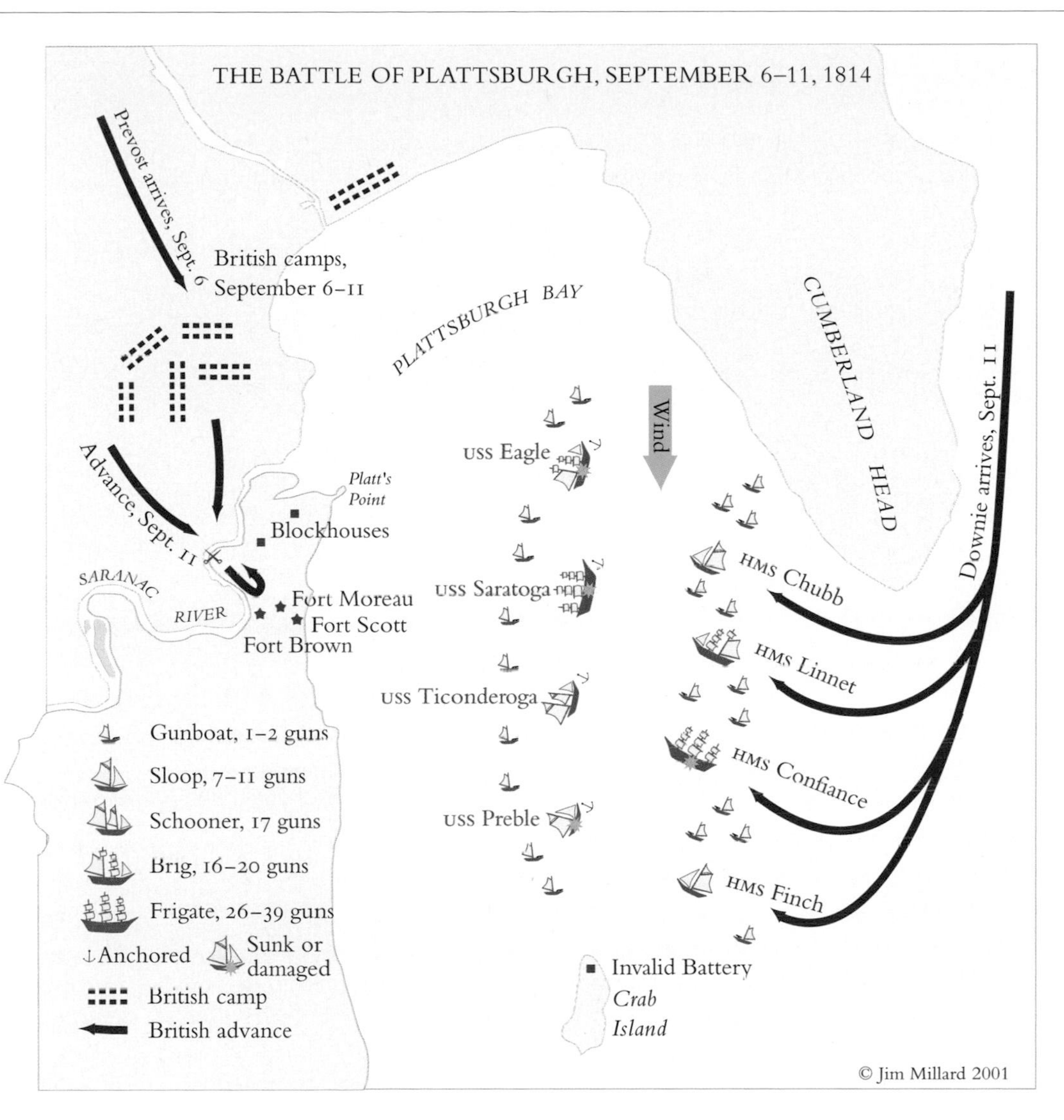

In June, 1814, the British general Sir George Prevost was ordered to invade Lake Champlain, possibly to provide a bargaining chip for the British to use in the stalled peace negotiations in Ghent. He had an army of about 14,000 troops and a small navy which he augmented by building several ships, in great hurry, at Île-aux-Noix. The largest, the 146-foot frigate *Confiance*, was only finished two days before the battle. She was the largest warship ever to sail Lake Champlain. Many of her crew, recruited just before the battle, had never sailed before. Some fifty of them died on their first morning at sea.

In the first week of September, Prevost marched his army down the west side of Lake Champlain and camped north of Plattsburgh. The American garrison at Plattsburgh consisted of 3,400 regular troops and 2,000 volunteers. Prevost could probably have taken the forts in a few hours of fighting. Instead he camped for five days and waited for his fleet to arrive.

The British fleet of four ships and twelve gunboats, commanded by Capt. George Downie, arrived at dawn on September 11. Facing it was an American fleet of four ships and ten gunboats commanded by a junior officer, Lieutenant Thomas MacDonough. Like the British, the Americans were barely ready: at the last minute MacDonough had to impress several military prisoners and six musicians to man his flagship, the *Saratoga*.

The battle that followed was, though short and small, a true naval battle, with all the suffering and fright that implies. According to the historian John Keegan, in almost no other situation in the history of war did men face such concentrations of artillery fire in such poorly protected situations as they did in battles between wooden ships.

The fighting began about 8 A.M. There was no wind, and the battle was a drifting match with guns. Downie died in the first salvo, as did a fifth of the crew of the *Saratoga*. MacDonough was hit by a falling spar and then knocked across the deck by the severed head of his chief gunner. He turned the *Saratoga* about and used her intact port battery to cripple the *Confiance*, which had become tangled in her own cables. The *Confiance* struck her colors, and the British fleet surrendered.

General Prevost, meanwhile, had advanced to the Saranac River and taken the river crossings. When he saw the rout in the bay he called it a day and retreated. His retreat was so sudden that it was some time before American defenders, greatly outnumbered and waiting for their positions to be overrun, discovered that the enemy was gone.

The war continued for another three months in the southern United States. Peace was signed in Ghent on December 24, 1814. The news traveled slowly, and, sadly, reached the U.S. several days after General Pakenham led 5,300 British troops into a pointless massacre in front of Jackson's guns at New Orleans on January 8, 1815.

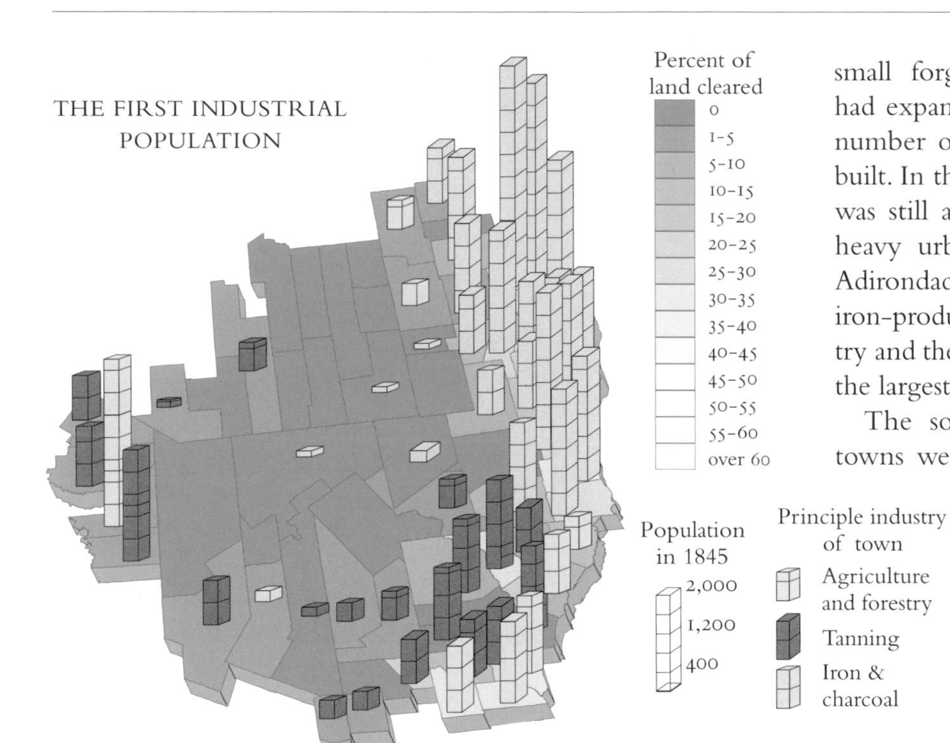

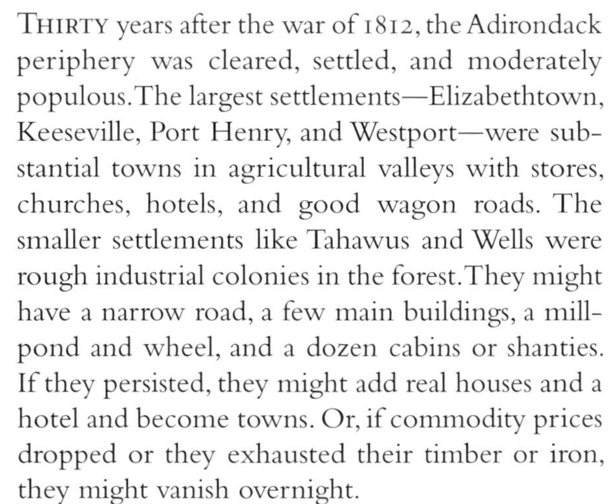

THIRTY years after the war of 1812, the Adirondack periphery was cleared, settled, and moderately populous. The largest settlements—Elizabethtown, Keeseville, Port Henry, and Westport—were substantial towns in agricultural valleys with stores, churches, hotels, and good wagon roads. The smaller settlements like Tahawus and Wells were rough industrial colonies in the forest. They might have a narrow road, a few main buildings, a millpond and wheel, and a dozen cabins or shanties. If they persisted, they might add real houses and a hotel and become towns. Or, if commodity prices dropped or they exhausted their timber or iron, they might vanish overnight.

The northeastern towns produced mostly iron, much of it in tiny settlements with names like Ironville, Upper Forge, and Ferrona. The small forges of fifty years before had expanded and multiplied, and a number of blast furnaces had been built. In the mid-1800s, iron making was still a rural craft rather than a heavy urban industry. The eastern Adirondacks were one of the major iron-producing regions in the country and the iron industry had become the largest Adirondack employer.

The southern and southeastern towns were built around tanneries, many of them quite large. All the Adirondack tanneries used hemlock bark to make their tanning liquor and so had to be built in or near the edge of the woods—close enough to the forest to get the hemlock they needed and near enough to good roads to be able to get hides in and leather out.

Tanneries and ironworks needed large amounts of bark and charcoal and tended to cut in concentrated areas and create large openings. A large tannery might cut all the hemlock on a thousand acres in a year; a large ironworks might take all the hardwoods from two thousand acres or more. The aggregate cutting for charcoal and tanning in the industrial periphery probably exceeded 25,000 acres per year, equaling or exceeding what the softwood loggers in the interior could do at their most rapacious. Sustainability was not a big issue in those days. You kept your capital investment small and when the wood ran out you closed and moved. In an expanding country with a thousand-mile frontier, there was always some place to go.

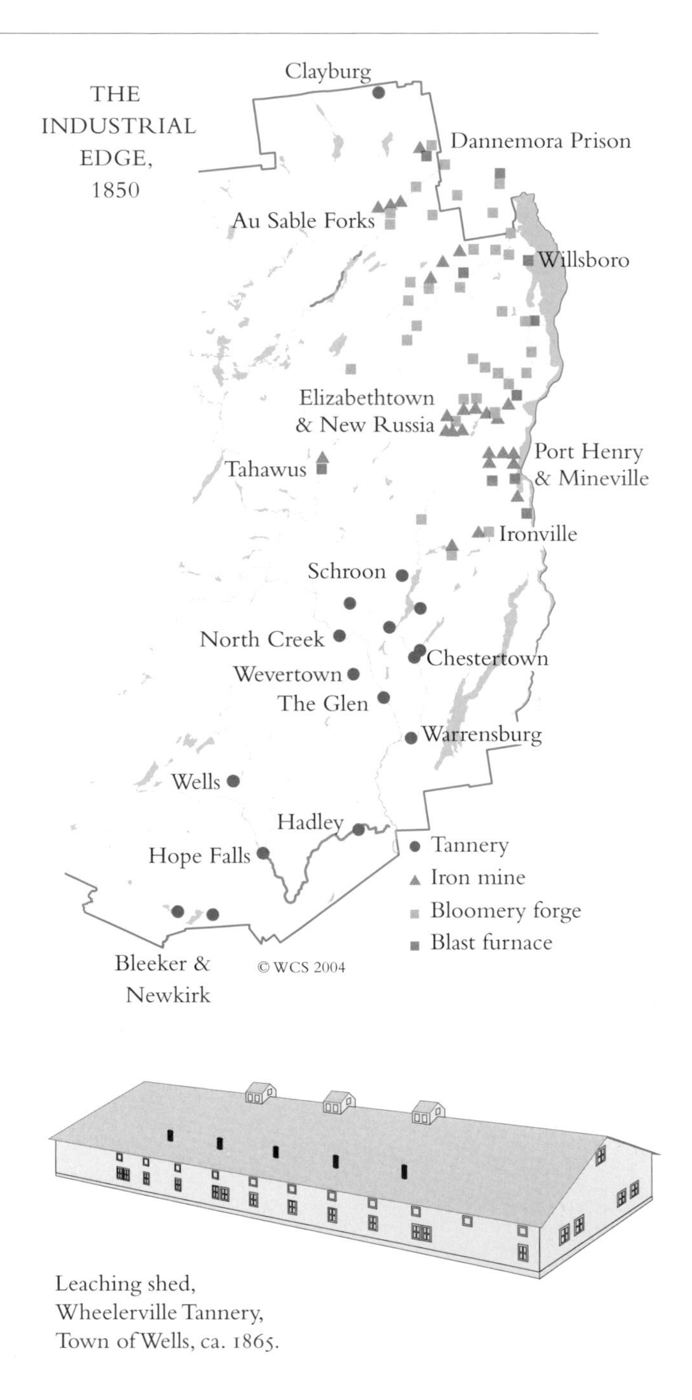

Leaching shed, Wheelerville Tannery, Town of Wells, ca. 1865.

IN 1850, people went to the Adirondack interior either to hunt for fun or cut trees for money. Sport hunting was a relatively small business but was already supporting hotels and guides in a number of towns. It would grow in popularity as the century progressed and game became scarce everywhere else.

Logging was the big business. By 1850 the serious cutting of virgin spruce had begun. The Adirondacks were the leading timber producing area in the country. Albany, with over a mile of lumber docks, was the country's leading lumber port. The lumber on those docks, and the fortunes that the lumber was generating for their owners, came from Adirondack logs.

Inside the Adirondacks there were no fortunes yet; just scattered hotels, lumber camps, and small settlements. Venison, logs, lodgings, and fur were the main commodities. A small resident population of innkeepers, market hunters, and guides served a much larger transient population of loggers, teamsters, and guests. The residents, from what we know of them, seem to have been an interesting lot—much the same kind of able and enterprising expatriates found on other frontiers today.

The map, appropriately sparse, shows the known components of this economy: the railroad terminals at the park edge, the stage roads radiating from them, the log-drive rivers and dams, the early hotels, the few small villages, and the single industrial site at Tahawus. Many other components—the small farms, the thousands of camps, paddocks, cookhouses, shanties, and smithies, the splash dams on small streams—have been swallowed up by the woods and are all but undiscoverable now.

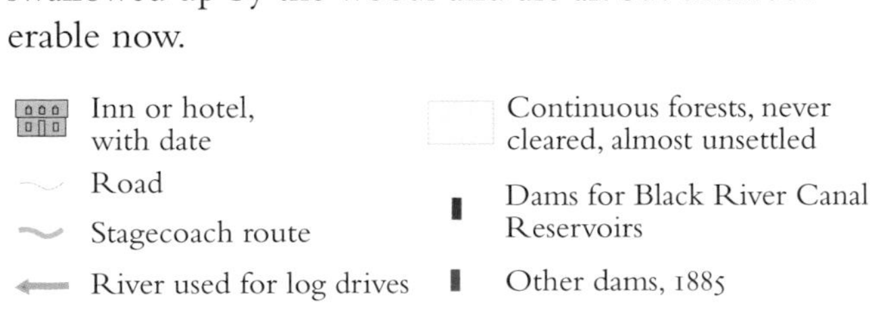

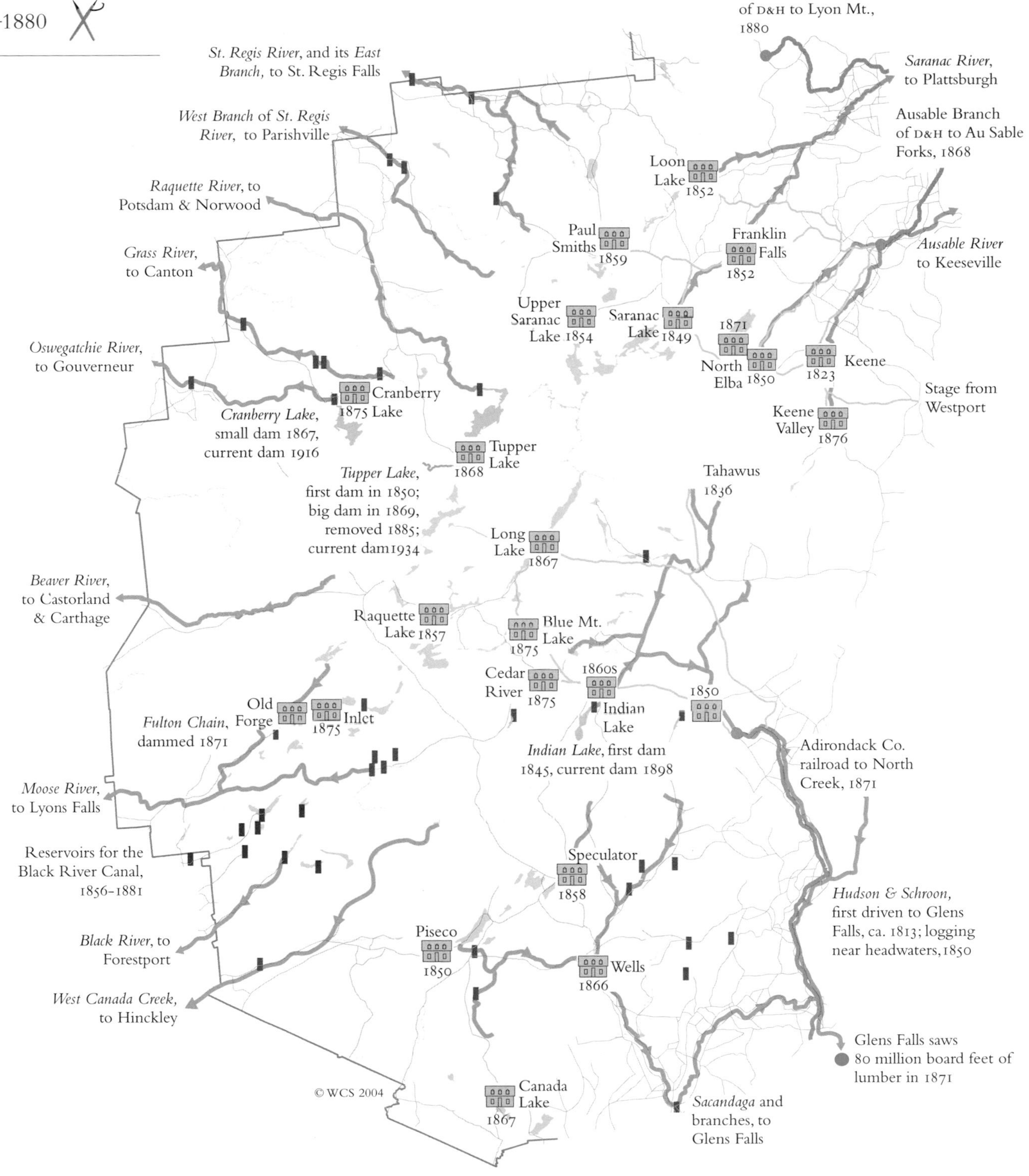

RAILROADS made possible the commercial and industrial development of the interior Adirondacks. By 1900, most of the major Adirondack towns had rail service, and almost every major Adirondack mill, mine, and hotel moved its products and clients by rail.

Today, when rail service is mostly limited to interurban freight and farming has all but vanished, it is easy to forget how vigorous the rural economy once was and how much it depended on rail. A 1900 railroad map of New York looks like a map of paved roads today. There were at least ninety-eight main-line railroad stations in northern New York and perhaps five to seven hundred in the rest of the state. Most of these shipped the products of local farms and industries, and most had passenger stations as well.

The Adirondack rail system developed quickly and vanished equally quickly. In 1850 the only railroads were interurban lines in the major river valleys. By 1875 main lines had completely encircled the Adirondacks, and four spur lines had entered the park. By 1900 railroads entered the park at nine points, four of which were interconnected by a main line that crossed the interior.

After 1900 no new main lines were built. A number of logging railroads were built between 1900 and 1930. The last new tracks in the Adirondacks were laid in 1944 to connect the D&H main line at North Creek to the National Lead titanium mine at Tahawus. After that, massive public investment in highways and air travel doomed the railroad system. Passenger service was largely gone by 1960, and most freight service by 1980. The New York Central tracks through the center of the park are still maintained but are unused except for short segments used by excursion trains.

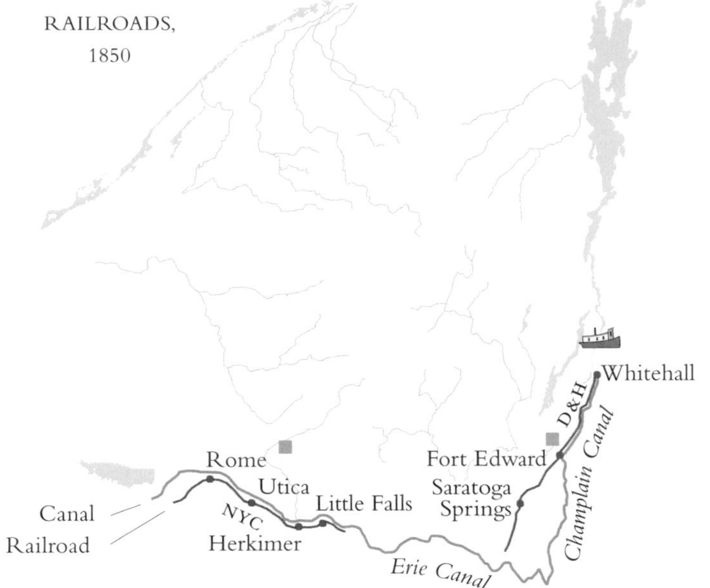

- ■ Mining town
- ■ Major sawmills
- ▲ Paper or pulp mills
- ● Resort town
- Steamboat connections
- Canal
- New railroad
- Existing railroad

By 1874, five main-line railroads circled the Adirondacks, and two branch lines penetrated the settled eastern Adirondack valleys. Major sawmills had developed in the St. Lawrence and Black River Valleys. The North Creek and Ausable Branches of the Delaware & Hudson provided local service in the eastern Adirondacks. Industrial railroads served the mines at Clifton Mines, Ironville, and Mineville. The Lake George Branch, a short spur from the D&H mainline, connected the steamboat lines on Lake George to those on Lake Champlain. The Black River Canal (BR), though only in operation for twenty years, was already an anachronism.

Like the superhighways of the 1950s, the early railroads were all in previously developed corridors. The first railroads in northern New York were the New York Central, paralleling the Erie Canal, and the Delaware & Hudson, paralleling the Hudson River and Champlain Canal. By 1850 steamboat service had begun on Lake Champlain, but there was still no mechanized transport in the Adirondack Park or in the Black River and St. Lawrence Valleys.

AB	Adirondack Branch of D&H
AuB	Ausable Branch of D&H
D&H	Delaware & Hudson
NYC	New York Central
R	Rutland Railroad
RW&O	Rome, Watertown, and Ogdensburg
SL	St. Lawrence Branch of New York Central

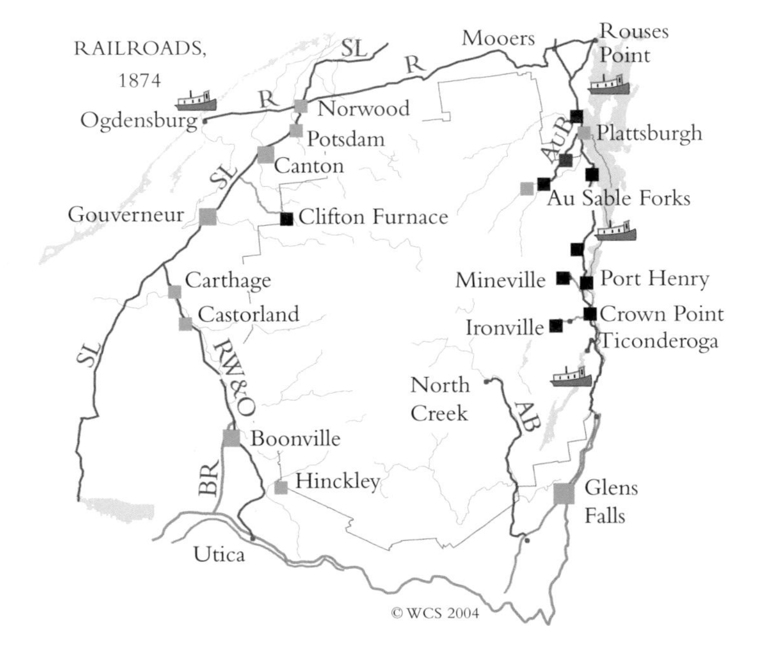

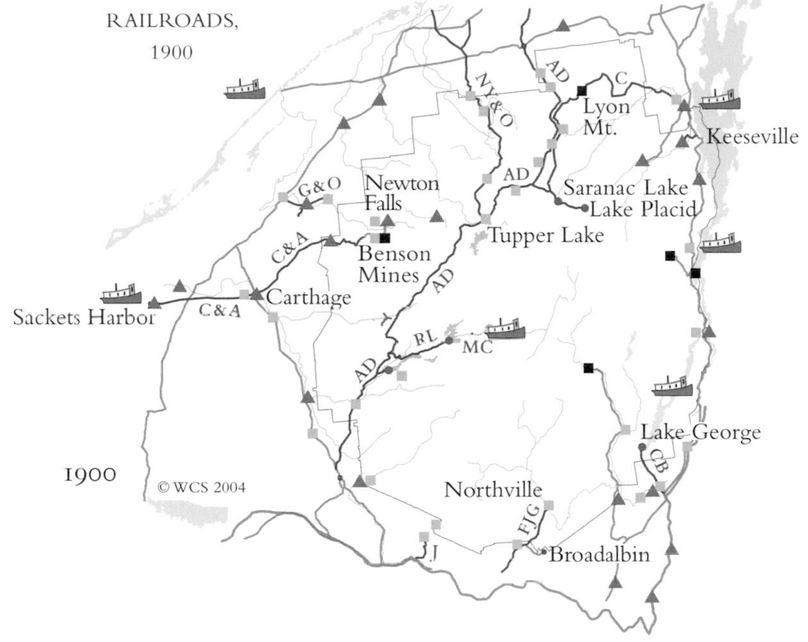

AD Adirondack Division of New York Central
C Chateaugay
C&A Carthage & Adirondack Railroad
CB Caldwell Branch of D&H
FJG Fonda, Johnstown & Gloversville
G&O Gouverneur & Oswegatchie
J Jerseyfield
MC Marion Carry
NY&O New York & Ottawa
RL Raquette Lake

In 1900 a period of industrial consolidation was beginning. Most of the small mines and several of the mining railroads associated with them were gone. The Erie Canal was about to undergo a major rebuilding. The Black River Canal was almost unused and about to be abandoned. The tourist trade was increasing, and there were now three short railroads built solely to carry passengers to summer colonies and resorts.

By 1893 all the main Adirondack railroad lines were in place, and a number of new railroad-served industries, particularly sawmills and paper mills, had developed within the park. The Chateaugay railroad had reached the iron mines at Lyon Mt. in 1880 and the resort villages of Saranac Lake and Lake Placid in 1887 and 1893. The Carthage & Adirondack had reached the iron mines at Benson in 1889, and the paper mills at Newton Falls in 1896. The New York and Ottawa had reached Tupper Lake in 1890 and triggered the construction of over a hundred buildings in one year. The Adirondack Division of the New York Central, the only railroad ever to cross the park, connected Herkimer with Malone in 1892.

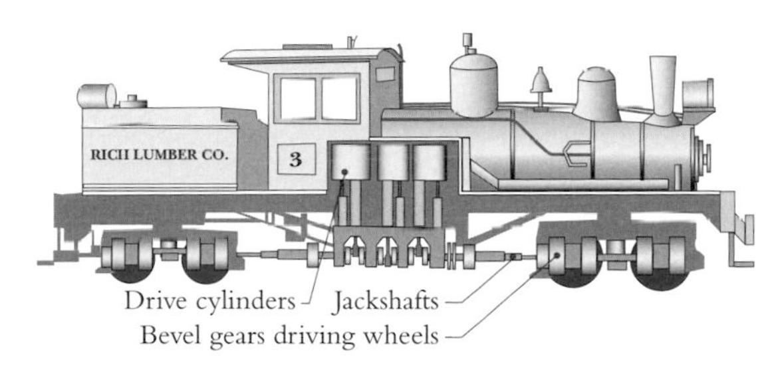

Forty-Ton Shay Logging Locomotive, produced by the Lima Locomotive Works in Lima, Ohio. Shay locomotives were the first logging locomotives suitable for steep terrain. They were driven by vertical cylinders just forward of the cab, connected to the wheels by jackshafts and bevel gears. The gearing allowed them to pull hard at slow speeds. The three-cylinder Shay shown here, although a relatively small locomotive, had eight tons of traction and could pull loads up a 14% grade. Shays were introduced to the Adirondacks in the 1880s and were used by many Adirondack logging railroads.

Logging Railroads. Between 1890 and 1940 at least fifteen major sawmills, three paper mills, and several hundred miles of logging railroad were constructed in the northwestern Adirondacks, starting a cutting spree that lasted until all the virgin timber on private lands had been cut. Because the loggers were cutting so fast and moving so often, most logging railroads were short-lived. The North Tram was the only railroad to operate for over twenty years. Many others didn't make ten.

Industrial Railroads

- Brooklyn Cooperage (BC)
- Bay Pond (BP)
- Everton (E)
- Grasse River (GR)
- Hannah Ore (HO)
- Horseshoe Forestry (HF)
- Newton Falls Paper (NF)
- North Tram (NT)
- Oval Wood Dish (OW)
- Post & Henderson (PH)
- Rich Lumber (RL)
- Watson Page (WP)

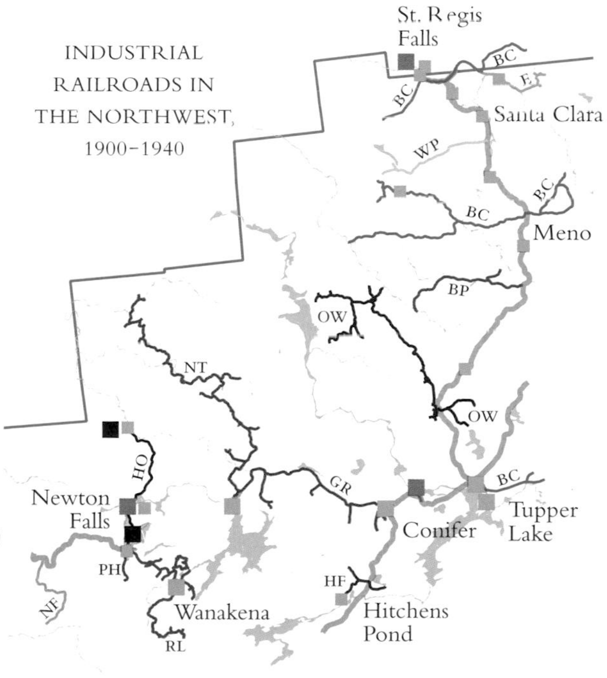

BY the 1880s, quite unexpectedly, the Adirondacks had become fashionable. People wanted to visit them, explore them, and own them. To do this fashionably meant service and luxury: resort hotels with ballrooms and bowling alleys, railroads and steamboats to get you there, guides and chefs to take care of you after you got there. And, of course, the highest fashion was private wealth: private steamboats and railroad cars, private train stations, private shores and islands, and finally private town-sized parks.

With the private parks came a new haute rustic style of building and decor, pioneered in the "Great Camps" built by W.W. Durant and his peers. It was a remarkable style, half Scandinavian, half chalet-inspired, at its best extremely sensitive to setting and mood. It produced many genuine beauties and perhaps even a few masterpieces. It began, the books say, on the east shore of Raquette Lake in 1877.

A cottage at the Antlers, a hotel run by the legendary Charley Bennett, an Adirondack guide turned hotelier and chef. Bennett traveled Europe in the winters, observing hotels and collecting recipes.

Inman Camp, Round Island, Horace Inman, ca. 1880. The original camp was a simple frame cabin with an attached lean-to. Later, as the camp grew, they added a floating vegetable garden and a dance floor built out over the water.

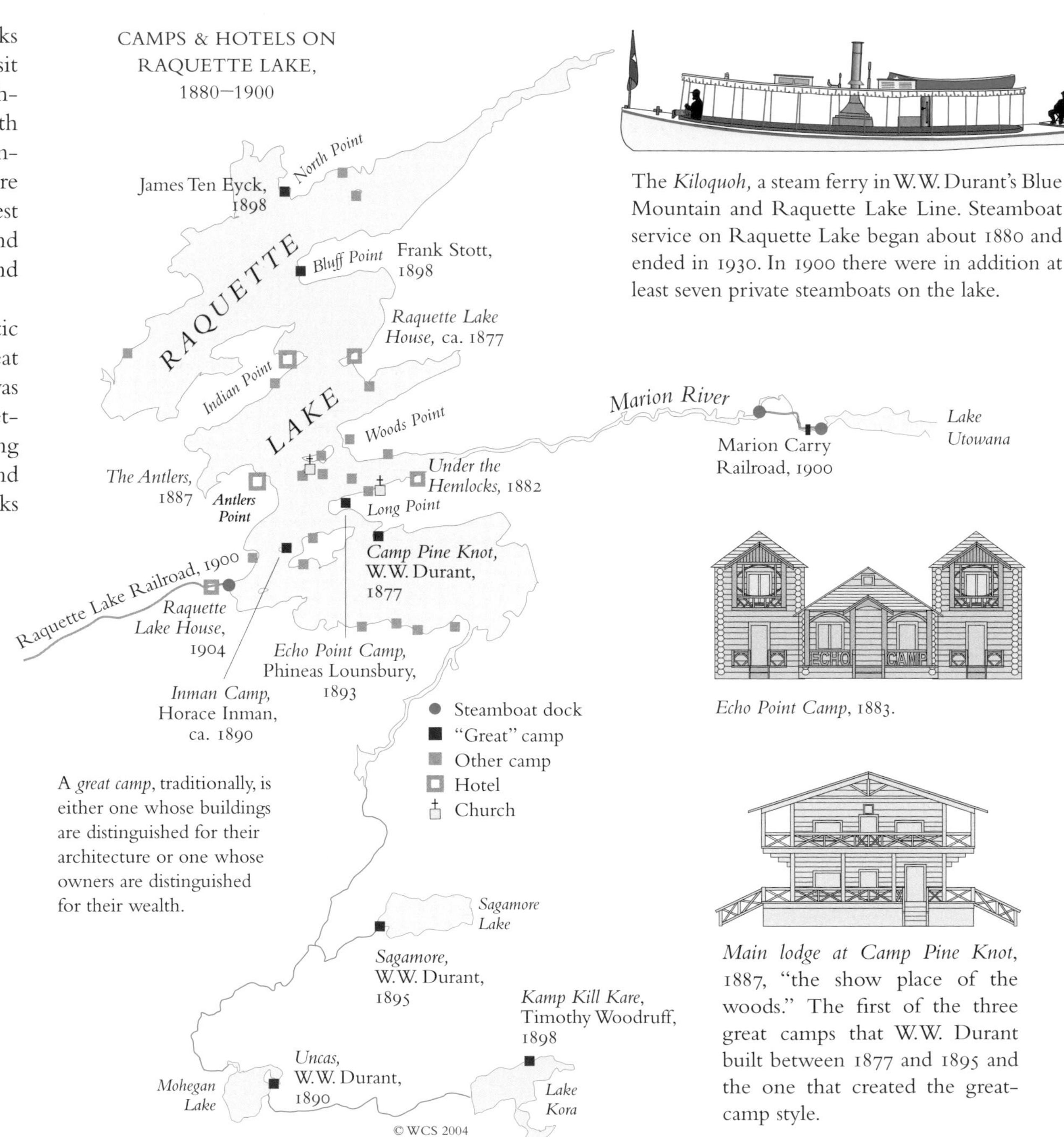

A *great camp*, traditionally, is either one whose buildings are distinguished for their architecture or one whose owners are distinguished for their wealth.

The *Kiloquoh*, a steam ferry in W.W. Durant's Blue Mountain and Raquette Lake Line. Steamboat service on Raquette Lake began about 1880 and ended in 1930. In 1900 there were in addition at least seven private steamboats on the lake.

Echo Point Camp, 1883.

Main lodge at Camp Pine Knot, 1887, "the show place of the woods." The first of the three great camps that W.W. Durant built between 1877 and 1895 and the one that created the great-camp style.

By 1900, the railroads had created the first major Adirondack travel corridor, along which there were about a dozen large resort hotels and a million acres of private parks and preserves. Many of these belonged to the railroad plutocracy themselves. In 1902, for example, the directors and stockholders of the one-mile-long Marion Carry Railroad included J.P. Morgan, W.S. Webb, R.C. Vanderbilt, H. Huntington, and H.P. Whitney. All owned land nearby.

The *Prospect House,* Blue Mt. Lake, 1880, "the largest and most luxurious hotel in the woods," located thirty miles and so nine hours by stage from the nearest railroad. It was two hundred fifty feet long, with space for five hundred guests. It had a bowling alley, a shooting gallery, a telegraph office, and, almost unique at the time, steam heat, running water, and electric lights in every room. Its remote location was at first fashionable, then disastrous. By 1894, two years after Webb built his railroad and competing hotels like the Childwold and the Forge House got railway service, it was insolvent and mortgaged; eight years after that it closed.

The *Forge House,* Old Forge, ca. 1896.

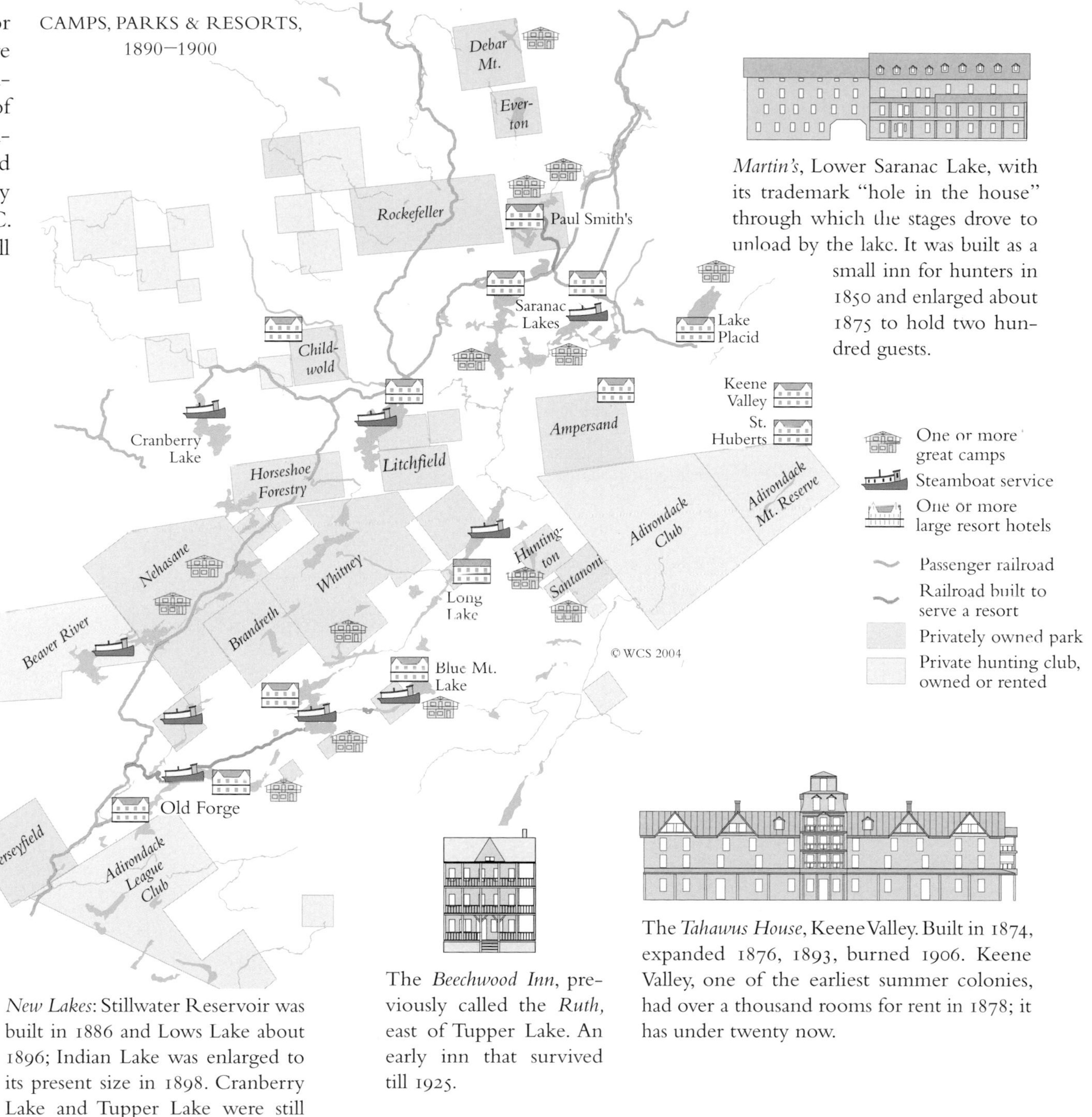

Martin's, Lower Saranac Lake, with its trademark "hole in the house" through which the stages drove to unload by the lake. It was built as a small inn for hunters in 1850 and enlarged about 1875 to hold two hundred guests.

New Lakes: Stillwater Reservoir was built in 1886 and Lows Lake about 1896; Indian Lake was enlarged to its present size in 1898. Cranberry Lake and Tupper Lake were still much smaller than at present.

The *Beechwood Inn*, previously called the *Ruth,* east of Tupper Lake. An early inn that survived till 1925.

The *Tahawus House*, Keene Valley. Built in 1874, expanded 1876, 1893, burned 1906. Keene Valley, one of the earliest summer colonies, had over a thousand rooms for rent in 1878; it has under twenty now.

At the end of the nineteenth century, some Adirondack industries were vanishing, and others were becoming larger and more concentrated. Tanning and iron-making, mainstays of the park economy for a hundred years, were almost gone. Most small mines had closed; the four remaining ones were large and mechanized. Large sawmills and paper mills had developed in the northern Adirondacks and were consuming virgin timber at an unprecedented rate.

The problem with transitions is that they keep going. The turn-of-the-century transition to larger industry led not only to the industrial peak shown in this map but also to the deindustrialization of the park shown in *Map 5-13*. While big resource-extracting industries can usually out-compete small ones, they also compete with each other in larger and tougher markets, and when they lose money they lose a lot of it. While the virgin timber, rich shallow ores, and cheap labor lasted, Adirondack heavy industry was very competitive. When they finally ran out, it was gone.

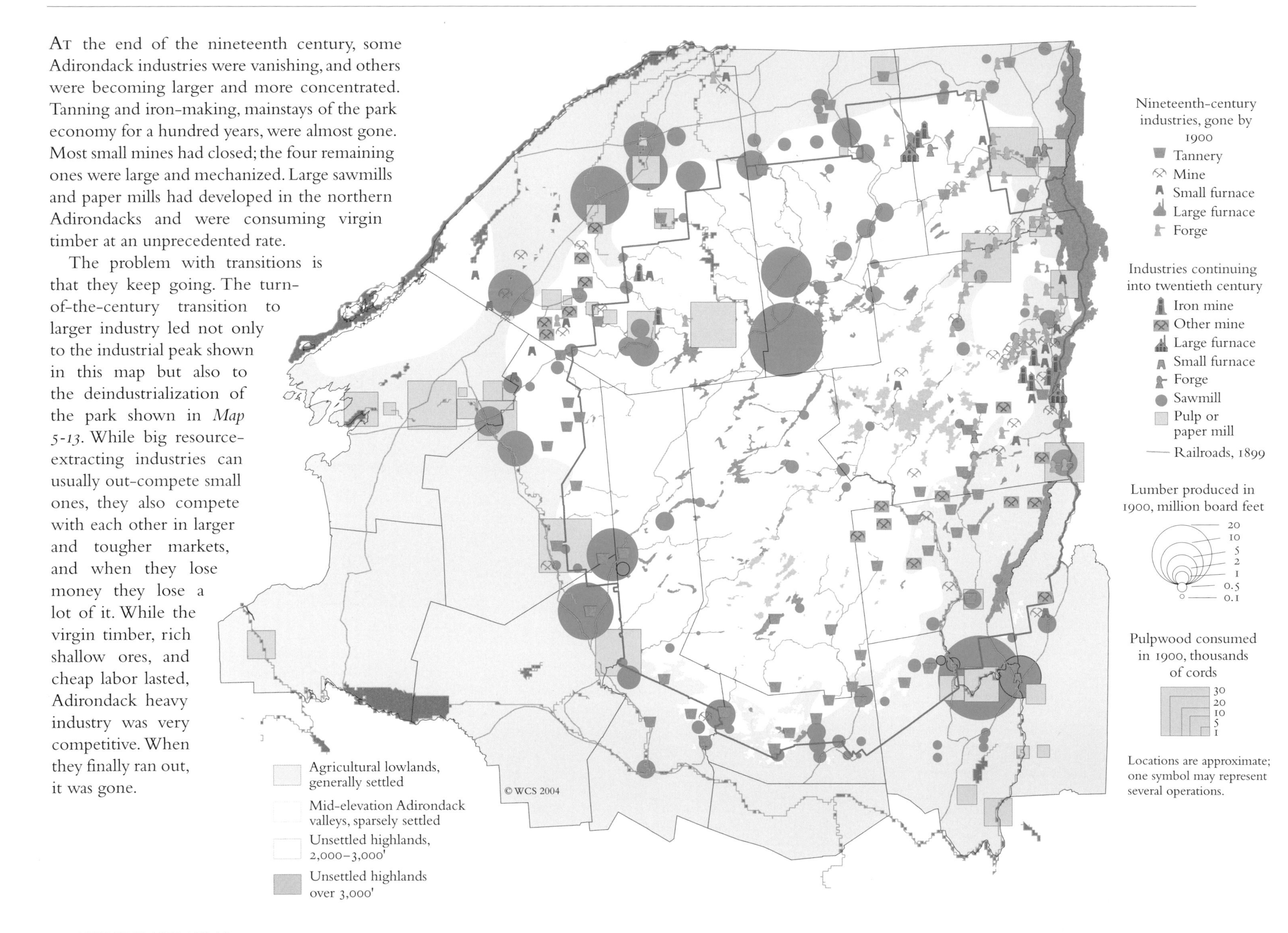

MUCH of the geography of the turn-of-the-century Adirondacks was defined by where the visitors wanted to go and how they got there. Here is a traveler's map, showing 143 railroad stations and steamboat docks in about 1900. All had post offices and were, at least potentially, important points in the future geography of the park.

The arrival of paved roads and automobiles changed, quickly and dramatically, both the geography of transport and the geography of settlement. A hundred years after the period shown in the map, less than half of these railroad and steamboat stops have survived. About fifty-seven, shown in red and blue, are still settled enough to have a post office. A few others are still settled but get mail from elsewhere. The rest have all but vanished.

Electric Trolly Car which operated between Lake Clear Junction and Paul Smith's Hotel from 1907 to about 1928. The car ran on 5,200-volt alternating current, originally supplied from Paul Smith's generator at Keese Mills. A motor-generator within the car converted this to 600-volt direct current, which ran four motors attached to the axles.

THE fifty years from 1900 to 1950 were peak years for Adirondack industry. The map shows 14 paper mills, 11 large sawmills, 12 logging railroads, 4 main surface mines, 2 big underground mines, and 2 blast furnaces. All of them were truly heavy industries, with outputs measured in millions of tons of ore and tens of millions of board feet of lumber. If the emblems of the previous industrial period had been the crosscut saw and the Shay engine, those of this period were the steam drill, the electric hoist, the paper machine, and the coal-fired furnace.

Little of this lasted much past World War II. The logging railroads and the big sawmills vanished as soon as the virgin timber was gone. The first generation of paper mills, uniformly dirty and inefficient, closed when the supply of spruce ran out or when they were out-competed by cleaner and more modern mills elsewhere. Only four paper mills are currently in operation, and only two of these still buy logs and make pulp. The four big iron mines, all of which had greatly expanded production during World War II, struggled against rising production costs and foreign competition for twenty years before closing in the sixties and seventies.

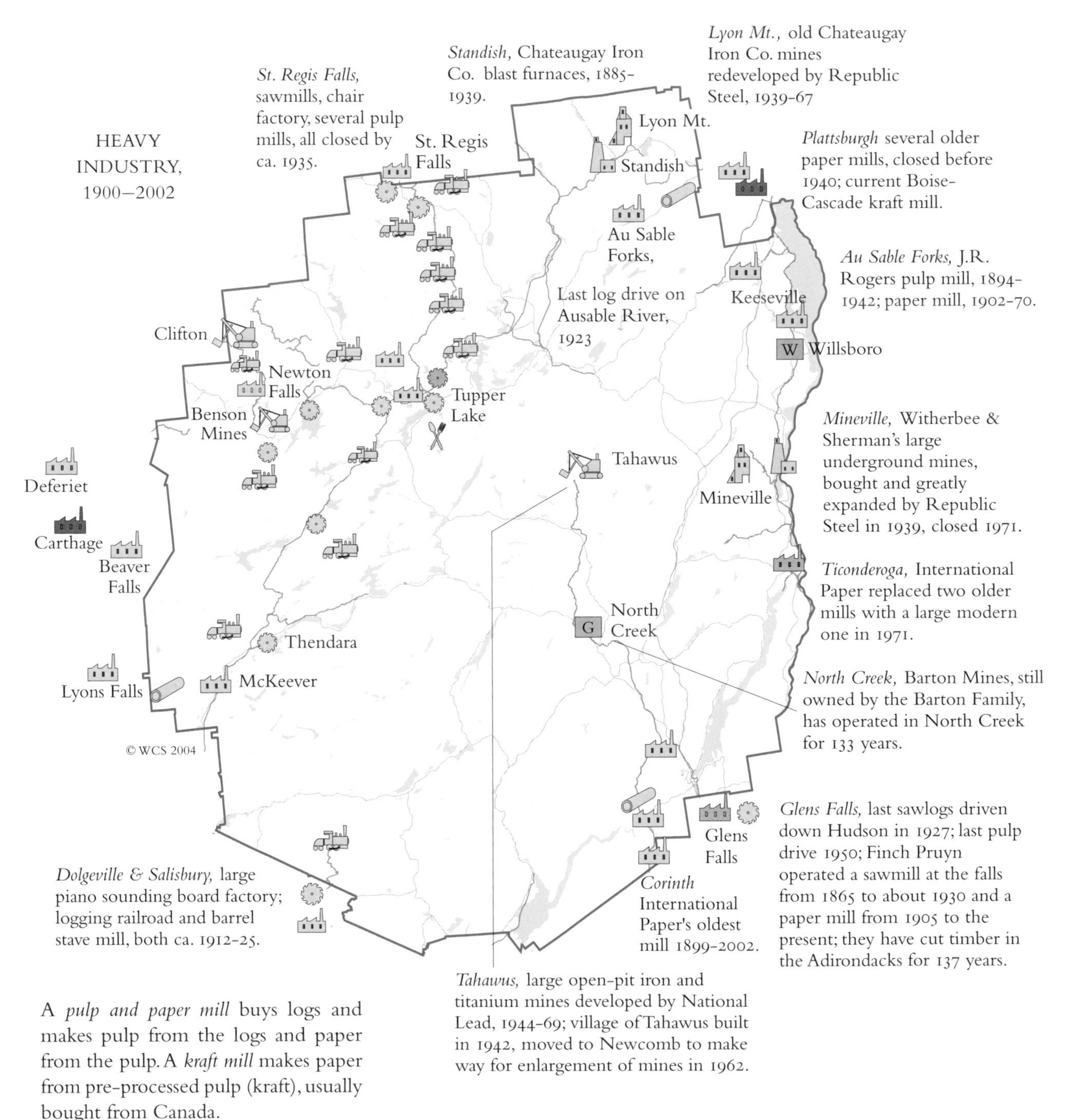

Tahawus, large open-pit iron and titanium mines developed by National Lead, 1944-69; village of Tahawus built in 1942, moved to Newcomb to make way for enlargement of mines in 1962.

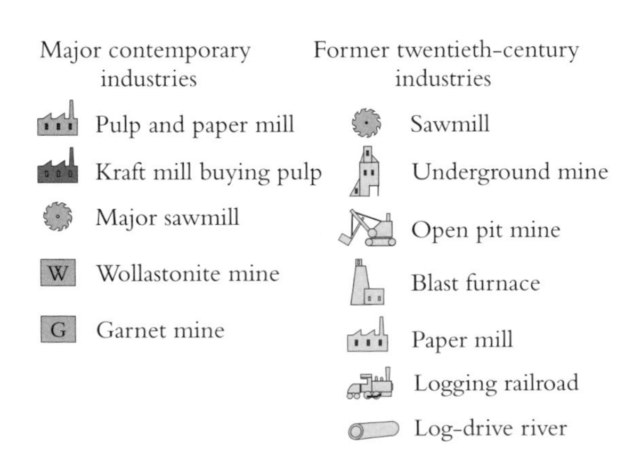

A *pulp and paper mill* buys logs and makes pulp from the logs and paper from the pulp. A *kraft mill* makes paper from pre-processed pulp (kraft), usually bought from Canada.

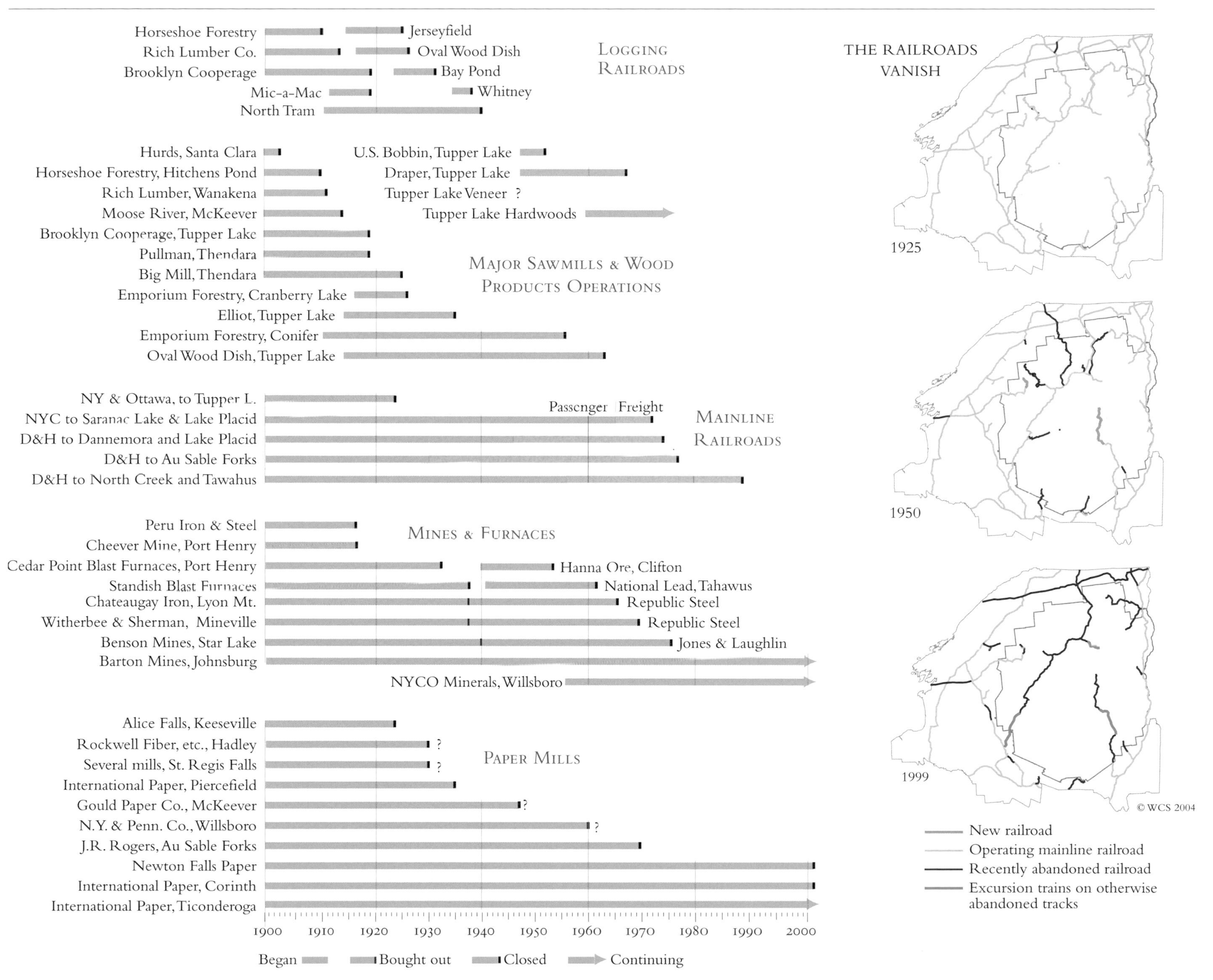

Logging Railroads
Horseshoe Forestry
Rich Lumber Co.
Brooklyn Cooperage
Mic-a-Mac
North Tram
Jerseyfield
Oval Wood Dish
Bay Pond
Whitney
Major Sawmills & Wood Products Operations
Hurds, Santa Clara
Horseshoe Forestry, Hitchens Pond
Rich Lumber, Wanakena
Moose River, McKeever
Brooklyn Cooperage, Tupper Lake
Pullman, Thendara
Big Mill, Thendara
Emporium Forestry, Cranberry Lake
Elliot, Tupper Lake
Emporium Forestry, Conifer
Oval Wood Dish, Tupper Lake
U.S. Bobbin, Tupper Lake
Draper, Tupper Lake
Tupper Lake Veneer ?
Tupper Lake Hardwoods
Mainline Railroads
NY & Ottawa, to Tupper L.
NYC to Saranac Lake & Lake Placid
D&H to Dannemora and Lake Placid
D&H to Au Sable Forks
D&H to North Creek and Tawahus
Passenger
Freight
Mines & Furnaces
Peru Iron & Steel
Cheever Mine, Port Henry
Cedar Point Blast Furnaces, Port Henry
Standish Blast Furnaces
Chateaugay Iron, Lyon Mt.
Witherbee & Sherman, Mineville
Benson Mines, Star Lake
Barton Mines, Johnsburg
Hanna Ore, Clifton
National Lead, Tahawus
Republic Steel
Republic Steel
Jones & Laughlin
NYCO Minerals, Willsboro
Paper Mills
Alice Falls, Keeseville
Rockwell Fiber, etc., Hadley ?
Several mills, St. Regis Falls ?
International Paper, Piercefield
Gould Paper Co., McKeever ?
N.Y. & Penn. Co., Willsboro ?
J.R. Rogers, Au Sable Forks
Newton Falls Paper
International Paper, Corinth
International Paper, Ticonderoga
1900 1910 1920 1930 1940 1950 1960 1970 1980 1990 2000
Began
Bought out
Closed
Continuing
THE RAILROADS VANISH
1925
1950
1999
© WCS 2004
New railroad
Operating mainline railroad
Recently abandoned railroad
Excursion trains on otherwise abandoned tracks

THE Cold War was America's strangest and most frightening war. Strange because it was over ideas and intentions rather than actions. The U.S. and the U.S.S.R had in fact fought as allies in the two world wars but had since become bitter rivals, each of whom sought dominance over the other. And frightening because the power of the weapons and the deadliness of the radioactivity they created were unparalleled. One estimate suggested the first few seconds of a nuclear war would release more explosive energy than all of World War II.

The rate at which the arms accumulated was also unparalleled. In 1947, the U.S. had fifty-six atomic bombs and was planning to build a hundred more. Russia was building its first. By 1960, the period shown on the map, the U.S. had over ten thousand nuclear weapons and Russia perhaps three thousand. At the peak of the Cold War in the 1980s, when Plattsburgh had more aircraft than any other Strategic Air Command base in the country, each side would have about twenty thousand nuclear weapons.

Both New York and the Adirondacks were deeply involved in the Cold War. In the early sixties, New York was one of the most heavily armed places in the world. Its offensive weapons were two bomber squadrons and one missile squadron, probably with several hundred hydrogen bombs total. Its defensive weapons were one fighter squadron and nineteen surface-to-air missile bases, also with significant number of atomic weapons.

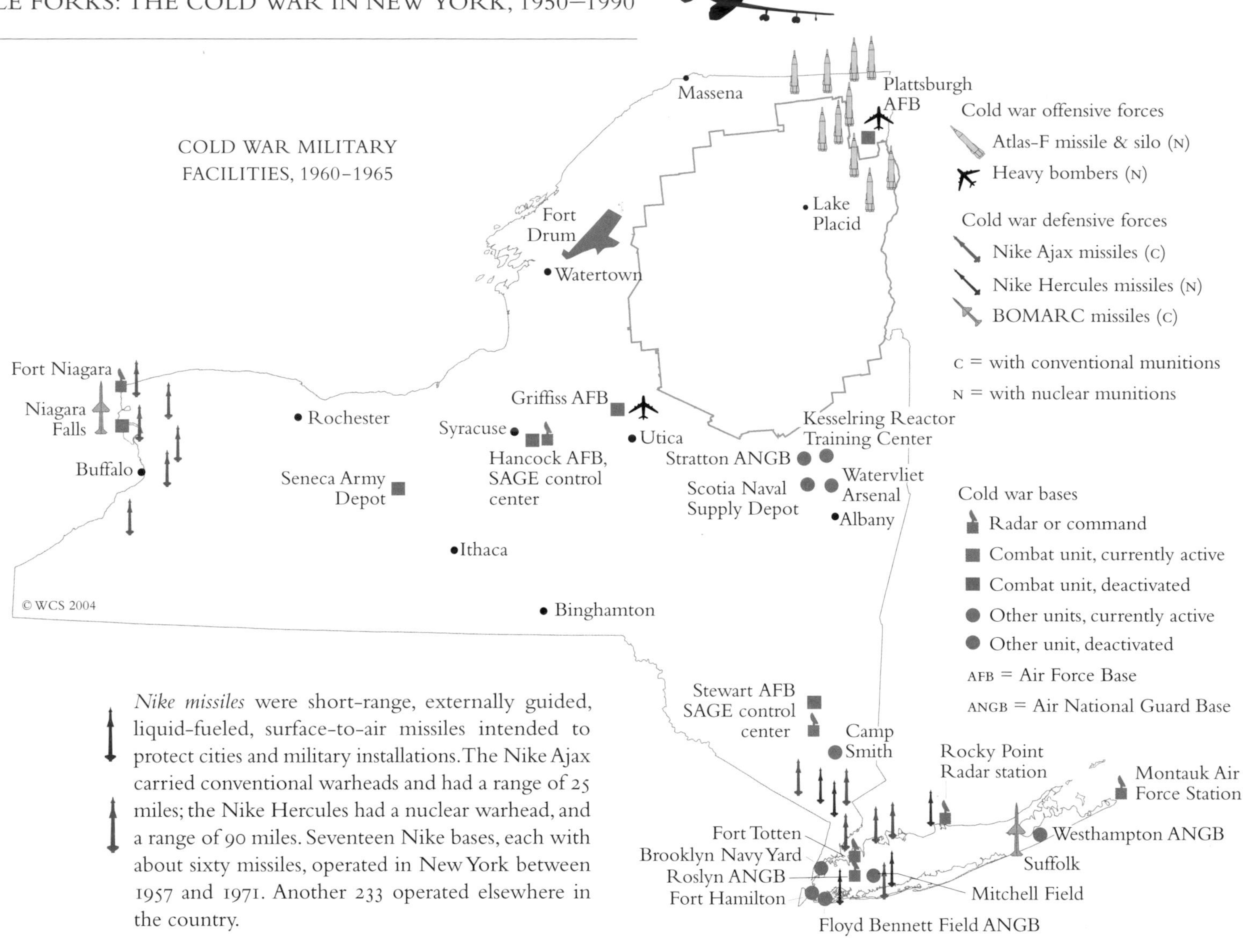

Nike missiles were short-range, externally guided, liquid-fueled, surface-to-air missiles intended to protect cities and military installations. The Nike Ajax carried conventional warheads and had a range of 25 miles; the Nike Hercules had a nuclear warhead, and a range of 90 miles. Seventeen Nike bases, each with about sixty missiles, operated in New York between 1957 and 1971. Another 233 operated elsewhere in the country.

BOMARC missiles, which operated in the same period as the Nikes, were medium-range, subsonic cruise missiles, 43 ft long with an 18-ft wingspan. They were the first missiles that could track a target themselves. There were 56 missiles at each of the two bases in New York and another 28 bases elsewhere in the country.

SAGE control centers were the first computerized missile-defense command posts. Each was a building the size of a football field that contained a 275-ton computer with 55,000 vacuum tubes. The computers required 3 million watts of power, burned out about a hundred tubes a day, and had 256 kilobytes of random-access memory, about half that of a 1980s personal computer. There were thirty SAGE centers in the U.S. and two in New York.

By the early 1960s the rapid deployment of long-range missiles was making defensive weaponry obsolete and changing our ideas about what bombs were good for. In the fifties, when we could only be reached by Soviet bombers on suicide missions, it was still possible to talk about defense against air attacks and even to speculate that we might be able use our missiles preemptively (or at least brandish them threateningly from time to time) without risking retaliation. By the mid-sixties, when we and the Russians had thousands of missiles on submarines and in hardened silos, neither defense nor preemptive use made any more sense. The defensive missiles were deactivated, the first-strike plans mostly put away, and we settled down to the "mutual assured destruction" strategy that, in one form or another, has been our basic nuclear posture ever since.

The Adirondacks' part in the story centered around Plattsburgh Air Force Base. The 380th Bomber Wing operated medium and heavy bombers there from 1956 to 1971 and fighter-bombers from 1970 to 1991. All were armed with nuclear weapons. The 380th Air Refueling Squadron operated tanker aircraft from 1956 until 1995. And for five years the 556th Strategic Missile Squadron maintained a dozen Atlas-Fs, the only ICBMs ever based east of the Mississippi, in hardened silos in the Adirondacks and Vermont.

When the Berlin Wall fell in November 1989, the Cold War was nearly over. Plattsburgh's part in it lasted only a few years longer. The last FB-111A fighter-bombers left in July 1991, and the last KC-135 air tankers in September 1994. When the base closed in September 1995, a long chapter in Lake Champlain history closed with it. For the first time in 329 years, since the building of Fort St. Ann in 1666, there were no military facilities of any kind—active or inactive, European, Indian, or American—anywhere in the Champlain Valley.

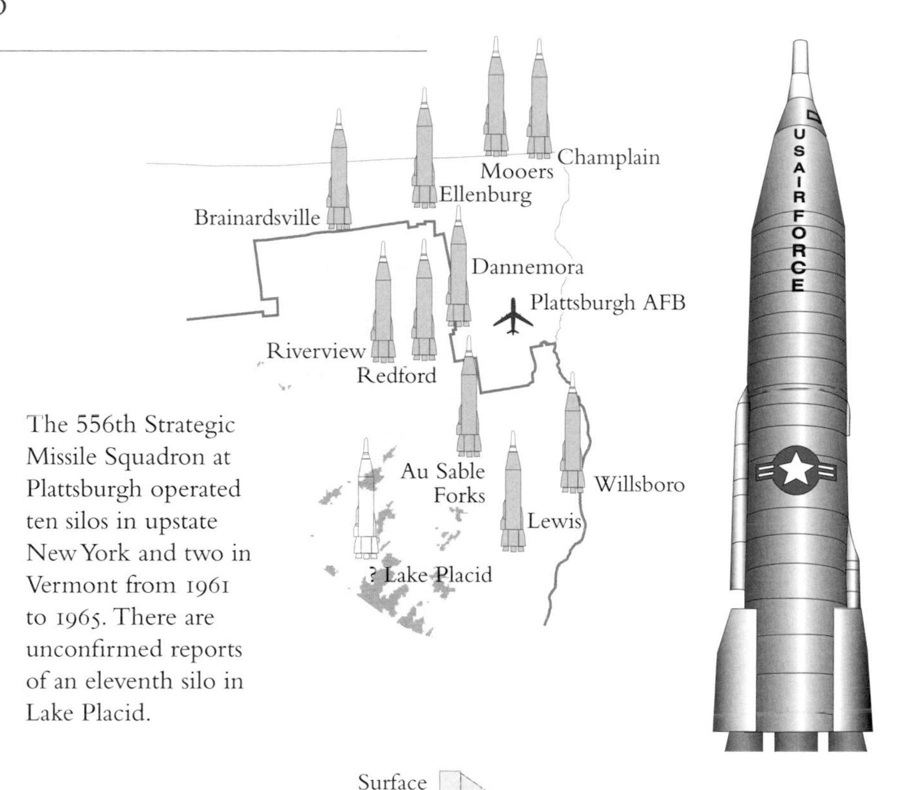

The 556th Strategic Missile Squadron at Plattsburgh operated ten silos in upstate New York and two in Vermont from 1961 to 1965. There are unconfirmed reports of an eleventh silo in Lake Placid.

The *Atlas-F* was the last and most sophisticated of the United States' first generation of intercontinental missiles. It was a liquid-fueled, inertially guided rocket 85 ft. tall, 10 ft. in diameter, and weighing about 125 tons. It had three main engines and carried a single thermonuclear warhead. It had a range of 6,300 miles; a missile fired from the Adirondacks could reach any of the major cities or military installations of western Russia.

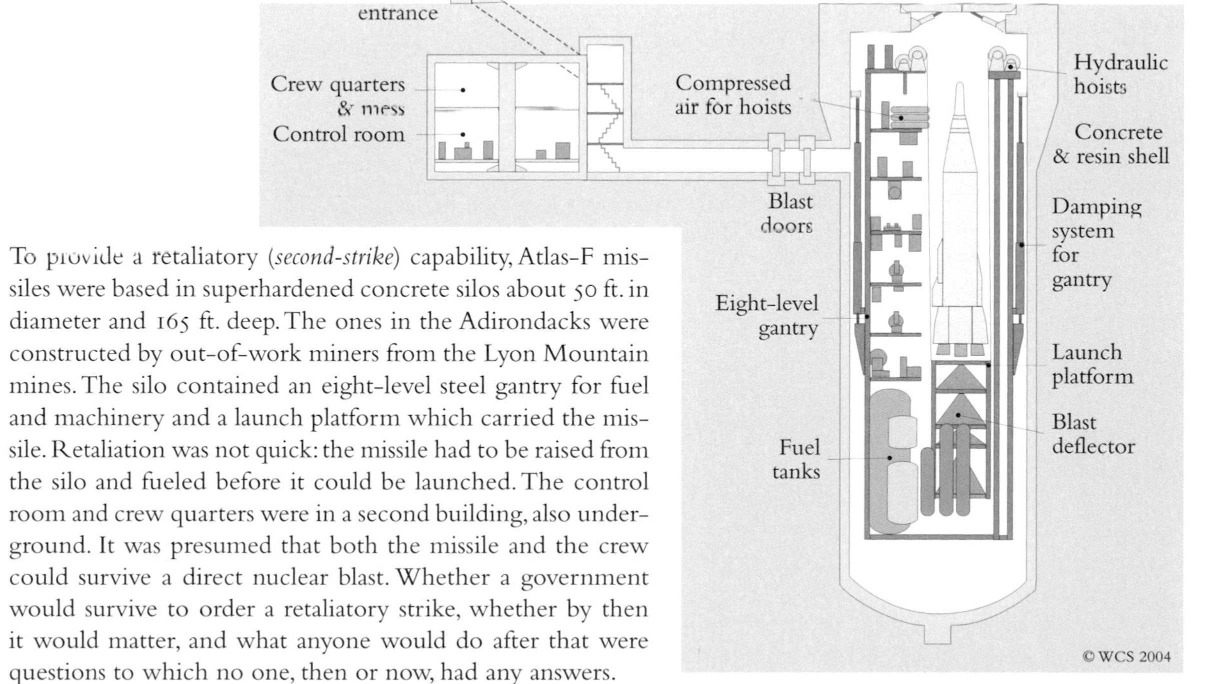

To provide a retaliatory (*second-strike*) capability, Atlas-F missiles were based in superhardened concrete silos about 50 ft. in diameter and 165 ft. deep. The ones in the Adirondacks were constructed by out-of-work miners from the Lyon Mountain mines. The silo contained an eight-level steel gantry for fuel and machinery and a launch platform which carried the missile. Retaliation was not quick: the missile had to be raised from the silo and fueled before it could be launched. The control room and crew quarters were in a second building, also underground. It was presumed that both the missile and the crew could survive a direct nuclear blast. Whether a government would survive to order a retaliatory strike, whether by then it would matter, and what anyone would do after that were questions to which no one, then or now, had any answers.

FREQUENCY OF WITNESS TREES IN PRESETTLEMENT SURVEYS, 1700-1830

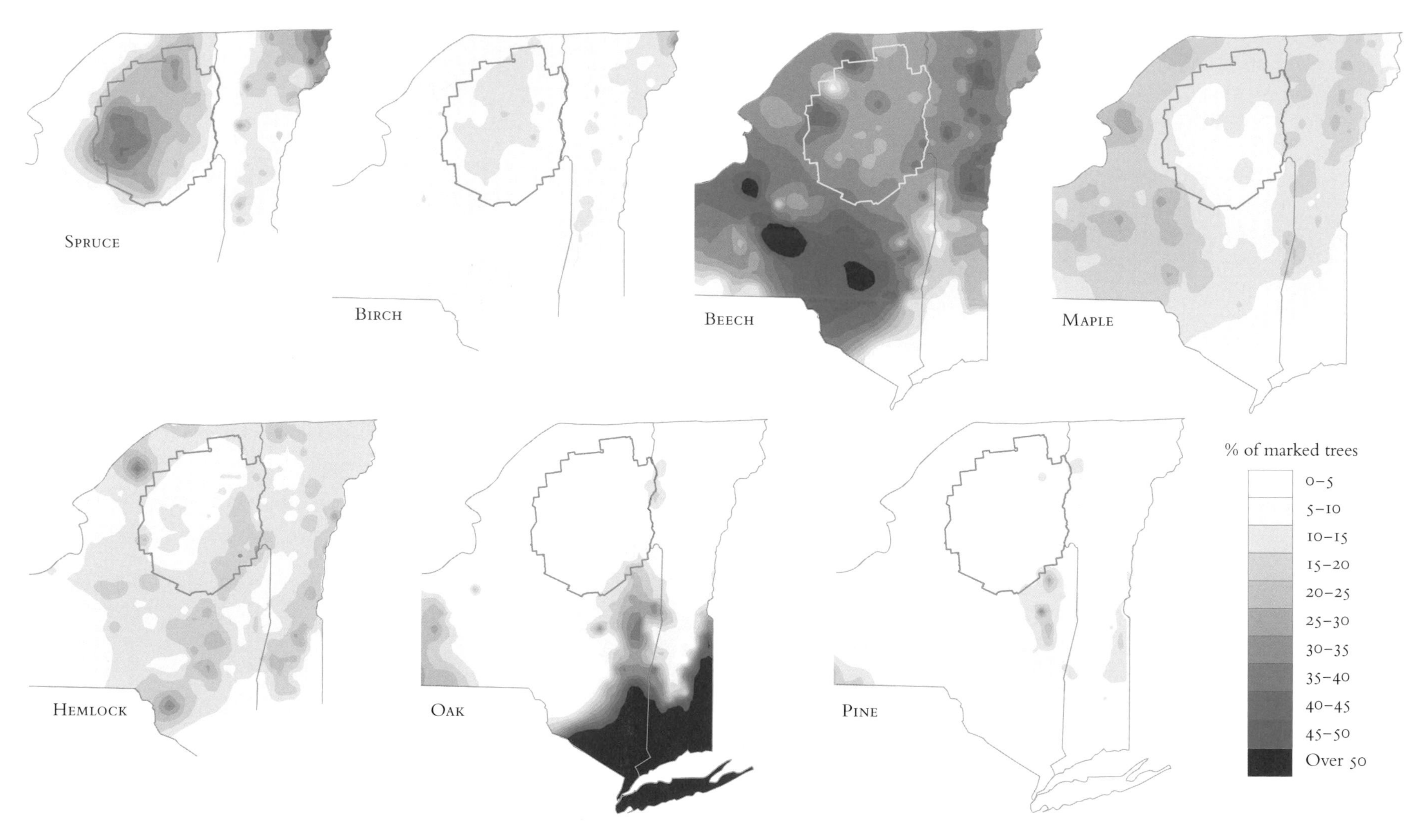

After the boundaries of a town had been run, the surveyors then ran the major *lot lines* (*Map 5-7*) and marked corner points roughly forty chains (a half mile) apart. At each corner they placed a stake or a stone marker and then marked the nearest tree as a *witness tree* and recorded it in their log. The records of these witness trees, compiled in the *returns of survey* in state and county archives, are in fact a sort of sparse forest inventory that can be averaged to produce a picture of the composition of the presettlement forest. The maps, from data compiled by Charles Cogbill, show show the percentage of the witness trees belonging to each species or species group. They have a recognizable ecological logic, plus some surprises. Note the remarkable abundance of beech northwards and oak southwards; the much lower abundance of maple and hemlock; and the restriction of spruce-dominated stands to the part of the western Adirondacks we have called the boreal core.

6 Forest Change

THIS chapter tells the story of how industry, fires, and storms have changed the woods. It is a story of very thorough change but not of exceptionally big changes. Two hundred years after logging began there are still the same kinds of trees in roughly the same places. But they are, on average, smaller and younger than their ancestors were and, perhaps again on average, less healthy as well.

Forests are large and complex, and detailed accounts of their composition and condition are rare. In the Adirondacks we are lucky to have four such accounts. Two are remarkable historical maps—the Sargent Commission Map of 1885 (*Map 6-1*) and the Forestry Commission Map of 1916 (*Maps 6-2, 3*). The other two are equally remarkable historical reconstructions—Charles Cogbill's assembly of witness tree records shown on the opposite page and Barbara McMartin's map of early Forest Preserve acquisitions (*Map 6-3*). Together these four accounts give the most detailed picture of the forests of a hundred years ago that we have for any place in the Northeast and form the basis of this chapter.

The chart on this page shows four periods of forest history and the approximate percentages of forests of different types in each period. The total area of the park is 6 million acres; one percent of the park is about 60,000 acres.

The story begins in 1800, when settlement and commercial logging are just beginning around the Adirondack edge. About ninety percent of the current park was virgin forest. By today's standards it was a very impressive forest, with more spruce and beech than today and many trees two to three feet in diameter and over a hundred feet high.

In the first period (*Maps 5-9, 6-1*), the valleys of the Adirondack periphery were cleared for farming, and the adjacent hills heavily cut for fuel-wood for forges and bark for tanneries. The interior was cut, extensively but very selectively, for big spruce. By the end of this period, in 1885, roughly one-third of the park had been cleared. Another third was still virgin forest. The remainder had been logged but fairly lightly.

The second period (*Maps 5-10, 5-12, 6-2, 6-3*) began about 1880, with the arrival of the railroads, paper mills, and big steam-powered sawmills. The mills took all the softwoods they could get, and the new railroads could get them from anywhere in the park. This was a period of big heavy cuts and big burns but also, paradoxically, a period in which the timber owners discarded many virgin but inaccessible lands which are now the jewels of the Forest Preserve.

In the third period, beginning about 1920 and extending till about 1940, logging first left the lowland corridors and moved onto the uplands and mountain slopes (*Maps 5-13, 6-4*). Industrial logging was extended to the hardwood forests, and these were cut almost as fast the softwood forests had been. But the cuts were less heavy because the hardwood mills didn't want small logs, and there was less clear-cutting and far fewer fires. Much land revegetated, and the total acreage of standing timber increased, though the quality of much of the new timber was low.

The final period (*Maps 6-4, 6-6*) began when the last virgin forests on private lands were gone, and with them the big sawmills and paper mills that had depended on them. It has been, for forests and foresters alike, a very unsettled time: new diseases, big storms, big changes in how timber is owned, cut, and sold, acid rain as an ongoing problem, climate change as a coming one. The trees, as always, are simply trying to hold on; the foresters are trying, for the first time, to produce sustained yields. Neither, as the new century begins, seems to be succeeding.

A GENERALIZED ADIRONDACK LAND-USE HISTORY

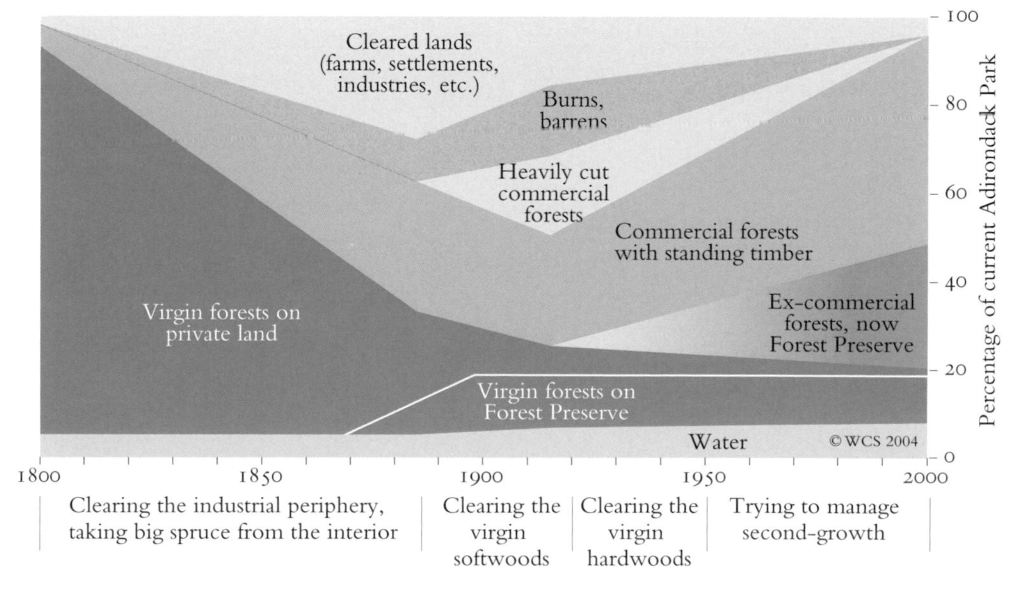

THE first assault on the Adirondack forests began about 1800 with the growth of settlements and a few small industries along the south and east edges of the park. It accelerated in 1813 when log-driving began and even more markedly when the tanning and iron industries expanded at mid-century. It ended sometime in the 1880s when the railroads and paper mills arrived and many of the early mines and industries closed.

The first assault was *eotechnic,* using only hand tools and simple machines and powered mostly by muscles and water. In the woods the emblems of the eotechnic period were the stable, the bunk-house, the smithy, the sled, and the axe. On the river they were the splash dam, the jam boat, and the peavey. In the mill towns they were the wheel pit, the vertical saw, the tanning vat, the kiln, the trip hammer, and the open forge.

Eotechnic industry was limited in what it could do and where it was able to do it. It could not mine more than a few hundred feet into the ground or haul logs or charcoal or bark overland more than a few miles. It could not move softwood logs any distance up hill and could barely transport or saw hardwood logs at all.

But what it could do it did well and surprisingly fast. By the end of this period at least sixty different companies were registered to drive logs on Adirondack rivers. In 1873, a peak year, the Adirondacks were the third largest lumber-producing region in the country. Over a million logs were floated down the Hudson alone, and the mills at Glens Falls made nearly two-hundred million board feet of lumber from them. About the same time thirty forges with a hundred and twenty fires were producing between thirty thousand and forty thousand tons of iron a year. To do this they used about four million bushels of charcoal a year, and clear-cut perhaps five thousand acres of hardwoods a year to produce it. And whatever the loggers and charcoal-makers could do was more than matched by the hundred-plus Adirondack tanneries, which removed much of the hemlock from between a million and a million and a half acres in eighty years.

The map below shows the aggregate effects of the harvests. The loggers, working in the light green areas, cut about 1.7 million acres but probably took only about a quarter of the total volume of timber. Today that would be called a light thinning. The farmers and tanners and iron-makers, working in the white areas, cut a little less, about 1.6 million acres, but cut much more heavily and created much larger and longer-lasting openings.

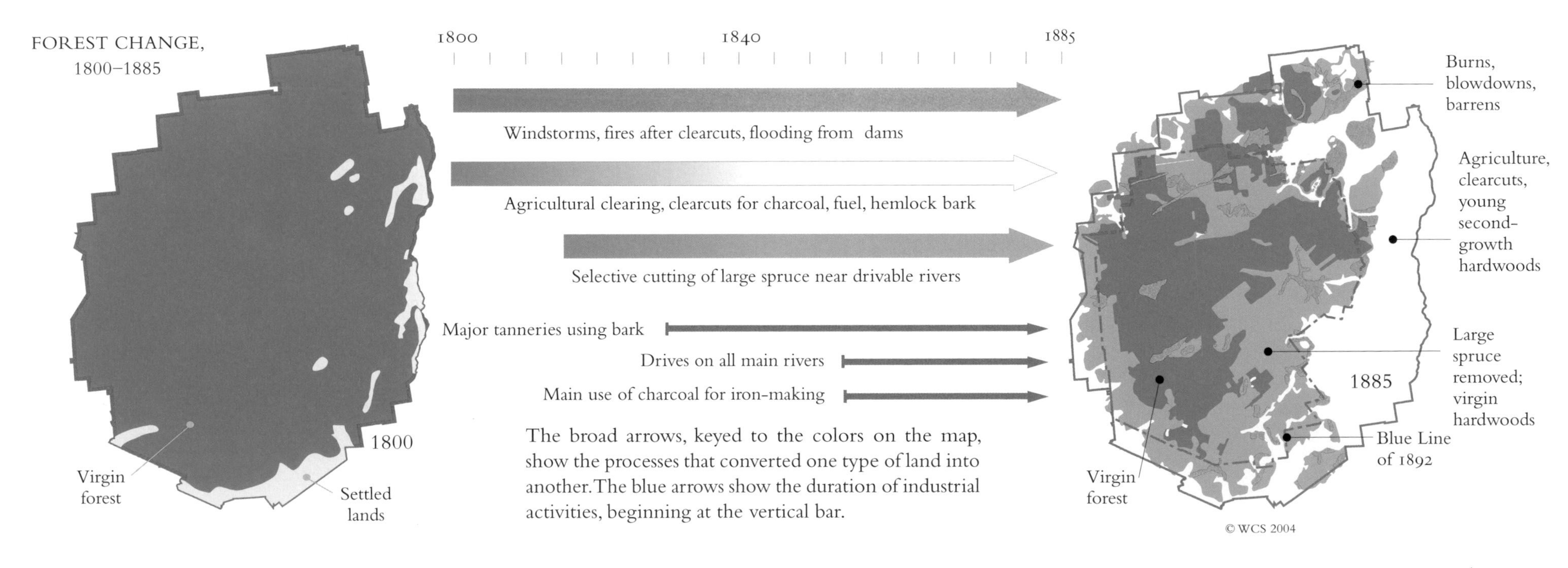

The broad arrows, keyed to the colors on the map, show the processes that converted one type of land into another. The blue arrows show the duration of industrial activities, beginning at the vertical bar.

By the 1880s there was a widespread perception that Adirondack forests were in trouble. Travelers in the eastern Adirondacks, where clearing for fuel and bark had been most extensive, were reporting clear-cuts and burns and wastelands. Legislative committees heard about timber thefts from state lands and unrestricted cutting on private ones. The sporting press, which hadn't figured out that deer love logging, magnified the reports into a general fear that the Adirondack wilderness was about to vanish and with it the immensely valuable supplies of Adirondack water on which downstate canals and mills depended.

The alarms were not unfounded. Big pieces of the wilderness had already been cut, and there were already many burns and barrens and newly made lakes filled with dead trees. But they were overly general. In fact, about a third of the park was still virgin and another third only lightly cut.

Whether justified or not, the alarms were heard in Albany and had two consequences of importance to our story. The first was that a legislative commission, investigating deforestation, produced the map we show here. And second was that the legislature, following this map, created a Forest Preserve and an Adirondack Park and gave them constitutional protection.

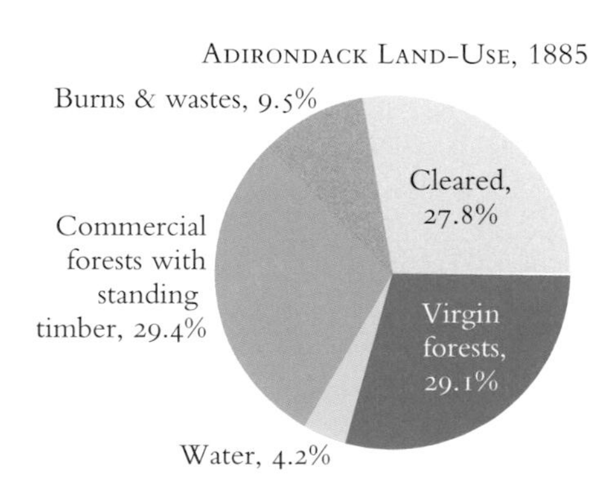

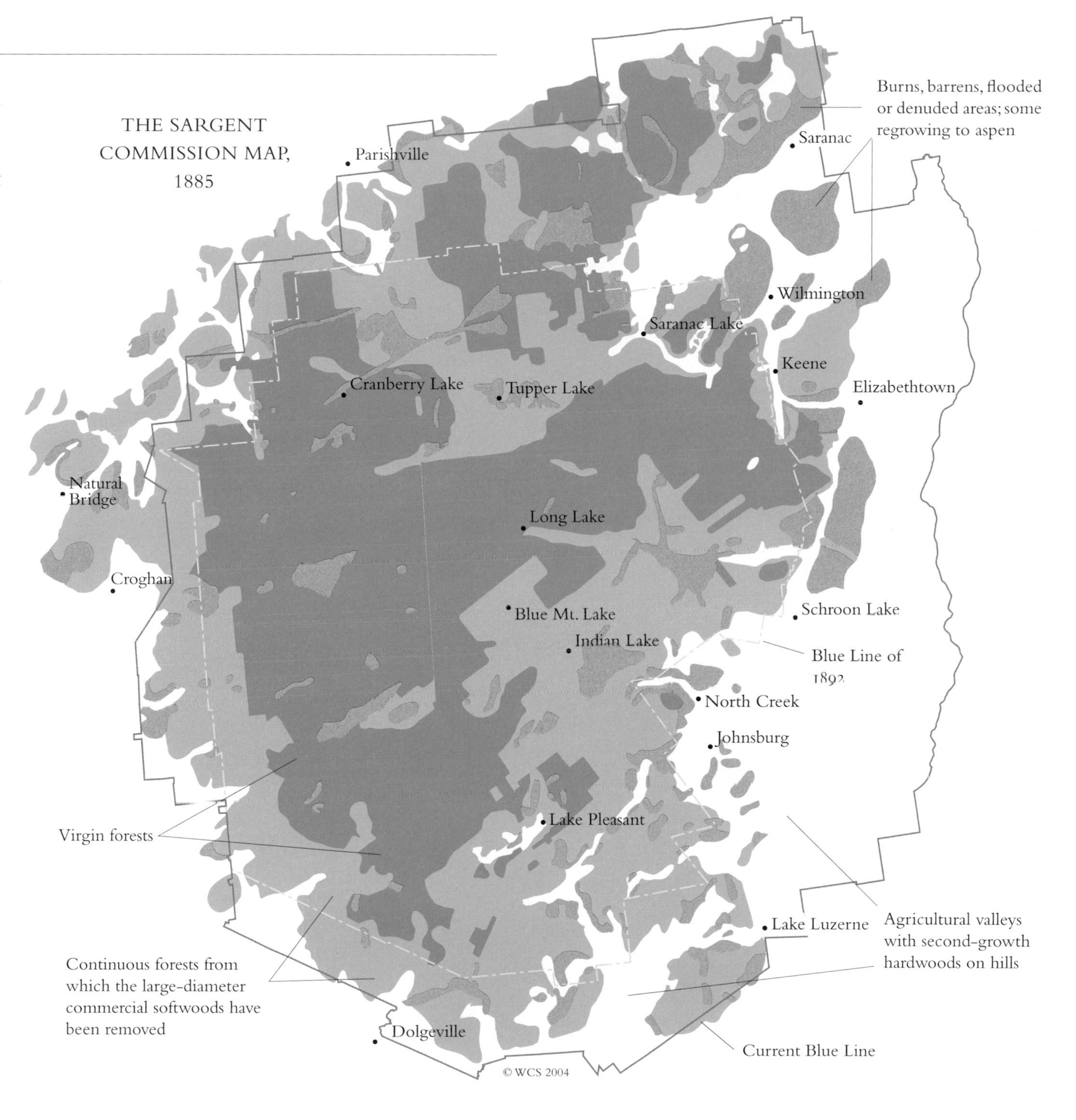

In the 1890s the railroads arrived. The first large, railroad-based sawmills and paper mill were built in the park (*Map 5-10*), and for the next forty years Adirondacks were treated to mechanized logging at its most ferocious. The forests survived it but at a considerable cost: by the time the big trees ran out and the loggers went west, 90% of the virgin softwoods were gone, and nearly a fifth of the park had burned in logging-related fires.

When the period began, Adirondack softwood production had peaked at 300 million board feet a year and was starting to decline as the supply of spruce near the rivers ran out. The paper mills had just arrived and were already using 50 million board feet of pulp a year. In a few years they would be using six times that amount.

Together, the railroads and paper mills were a timber-baron's dream. The railroads make it possible to get the last virgin spruce out of the swamps and off the mountains. The paper mills make it profitable, for the first time, to cut the small softwoods that had previously been ignored.

The result was an industrial jamboree that lasted until the virgin softwoods were gone. There were rivers full of logs, logging railroads everywhere, hundred-man woods crews, five-mile-long flumes, donkey engines and high-wires on the mountain slopes. And of course a lot of mess: clear-cuts and slash, big hot fires set by wood-burning engines, and rivers full of the bark and wastes from the mills.

Ironically, it was probably the messiness of the new industrial logging that led to the protection of the Forest Preserve. Up to this time the state had thought of its forest lands as temporary holdings, a sort of land bank into which it put lands when their owners didn't want them and from which it could withdraw lands to log itself or sell off as the timber on them matured. But as the condition of the private lands got worse and as the state suffered more timber theft from the Forest Preserve, the conservationists came to believe all logging was inherently destructive and that the only way to protect the Adirondacks was to create a core of state lands that could not be cut.

The result was the series of legislative acts (p. 27) that protected the Adirondack Park and began the regrowth of the burns and barrens. The new forests grew quickly, and by 2000 Adirondack forests were more continuous than they had been for 150 years. But, because it has been a difficult century for trees, they were not as big or as grand. In the last fifty years many old trees have been lost to disease, pollution, and wind. The young trees are doing their best but face damaged soils and a changing climate; it seems unlikely that any of them will ever attain the size of their ancestors.

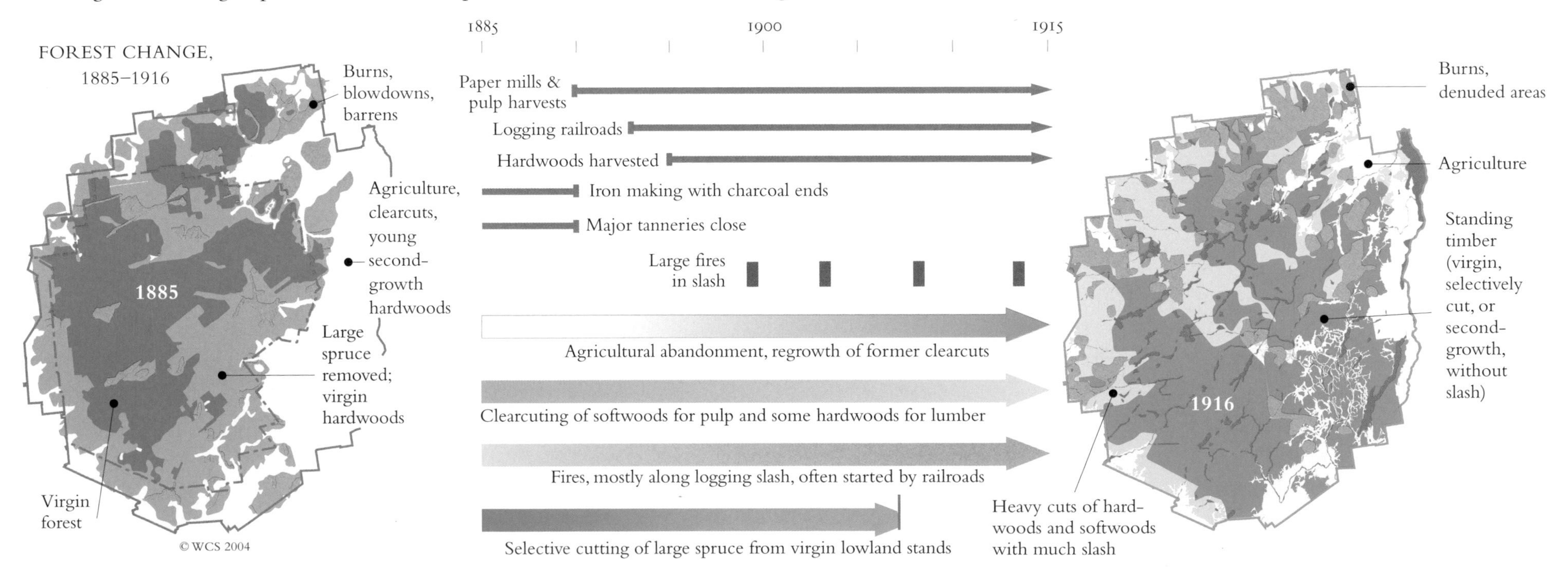

BETWEEN 1888 and 1915 a million acres in the Adirondacks burned. The fires were almost all in heavily cut lands and were especially common along the railroads in the northwest. All occurred, as wildfires always do, in dry periods. They varied greatly in size and intensity. Some were hot and quick-moving, others low and slow-moving. A few were fought energetically. Many others were left to burn themselves out. Most of the lands that burned were worth very little, and even for valuable lands no effective system of wildfire fighting yet existed.

The fires were blamed, depending on the blamer's prejudices, on droughts, fishermen, railroads, lightning, loggers, berry pickers, and the smoking public at large. But what was really important was not what started the fires but what fueled them. Almost all the big fires were the result of a particular style of forestry in which virgin forests were heavily logged and the slash left on the ground. Given wood-burning engines, enough slash and a heavy enough cut to allow it to dry out, fires were almost inevitable, with berry pickers or without.

New York State responded vigorously to these fires, creating, over the next forty years, an exemplary fire fighting system of fire towers, spotting aircraft, trucks, and trained crews. But ironically, while it was doing this, the regrowth of the forests was creating an even more effective fire-prevention system of its own. From about 1920 on, as the old cuts and burns regrew to young, green forests, fire frequencies declined. They have stayed low ever since. Wildland fire fighting is still an essential part of the state's mission and one at which it excels (*Map 12-1*). But with a different kind of logging in a different kind of forest, it is now a mission that the state undertakes less frequently and on a far smaller scale than it once did.

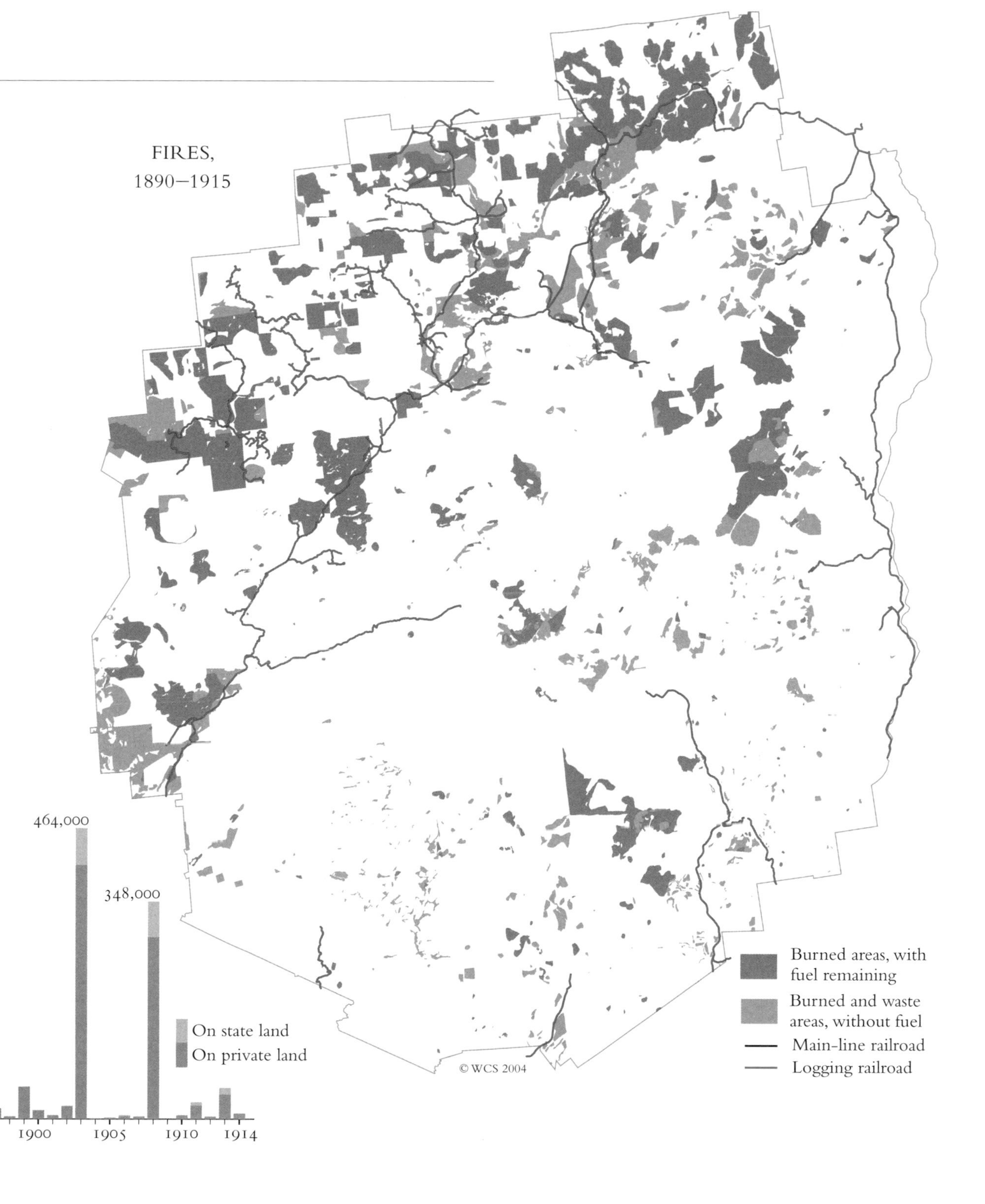

PERHAPS the oddest twist of Adirondack history is that the Forest Preserve lands that now constitute the largest collection of virgin forests in the eastern U.S. were acquired in scraps and protected by accident. The first half-million acres in the Forest Preserve, including everything acquired before 1891, were abandoned lands that the state bought at tax sales. The lands were abandoned because their owners wouldn't or couldn't pay the town taxes on them. Sometimes they were lands that had already been logged. More often they were lands that couldn't be logged, either because there was too little spruce or because what spruce there was was too far from the rivers. The towns seized the lands and put them up for sale. When there were no other buyers, the state stepped in, paid the back taxes that were owed to the town, and took title to the land.

The state, at first, thought of these acquisitions as a burden rather than a bargain. It had no idea that it was protecting the last virgin forests in the east and acquiring the lands that are now regarded as the heart of the Forest Preserve. It simply thought it was helping the towns out by taking unsellable lands off their hands.

The map is based on research by Barbara McMartin. It shows 1,575 parcels acquired in tax sales that were probably virgin forest when they were acquired and another 498 (in pink) purchased directly, many of which were either virgin or had been cut lightly for spruce. The total is a very respectable 1.3 million acres, 360,000 acres by direct purchase and 967,000 acres from tax sales.

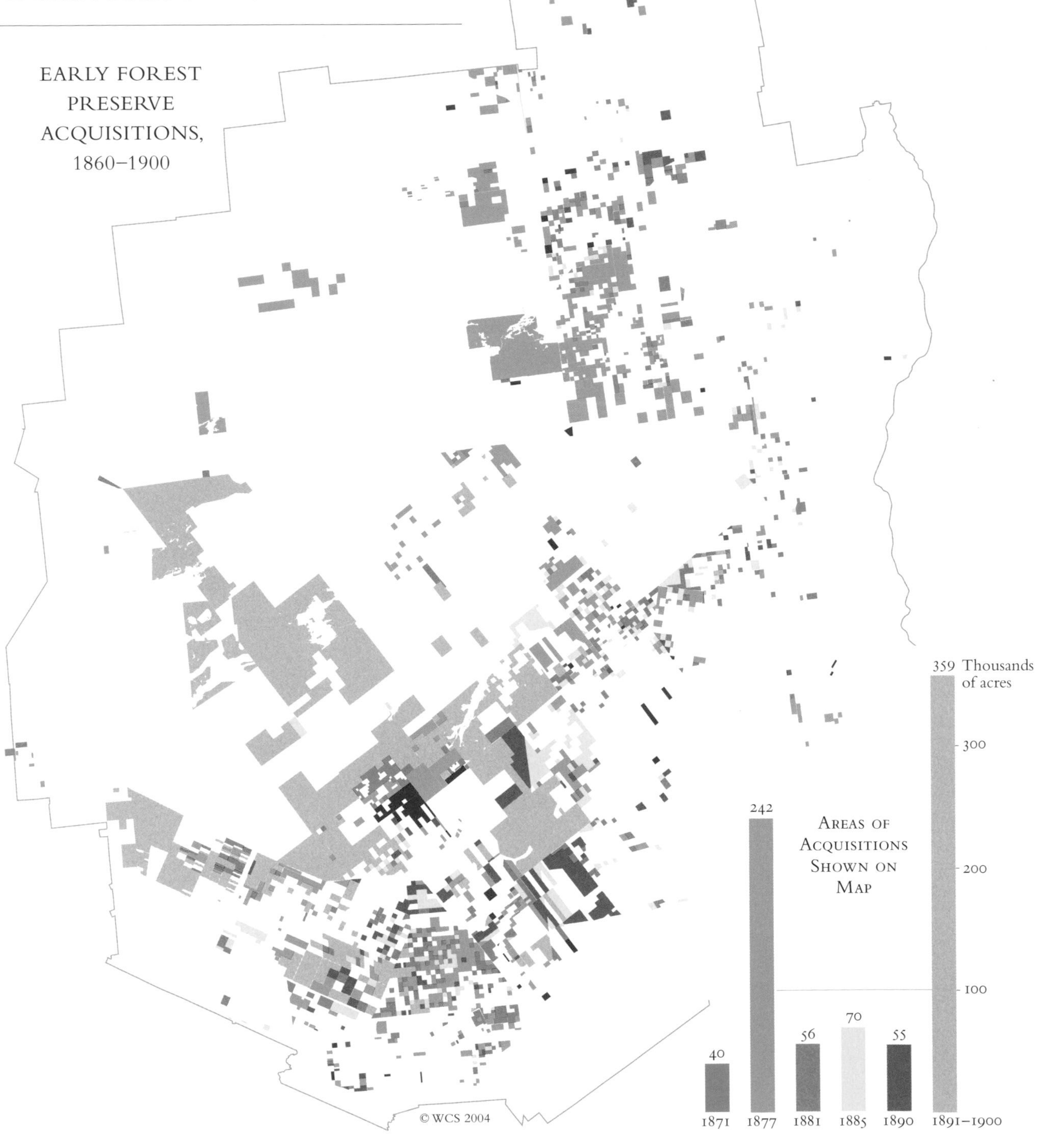

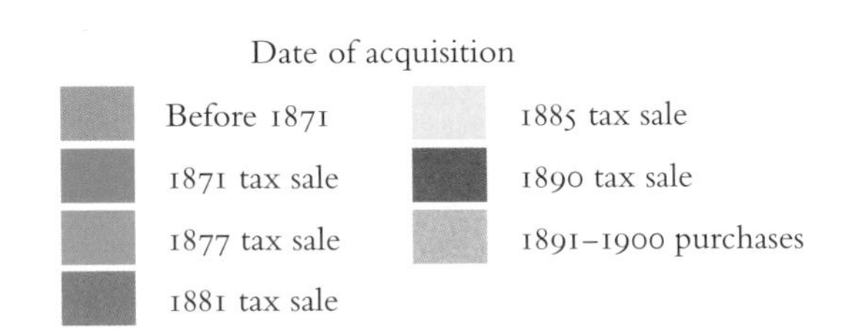

IN 1916, following fifteen years of bad fires, the Conservation Department produced a large-scale map showing what had burned and what might burn. This was the famous *Fire Protection Map*, by far the most detailed map of Adirondack forests that has ever been made.

The map shows the forests at the height of the railroad logging period, a singularly headlong time when industrial logging was rapidly eliminating its own resource base. A third of the Adirondacks had been recently cut or burned. Much of this was in areas that were mapped as virgin or lightly harvested in 1885. But even so, nearly half the park was still continuous forest and probably about half of this virgin or almost virgin. The largest cuts are in the center and northwest; their association with the railroads and large mills is clear.

This was a pivotal moment in the history of the forests. Most of the virgin softwoods had already been cut, but large areas of virgin hardwoods still remained. The state was trying to incorporate them in the park but had neither money nor a plan. The big sawmills had just started to cut them and had, unfortunately, lots of money and detailed plans.

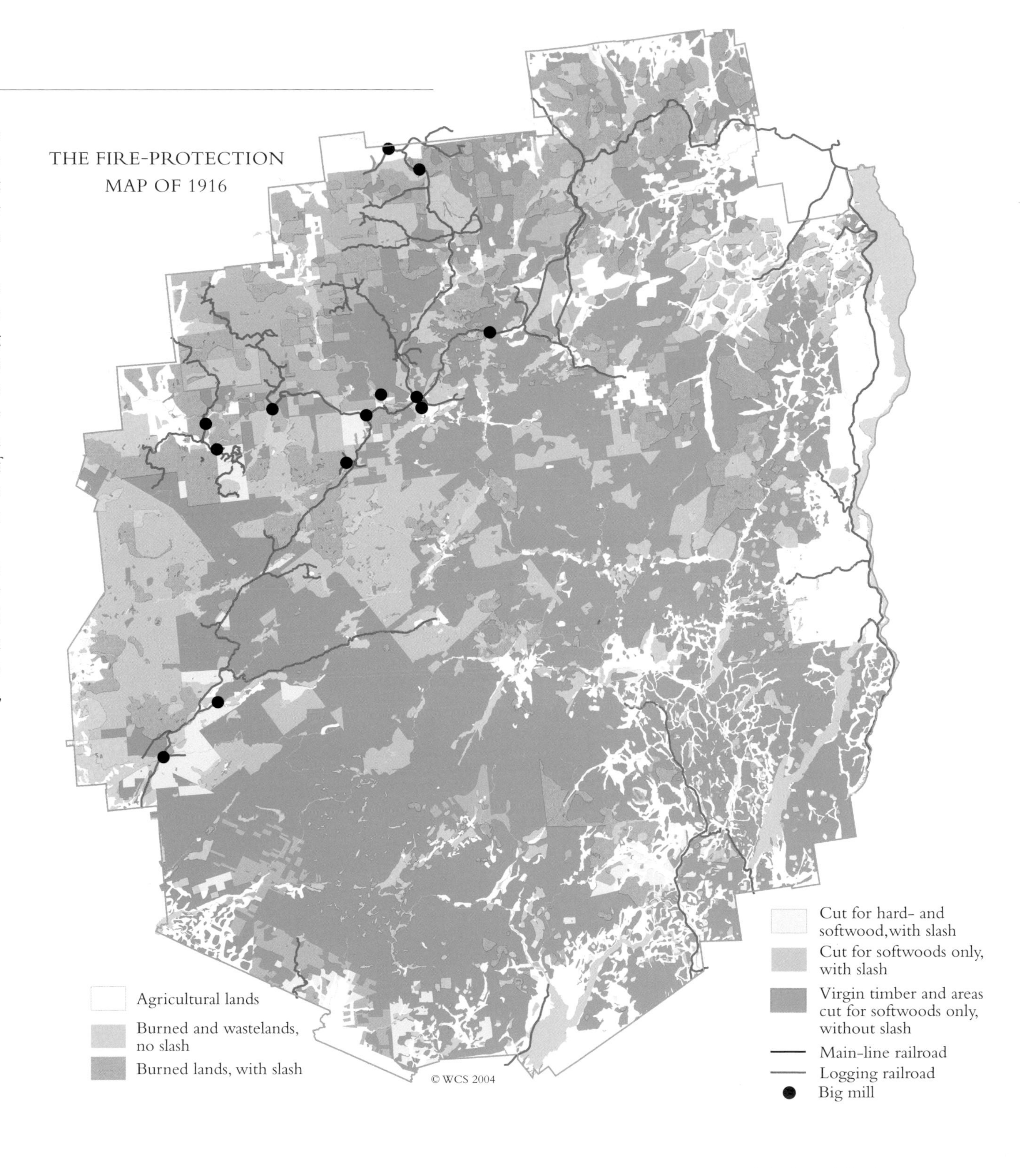

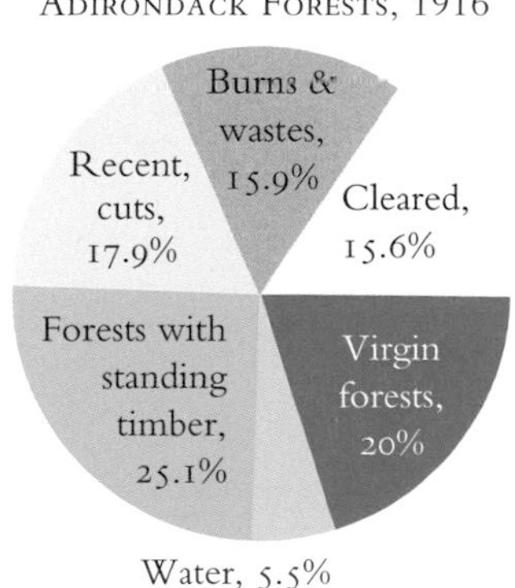

The amount of virgin forest is an educated guess; the true amount is somewhere between the 29% mapped in 1885 and the 15% incorporated in the Forest Preserve.

THE forest changes in the twentieth century are too complex, too poorly mapped, and probably too close to us to summarize easily. The central story is certainly the change from a high-speed, cut-and-run forestry based on virgin timber to a more cautious and deliberate one that works with second-growth and tries to produce sustained yields. But twined with this are other important stories: changes in ownership and machinery, changes in markets, big storms and frustrating salvage operations, and increasing evidence that pollution and climate change are affecting the basic biology of the forests. In lieu of a summary, we offer a generalized contemporary map and a few waypoints. The period begins about 1900, when the supply of lowland spruce is nearly exhausted.

1902 The Forest Commission, concerned about declining softwood supplies, establishes tree nurseries and begins planting white pine and red spruce on the Forest Preserve. Geared logging locomotives like the Shay engine (p. 89) are coming into wide use, allowing mechanized logging in steep terrain.

1910 The first steam tractors are used in the Adirondacks. Commercial-sized spruce on private lands is now restricted to higher elevations. A period of intense mountain logging begins. J.R. Rogers Company builds a seven-mile long flume to move pulp logs to the Ausable River.

ca. 1915 Finch Pruyn builds a log chute on Mt. Colden above Avalanche Pass.

ca. 1918 The era of the great fires is largely over. Large-scale cutting of virgin hardwoods begins. Four-foot pulp logs are included in river drives. Linn gasoline tractors are first used on the Moose River Plains

ca. 1925 J.R. Rogers harvests the John's Brook Valley and Big Slide Mountain. Clear-cuts reach an elevation of 3,100 feet on Giant Mountain.

1920 Seven thousand men are employed at 150 Adirondack logging camps. The demand for pulpwood is higher than the supply, and forty percent of the pulp used by New York paper mills comes from Canada.

1923 The last sawlogs are driven down the Ausable and Moose Rivers. Log trucks are coming into wide use, and year-round roads are first built in the woods.

1927 The last 13-foot sawlogs are driven down the Hudson. The last virgin spruce on private lands is cut, mostly in the mountains.

ca. 1938 Railroad and horse logging have largely ended; first-generation paper mills have closed at Piercefield, Plattsburgh, Keeseville, and St. Regis Falls.

1940 The first snowmobiles in the Adirondacks are used in logging operations on the Moose River Plains.

1945 Most virgin hardwoods on private lands are gone; foresters begin taking second cuts in many stands.

1950 The last Adirondack river drive: 10,000 cords of softwood pulp are driven down the Hudson to the Finch-Pruyn mill in Glens Falls. Chain saws and skidders come into wide use.

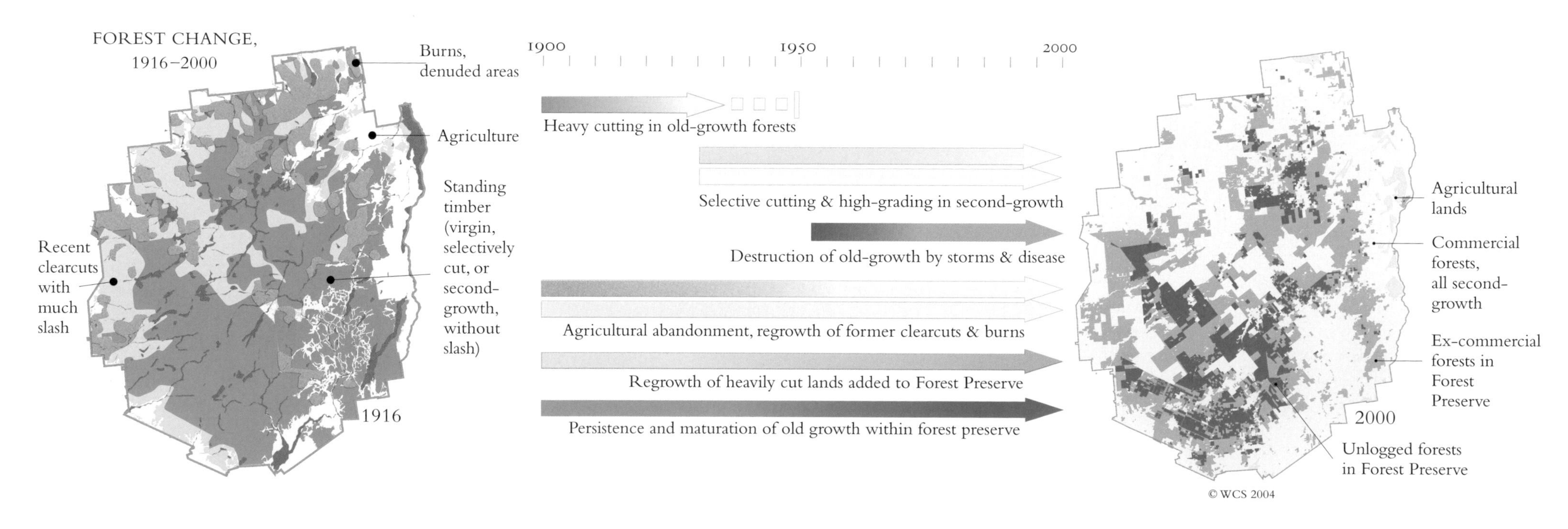

1950 The Big Blow (*Map 6-5*). An errant inland hurricane blows down timber on some 800,000 acres in the west and central Adirondacks; the state authorizes extensive and probably unconstitutional salvage operations in old-growth forests on Forest Preserve lands.

1955 Acid rain intensifies with rising fossil-fuel use. Many sugar maples in the western Adirondacks cease reproducing in this decade.

ca. 1960 Self-loading trucks (cherry pickers) come into wide use; paper mills in Willsboro and McKeever close; Emporium Forestry closes its mill at Conifer after fifty years of operation.

1967 The Oval Wood Dish mill in Tupper Lake burns.

ca. 1970 Beech-bark disease arrives; repeated high-grading of hardwoods has left many hardwood forests with much beech, which rapidly begins to sicken and die. J.R. Rogers shuts their paper mill at Au Sable Forks, the oldest in the Adirondacks.

1973 The Park Agency limits the size of clear-cuts.

1978 The first conservation easements on private lands are sold. Feller-bunchers appear in the woods.

1995 A derecho (*Map 6-5*) blows down timber on 160,000 acres in the west Adirondacks. Salvage of state lands is discussed but not authorized. Most adult beeches have died, and most young beeches over six inches in diameter are dying. Sugar maple reproduction is almost absent over large areas of the western Adirondacks.

1998 A multi-day ice storm damages three million acres of timber in northern New York, including eight-hundred thousand acres in the Adirondacks (*Map 6-6*).

2000-2002 Industrial forestry declines. Champion International sells all its Adirondack lands, partly to the state and partly to a forest investment group. The paper mills at Newton Falls, Deferiet, and Lyons Falls close and the market for pulp is very poor. International Paper sells several large parcels to conservation groups. Timber investment groups acquire large amounts of ex-industrial land. Domtar, the second-largest Adirondack landowner, puts all its Adirondack lands on the market.

LOGGING, though impressive, is not the only thing that changes forests. In a dry year at the peak of railroad logging, loggers and logging-related fires might affect several hundred thousand acres of the park. But on November 25, 1950, a hurricane blew down over 800,000 acres of timber in a single day. And in a five-day period in January 1998, the largest and most intense ice-storm ever recorded damaged about three million acres of forests in northern New York and a similar area in Ontario and Quebec.

The maps on the next four pages show the three biggest storms for which there are good records and then present a composite picture of their effects. The first is the "Big Blow" of November 1950. This was a cyclonic storm of hurricane force that tracked inland over the Appalachians, where hurricanes aren't supposed to go, and crossed the western Adirondacks. It damaged both softwoods and hardwoods, especially on north- and east-facing hilltops in the western Adirondacks.

The politics following the storm were as far off-track as the storm itself. Despite the constitutional prohibition on removing or selling timber from the Forest Preserve, the Conservation Department proposed, and the Attorney General and the legislature approved, a plan that would do just that. Conservation Commissioner Perry Duryea said that the fallen trees had created an unprecedented fire hazard and so must be removed from the woods. Attorney General Nathaniel Goldstein agreed and added that in the "grave emergency" (Korea) facing our nation the trees ought to be salvaged so that they could be used for national defense. There was of course no legal, ecological, or military logic in any of this. But the fear of a return of the great fires of 1905 prevailed. New roads were built in wilderness areas, some 40 million board feet of timber were removed from the Forest Preserve, and an attractive 1.1 million dollars of windfall profits added to the Conservation Department's budget.

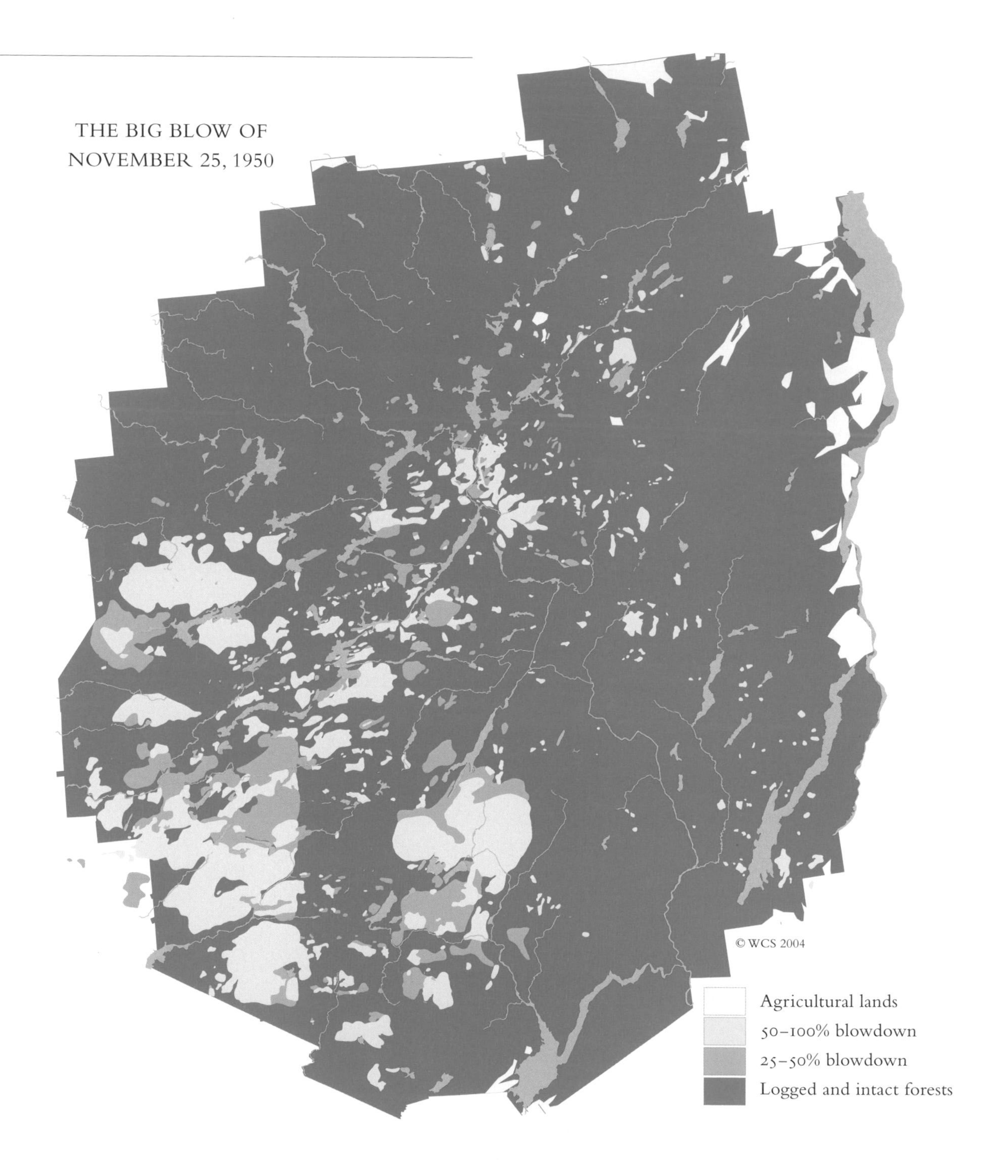

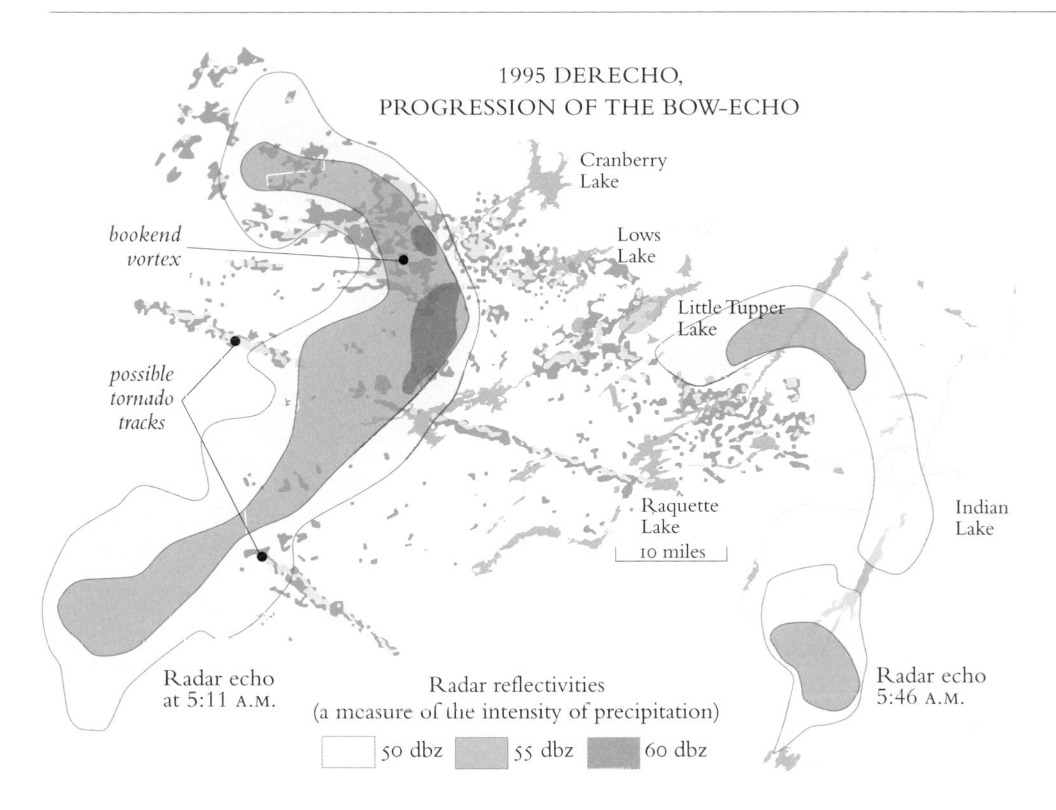

A *derecho,* meaning a straight-line storm, is an unusually intense squall line in which a group of thunderstorms moves together as a single unit, reaching speeds of over fifty miles per hour and creating a gust front of downdraft winds that can exceed two hundred miles per hour. The left map shows the progression of the radar echo, a measure of the intensity of precipitation, for the 1995 derecho. At about 4:50 A.M. the upper end of the squall line began to curl counterclockwise, generating a sharp bend, a *bookend vortex.* The vortex propagated faster than the rest of the line and, until it dissipated thirty minutes later, created a fifty-mile-long swath of intense damage. The two narrow lines of blowdown south of the main damage swath from the vortex are unexplained: they may have been caused by small tornados generated by shear along the gust front.

THE second of the three severe storms was a derecho that originated near Lake Ontario shortly after midnight on July 15, 1995. The atmosphere over the entire Northeast was extremely moist and unstable and so primed for a violent storm. The derecho crossed the Adirondacks heading southeast just after dawn and continued across southern New England and out to sea.

For most of its six-hundred-mile path and ten-hour life the storm did only scattered damage. But for a period of about a half-hour, while over the western Adirondacks, it developed the "bow-echo" curvature that is the signature of an exceptionally intense storm and produced downburst winds that were likely over a hundred and fifty miles per hour. In this half-hour it blew down about 130,000 acres of timber and damaged an equal amount or more. It blocked the Oswegatchie River and most of the trails in the Five Ponds Wilderness with fallen trees, requiring the DEC to mount its largest back-country rescue ever. About 5:30 A.M., as it began to dissipate, it crossed seven popular public campgrounds, knocking down trees, crushing cars, and killing three people. If it had maintained the intensity it had fifteen minutes earlier when it was devastating the Five Ponds, the toll might have been far higher.

THE last big storm was the both the largest and the most unusual. On January 5, 1998, a slow-moving rain storm overrode a layer of cold surface air, creating the largest ice storm on record for the Northeast. For the next four days ice built up on trees, buildings, and wires, creating a coating 1–4 inches thick and imposing loads of 2–5 pounds per foot on utility wires and 5–20 pounds per foot on tree branches. There was very little wind associated with the storm, but the weight of the ice by itself was more than trees or wires could take.

THE ICE STORM OF 1998

In northern New York, approximately three million acres of forests had moderate or heavy damage. Three hundred fifty maple sugar orchards had substantial damage, eight thousand electric poles snapped, and fourteen hundred dairy farms were without power and unable to milk for up to three weeks, losing several thousand cows to udder damage or mastitis.

Eastern Canada was, if anything, hit even harder: in addition to thirty-five thousand local breaks in electric lines, a hundred steel towers on the main Hydro-Quebec transmission line collapsed. The towers were two hundred and fifty feet tall and weighed some fifty tons each; it is likely that they were carrying more than their own weight of ice when they collapsed.

The storm was unique in several ways. Because of the temperature inversion, it was primarily a low altitude storm that concentrated its damage in the valleys and spared the mountains. Because it was slow moving it created the largest ice build-up of any storm on record. And because it was a frontal storm, the area damaged was exceptionally large, stretching from central Quebec south into New England, and from Ontario east to Maine.

As dramatic as single storms may be, they do not have great effects on forests. They knock some trees down, release others, and generate tip-up mounds that are colonized by a number of trees whose seedlings need mineral soil. But because they are always patchy, they don't have large overall effects on the age or composition of the forest.

It is when storms recur that they start to matter. Repeated storms limit the ages that individual trees can obtain, increase the amount of woody debris in the understory, increase the number of pits and mounds on the forest floor, and give early successional species like birch and aspen a permanent place in the forest. If they recur frequently enough, storms produce a forest that is lower, smaller, younger, and more diverse than an undisturbed forest, with more levels in the vegetation and a bumpier forest floor.

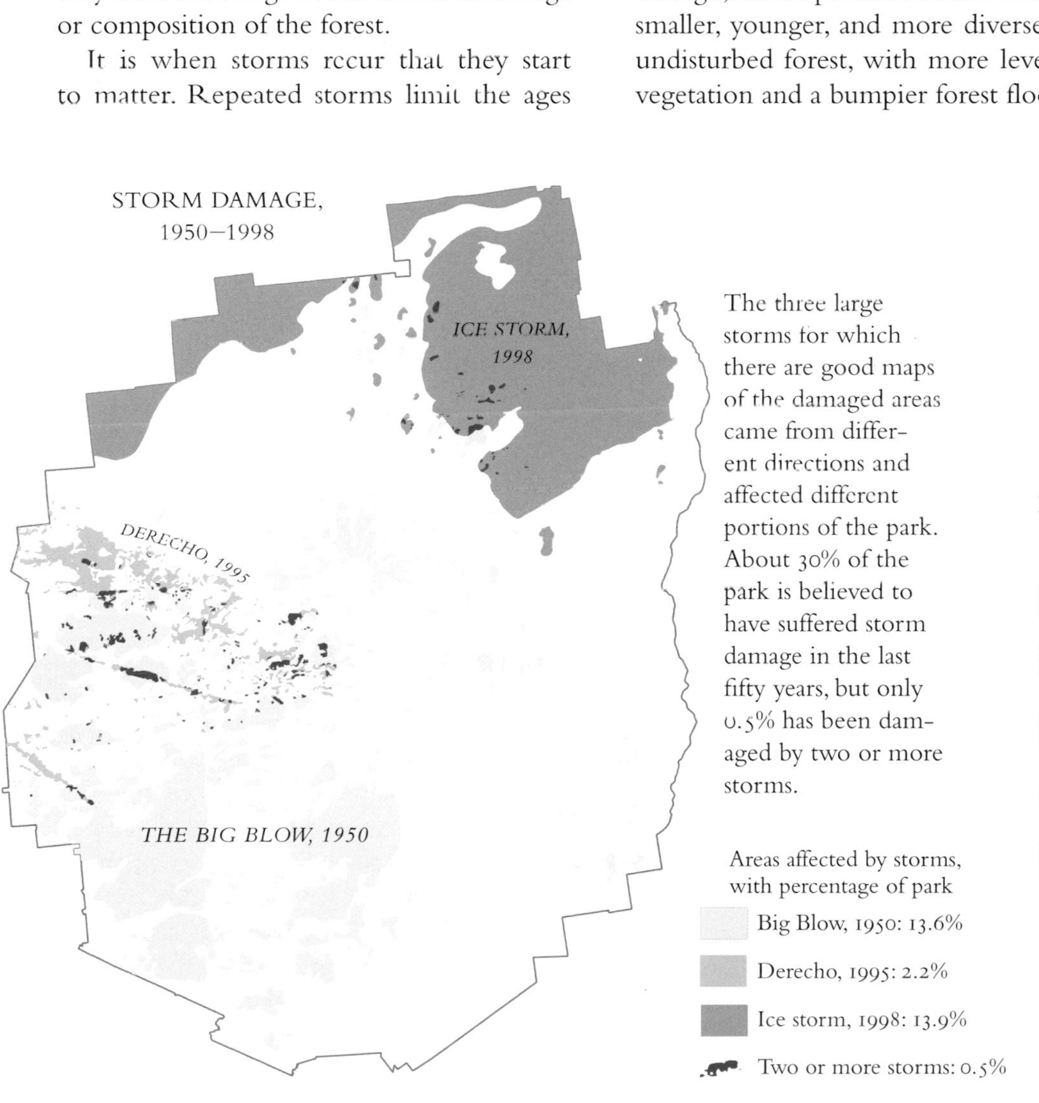

The three large storms for which there are good maps of the damaged areas came from different directions and affected different portions of the park. About 30% of the park is believed to have suffered storm damage in the last fifty years, but only 0.5% has been damaged by two or more storms.

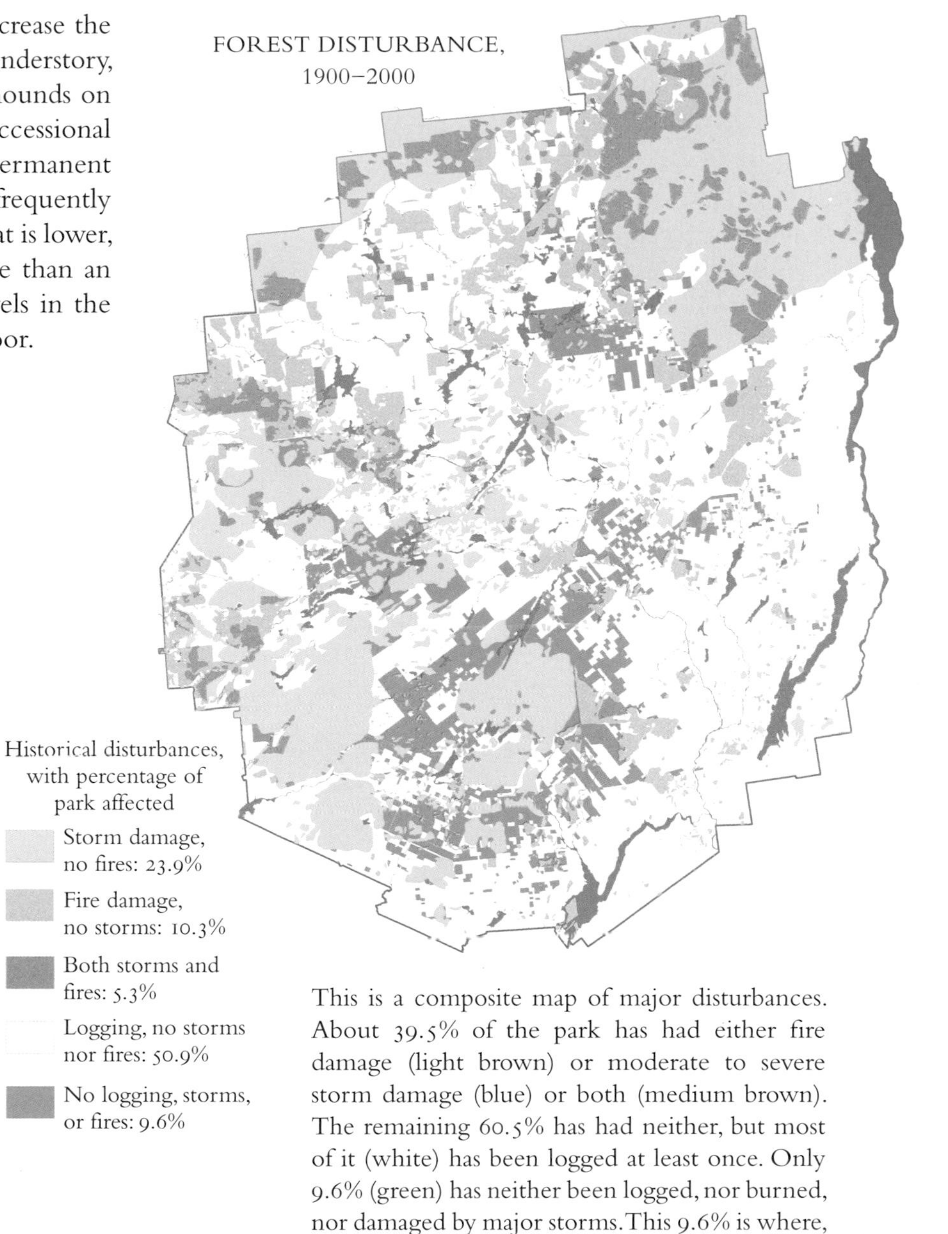

This is a composite map of major disturbances. About 39.5% of the park has had either fire damage (light brown) or moderate to severe storm damage (blue) or both (medium brown). The remaining 60.5% has had neither, but most of it (white) has been logged at least once. Only 9.6% (green) has neither been logged, nor burned, nor damaged by major storms. This 9.6% is where, hypothetically, the oldest trees in the park should be found.

THE ADIRONDACK POPULATION

7 *Vital Statistics*

THE next three chapters use state and federal statistics, particularly data from the U.S. Census Bureau, to give a picture of the residents of northern New York. The census data are by far the best picture of who lives here and what they do. But they have several important limitations. First, they are gathered only at ten-year intervals and so are usually somewhat out of date. Second, they only count the year-round residents, a major disadvantage in a park with numerous visitors and seasonal residents. And third, because the Adirondack boundary cuts through towns and census blocks, they give no precise statistics for the Adirondack Park as a whole.

Lacking parkwide statistics, the best we can do is use statistics from larger regions that include the park. The most convenient of these are the twelve-county *Adirondack region* and the ninety-two *Adirondack towns* that lie wholly or partially within the park. We use these repeatedly in the next several chapters.

Because the area of the ninety-two Adirondack towns is only 13% larger than the area of the park, it is tempting to use the ninety-two towns as a surrogate for the park. But because the Blue Line, by design, separates sparsely settled areas inside the park from densely settled ones outside it, this can be misleading. In 2000, for example, the total population of the ninety-two Adirondack towns was 235,885. Of these, only 111,047 lived in the sixty-one towns completely within the park. The remaining 124,838 lived in the thirty-one border towns that are partly in the park and partly outside. Thus the year-round population of the Adirondack Park—a much disputed number—is thus somewhere between 111,000 and 235,000. Because the developed parts of Gloversville, Glens Falls, and Plattsburgh lie outside the Blue line, the park population is probably closer to the former than the latter. Somewhere near 150,000 is a common guess but only a guess.

Old, Young, Nonwhite, White. The Adirondack population is a very typical northern U.S. rural population: about 29% young, 12% old, 96% white, 3% African American, 1.6% other races. The nonwhite population of about 7,000 is mostly in the prison towns and made up largely of prisoners. Dannemora, which houses many prisoners from New York City and its suburbs, has a racial composition similar to New York City. Statewide, nonwhites are concentrated in urban counties and, increasingly, in rural counties with large prisons.

Currently there are about 73,000 Indians in New York, perhaps twice what there were in 1650 when the Iroquois League was at the height of its power but now only 0.4% of the state population. They are mostly in the counties with Indian reservations.

The state-wide distribution of children is fairly uniform, though with noteworthy deficits in the career-oriented suburbs of Albany, Ithaca and New York City and also in our own fishing-oriented Hamilton County. The distribution of old people is much less uniform, with noteworthy concentrations in the Catskills and Adirondacks. Montgomery County, with 20.1% of the population over 64, has the largest percentage of old people in the state. Hamilton County, with 17.9%, is second.

THE 12-COUNTY ADIRONDACK REGION

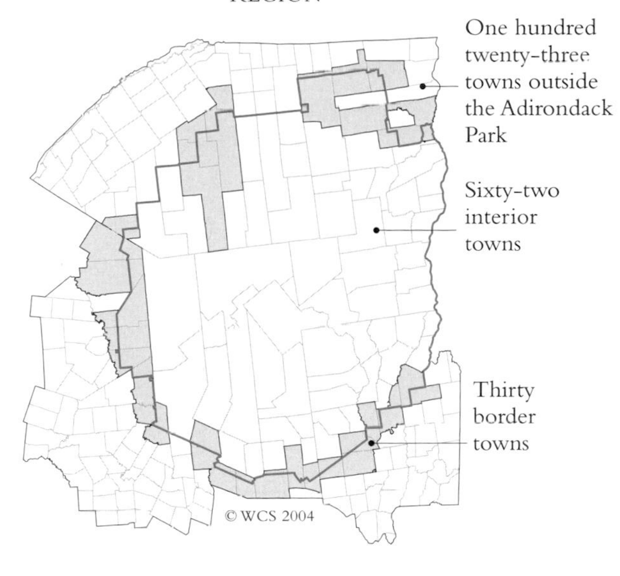

AREA, POPULATION, POPULATION DENSITY

Total 12-county area (sq. mi.)	16,520
Area outside park towns	6,056
Area of border towns	2,884
Area of interior towns	7,580
Total population (2000)	994,082
Outside of park towns	758,197
Border towns	124,838
Interior towns	111,047
Average density (persons/sq mi)	60.2
Density outside park towns	125.2
Density of border towns	43.6
Density of interior towns	14.4

DESPITE much cottage and second-home development and an expanding summer population, the year-round population of the Adirondacks has changed remarkably little in the last seventy years. The major increases are in the parts of the park within commuting distance of urban areas in the Hudson and Mohawk Valleys. Summer populations have probably increased much more but no numbers are available.

Like many rural areas, the Adirondacks have grown more slowly than the country as a whole. In the last seventy years, while the U.S. population has doubled and the New York population increased by fifty percent, the northern New York population has only grown by a relatively modest 35%, and the interior towns of the park by an unnoticeable 13%. In contrast, the suburban border towns of the park grew by a strapping 142%, faster than either New York or the U.S.

The population changes in the Adirondack region are the result of two trends: shrinking rural populations on the one hand and growing urban and suburban populations on the other. The result has been to increase the demographic contrast between towns and the surrounding countryside. The least settled areas have remained roughly constant or decreased in population; the areas that were more densely settled in 1930 have, in most cases, increased their population two to eight times.

Almost all of the major increases in resident population have been outside of the Adirondacks. Within the park the populations of the small interior towns have stayed the same or decreased, and even the populations of the resort towns grew fairly slowly, increasing by only 50-100% in sixty years. Only the southeast edge of the park, which has substantial amounts of ex-agricultural land available for development and is within commuting distance of Albany and Schenectady, has had widespread population increases of more than twofold.

Resident population (thousands)
1,000
800
600
400
200
0
1930 1940 1950 1960 1970 1980 1990 2000
NNY
POPULATION GROWTH
1930–2000
All park towns

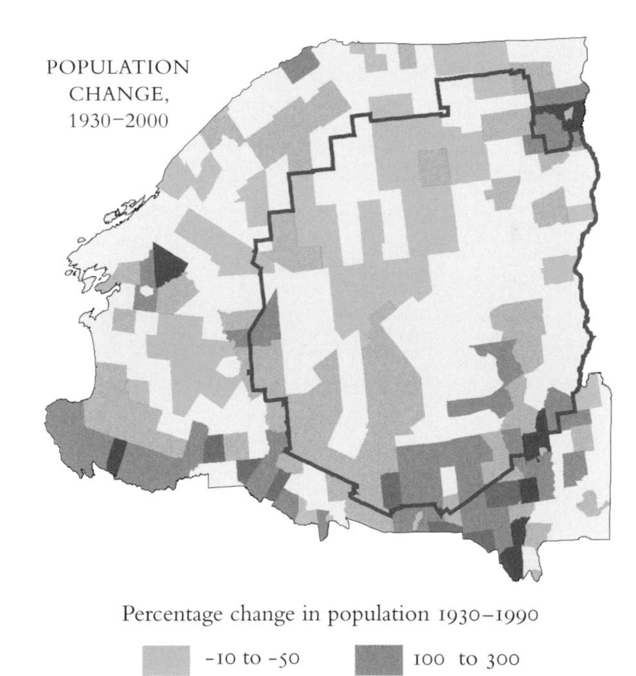

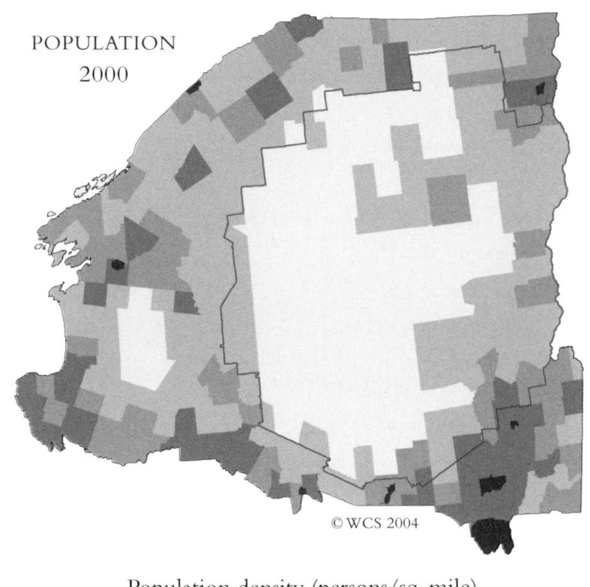

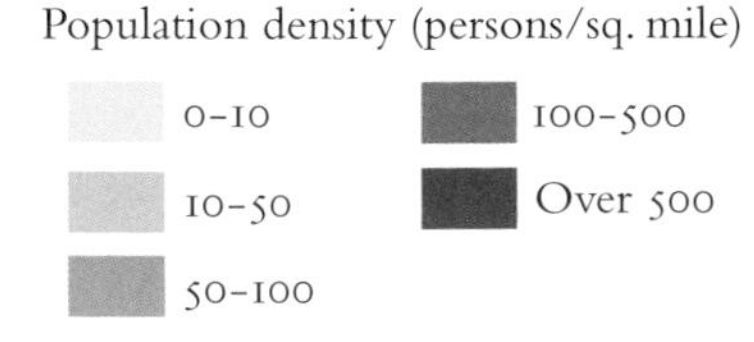

POPULATION growth in northern New York has been anything but uniform. While the total population has grown continuously from 1945 on, in every decade there have been some towns that lost population. No county has grown uniformly or continuously, and no center of growth has persisted for more than two decades.

In the 1930s, growth was very spotty. Population losses were as common as population gains, particularly in the agricultural counties and in the old paper and lumber towns near the west edge of the park.

In the 1940s, there was local population growth associated with war-time industry and military training in the St. Lawrence and Black River Valleys and in the main iron-mining towns in the park. There was also substantial growth in the south and southwest.

In the 1950s there was considerable growth in the north and northwest, some associated with the expansion of the aluminum industry in Massena and the development of a large Air Force base at Plattsburgh. In addition, the south edge of the park began to develop a commuter economy as a number of agricultural towns and summer colonies became bedroom towns for Albany and the other Mohawk Valley cities.

Commuter-related development continued for the next two decades, joined by a substantial amount of ski and resort development in the Lake Placid and North Creek corridors in the 1970s. The ski-related growth was only temporary, slowing in the 1980s and gone by the 1990s. The map of the 1990s looks strikingly different than those of the four previous decades, especially in the noticeable slowing of growth just outside the park. If this is a true change in pattern, rather than just a temporary fluctuation, it could have great significance for the park.

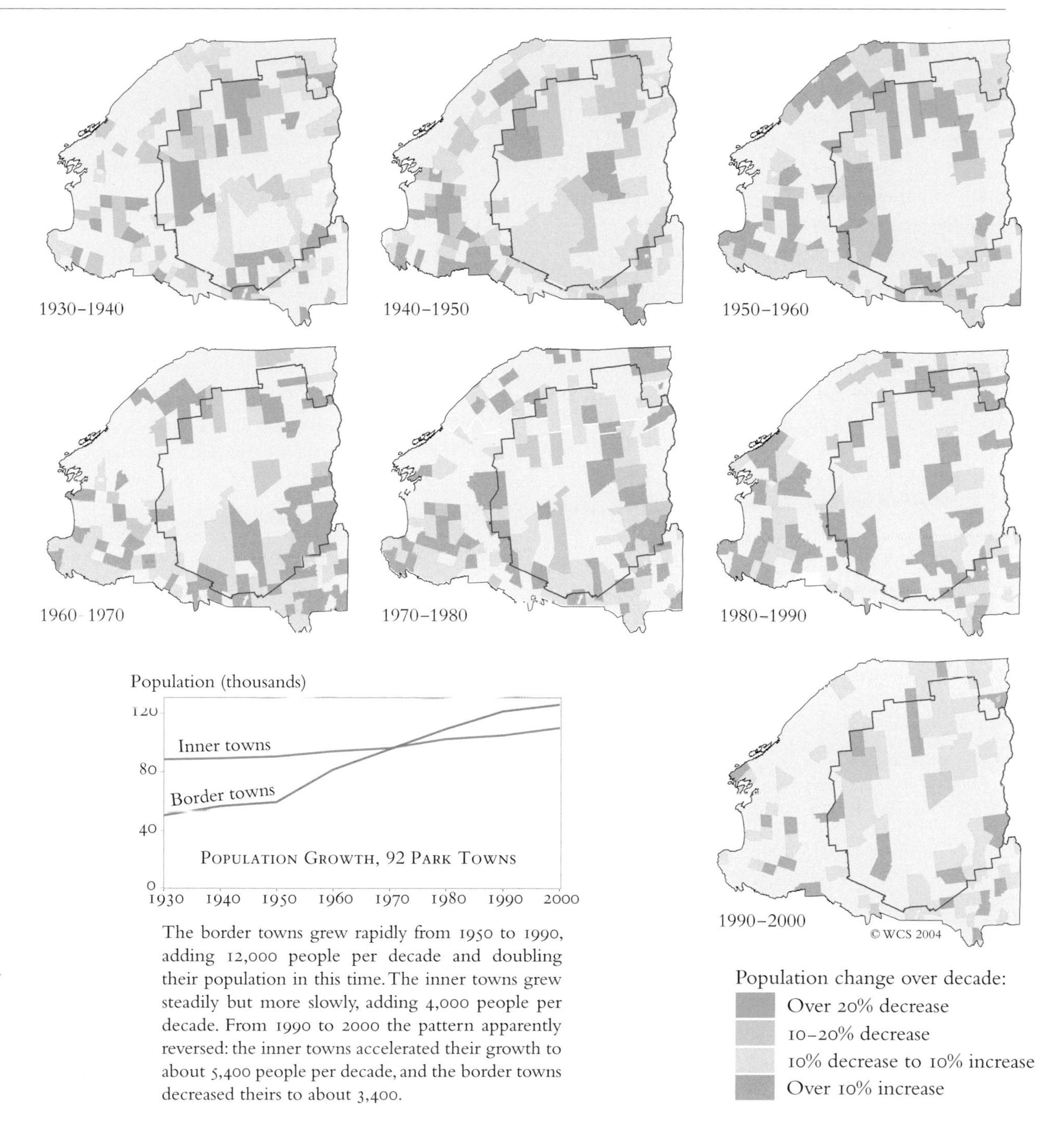

The border towns grew rapidly from 1950 to 1990, adding 12,000 people per decade and doubling their population in this time. The inner towns grew steadily but more slowly, adding 4,000 people per decade. From 1990 to 2000 the pattern apparently reversed: the inner towns accelerated their growth to about 5,400 people per decade, and the border towns decreased theirs to about 3,400.

BIRTHS and deaths are perhaps more important than anything else in this atlas because they determine how many of us there are and how crowded our world is.

In 2000, the New York State birth rate was 13.6 births per thousand persons per year and the death rate 8.5 deaths per thousand per year. The difference in these rates gives a hypothetical rate of *natural increase* of 0.5% per year, enough to double the population in about 140 years. This is not a particularly fast rate but it is still a significant rate: the current population of New York is nineteen million, and another nineteen million people will double both the density of settlement and the already large impact of humans on the natural world, no matter how slowly or quickly they come.

The actual New York State growth rate depends both on natural increase and migration into and out of the state. Currently, and perhaps accidently, it is also fairly close to 0.5%.

The geography of birthrates and death rates is odd. Both vary substantially from county to county but do not correlate with population, development, or, in fact, any other factor we have mapped. Birthrates are high along the Lake Ontario shore and in the New York City suburbs; death rates are high near Lake Erie and in a north-south corridor just west of the major mountains.

Some of the variation is related to the average age of the population. Northern and western New York is, on average, younger than eastern and southern New York and so should have a higher birthrates and a lower death rate. But, except that dead people don't have children, cause and effect are not very clear. Does a young population cause a high birthrate or result from it? And what makes the idiosyncratic differences in the pattern? Why are there more births in Brooklyn and the Bronx than in Queens, and why do so few people die in Putnam and Saratoga counties?

The rate of natural increase is even more variable than either the birth rate or death rate but seems to show a much clearer geographic pattern. It is uniformly high in the urban and suburban counties and also in Jefferson County where it may reflect the young military families at Fort Drum. It is quite close to zero in the more rural counties and actually below zero in forested Hamilton, Franklin, and Delaware Counties. All three are quiet, woodsy places where people like to birdwatch and fish. Children may just not fit the lifestyle or may wander away among the trees.

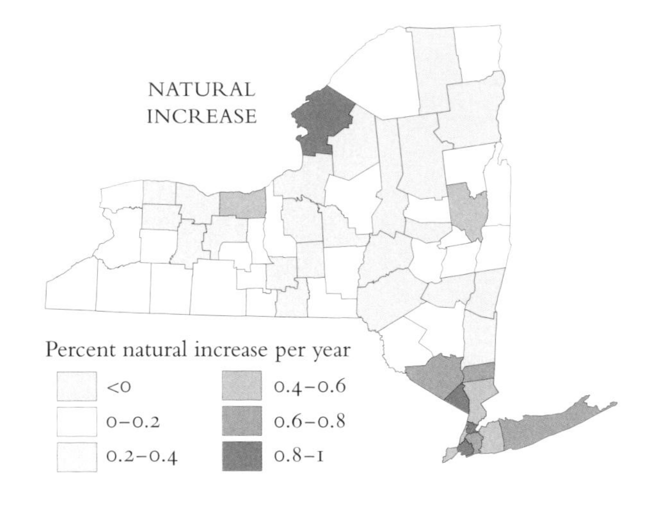

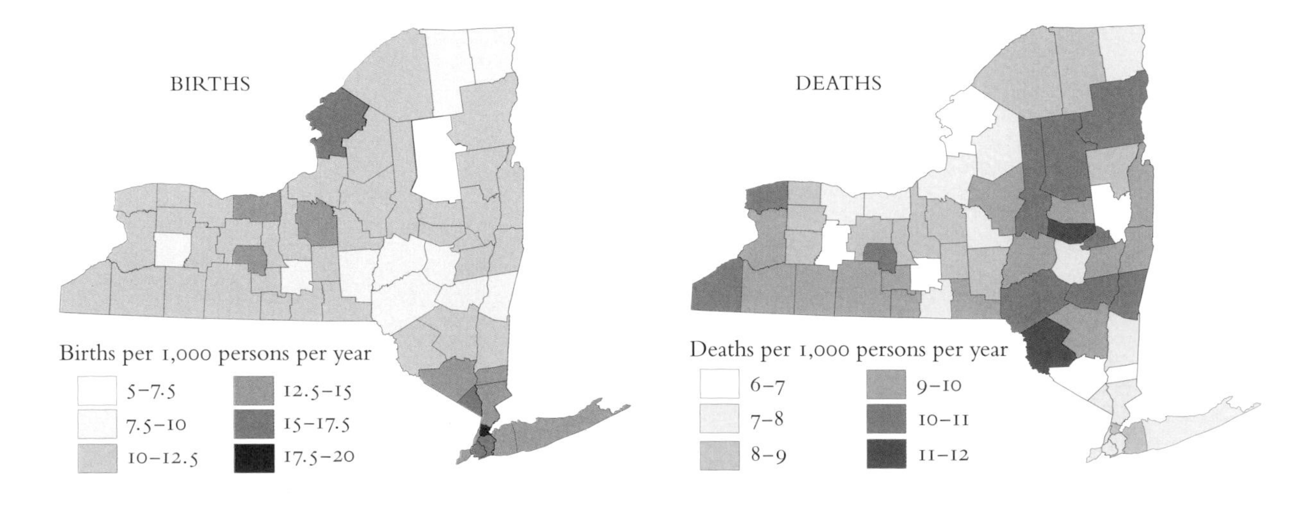

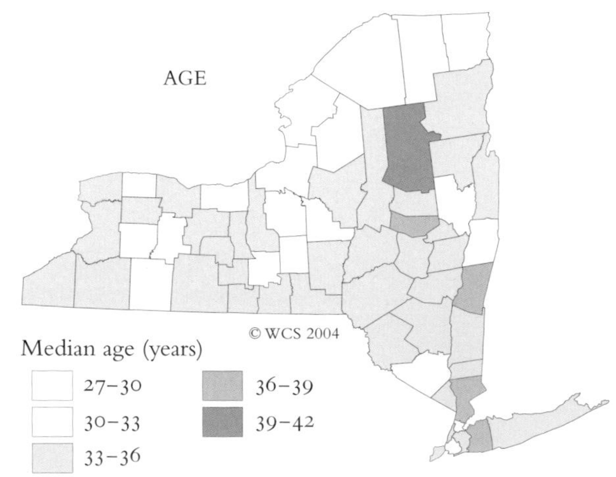

BOTH birth and death rates have decreased dramatically in the last 150 years and are now less than half of what they were in 1850. This *demographic transition*, which has happened in all affluent countries but stubbornly refuses to happen in many of the nonaffluent ones, has slowed population growth and produced an older, less fecund, and dramatically longer-lived population.

The U.S. death rate, which was about 1.9% per year in 1860, started to fall about 1880 and reached 1.0% in the 1940s. Much of the decrease was because of improvements in environmental quality and public health—nutrition, vaccination, and sanitation—that affected the survival of children and young adults. The dramatic medical advances since 1940 have resulted in great increases in longevity but, because a five-year-old who doesn't die of measles postpones his or her death far longer than a sixty-year-old who gets a bypass, caused only small changes in the death rate.

The U.S. birthrate was an impressive 5.5% in 1820. By 1935 it was slightly below 2%. It rose to a modest peak of 2.4% in the 1945-1960 baby boom—which only looked like a boom because the rates in the Depression were very low—and then fell to 1.5% in the 1970s. It climbed slightly until about 1990 as boomers began having boomerettes and now has fallen again.

Within New York there is considerable county-to-county variation in age-structure and birthrate. Jefferson County, for example, has a large percentage of twenty-year-olds who are producing a small baby boom of their own. Hamilton County has more sixty-year-olds than thirty-year-olds and, despite many of them doing their best, only half the birthrate of Jefferson County.

Because New York's population is large to start with, the rate at which it is adding people is very important. The current 0.5% annual increase in a population of 19 million is 85,000 people, or the average population of a northern New York county per year. Providing food, medicine, education, homes, and jobs for this many new people every year is not an easy task for any society and especially not for one with as large a gap between rich and poor as ours. Many of the new New Yorkers are born into poverty, and some of them will remain poor all their lives.

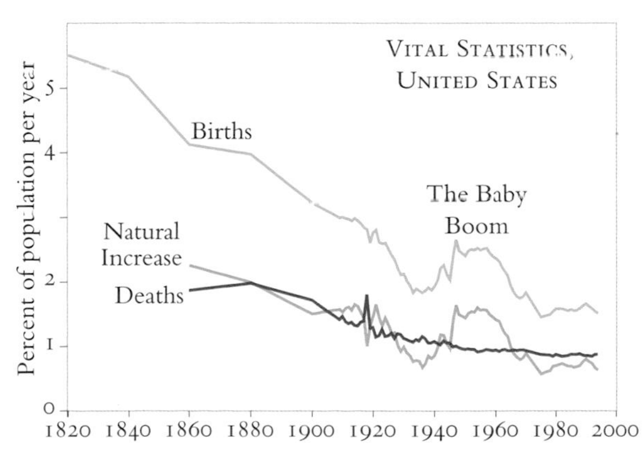

U.S. death rates are not available for 1860 and 1880. Massachusetts death rates, which are likely smaller, have been substituted. There are no death-rate increases in either of the world wars, but there is a strong spike associated with the deadly "influenza" epidemic of 1918.

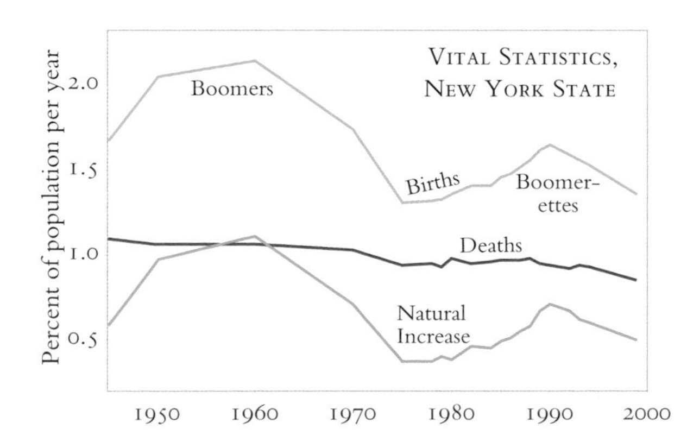

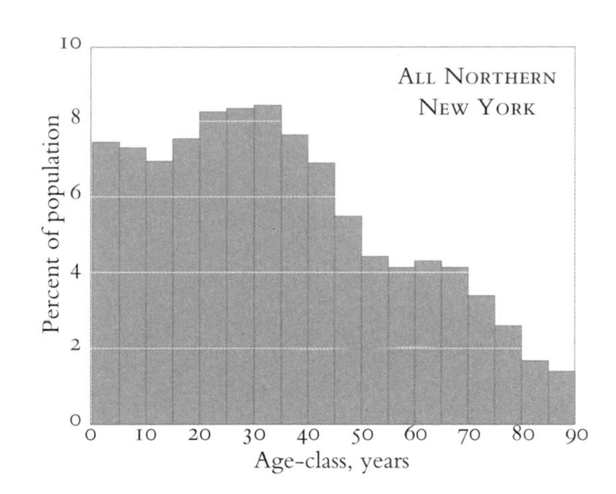

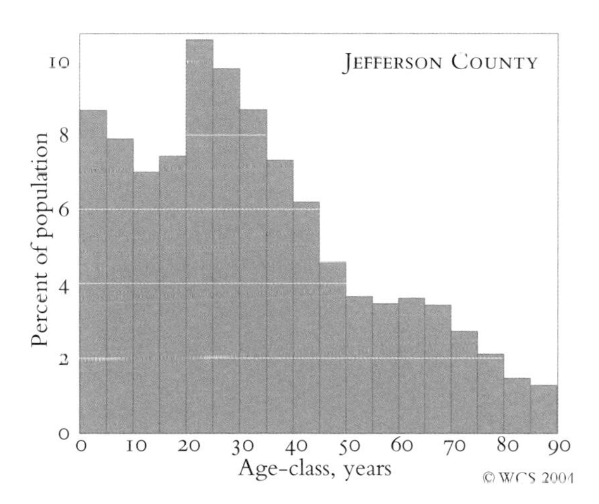

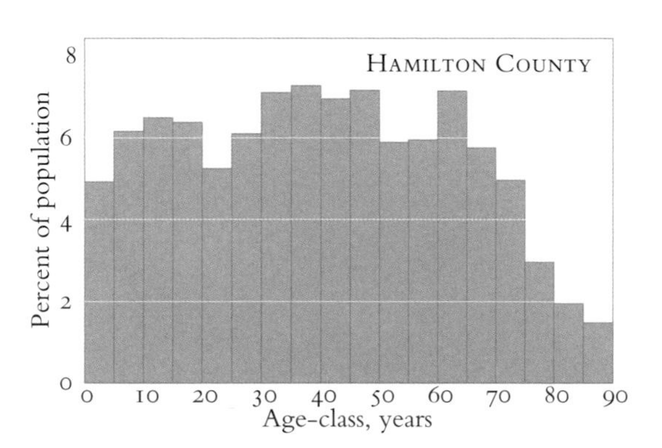

THE marriage rate in New York State is fairly uniform, city and country, at about 6-10 marriages per 1,000 persons per year. Love, sex, and parental pressure are apparently geographic universals. The state averages around 150,000 marriages, involving 1.8% of the population each year. The marriage rate is a little lower in some urban and suburban areas but nowhere falls below 3 per 1,000; there are no true bachelor havens. It is locally high in the central and eastern Adirondacks and highest of all, a preternuptial 15.6 per thousand, in Manhattan.

Ages of Brides and Grooms. In New York the forty years between twenty and sixty are the marrying years; only 4.5% of the men and 6.9% of the women who marry at all marry outside of these ages. The peak for both men and women is in their late twenties: 29.9% of all men and 29.6% of all women who marry are between twenty-five and twenty-nine.

Variety and adventure may matter more than is generally believed. Both sexes typically choose partners close to their own age, fully 60% percent of all marriages are between partners more than five years apart in age. And while it is commoner for a man to marry a younger woman than vice versa, it is not uncommon for women to marry younger men. In 1997, fourteen thousand upstate New York women married men five years or more younger then they. Five of these, if the *Vital Records* of New York State are to be believed, married men forty to fifty years younger than they.

Divorces. Change the bride with spring, Harun al-Rashid said, and in 1997 some 124,000 New Yorkers did just that. Most of them had been married less than ten years. The median duration of their marriages was about eight years. The divorce rate is low in the first year, rises rapidly to a peak around year four, and then remains fairly constant

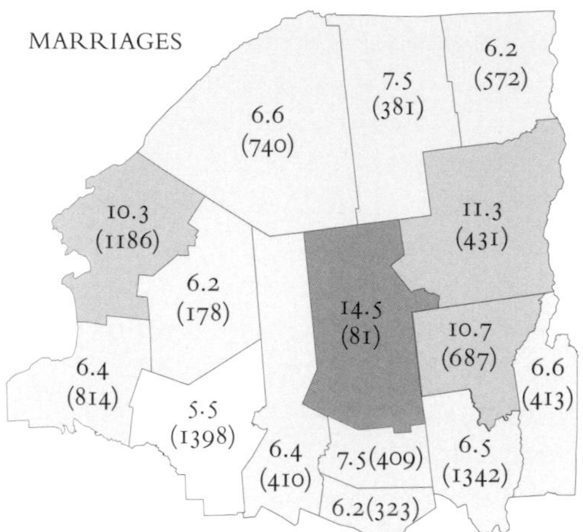

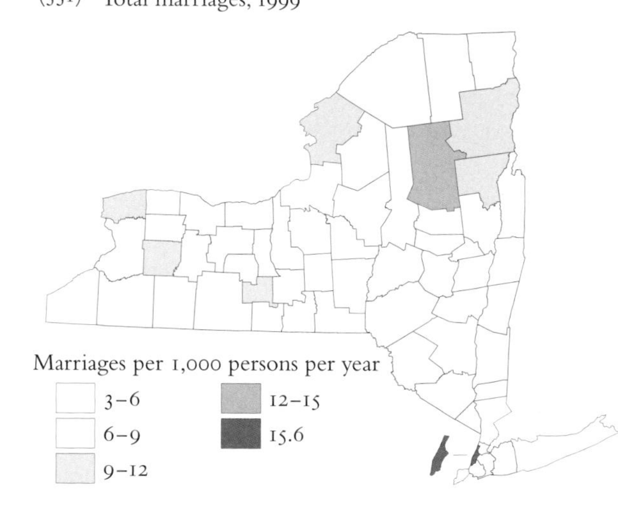

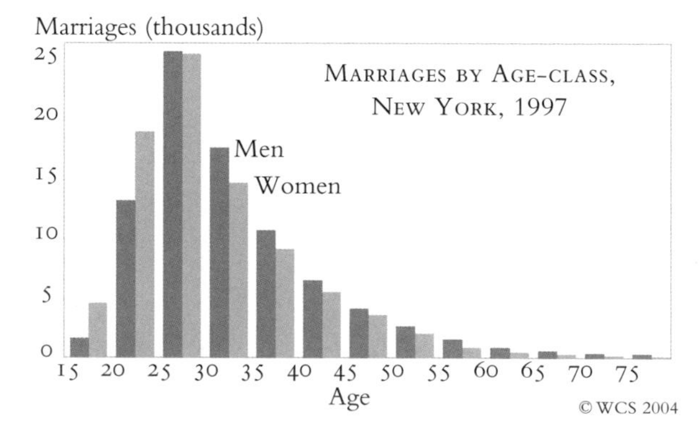

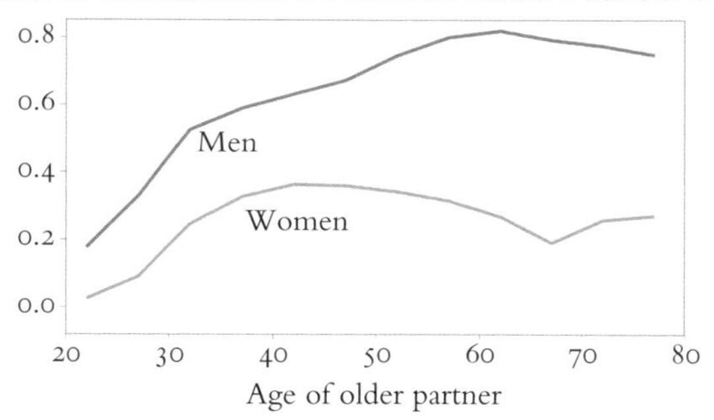

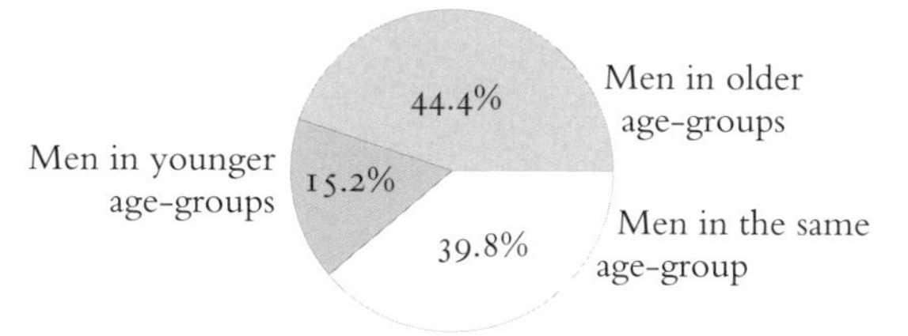

Age Differences. Although women marry earlier than men and so often marry men older than they, the majority of both men and women in their twenties choose partners near their own age. Older men and women, perhaps for reproductive fitness or perhaps for fun, often choose partners an age-class or more younger than they. Fully 80% of the sixty-year-old men who marry choose younger partners, as do an interesting, and less often noticed, 35% of the women in their forties and fifties.

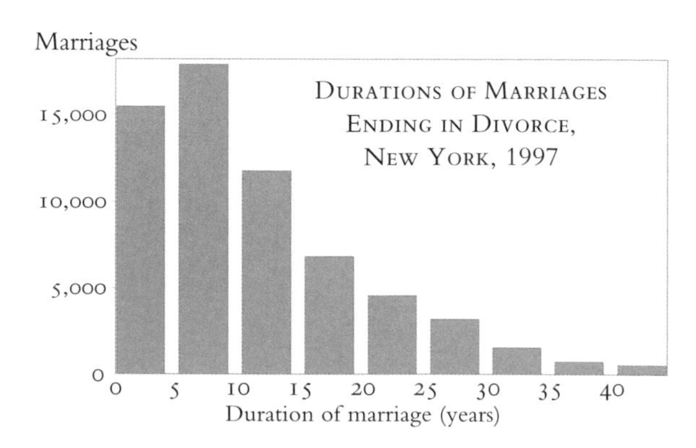

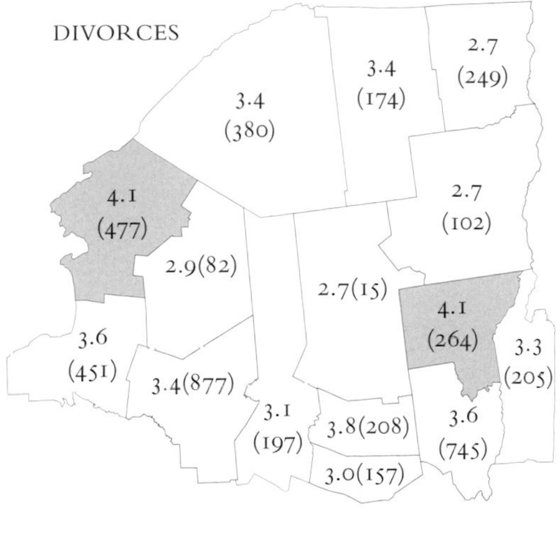

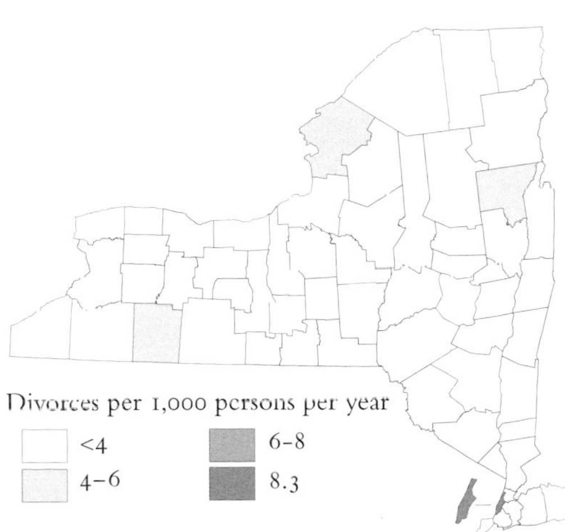

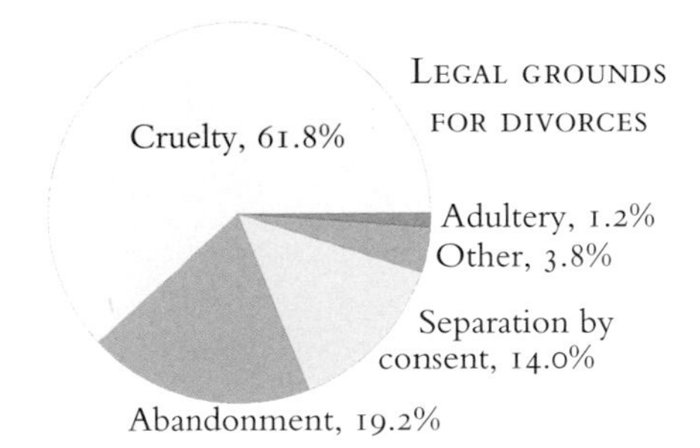

for the next forty years. Apparently, after about five years of marriage the chance of divorce doesn't change much. Fewer couples separate after thirty years of marriage than after ten but only because there are fewer left to separate. The likelihood is the about same for both.

Per capita divorce rates are even more uniform than marriage rates, running at about 3–4 per 1,000 persons per year. Since death rates are two to three times this, it appears that, if death rates are the same for the married and the single, somewhere between a half and two-thirds of all marriages will be parted by death and the remaining third to a half by a judge.

Extra Marriages. The *excess marriage rate* is the amount by which the marriage rate exceeds the divorce rate. It is uniform at a nonthreatening 2–4 excess marriages per 1,000 persons in much of the state but rises, strikingly, to three times that in the recently discovered *bliss belt* that arches across the Adirondack region from Jefferson County to Essex County. At the apogee of the bliss belt in Hamilton County, the married population is increasing, though perhaps not sustainably, at a striking 11.8% per decade. Having half the population over forty may not be as dull as it first appears.

Manhattan is, as always, a special case. It combines a record 15.6 per 1,000 marriage rate with a scorching 8.3 per 1,000 divorce rate to produce a 7.3 per 1,000 excess of marriages. A busy place indeed.

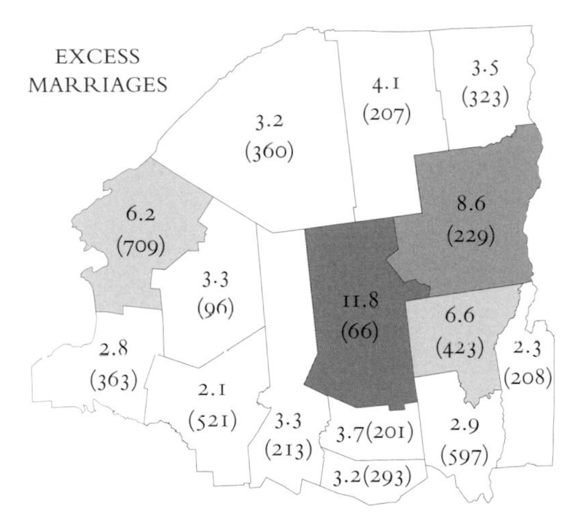

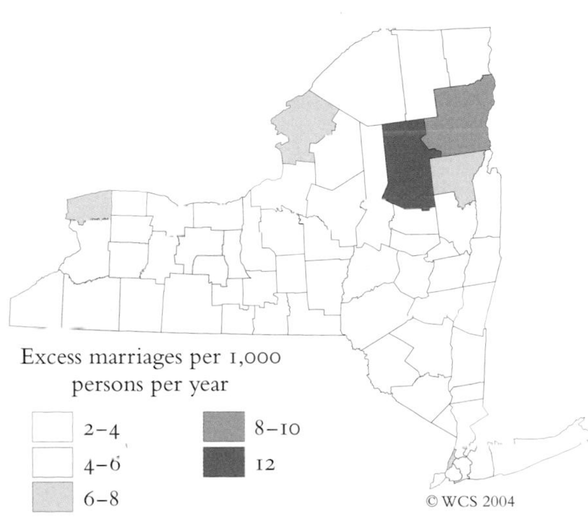

In the upper figures on this page, the top number in each county is the rate of divorces or excess marriages per 1,000 persons per year in 1999. The number in parenthesis is the total number of divorces or excess marriages in that county in 1999.

DIFFERENT TYPES OF EMPLOYERS

The pies show the percentage of the workforce in each county that work for themselves, that work as unpaid employees of family businesses, and that work for each of the five other main types of employers. Note that the employers may be in different counties from the workers they employ.

Examples of Employers

Private For-profit
Accounting firm
Bank
Ferry operator
Hotel
Retail store
Real estate agency
Sawmill
Trucking company
Quarry

Private Nonprofit
Church
College
Hiking club
Hospital

Government
Highway department
Military base
Police department
Prison
Public school system
State agency
Town board

Self-employed
Boat-builder
Consultant
Doctor
Farmer
Plumber
Writer

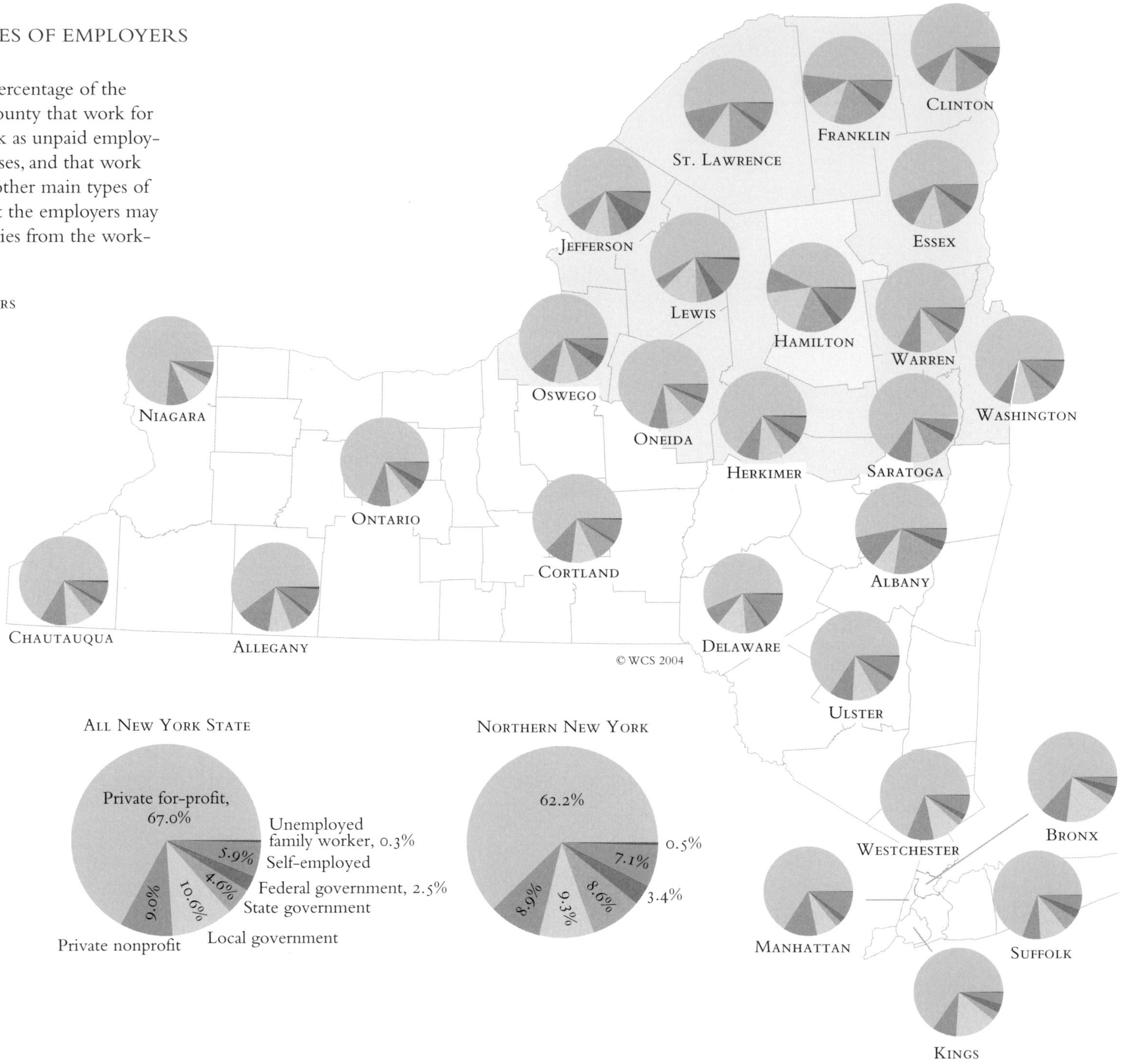

Clinton
Franklin
St. Lawrence
Essex
Jefferson
Lewis
Hamilton
Warren
Oswego
Washington
Niagara
Oneida
Herkimer
Saratoga
Ontario
Cortland
Albany
Chautauqua
Allegany
Delaware
Ulster
© WCS 2004

All New York State

Private for-profit, 67.0%
Unemployed family worker, 0.3%
5.9% Self-employed
4.6% Federal government, 2.5%
State government
10.6% Local government
9.0% Private nonprofit

Northern New York

62.2%
0.5%
7.1%
3.4%
8.6%
9.3%
8.9%

Westchester
Bronx
Manhattan
Suffolk
Kings

8 Employers, Jobs, & Income

THIS chapter describes Adirondack workers and employers: who they are, what they do, and how much they make or don't make doing it.

Because every job involves both a worker and an employer, mapping jobs is complicated. Someone may live inside the park but work outside of it. Further, she or he may work for a hospital, and thus in a service industry, but do plumbing and electrical work for them and so be classified as a craftsman rather than a service employee. To reflect this complexity, we give separate maps of *occupations*, telling what the worker do, and *industries*, telling what their employers do.

JOBS WITHIN THE ADIRONDACK PARK, BY INDUSTRY, 1997

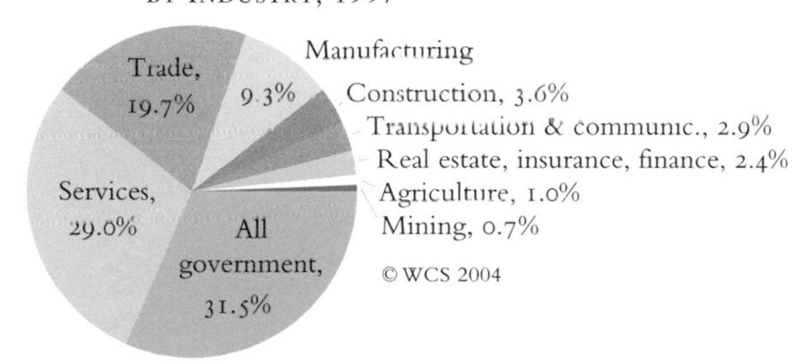

A further complexity is that most of the publicly available data count workers rather than jobs. Thus the map on the left page, which is based on census data, shows what percent of the *workers* in each county work for each sort of employer. It does not—and this is an important distinction—show what percentage of the *jobs* in the county each sort of employer provides. As a result we know much about what the people in each county do for their livings but little about where they go to do it.

Starting with the park, the New York Department of Labor estimates that there were 37,681 full-time jobs within the Blue Line in 1997. Over half were in government or in service industries like health, education, and tourism. About a quarter were in trade and manufacturing. Everything else—farming, mining, building, forestry, finance, etc.—employed less than an eighth of the work force. Unlike a century ago, if you work in the Adirondacks today, the chances are high that you will work indoors and at a white-collar job.

PERCENTAGE OF HAMILTON COUNTY WORKERS EMPLOYED BY EMPLOYERS OF DIFFERENT TYPES

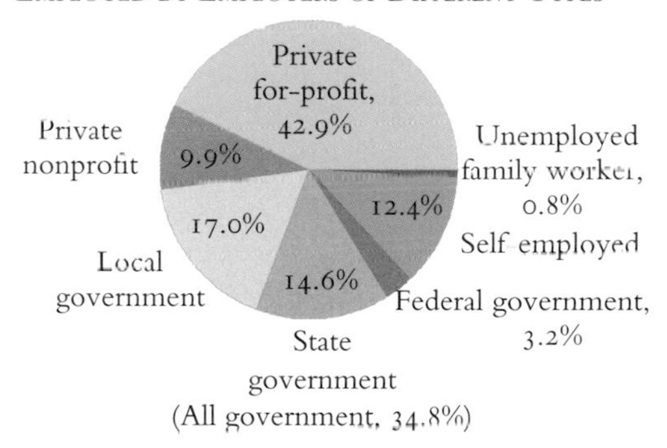

Further, it is more likely that you will be self-employed or working for a government or a nonprofit organization than working for a private, for-profit business. The first three employ a remarkable 56.3% of the work force in Hamilton County. Nothing like this occurs in southern New York. But its meaning is unclear. Is Adirondack government unusually large, or the Adirondack for-profit sector unusually small, or possibly both?

While we know the numbers of *jobs* in the park with some accuracy, because the Blue Line cuts through census districts we do not know much about the size or composition of the Adirondack workforce. The best we can say is that the census shows a work force of 92,860 in the ninety-two Adirondack towns, that somewhere between half and two-thirds of this work force lives in the park, and that their occupations are as shown in the graph below.

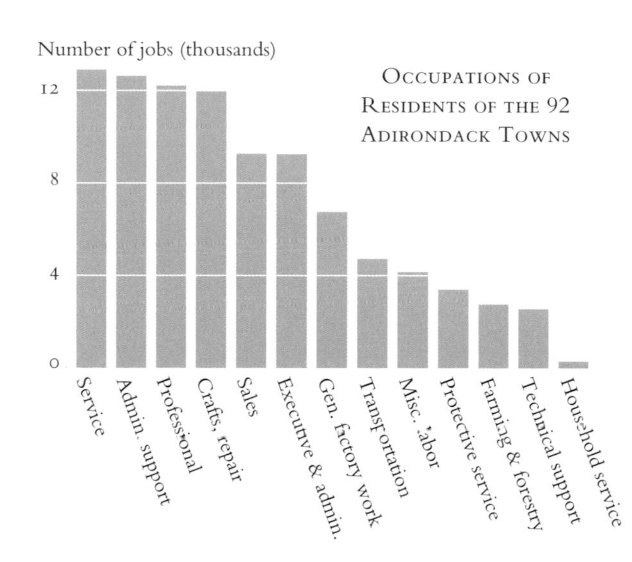

By the standards of the old Adirondacks, centered on mining, forestry, and manufacturing, this is a surprising work force: 23% professional and 67% white collar but only 17% unskilled labor and 3% natural resource dependent. The iron-furnace and log-drive days were not long ago, but they are long gone.

An industry, to the census, is any legal economic activity with a more-or-less classifiable purpose. Industries may be large or small and corporate or personal. Tools, factories, and loading docks are not necessary. Banks, lemonade stands, paper-mills, armies, legislatures, funeral parlors, and jazz bands are all industries to the census.

The maps give the percentages of the workforce employed in different industrial sectors. Note again that this classification depends on what the business does and not what their individual workers do. In these maps, an accountant employed by a builder works in the construction sector, and a janitor at a brokerage works in finance.

Examples of Industries and Industrial Sectors

Agriculture & Forestry
- Dairy farm
- Logging contractor
- Tree farm

Mining
- Gravel pit
- Iron mine
- Slate quarry

Construction
- House-building
- Road-paving

Manufacturing
- Boatbuilder
- Concrete company
- Paper mill
- Print shop
- Sawmill
- Sheltered workshop

Transportation & Communication
- Fuel-oil dealer
- Phone company
- Television station
- Trucking company

Trade
- Food service
- Gas station
- Hardware store
- Restaurant

Finance, Insurance, & Real Estate
- Bank
- Brokerage
- Insurance agency

Services
- Auto-repair shop
- Hotel
- Hospital
- Law office
- Membership organization
- Museum
- Private school
- Ski area
- Summer camp
- Theatre

Government
- Elected official
- Government agency
- Highway department
- Military base
- Police department
- Post office
- Prison
- Public school system

Also note that, unlike service *jobs*, which tend to be unskilled, the service *industries* include law, medicine, architecture, and a number of other professions which have highly skilled employees.

Within the Blue Line, the largest and most rapidly growing industries are government, service, and trade. Manufacturing is a distant fourth. Together, these four provide about 33,000 jobs, or 89% of the jobs within the park. Construction, transportation, real estate, and finance provide another 3,400 jobs. Agriculture, mining, and forestry, which fifty years ago were some of the largest Adirondack employers, now provide only 700 jobs, less than 2% of the jobs in the park.

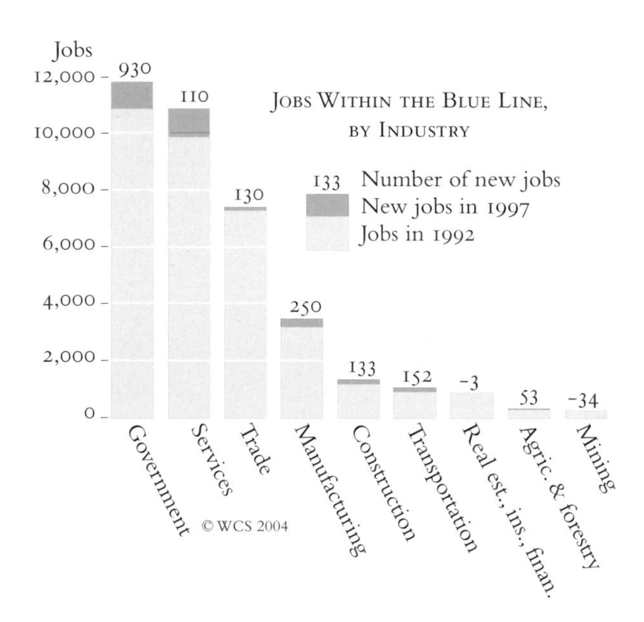

Elsewhere in northern New York, manufacturing and trade are more important and government and services slightly less so, but service and government are the dominant industries everywhere.

In New York as a whole, it is striking how little the proportion of the workforce in different industrial sectors varies from place to place. Retail

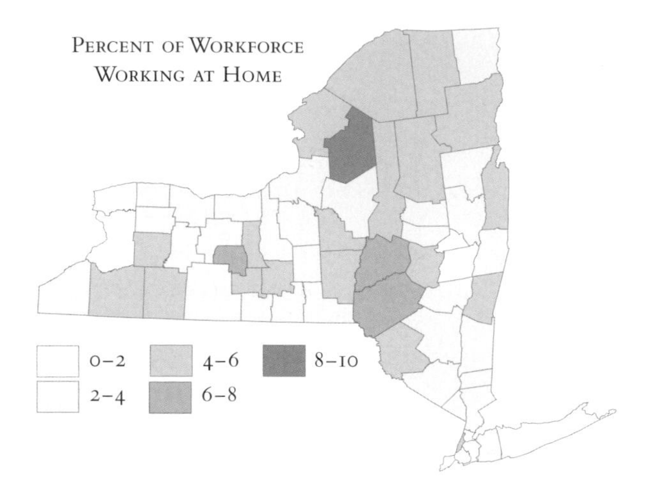

trade employs about 20% of the work force statewide, hardly varying between forested Hamilton County and agrarian Cortland or between affluent Westchester and the impoverished Bronx. Agriculture and forestry are remarkably small everywhere, especially for a state that is still mostly fields and woods.

The real story told by the map on the opposite page is that the northern New York economy has at least three levels. The first level—extracting resources and making things out of them—supplies a little over a quarter of all jobs. The second level—moving people and things about and selling things to people—supplies about another quarter of the jobs. The third level—moving money and information, providing professional services, and managing the other levels—supplies a little less than half of all jobs.

Worldwide, only very primitive or very subjugated economies are dominated by the first level, and only very wealthy ones by the third level. By world standards, and even in comparison to much of the U.S., northern New York, rural and urban, has a very wealthy economy.

The shares of the different industrial sectors in the northern counties are more alike than different. The entertainment-recreation sector is larger in the central park counties—Hamilton, Essex, Warren—than in the peripheral ones. Manufacturing varies in the reverse direction and is stronger outside the park than in it. Public administration is more important in the far northeastern counties (Franklin, Clinton, Hamilton, Essex) than in most of the rest of the state. Hamilton County, with many trees and only a few people, has a surprisingly large construction sector and, compared to the other northern counties, a very small agriculture-forestry sector. This is because cows are milked every day, while forests are only cut every ten or twenty years. Thus while there will be a farm job for every few hundred acres of farmland, there will only a forestry job for every five or ten thousand acres of forest.

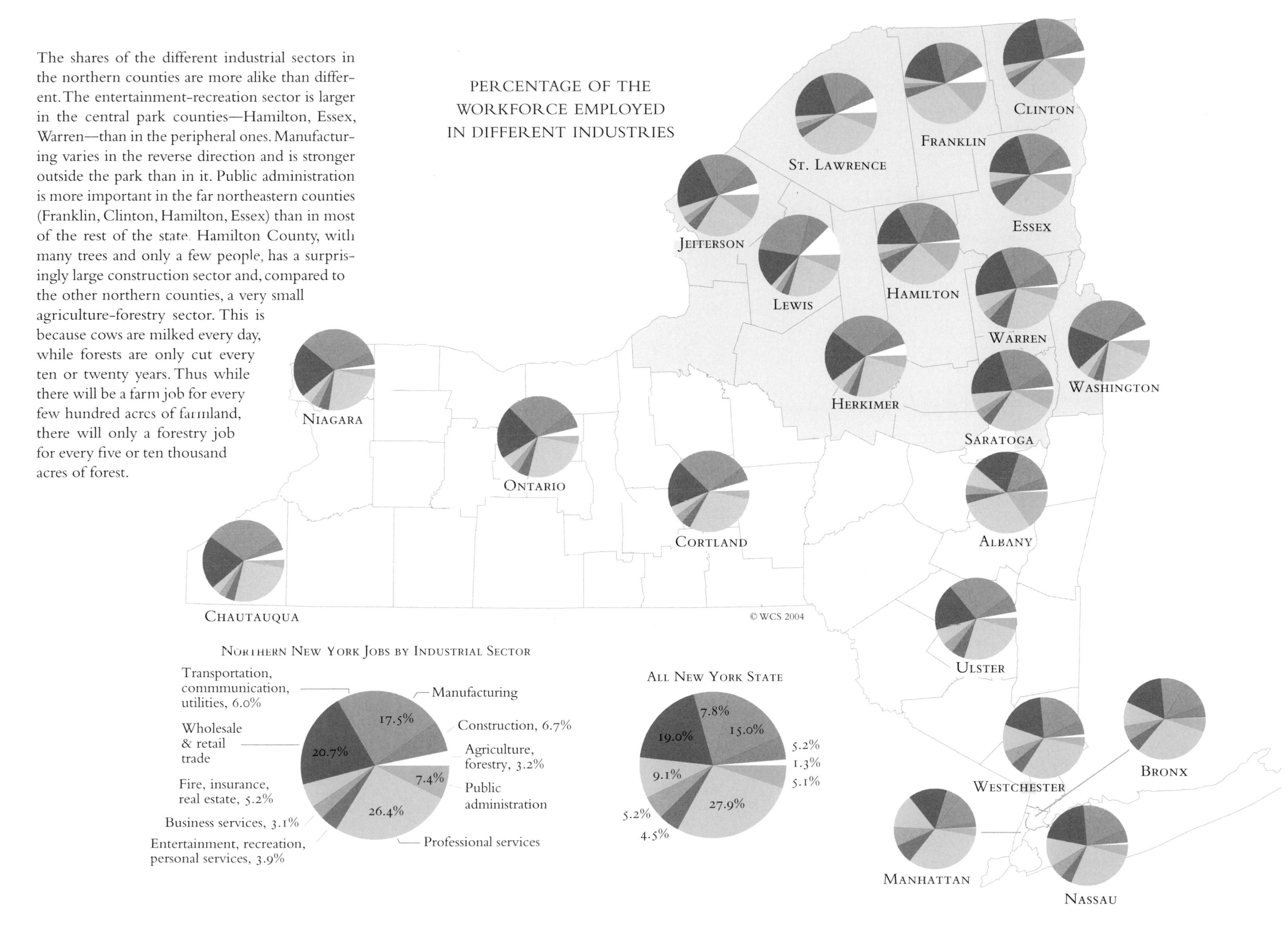

By an *occupation* we mean what an individual worker does, regardless of what his employer does. Note that in this classification professionals like teachers and lawyers are not considered to hold service jobs, even though they work in service industries.

In northern New York about 30% of the work force works with their hands and the remaining 70% manages, offers services, or generates and processes information. Only 3% of the workers farm or work directly in forestry, though many other workers support or profit from these occupations.

Statewide the pattern is similar: the downstate counties have are fewer farmers and blue-collar workers and more managers and professionals, but the differences are only a few percent either way.

The important thing to notice is that, contrary to common belief, the dominant occupations in northern New York are neither resource extraction nor service. Rather, like the rest of the state, two-thirds of the people work with information or other people, a little over a quarter make things and move things and fix things, and only 3% work directly with the soil and the forests.

The maps of occupational groups bear this out. All of the occupations except forestry and farming are widely distributed throughout the state. The centers of concentration are about where you would expect them: white collar jobs near the cities, blue collar jobs in the old industrial areas, and service, precision assembly, crafts, and repair pretty much everywhere. Farming and forestry, on the other hand, have a very spotty distribution. They are almost absent from urban and suburban areas, uncommon in most rural counties, and employ more than 5% of the work force in only 11 of the 62 counties in the state, and more than 10% in only one county.

Examples of Occupations

Professionals	Secretaries	Carpenters
Athletes	Tellers	Linemen
Doctors	*Service Providers*	Machinists
Embalmers	Bailiffs	Printers
Foresters	Barbers	Plumbers
Musicians	Bartenders	*Operators, Laborers,*
Nurses	Guards	*Transporters*
Teachers	Guides	Axemen
Managers	Janitors	Chainmen
Executives	Policemen	Deckhands
Principals	Orderlies	Excavators
Proprietors	Sextons	Drivers
Sales & Support	Waiters	Lockkeepers
Clerks	*Crafts & Repair*	Mill hands
Salesmen	Boatbuilders	Pilots

Occupations in the 92 Adirondack Towns

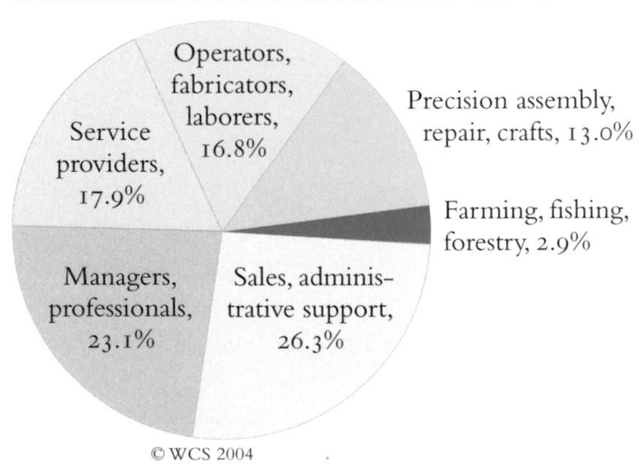

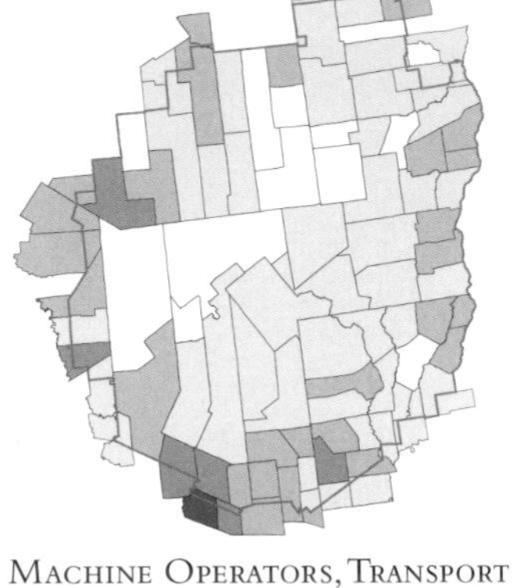
Machine Operators, Transport Workers, General Laborers

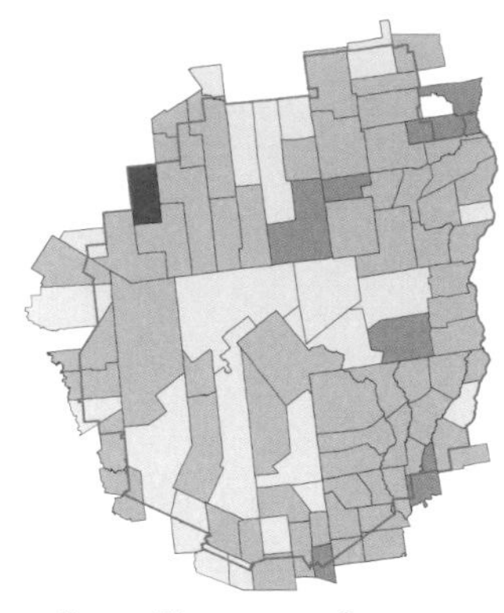
Sales, Technical Support, Clerical

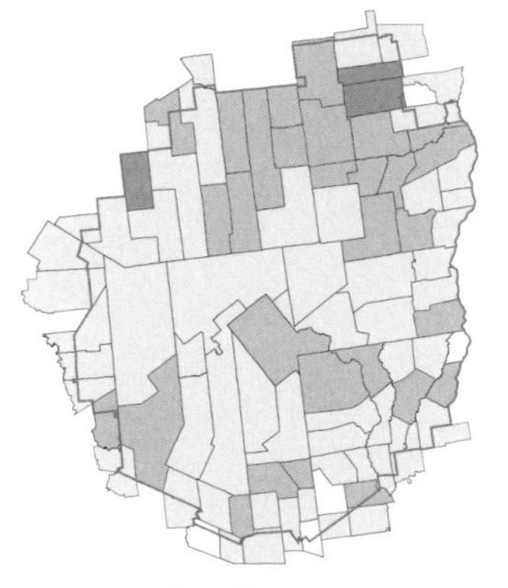
All Service

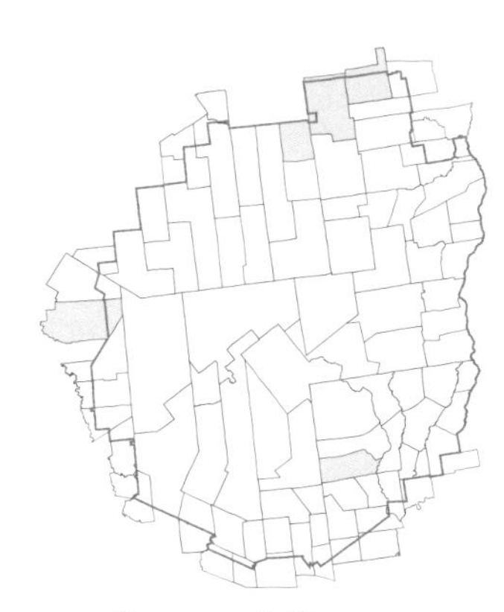
Farming & Forestry

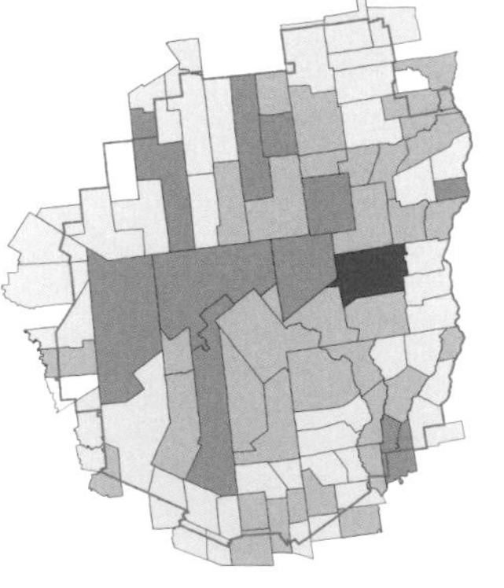
Managerial & Professional

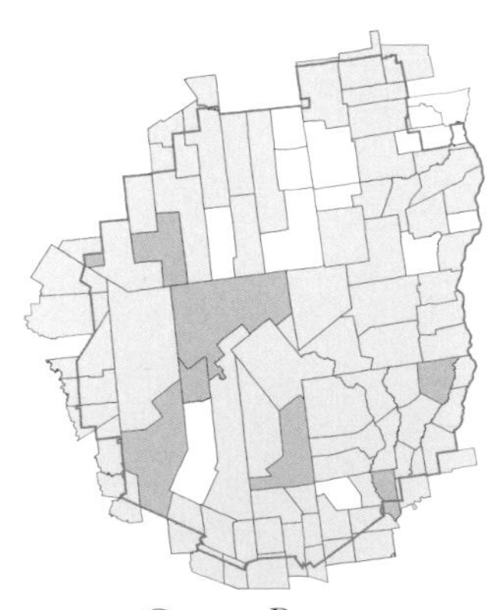
Crafts, Repair, Precision Assembly

Percent of labor force employed
- 0–10
- 10–20
- 20–30
- 30–40
- 40–50

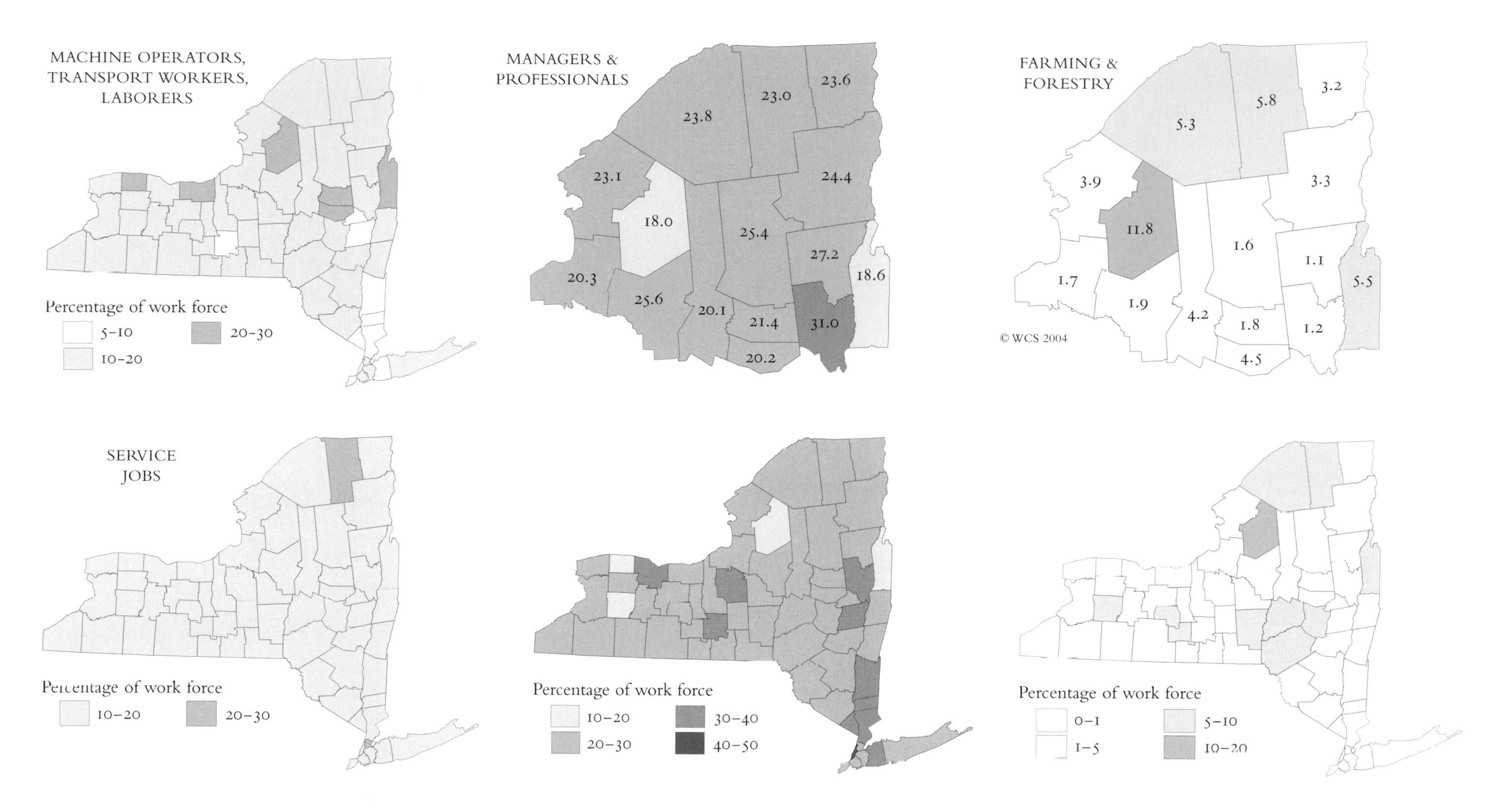

Factory work and ordinary labor are found throughout the state except the suburban counties but exceed 20% only in the old industrial towns along the rivers and canals. *Service Jobs* are very uniformly distributed; the only exceptional concentration is in the prison belt in Franklin County.

Managers and professionals, though most concentrated in the cities, are a significant and fairly uniform proportion of the work-force everywhere in the state. In rural St. Lawrence County, once one of the famous agricultural regions of the East, there are now four of them for every farmer.

Farming and forestry are absent from the suburban counties and uncommon, except for scattered concentrations, in the rest of the state. In northern New York there are still moderate numbers of farmers in Washington, Franklin, and St. Lawrence Counties and along the Black River in Lewis County. But the days when 90% of upstate New Yorkers farmed are long gone.

THESE graphs show the total wages paid by various northern New York industries from 1965 to 1995. The total wages, while less than the total local expenditures of the industry, are the best measure of what the region receives from an industry in return for the use of its land, workers, and resources.

The graphs document a period of remarkable changes. Overall, the resource-based industries have decreased, and trade, government, and services have increased. Farming, which was already in decline by 1965, has declined another 60% since then, creating a rust-belt of unpainted barns and derelict machinery. Many farms that were showplaces thirty years ago are now struggling or bankrupt. The basic cause has been persistently low commodity prices, which have forced farmers to sell at or below their production costs.

Mining and manufacturing have also declined, less abruptly than farming but with frightening steadiness. High operating costs, capital flight, and foreign competition are the main causes. Mining, once the largest industry in northern New York, now employs less that 0.5% of the labor force.

Transportation (which includes communications), construction, and finance have been steady or shown modest growth. If our future is really in housing developments, online brokerages, and cell phone towers, we may be pleased to note that it isn't coming very fast.

All of this says, in effect, that while oil remains cheap we can import goods and raw materials from anywhere in the world but will still need services and government close to us. If you quarry or chop or weld or sew or farm, you may have to go elsewhere—perhaps a long way elsewhere—to find work. But if you teach or cook or manage or advise or guard or cure, there is very likely a job for you nearby.

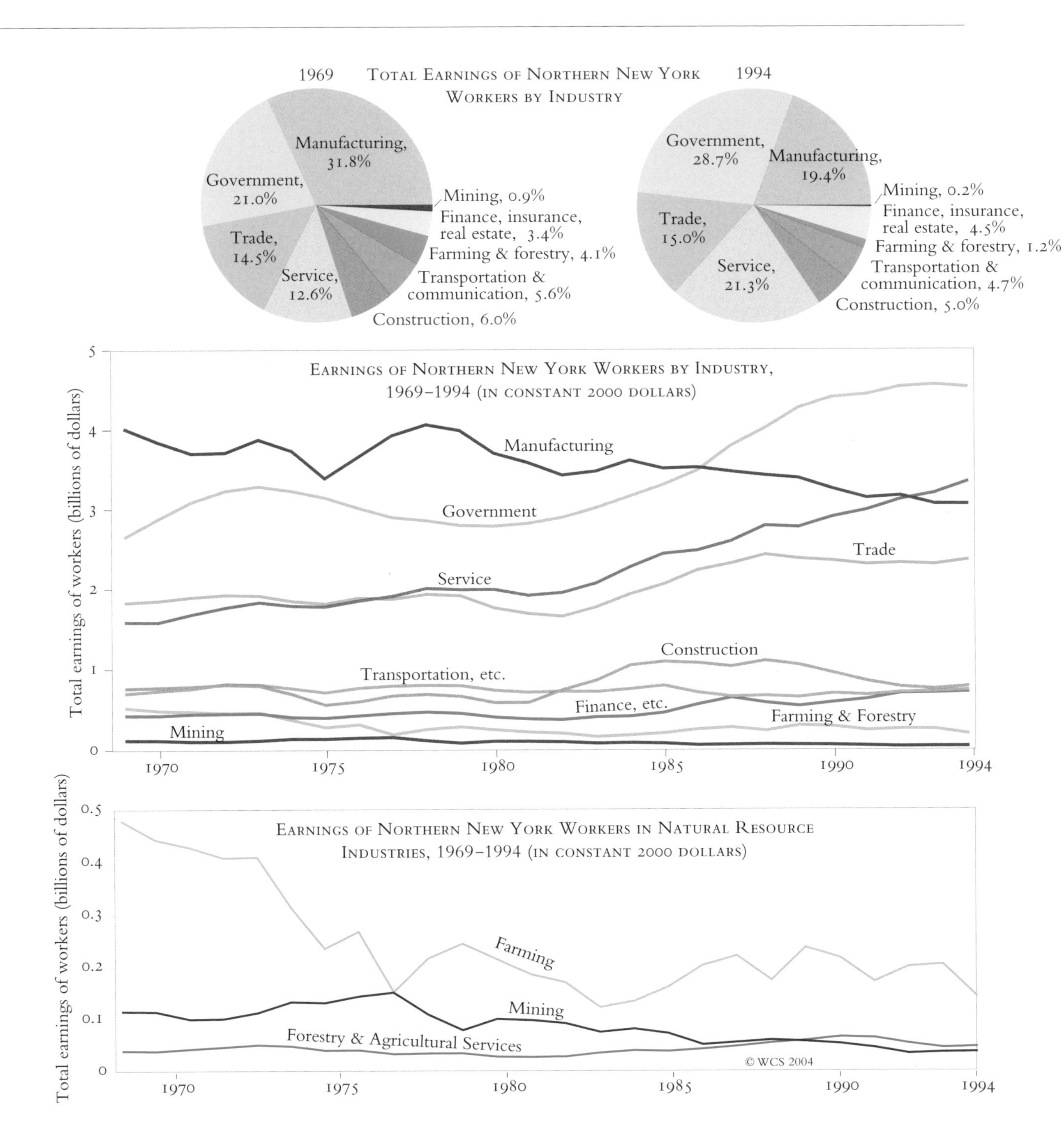

GOVERNMENT is our largest and most steadily growing industry. The several hundred northern New York governments currently employ about 120,000 people and contribute 4.3 billion dollars to the local economy each year. The number of government employees increased 20% between 1969 and 1994, and their total earnings, even after adjusting for inflation, nearly doubled.

Almost all of the growth in jobs came from increases in state and local government. Many new state facilities—including seventeen prisons—were built in this period. Almost no new federal facilities were built, and a major one closed.

Most of the increase in the total government payroll resulted from increases in employment rather than increases in wages. State wages, high in the 1970s, have actually fallen about 20% in real value since 1980. Federal wages, lower to start with but still very good by local standards, have risen 30–40% in the last 20 years and are now close to state wages.

There are ironies in all of this. While almost everyone wants the the federal and state governments to be small, almost no one seems to mind having big county and local governments, especially when thay are educating the kids and providing secure jobs with good benefits. And even if some government jobs are difficult or dangerous, they are probably less so, and often better paying as well, than many of the mine and forest jobs they have replaced.

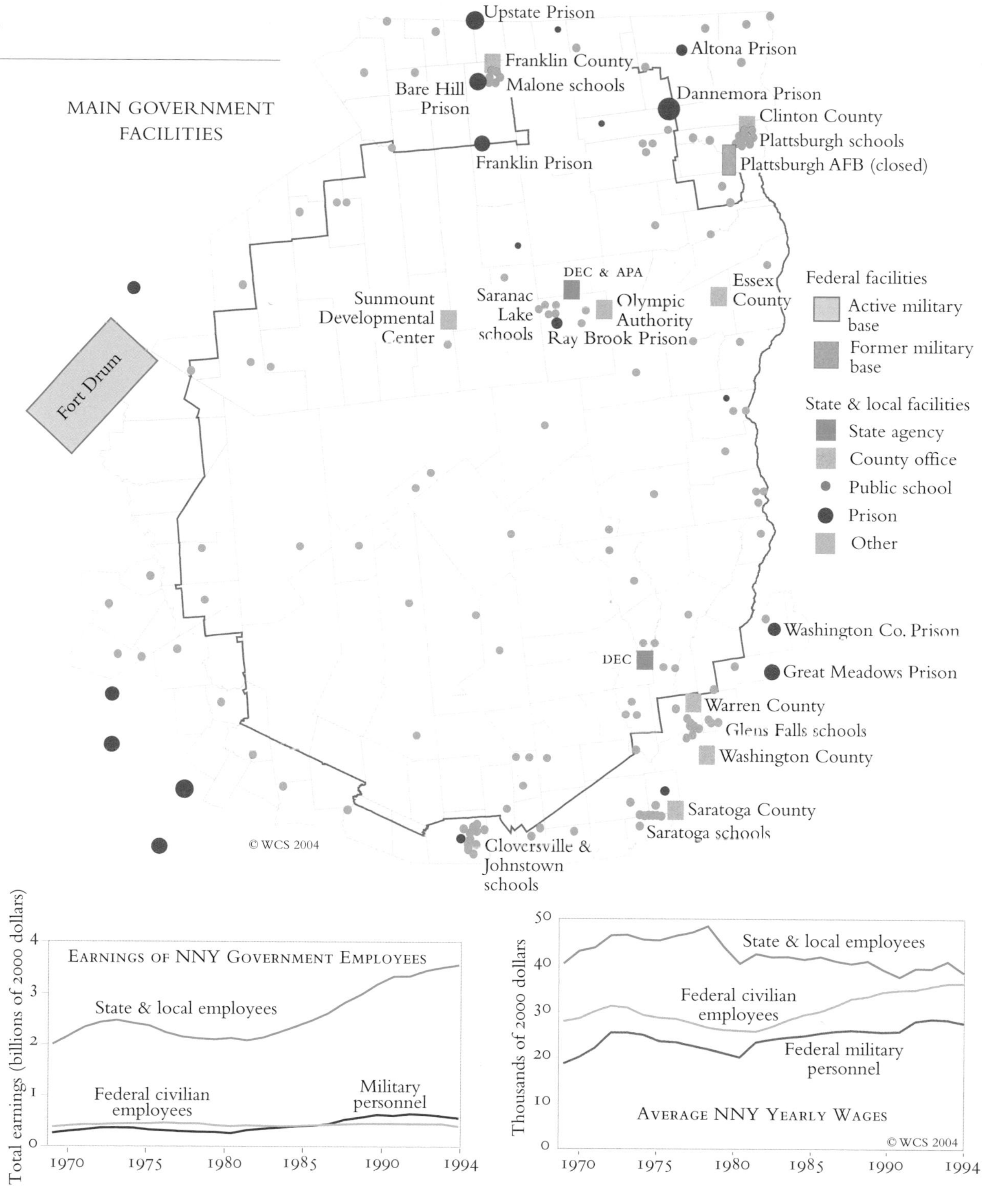

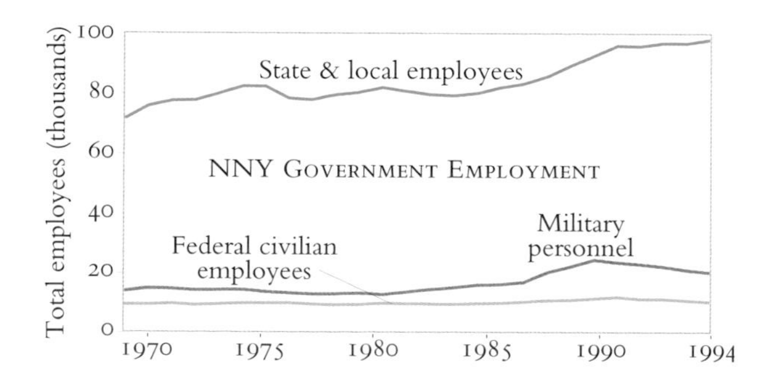

Per capita income, which includes both wage and nonwage income, is a good measure of both the average wages of workers and the overall wealth of a town. Regionally, it is highest in New York City and the wealthier suburbs, moderately high in the upstate cities and suburbs, and low in the rural counties and the poorest parts of the inner city. But in the Adirondacks the pattern reverses, and the wilderness center is, interestingly, better paid on average than the northern and western periphery. The most prosperous inner towns, with incomes in the \$17,000 to \$20,000 range, are comparable to many upstate suburbs. The poorest peripheral towns, with incomes under \$15,000, are some of the poorest places in the state.

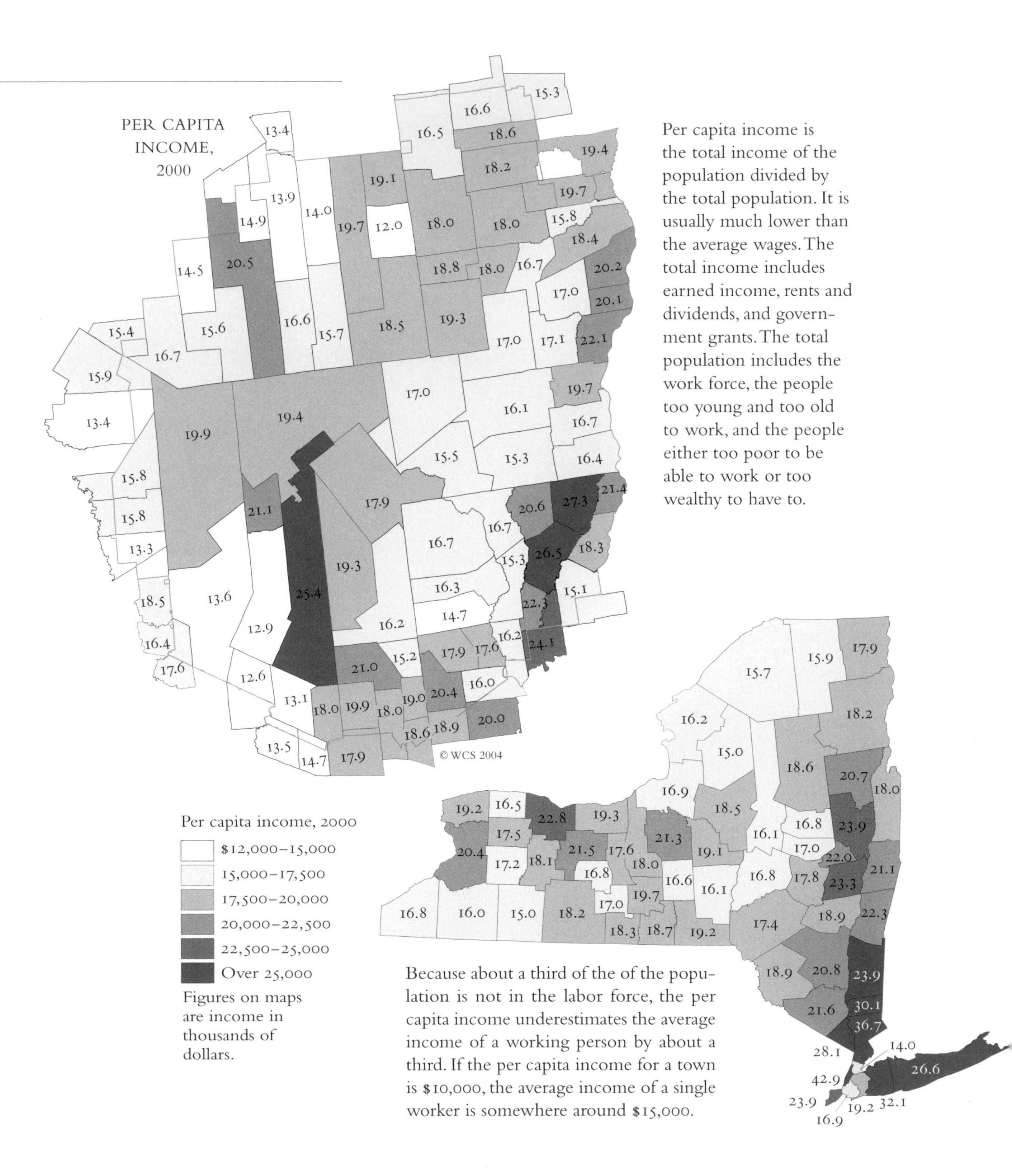

Figures on maps are income in thousands of dollars.

Per capita income is the total income of the population divided by the total population. It is usually much lower than the average wages. The total income includes earned income, rents and dividends, and government grants. The total population includes the work force, the people too young and too old to work, and the people either too poor to be able to work or too wealthy to have to.

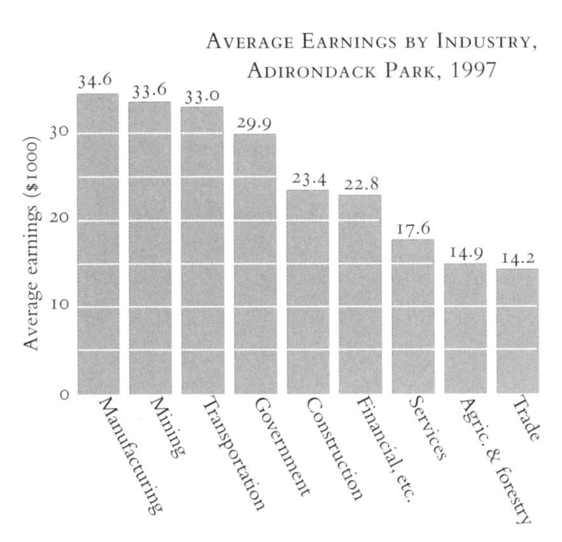

In the Adirondacks, as elsewhere, wages are high in manufacturing and transportation and low in services, agriculture, and trade. Mining and government wages are also quite high. The losses of mining jobs in the 1970s hit some communities very hard, and the growth in government jobs since then has been a boon to others.

Because about a third of the of the population is not in the labor force, the per capita income underestimates the average income of a working person by about a third. If the per capita income for a town is \$10,000, the average income of a single worker is somewhere around \$15,000.

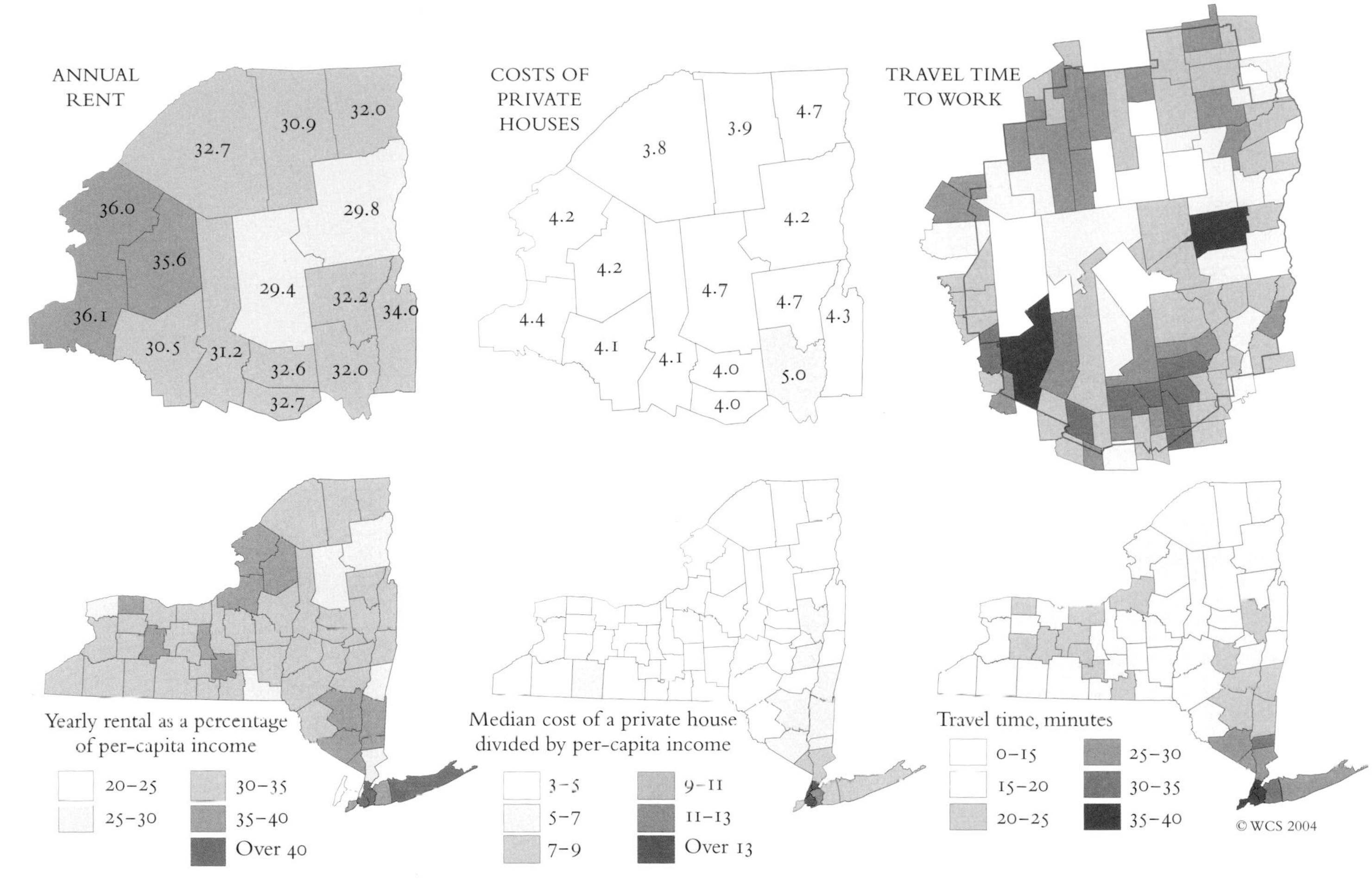

Median yearly rents tend to be about a third of yearly income in much of the state. Landlords operate in a market where demand is high and probably adjust prices fairly carefully to what people can pay. Rents are relatively high in some of the suburbs and in Lewis and Jefferson Counties where Fort Drum creates a large demand for housing. They are surprisingly low, at least relative to the high average income, in Westchester and Manhattan.

The *median private housing cost* divided by per capita income, is an approximate measure of how many years of income it takes to buy a house. It averages four to five years in much of rural New York, nearer five in the most developed counties and nearer four in the most rural ones. It rises sharply in the eastern suburbs, reaching ten years of income near New York City and twenty-three years of income in Manhattan. Very few people with average incomes own houses in Manhattan.

Average travel times to work vary greatly within the Adirondacks, tending to be low in the center towns and longer in the peripheral ones. They are even longer in the cities and suburbs. For many people, the price of suburban affluence is five hundred or a thousand hours a year of traffic, trains, and bad air. For some it is a fair price; for others, including many Adirondackers, it is almost inconceivable.

THERE is a definite economic penalty to being rural. Per capita incomes in the poorer towns of northern New York, like those elsewhere in rural New York, are only half to two-thirds of those in the cities and more affluent suburbs.

Just how much of a penalty this is depends on the cost of living. While a $15,000 per capita income is not going to buy many luxuries anywhere, it is quite possible that it goes as far, or farther, than $20,000 in an expensive suburb.

Unfortunately, because the standard cost-of-living indexes are all calculated for urban areas, there is no accurate way to compare the purchasing power of urban and rural incomes. But a few typical costs, shown in the maps on this page, suggest that the rural income penalties are not as bad as they look at first. Both housing costs, measured relative to income, and travel costs, which really represent unpaid working time, are less in rural areas than in the suburbs and cities. They are also somewhat less in the rural central and northern Adirondacks than along the more developed southern edge. If other living costs show similar differences, it raises the possibility that the state income map on the previous page badly misrepresents how well people actually live.

THE census estimates that in 2000 the ninety-two Adirondack towns contained 87,900 households with a total income of 3.3 billion dollars. If evenly divided, which incomes never are, each household would get $37,500. At least 14,000 of these households, 16% of the total, were humble or poor, with household incomes of $15,000 or less. About 870 households, 1% of the total, were affluent or rich, with household incomes of $200,000 or more. This top 1% had an aggregate income of $173 million, about 5% of the total income of the Adirondack towns and slightly greater than the total income of the 14,000 poorest households.

Poverty, with all the suffering, cruelty, and waste of human potential that it implies, is almost universal in rural America. In 2000, the highest poverty rates in New York were in Brooklyn and the Bronx where 25–30% of the population was below the poverty level. The next highest rates were in the northern Adirondacks and a few other poor rural counties. Fifty-eight of the ninety-two Adirondack towns had poverty rates of over 10%; twenty-one towns had poverty rates of over 15%.

Besides high poverty rates, the Adirondacks also have some of the highest unemployment rates in the state. In 2000, when the United States had an unemployment rate of 4.1%, twenty-two Adirondack towns had unemployment rates of over 10%, and six had rates of over 15%. Some of this unemployment may be part of an annual pattern in which people expect to work in certain seasons and not in others. But expecting not to work is not necessarily the same thing as choosing not to work. Both the strong similarities between the poverty map and the unemployment map and the stark fact that every Adirondack town of any size has an organization that distributes food to families who are unable to provide it for themselves suggests that for too many Adirondackers there is not enough work to go around.

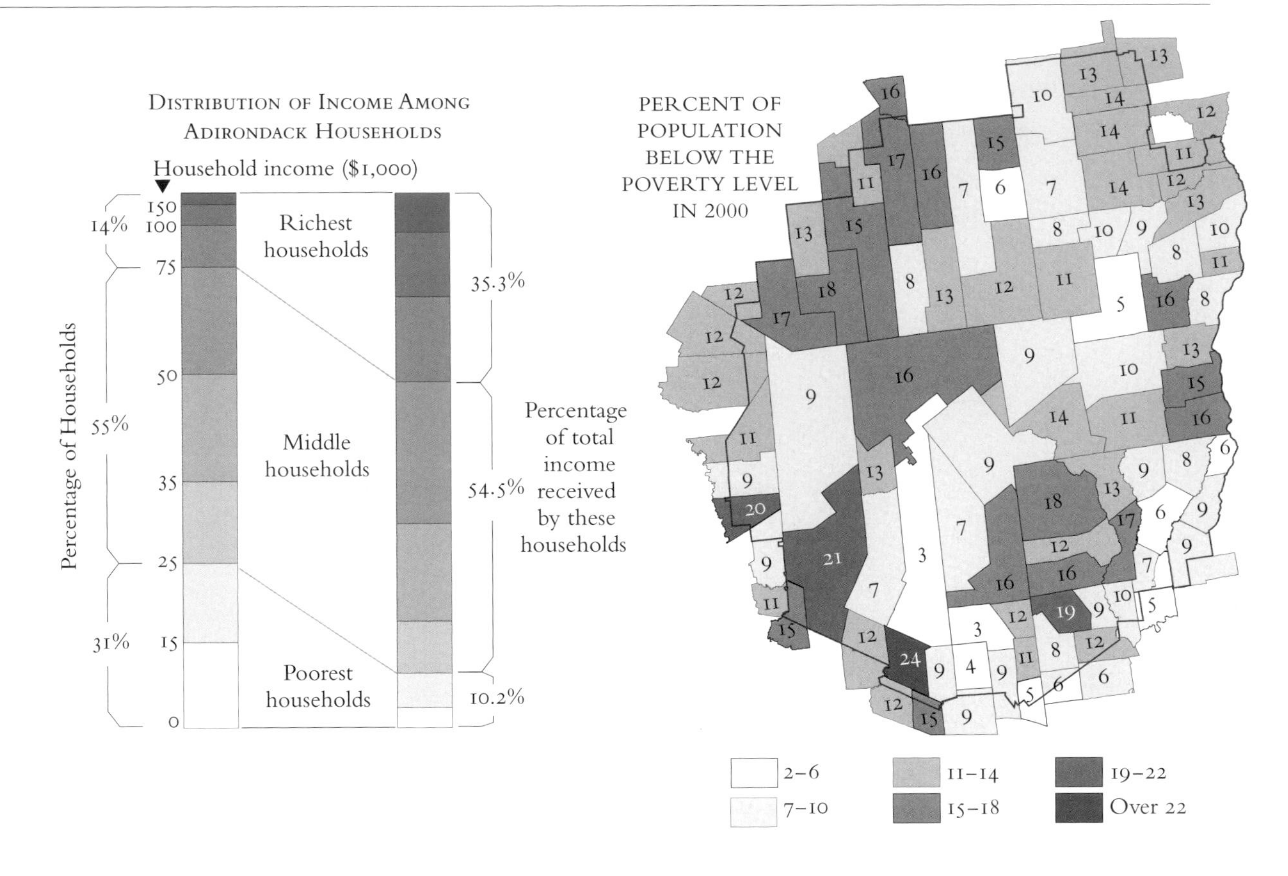

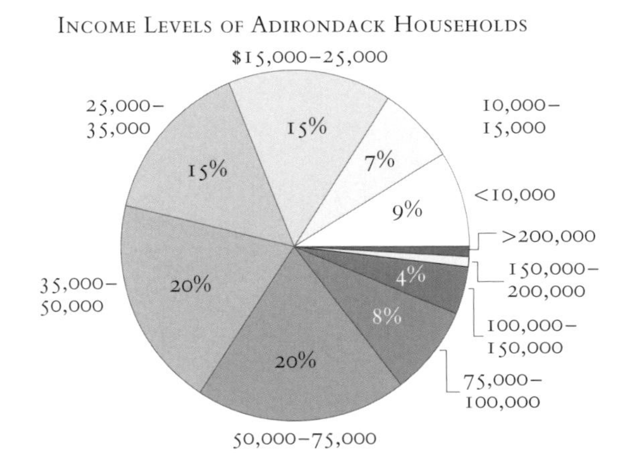

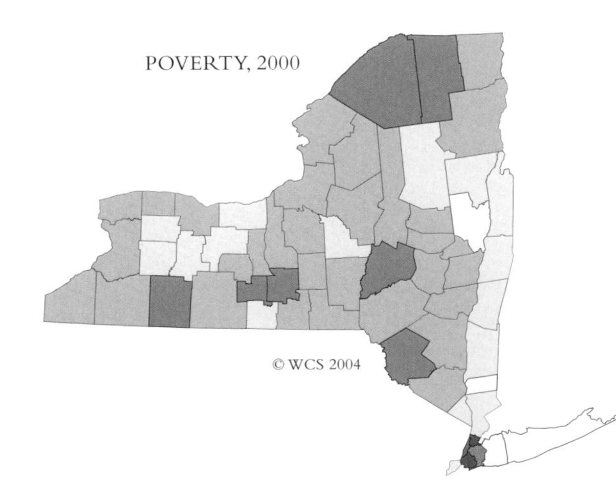

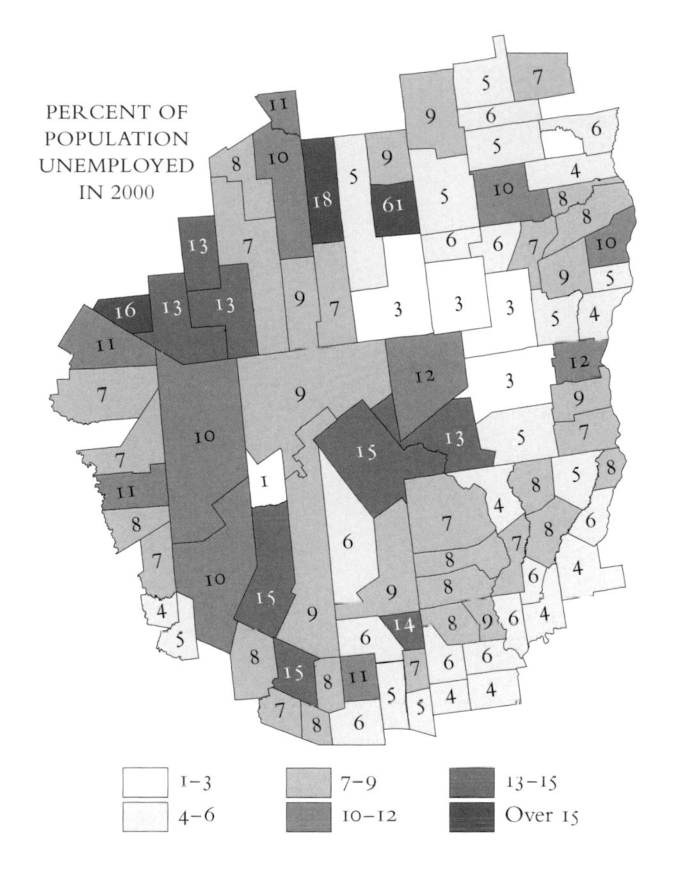

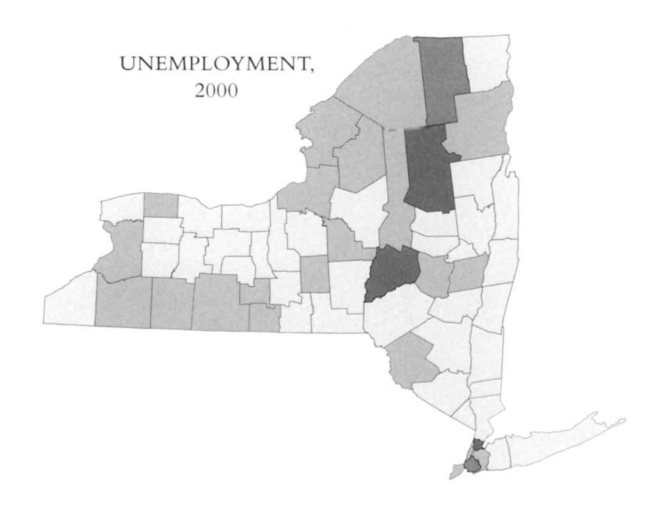

Social services are provided by both the county governments and private organizations. The counties operate two main county offices and twenty-one satellite offices within the park, which provide a variety of economic assistance, custodial, and protective programs. Private nonprofits and charities run fifty-one food pantries within the park and provide supplementary food to over 2,000 families regularly. The nineteen day-care centers are mostly run by private organizations but receive some public funding.

County social programs and food pantries are widely distributed in the park, though more scattered in the west than the east. Day care is limited to the larger eastern towns; there are only two programs in the whole west half of the park.

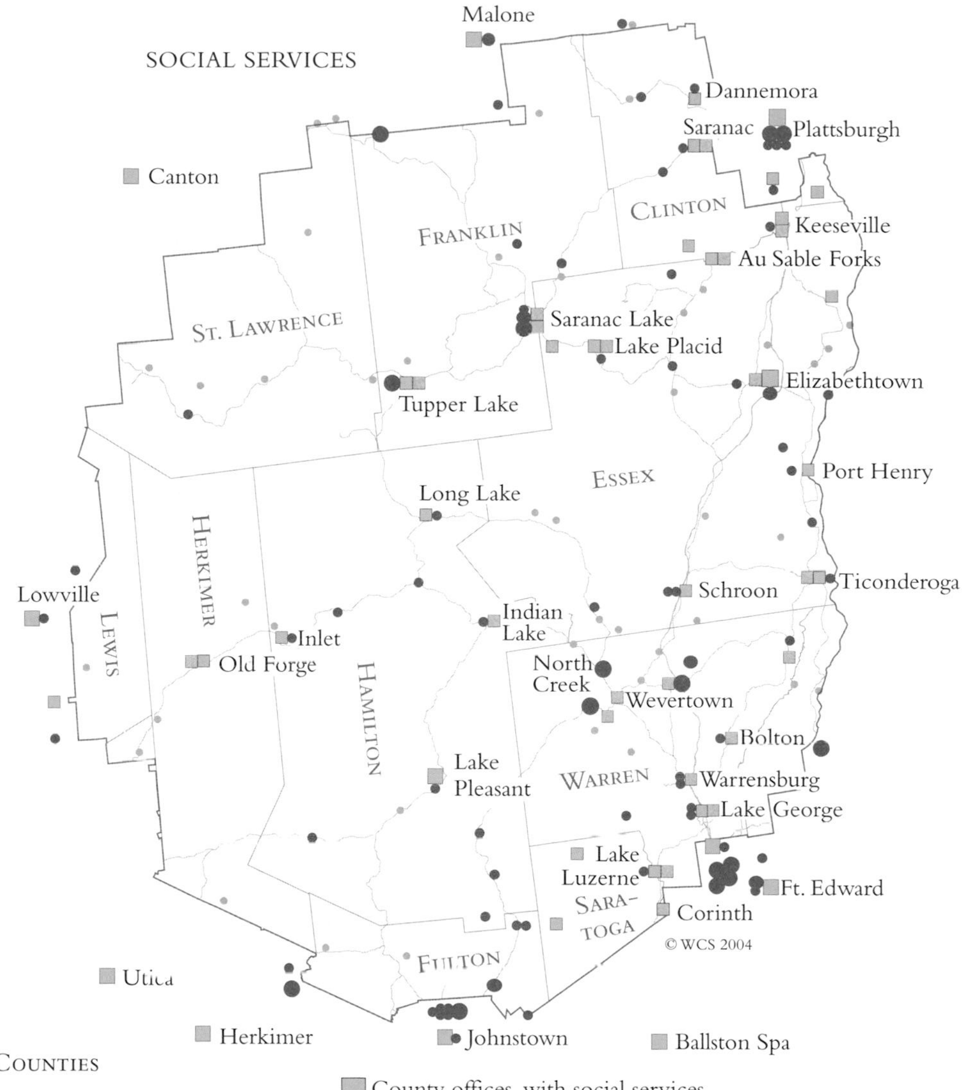

Social Services Provided by Counties

- Adoption
- Adult protective services
- Child preventive services
- Child protective services
- Child support
- Day care
- Employment
- Food stamps
- Foster care
- Home energy assistance
- Medicaid
- Public assistance
- Temporary assistance

MORTALITY RATES

Heart disease, cancer, and stroke are the three commonest cause of death in the U.S.. The upper maps show the total death rates per 100,000 persons in 1997. Each color represents a fifth of the counties; the white counties have the lowest death rates and the dark blue the highest. The lower maps show the spatially smoothed death rates for white men and women seventy years old in some eight hundred *health service areas.* The darkest blue represents the fifth of the service areas with the highest mortalities. The numbers on the upper maps are the total death rate; those below the lower maps are the range in mortality rates shown on the map. Note that while the darkest blue always represents the highest mortality rates on a particular map, the actual mortality rates differ greatly from map to map.

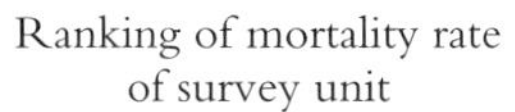

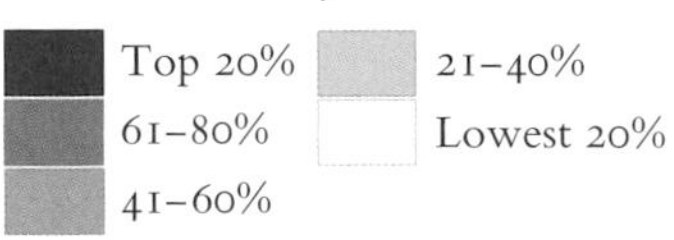

9 Death, Injury, Disease, & Crime

THIS is a tabloid chapter, about the things that hurt and kill us. Some, like falls and burglary, are human universals, and their geography doesn't tell us much except that we are mortal and fallible. Others, like suicide, robbery, and many fatal diseases are more selective and differ greatly in whom they strike and where they are most frequent. The ills in the second group make interesting maps and suggest important questions about why some people and places suffer more than others and what could be done to change this.

The commonest causes of death in northern New York, as in the U.S. as a whole, are heart disease and cancer. Strokes, lung disease, and pneumonia are also significant. Deaths not related to disease—suicide, accidents, and violence—are rarer but still consequential, totaling just less than 5% or about one death in twenty-two overall.

CAUSES OF DEATH, NORTHERN NEW YORK, 1997

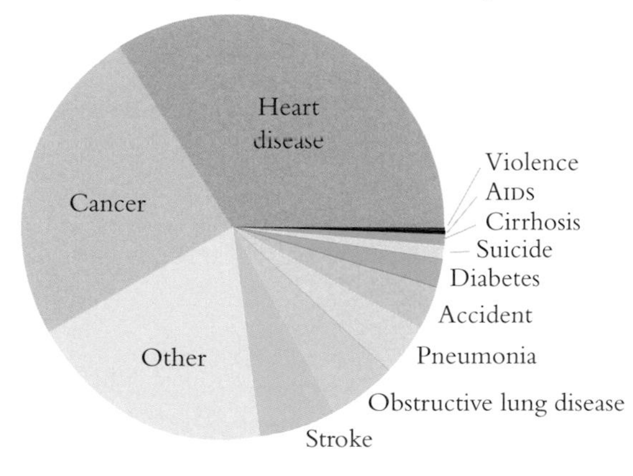

The maps on the opposite page show the relative rankings of the mortality from the three commonest causes of death. The mortality rates within New York are fairly uniform. Fourteen counties, for example, have heart-disease rates within 20% of the state average. In the United States as a whole they are much less uniform. New York and New England have relatively high rates of heart attacks and cancer. The West Coast has low rates of heart disease but higher rates of cancer and stroke. The mountain West and upper Midwest have low rates of everything, and the Ohio and Mississippi Valleys have high rates of everything. In North Dakota almost no one dies any more, and the morticians have moved away or are running pet cemeteries. Emptiness, wheat, and wind may be healthier than we think.

Injuries are common everywhere. Children and the elderly are the commonest victims, and males of all ages are injured more often than females. Falls are the commonest accident, city and country, and will likely remain so unless we cease to grow old or the government bans gravity. Assaults are the second commonest cause of injuries in New York City. Automobiles accidents are the second commonest elsewhere in the state. Deliberate self-inflicted injuries, somewhat surprisingly, are third, beating poisoning, bicycles, overexertion, and cutting tools easily; about one injury in fifteen, city or country, is self-inflicted. Firearm injuries are very rare in the country, where firearms are common, and three times more common in the cities, where firearms are rare.

HOSPITALIZABLE INJURIES, NORTHERN NEW YORK, 1990–1992

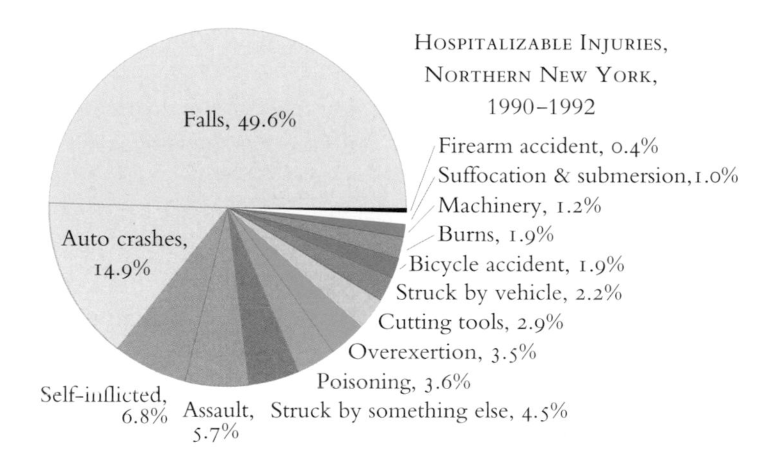

HOSPITALIZABLE INJURIES, NEW YORK CITY, 1990–1992

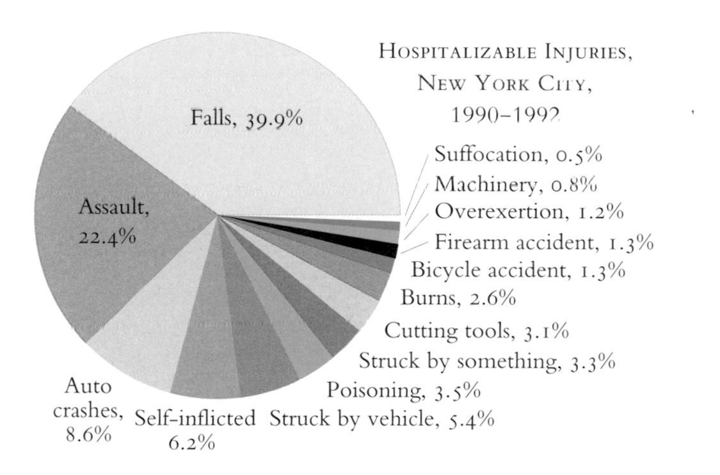

TOTAL ACCIDENT RATE

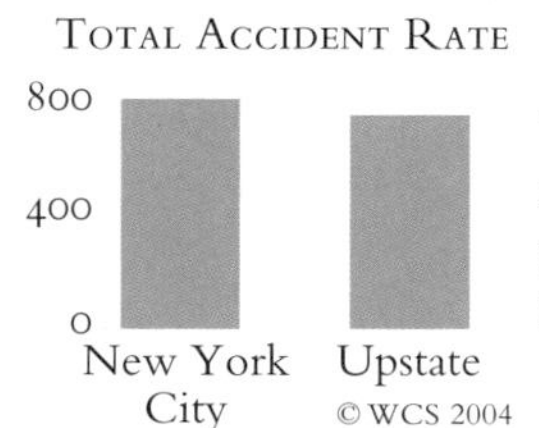

The accident rate is the number of hospitalizations per hundred thousand people per year.

CANCER is the second most common cause of death in New York, killing two New Yorkers out of every thousand per year. In some age groups, for example, women in their forties, it is the commonest cause of death. It is a complex and relatively modern disease related to environment, ancestry, and lifestyle. Its apparent incidence has increased over the last forty years, in part because of improvements in detection, in part because of decreases in other diseases, and possibly in part because of increased amounts of carcinogens in the environment (*Map 17-2*).

Cancer geography is of considerable interest because cancer rates differ greatly from place to place. There is a substantial—roughly one in four—chance of dying of cancer anywhere in New York state. But in some places the chance is almost twice this and in other places less than half.

The state maps, which show incidence rates, use data from 1993 through 1997. The U.S. maps, which show mortality rates, use data from 1988 through 1992. Both are age-adjusted and sex-adjusted, meaning that they compare the observed cancer rates for each survey area to an expected rate that is based on the number of people in each age-sex group and the overall cancer rate for that group. This is necessary to prevent us mistaking, for example, an area with 10% more women but an average rate of breast cancer for an area with a 10% higher rate of breast cancer.

Much of the variation shown on the maps is probably accidental: if a small study area only expects a few cases of lung cancer a year, a few cases more or less in a five-year period will make a large change in the apparent rate. But some of the concentrations, indicated by the circles on the state maps, are statistically significant and unlikely to have resulted from chance.

Even with only five years of data there are significant regional differences in the incidence rates for particular cancers. The high concentrations of lung and colorectal cancers in northern New York are particularly striking and particularly important to us.

The reasons for these concentrations are unknown; they could be something in the environment—acid rain, radon from bedrock, dusty workplaces, toxics from industry. Or they could be something in people's lives—high rates of smoking, the wrong grandparents, too much indoor air, or too few green vegetables in the diet.

Considering how important cancer is and how specific its geography seems to be, our inability to say more about its environmental causes is frustrating. Part of the difficulty is that unraveling the causes of any disease is hard, and especially so for a disease with multiple risk factors in which the onset of the disease may occur long after exposure to the risk. Another difficulty is that, although there has been much research on the toxicity of specific carcinogens, there has been much less research on the general geography of cancer. Thus, for example, surprisingly little is known about the overall health effects of carcinogens like PCBs and strontium 90 that were once widely produced by industry and government.

The disquieting result is that we know very little more about cancer geography than we did thirty years ago. All we can really say is that for some reason, perhaps connected with their environments and perhaps connected with their lives, northern New Yorkers are at significantly higher risk of contracting two of the most common fatal cancers than people elsewhere in the state. What the reason is, and what we should do about it, are still unknown.

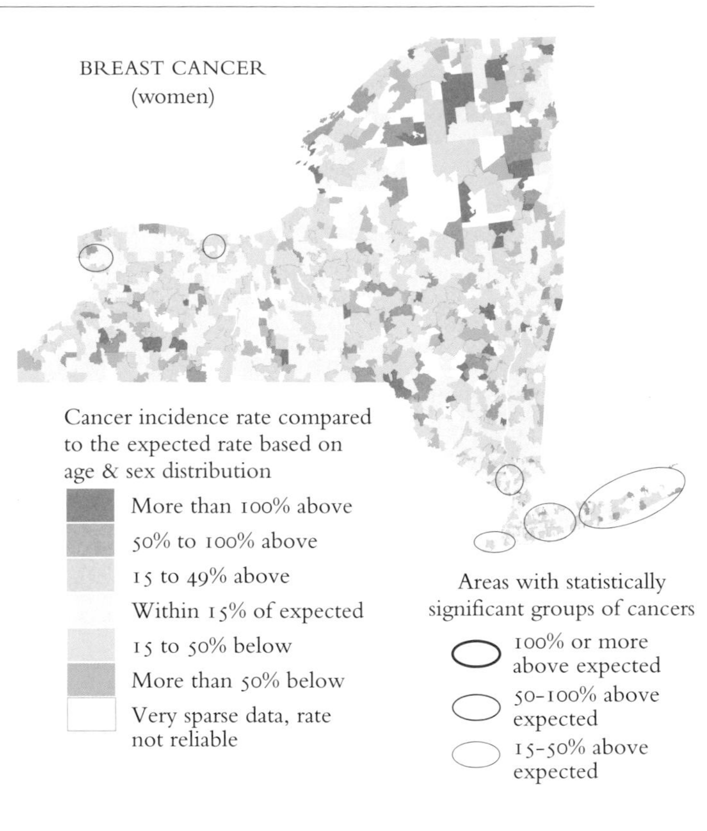

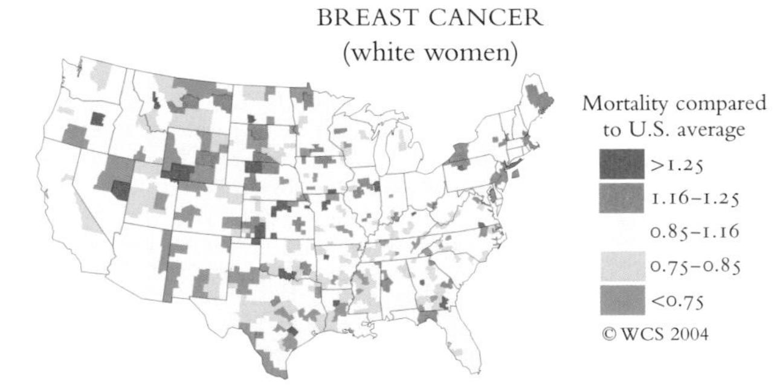

Breast cancer, which killed 3,300 women in New York in 1997, is very uniformly distributed in both the U.S. and New York. Significant concentrations occur only on Long Island and at a few other scattered sites. Apparent concentrations, of unknown significance, occur in the Adirondacks.

Lung cancer, the commonest form of cancer in New York, is especially common in rural areas upstate. The mortality rates in much of the Adirondacks for both men and women are one-and-a-half to two times the state average.

Colorectal cancers, which killed 4,400 New Yorkers in 1997, are also significantly clustered in the Adirondacks, sometimes but not always in the same zip codes where lung cancer is high. The New York pattern seems an extension of a national one; this is one of the few cancers that is commoner in the northern U.S. than the south.

AIDS is both our newest pandemic disease and one of the most fearsome that humans have yet encountered. It appears to be derived from an animal disease caused by a simian immunodeficiency virus that has mutated and spread to people. Like rabies (*Map 4-14*) it originated somewhere in the interface where human populations and wild animal populations overlap. Given the expanding human and animal geographies of the last fifty years, it is unlikely to be the last important disease to come from there.

AIDS is caused by the human immunodeficiency virus (HIV) and usually develops five to ten years after infection with the virus. Current therapies attempt to suppress or eliminate the virus and prevent the disease from developing. The therapies, involving a mixture of drugs, are quite effective for those for whom they are available.

Since 1981, when the first cases were recognized, there have been 125,000 known AIDS cases in New York or about seven cases for every thousand persons. Eighty-one percent of the cases were in New York City. Originally the fatality rate was very high: ninety percent of the people who contracted AIDS in the 1980s are now dead. Recent advances in treatment have brought the fatality rate down dramatically; over sixty percent of the people who contracted AIDS between 1993 and 1997 are currently alive. AIDS is still a very dangerous disease but no longer a death sentence.

While still concentrated in the cities and commonest in men, AIDS is now found in both men and women throughout New York. Through 2000, the last year for which data are available, there had been 926 cases among northern New York residents, plus another 2,379 cases among prisoners, making it the commonest potentially fatal infectious disease in northern New York.

Fortunately, in New York the worst of the AIDS epidemic seems to have passed. In 2000 there were 6,094 new cases in New York, less than half the number of new cases reported at the peak of the epidemic in 1994.

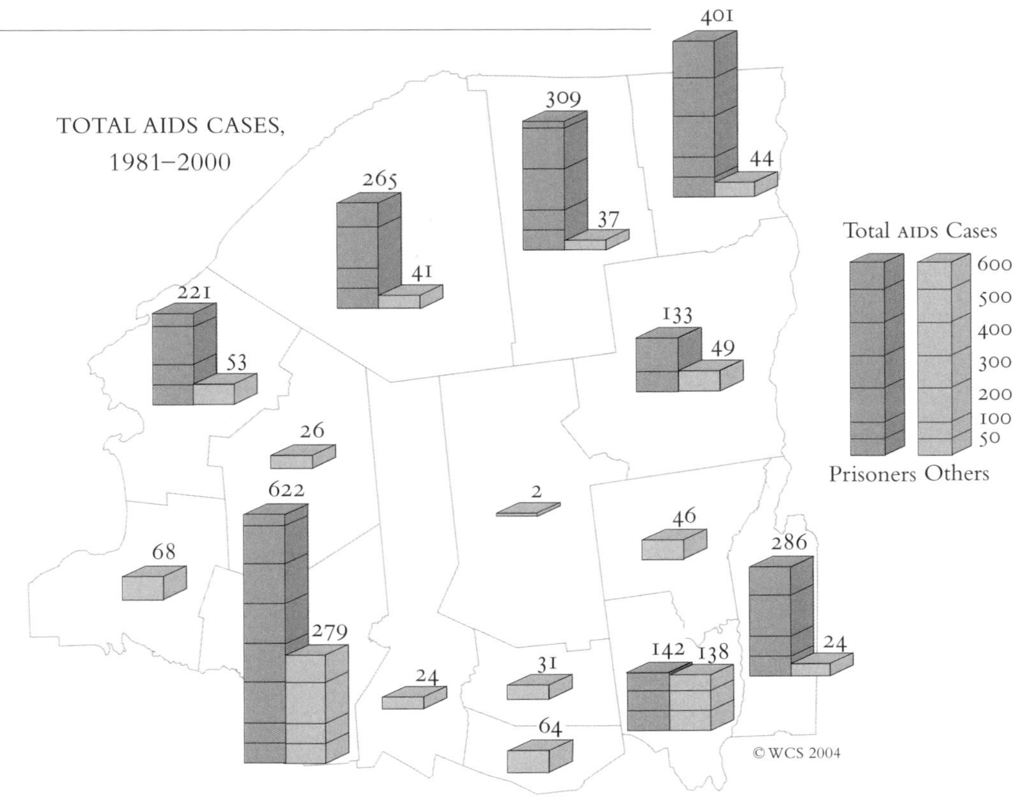

Infection Rates. HIV infection rates, the best measure of the future course of the AIDS epidemic, are only known from a few groups for whom testing is mandatory. Currently in upstate New York, 0.08% of all pregnant women, 0.1% of all newborn babies, 0.04% of all military recruits, and 0.6% of all prison inmates are HIV-positive. The rates in New York City are substantially higher.

While AIDS is currently our most threatening disease, our infection rates, likely somewhere around 0.1–0.2% of the total population, are very low compared to countries where the infection is severe. In some African counties 30% or more of the adult population and 50% or more of the newborns are HIV-positive.

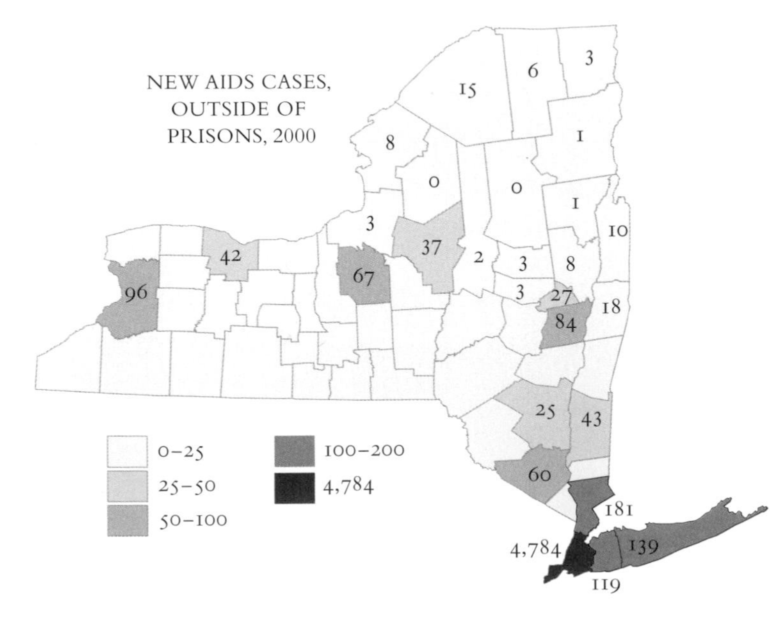

LYME DISEASE, named for Lyme, Connecticut, where it was first recognized is, like AIDS, a disease of wild animals that has only recently migrated to humans. Thirty years ago it was unknown; ten years ago it was confined to Long Island and the lower Hudson Valley; now it is found throughout the U.S. and has reached southern Canada and Alaska. There were 17,730 reported cases in the U.S. in 2000, nearly twice the number in 1990.

Lyme is an outdoor disease that depends on animal vectors. The disease organism, a spirochete, accumulates in populations of deer mice and is then transmitted by a tick, which in turn is dispersed by deer.

Lyme disease began invading northern New York and the Adirondacks about five years ago. It is now present, though still uncommon, in most northern New York counties. A warming climate, which has allowed ticks to spread north, increasing deer populations, and the widespread movement of livestock and pets have contributed to its spread. Since neither the climate, the deer, nor the ticks show any signs of stopping, it may continue to increase for some time.

Giardia is a thoroughly unpleasant but rarely serious intestinal disease, typically producing nausea, loss of appetite, cramps, and diarrhea. It is caused by a protozoan that infects many mammals and is often contracted from contaminated water or food. While it is called beaver fever and warned against by every book about safe camping, it is in fact far more common in urban areas and institutional settings than in rural and wilderness ones. You can certainly get it the Adirondacks from moose or beaver or other campers. But you can also get it from a restaurant, a summer camp, an army base, or your own water supply. And, as the maps suggest, it is far more likely that you will.

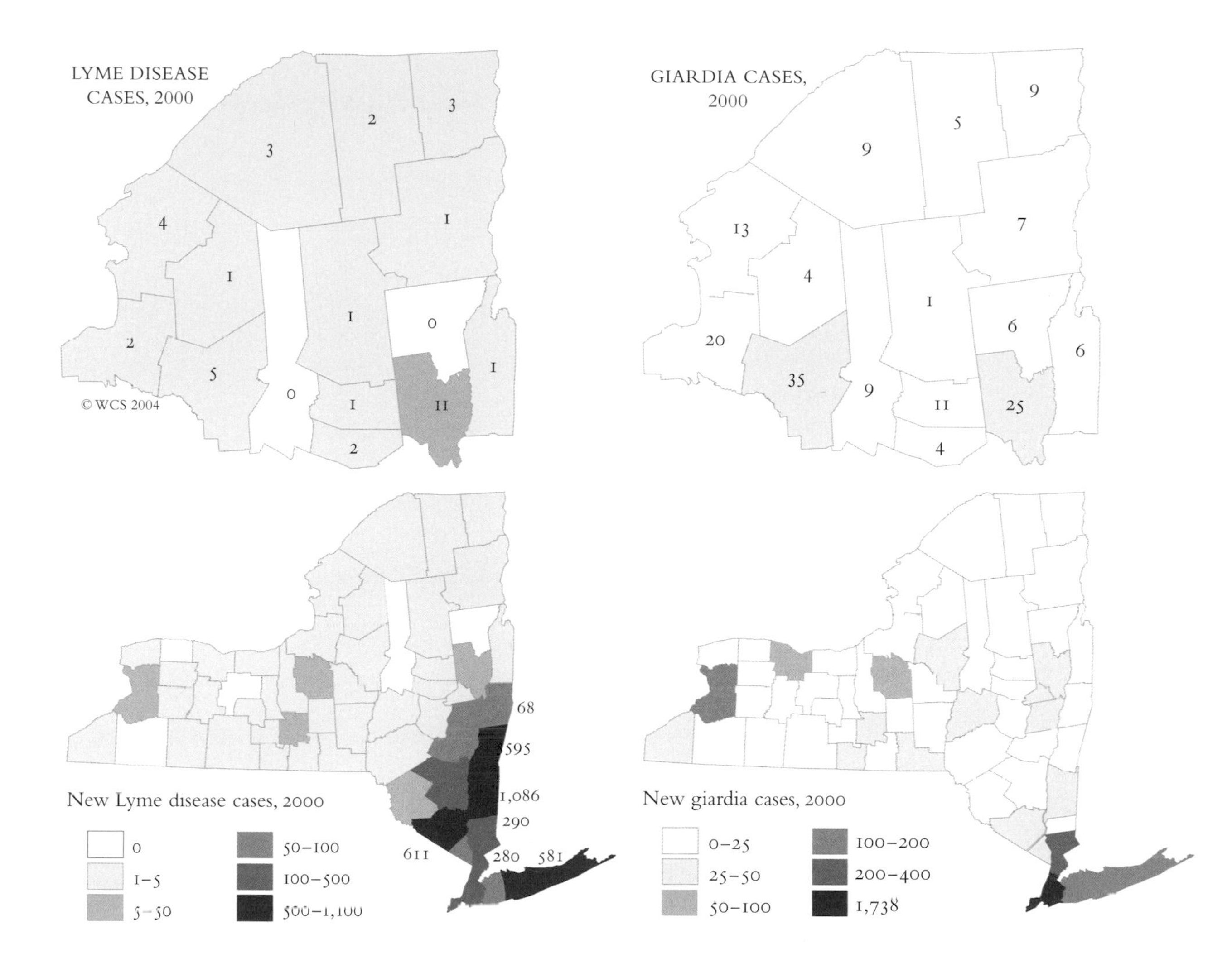

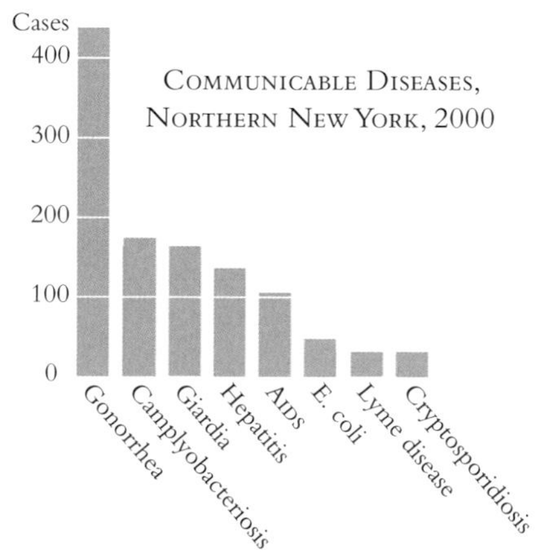

In 2000, *Lyme disease* (above left) was still concentrated in the lower Hudson Valley but was becoming more frequent to the west and north. Since then it has increased substantially in the lower and western Adirondacks. *Giardia* (above right) was commonest in the most populated areas and actually rare or uncommon in the interior Adirondacks. There were, for example only 164 cases of giardia in all northern New York, but 1,738 in New York City.

BECAUSE children are in many ways defenseless, their health, much more than that of adults, depends on the protection they receive and the environments provided for them. When we provide good care and a good environment, they prosper; when we don't they sicken, suffer injuries, and die.

Childhood diseases and deaths were frighteningly common a century ago and remain significant today. Out of the 5 million children nineteen years old or less in New York, every year about 4,000 die, 13,000 are suspected victims of abuse, 45,000 are hospitalized, and 200,000 are suspected to be victims of maltreatment or neglect.

The geography of childhood health, shown in the aggregate map on the opposite page, is to a great extent the geography of affluence and poverty. Families in the poor counties have more uninsured and unmarried mothers and are less likely to have prenatal care in the first trimester. In consequence, they have more miscarriages, more low-birthweight babies, and more infant deaths. When in addition, as in the New York City area, their environments are dangerous or degraded, the rates of childhood disease and violent crime increase substantially.

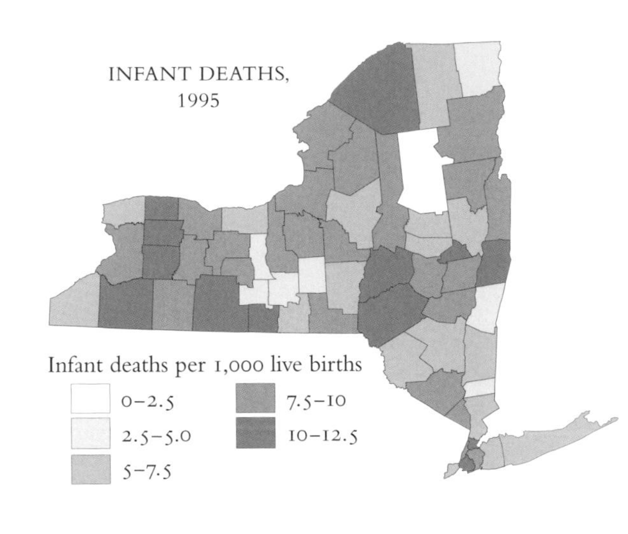

The New York death rate for infants from birth to one year averages 7.9 per thousand, quite comparable to the overall death rate of 9 per thousand. The infant death rate is high in the poor cities and poorer rural areas and low in the prosperous suburbs. Poverty, an affliction for adults, can be fatal to children.

Deaths rates for children between one and fifteen are much lower, about 2-3 per thousand, but still consequential. About two-thirds of these deaths are the result of congenital abnormalities and disease, the remainder the result of accidents and violence. The death rate triples in late adolescence, almost entirely because of an increase in accidents and violence. The adolescent death rate is highest in New York City where the homicide rates are high. It is also surprisingly high in upstate New York where, although the homicide rate is much lower, the motor vehicle accident rate is much higher than in the city.

The true extent of child abuse and maltreatment is unknown. A compilation of reports by the Office of Temporary and Disability Assistance suggests an overall rate of 46 per thousand (one child in twenty) and an upstate geography, with concentrations in small upstate cities and some of the poorest rural counties. This may, however, be an artifact of poor reporting in the cities and poor suburbs. If the abused children rarely see a doctor or a social worker, or if, sadly, the maltreated children in a poor school don't look too different from the rest of the children, then they will not appear on the map.

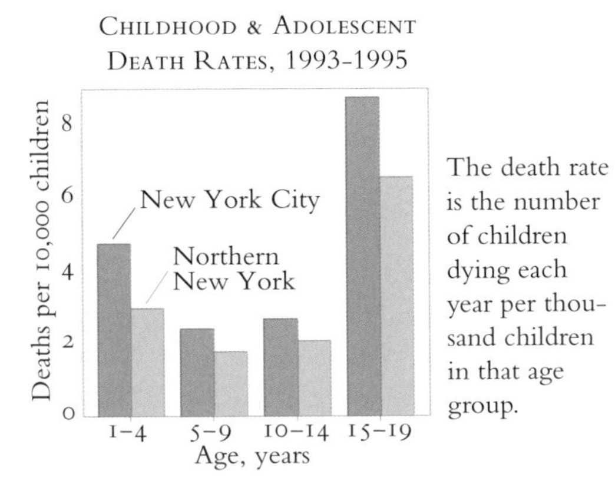

The death rate is the number of children dying each year per thousand children in that age group.

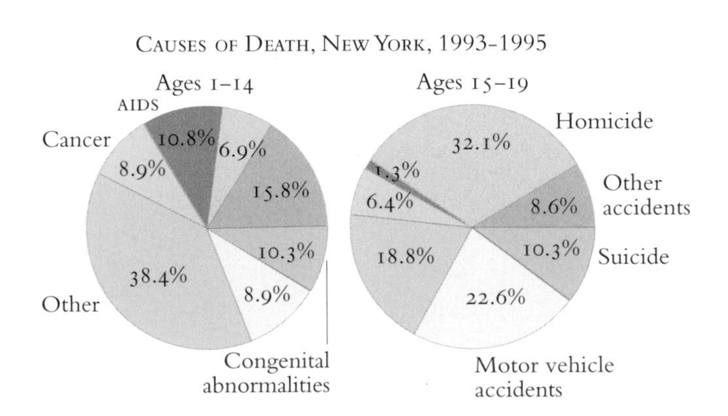

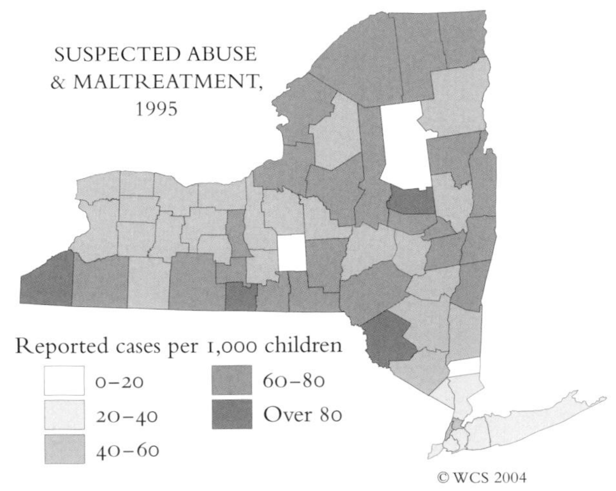

CHILDHOOD RISK FACTORS & OUTCOMES

THE graphs show a *risk-and-response spectrum*, comparing seven risk factors and seven measures of infant and childhood health for selected counties. Brown bars indicate that the frequency of risk factors or adverse outcomes is greater than the upstate New York average; green bars indicate the reverse.

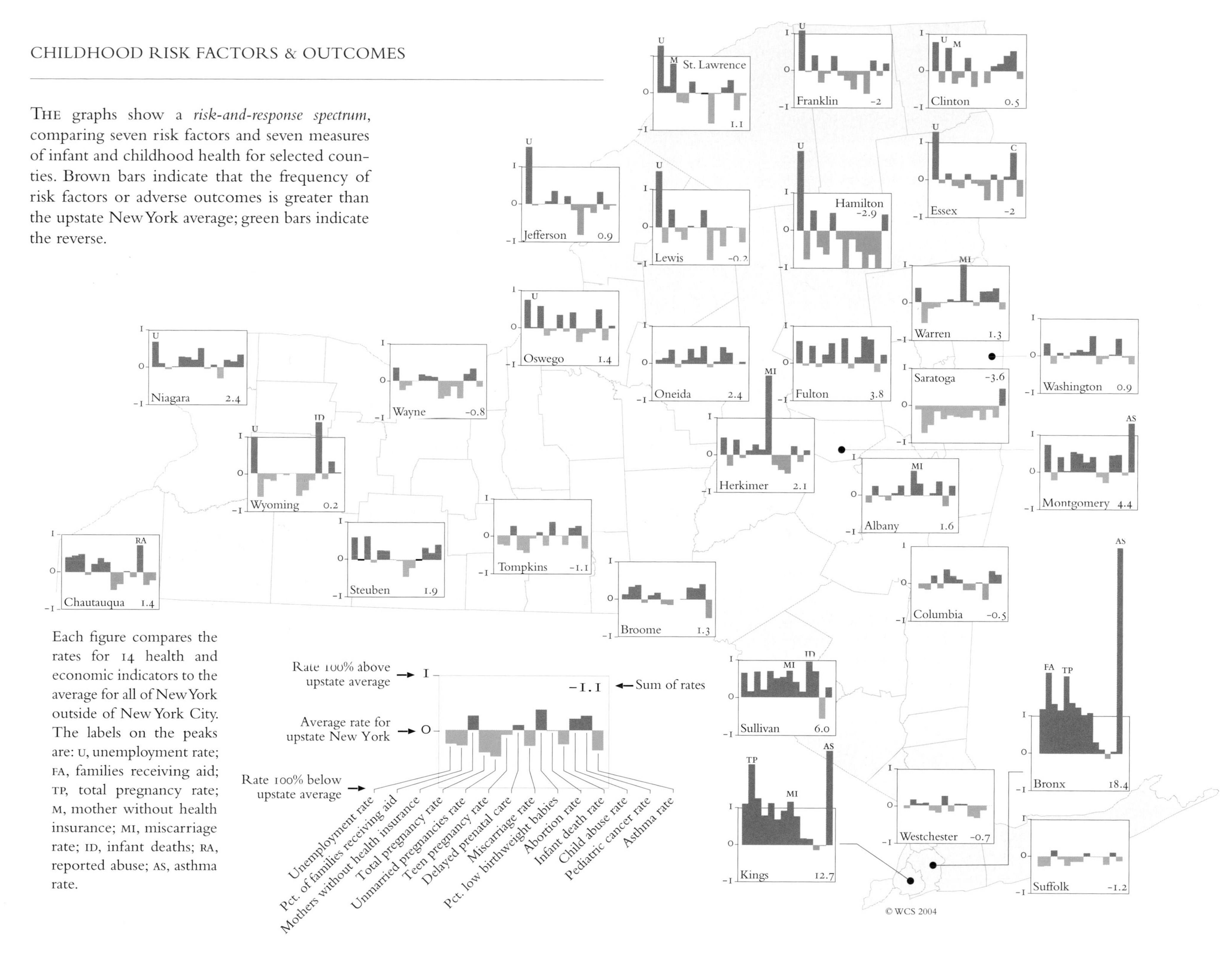

Each figure compares the rates for 14 health and economic indicators to the average for all of New York outside of New York City. The labels on the peaks are: U, unemployment rate; FA, families receiving aid; TP, total pregnancy rate; M, mother without health insurance; MI, miscarriage rate; ID, infant deaths; RA, reported abuse; AS, asthma rate.

THE next four pages, dealing with crime, are based on statistics from the New York Department of Criminal Justice and the U.S. Census. Please be aware that crime statistics, whatever their source, are inherently unreliable, both because the reporting process is complex and fallible and because many crimes are never reported.

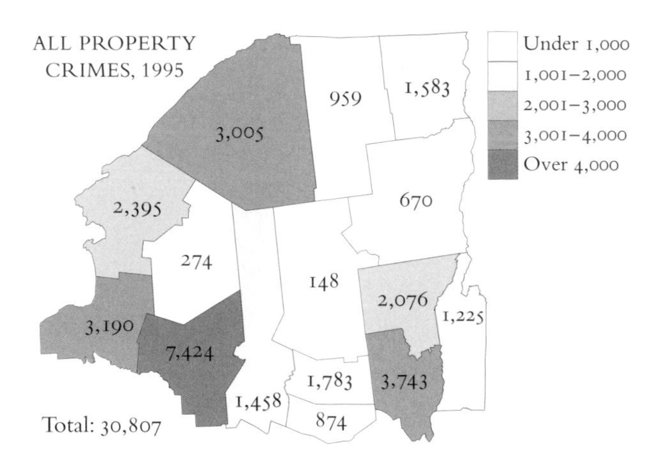

Property crimes, including burglary, auto theft, and larceny theft, are crimes in which property is taken without its owner noticing. By definition, no violence against persons is involved, though windowpanes, dead bolts, and safe hinges may suffer in the process.

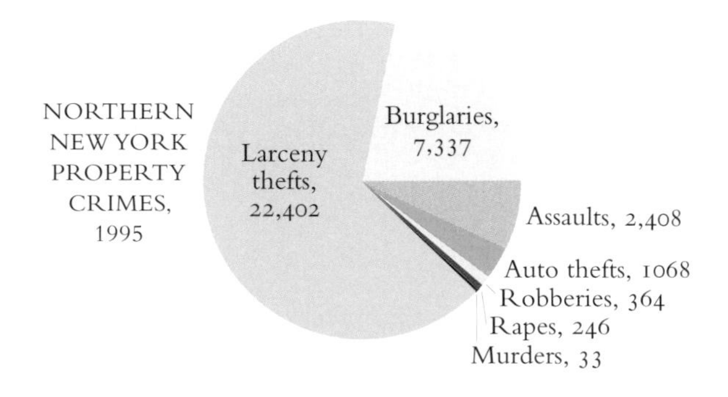

Property crimes are by far the most common type of crime, making up 90% of the reported crime in northern New York. They are committed by people of all ages and degrees of criminality: old and young, amateurs and professionals, one-timers and habitués. The causes are as varied as their participants. Poverty and inequality certainly figure in (my rich neighbors, I am virtually sure, do not steal from my poor ones), but so do adventure, misanthropy, temptation, addiction, and probably good old American lawlessness.

In 1999 there were 188,704 reported property crimes in upstate New York, down a healthy 34% from the 285,718 of 1980. Burglary has fallen steadily over this whole period and is now at half its 1980 level. Larceny, which now dominates the property crime figures, fell in the early eighties, then rose abruptly and now has fallen equally abruptly. Auto theft, the rarest of the property crimes, has been fairly constant.

Burglary is both one of the commonest and the most equably distributed of all crimes. Statewide, it is high in the cities, low in the suburbs, and moderately high again in some rural areas, especially those with many second homes. It is the only category of crime in which Fulton, an Adirondack county, has a rate similar to that of the Buffalo area and in which Sullivan, a rural county with a population of 70,000, has the highest rate in the state.

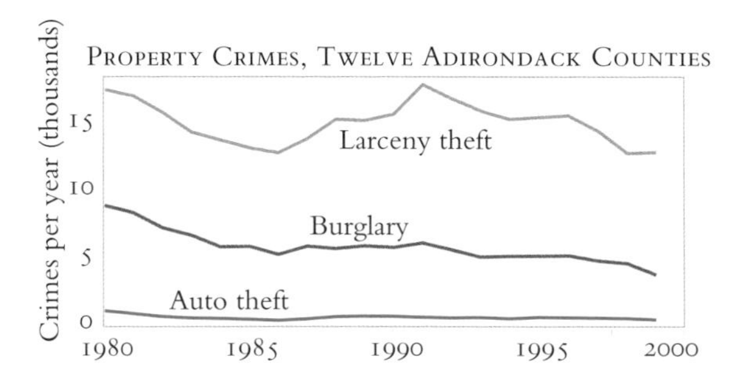

In 1995 New York State had 141,719 burglaries or 390 burglaries per day. Even the poorest rural areas still managed rates of around 400 burglaries per 100,000 persons per year or one burglary per 250 persons per year.

If burglars, like other traveling professionals, need nonpaying time for lying low, casing jobs, fencing booty, and so on, then these rates sug-

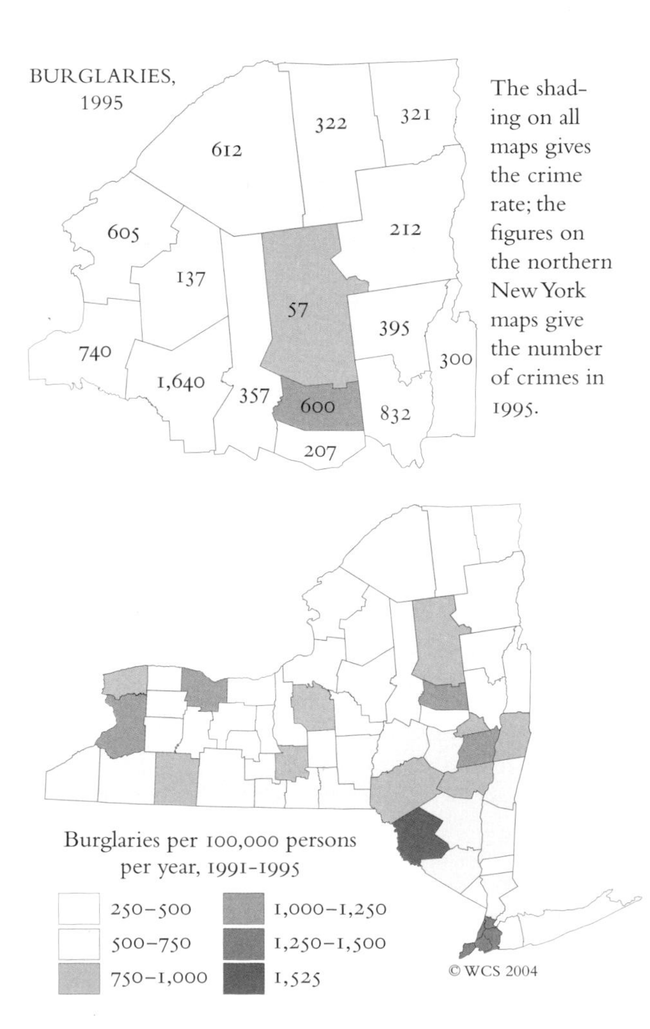

gest a substantial burgling population. If full-time burglars average a job or two a week, then New York has somewhere between one and three thousand professional burglars. And if even 10% of the burglaries are by amateurs who only do a job or two a year, then there may be 10,000 or more occasional or journeyman burglars outside the professional ranks.

Auto theft is the commonest open-air felony in New York. It is almost as common as burglary in the cities but much less so in rural areas. Although in Washington County this is believed to be the one salutary effect of road salt, it may also reflect differences in logistics and markets. It is probably easier to steal a car off a city street than out of a driveway. At a minimum there are more to choose from, and the owner is less likely to be looking out the window and phoning the sheriff. And disposing of stolen cars, which after all are larger and more identifiable than chain saws and cuts of venison, is probably much easier in the city than in the woods.

Robbery, though by definition a coercive or violent crime, is still more a crime of money than of passion. It is, in any event, a very urban crime. There are more places to steal from in cities, and even more important, probably more desperate people for whom the prospects of small amounts of cash are worth the risk of being shot or jailed. While robberies occur even in the most rural counties—Clinton County averaged five per year from 1991 to 1995—the mean rates are less than a fifth of what they are in the upstate cities and less than a twentieth of what they are in New York City.

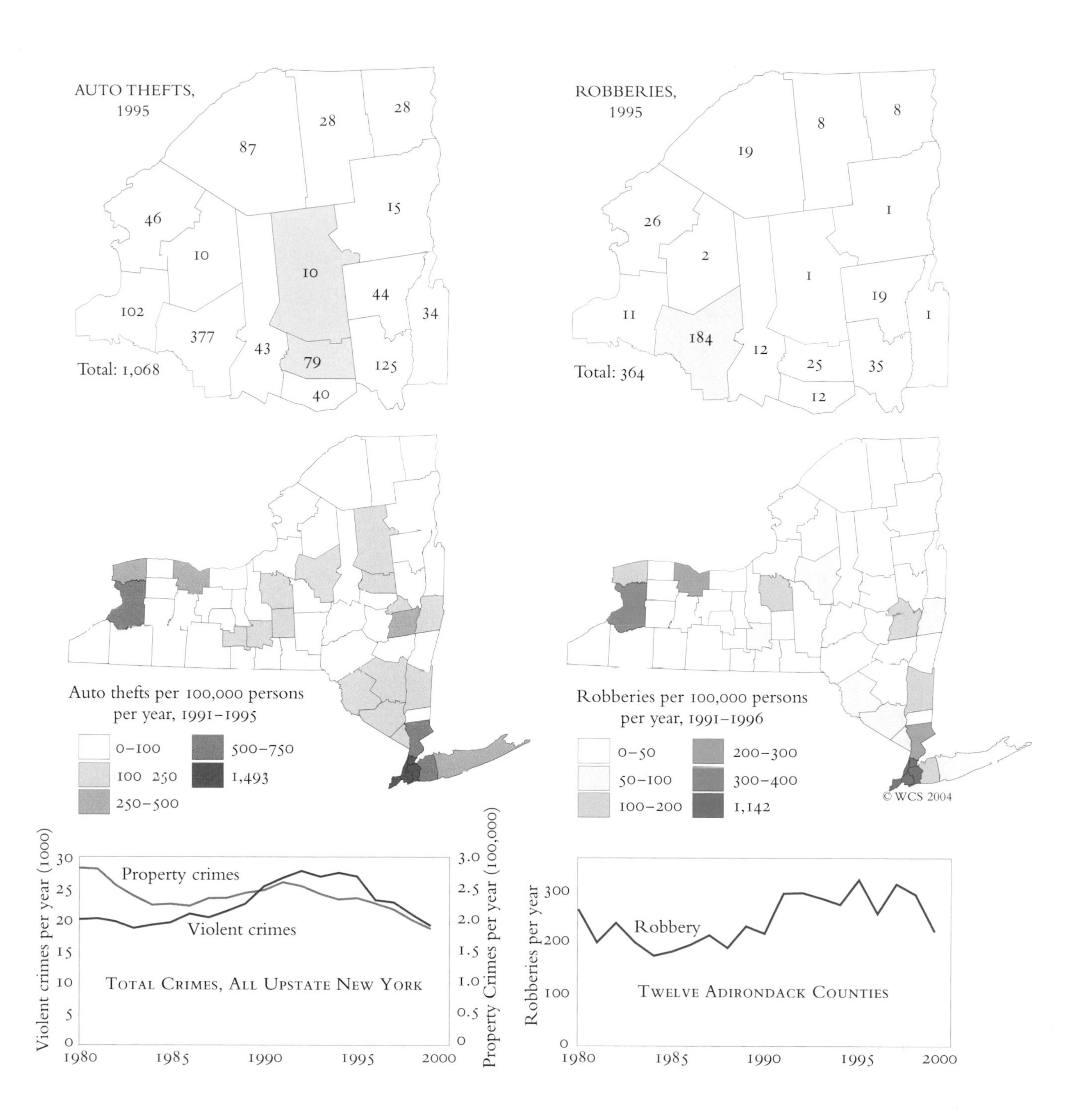

AMONG first-world countries, America's high rate of violent crime is its peculiarity and shame. Northern New York, which lacks ethnic, racial, or sectarian violence and has not had a war on its soil for nearly 200 years, had 32,254 reported violent crimes between 1981 and 1995. When we consider further that many violent crimes are never reported, that crimes affect many people besides the victims and the perpetrators, and that violence perpetuates itself in families and communities we start to get some sense of how high the human costs of violent crime really are. We can be grateful that the rates are lower here than elsewhere but should be mindful that they are much higher than they might be and, indeed, in many cases higher than they were twenty years ago.

Aggravated assault, which is assault intended to cause serious harm, is the commonest violent crime in New York. There are about 2,500 assaults in northern New York each year and about 75,000 in the state as a whole.

Like burglaries, assaults happen everywhere. The highest rate in the state, 838 assaults per 100,000 persons per year, is in New York City. But some rural counties also have high assault rates. In the Adirondacks, Warren County has an average assault rate of 442, half of that of New York City.

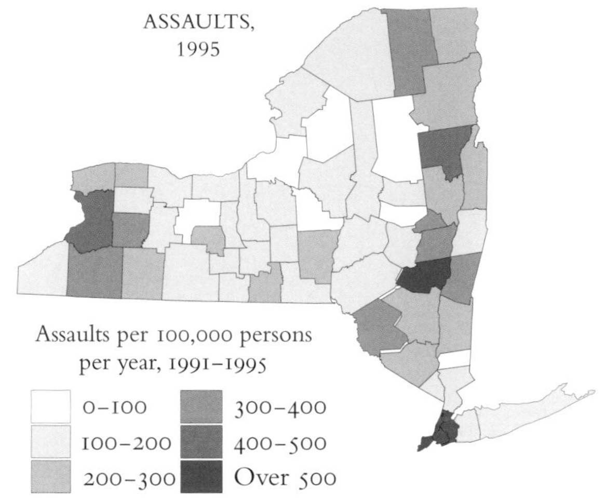

In the Catskills, Greene County has an assault rate of 525, almost two-thirds that of New York City.

The rates of *forcible rape* are high throughout New York. The statewide rate is about 23 rapes per 100,000 persons or 1-3 rapes per year in a town of 10,000 persons. The highest rates, like those of other violent crimes, occur in the inner cities. But rape is not an urban specialty like robbery or auto theft and can and does happen everywhere. The rape rates in small cities are as high, or in some cases higher, than those in New York City; the rates in many rural areas, including parts of the Adirondacks, are as high as those in many cities.

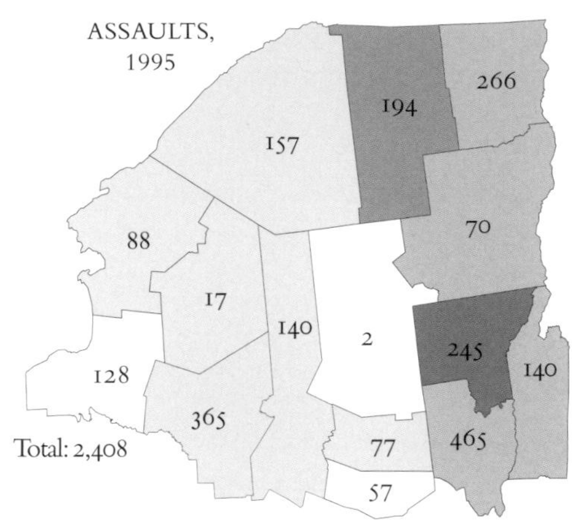

The shading on all maps gives the crime rate; the figures on the northern New York maps give the number of crimes in 1995.

The official numbers of forcible rapes greatly underestimate the frequency of sexual crimes. Many forcible rapes are unreported, as is a large percentage of child abuse and almost all unforced rape. In Washington County, with 60,000 persons, the official 1991-95 figures ranged from two to five rapes a year. But in 1999 alone the Sexual Trauma Center at Mary McClellan Hospital in Washington County provided services to 72 rape victims and 297 victims of child abuse. Even assuming that some of these cases come from other counties, the actual rates still appear to be many times higher than the reported rates.

The causes of high rape rates are debated and uncertain. Individual sociopathy, general societal disrespect for women, and innate or culturally encouraged male aggressiveness probably all play roles. The geography of the rates is less equivocal: rape is a common and, regrettably, almost universal crime in our society. No part of New York does not suffer violent sexual crimes; no cultural group does not suffer them, and probably few if any cultural groups do not commit them.

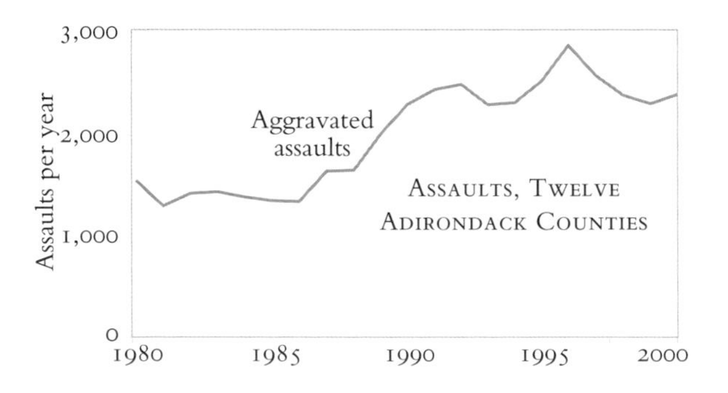

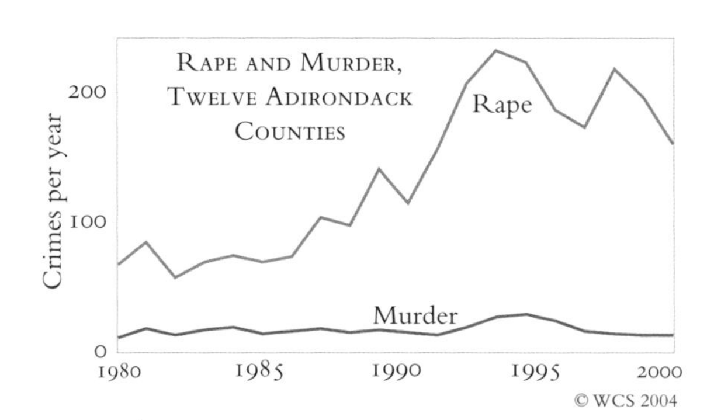

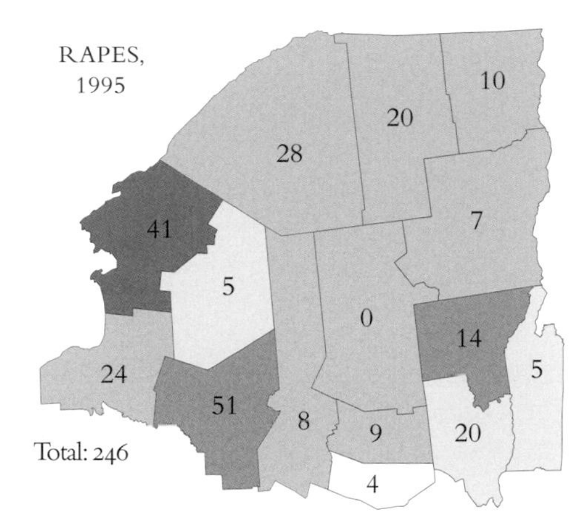

Murder is the rarest of the violent crimes, occurring at rates of less than 4 per 100,000 persons in much of the state, for a total of about 2,000 murders a year. Even so, it is a surprisingly common way of dying. Murders kill half as many people as accidents and two-thirds as many people as AIDS. One percent of all New York deaths, seven percent of the deaths of children between one and fifteen, and thirty-two percent of the deaths of adolescents between sixteen and nineteen are murders.

In New York the highest murder rates are associated with urban violence. But nationwide the pattern is quite different. In the U.S. as a whole rural murder is very common, especially in a belt across the south and west. Disturbingly, murders of women are far more widespread than murders of men. Men, who commit almost all the murders, only murder substantial numbers of other men in the south and far west. But they murder women, at significant rates, not only in the south and west but in the northeast, midwest, and intermountain west as well.

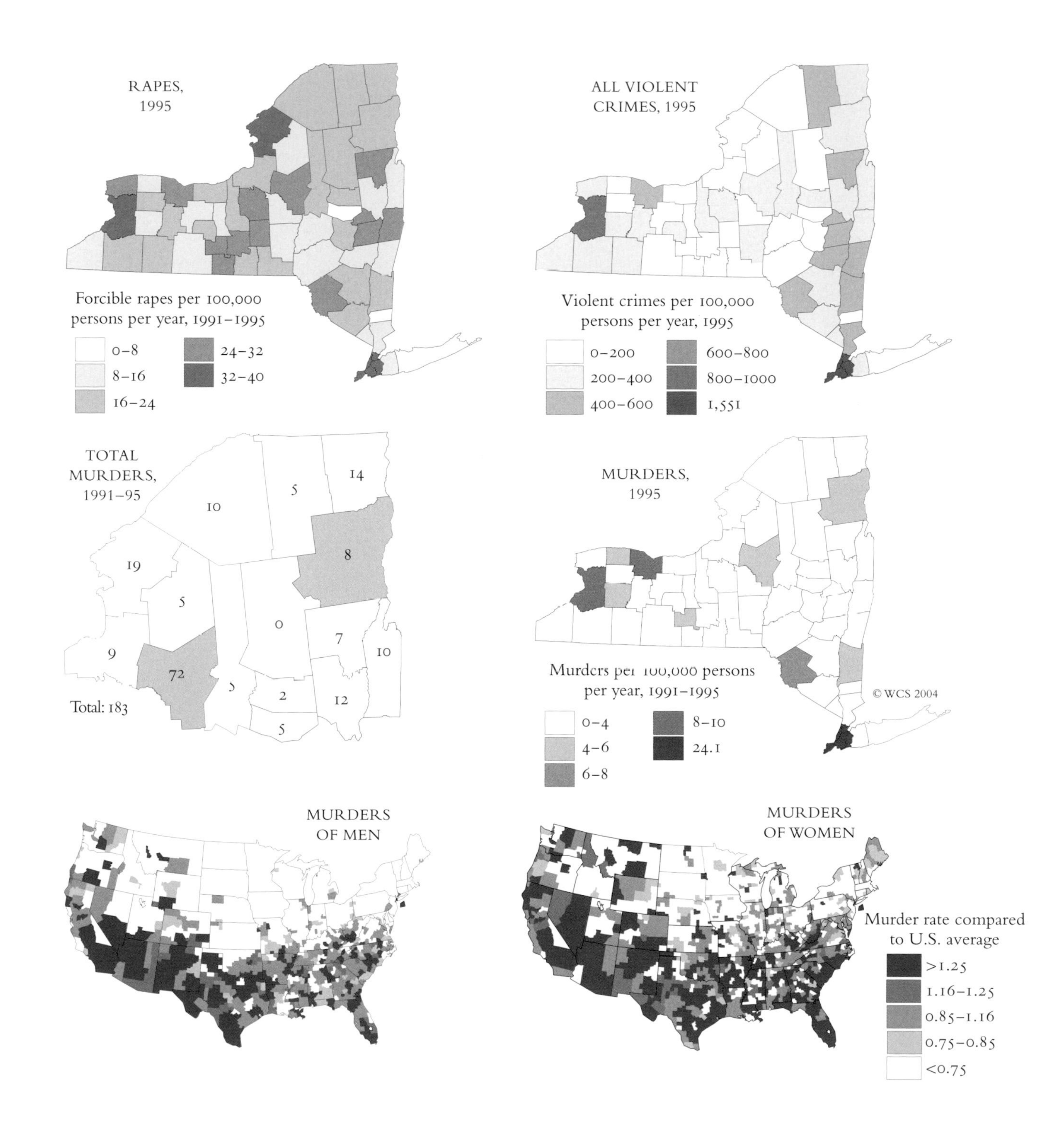

The New York state prison system consists of 70 facilities that on January 1, 2000, housed 71,466 long-term prisoners at an annual cost of about $2,000,000,000. An unknown but large number of additional prisoners were housed, for shorter terms, in city and county jails at an additional cost of $1,900,000,000.

The northern New York portion of the system consists of twenty-three prisons holding 26,000 inmates. Three are minimum security, seventeen are medium-security, and three maximum security. One of the maximum security prisons, Upstate, is a "maximum control" or "24-hour lockdown" facility, and another one, Clinton, has prisoners on death row.

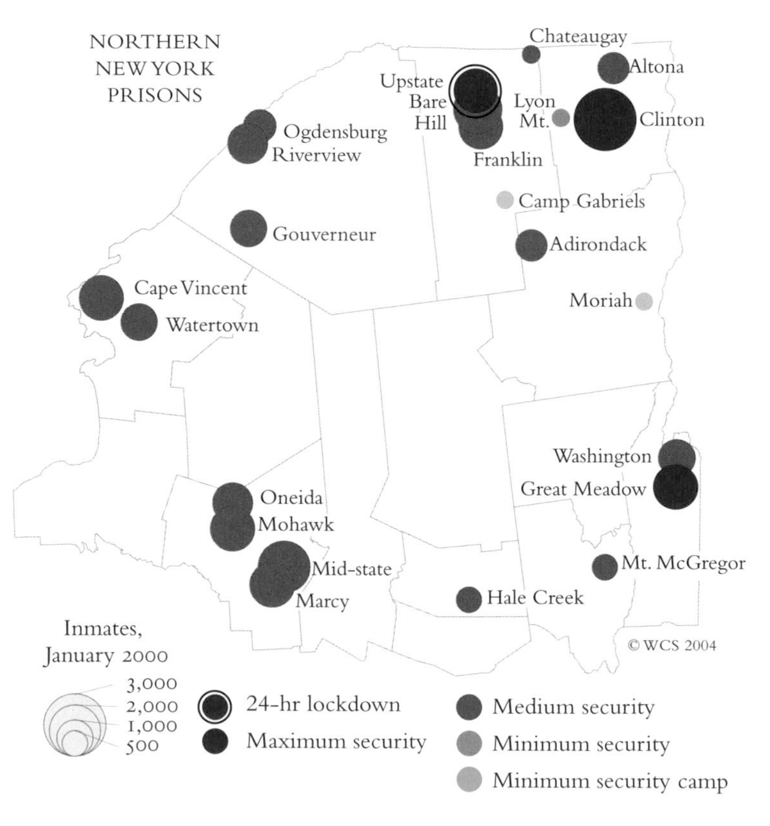

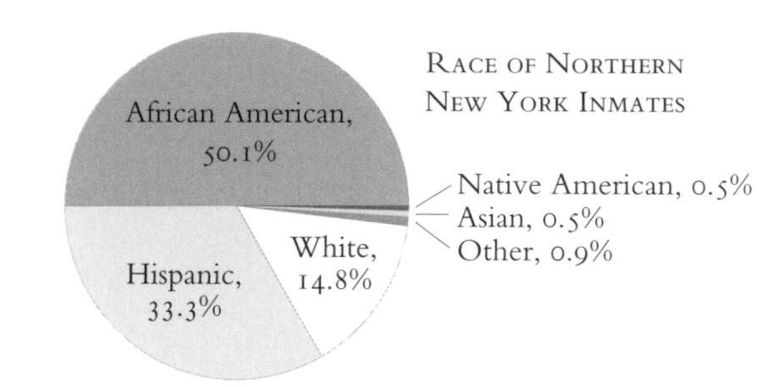

The New York prison population is largely urban, poor, nonwhite, male, and quite experienced in crime. Eighty-three percent are African American or Hispanic. Ninety-five percent are male, 59% percent have had previous jail time, and 54% have committed two or more felonies. Only 42% have completed high school, and only 25% are married, though 60% have children. Fifty-five percent report having used drugs, and 30% have been diagnosed as having problems with alcohol abuse.

Most New York prisoners are serving sentences of between two and ten years. The median minimum sentence is four years three months, and the median time already served twenty-four months. Over half, if the national averages are any guide, will commit other felonies after they leave prison.

The prisoners held in northern New York are typical of those in the state as a whole. Over half are between 25 and 40; 85% are nonwhite; 35% were convicted of drug sale or possession, and many of the rest, including the 45% that have committed violent crimes, were involved in one way or another with drug use or the drug trade.

Most northern New York prisoners are young and will be free in two to six years. A few are older and have a bleaker future. Two hundred and ninety are sixty or older; fifteen, all in Clinton, have been sentenced to life without parole; and five, if their convictions and the constitutionality of their sentences are upheld, will eventually be executed in the death house at Greenhaven Prison.

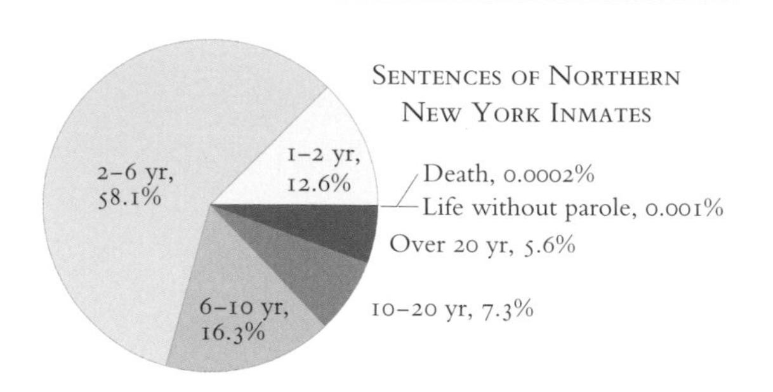

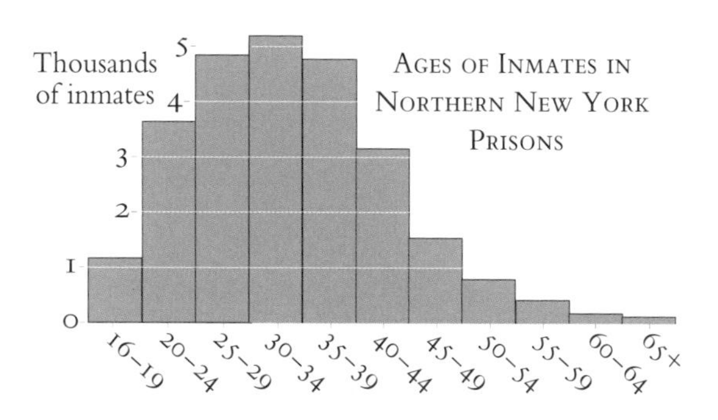

In 1975 northern New York had two prisons and under six thousand prisoners. In the last twenty-five years it has built twenty-one new prisons, one every fourteen months, and has more than tripled the number of prisoners. The prisons were not added because of an increase in crime but because of new laws that mandated longer sentences for the same crimes. It was originally hoped that these laws, particularly the so-called *Rockefeller drug laws* and the *second-offender law,* would reduce crime and decrease drug use. Only later was it realized that they failed to distinguish between minor and serious offenses and that they were in fact aggravating the drug problem by sending nonviolent offenders, particularly African Americans and Hispanics into prisons rather than into treatment programs.

The question of racial biases, either in the laws themselves or the way they are enforced, is a serious one. Studies by the FBI and the National Institute for Drug Abuse suggest that the majority of drug users are white. But currently African Americans and Latinos, who make up 25% of the New York population, make up 92% of the people in New York prisons for drug offenses. In 1980, approximately two African Americans and Latinos were sent to prison on drug charges for every white. Now seventeen are.

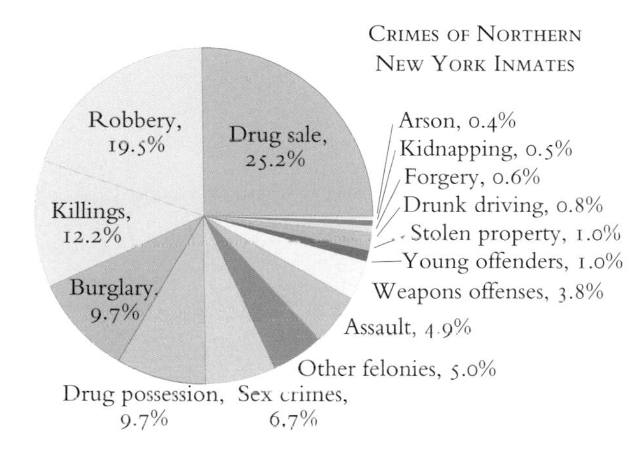

Whether necessary or cruel, just or unjust, the increase in the number of prisoners since 1980 has been dramatic. Parallel changes in federal and state laws elsewhere have tripled the total number of prisoners and have given the U.S. one of the highest imprisonment rates in the developed world. Before 1990, no developed country had an imprisonment rate of over 400 per 100,000. Currently, the U.S. imprisons about 520 people out of every 100,000. New York, a bit below the national average, imprisons 394. Canada, by way of contrast, imprisons 120. Most other western democracies are under 100; only South Africa with 370 is comparable to the U.S., and only Russia with 550 exceeds it.

Issues about the cost, necessity, and the fairness of the prison system are very important for the Adirondack region which contains a third of the state's prisons, and profits from the 10,000 jobs and $465,000,000 annual payroll these prisons produce. If our high imprisonment rates are necessary to prevent crime and reform criminals, then our prisons, however somber, are necessary and honorable institutions. If the high imprisonment rates do neither, then the prisons are neither. And if, as many suggest, there are fundamental inequities in the laws or law enforcement of a state that has more African Americans and Latinos in the state prisons than in the state universities, then the Adirondacks, however far from the places where the drugs and crimes flourish, are benefiting from a racist system and need to think deeply about their part in it.

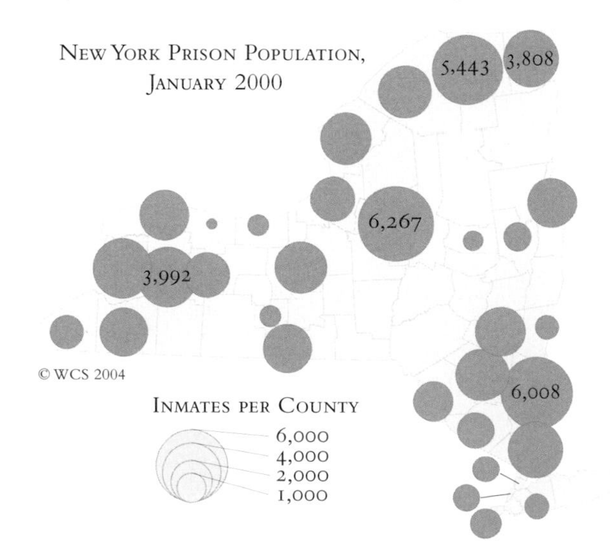

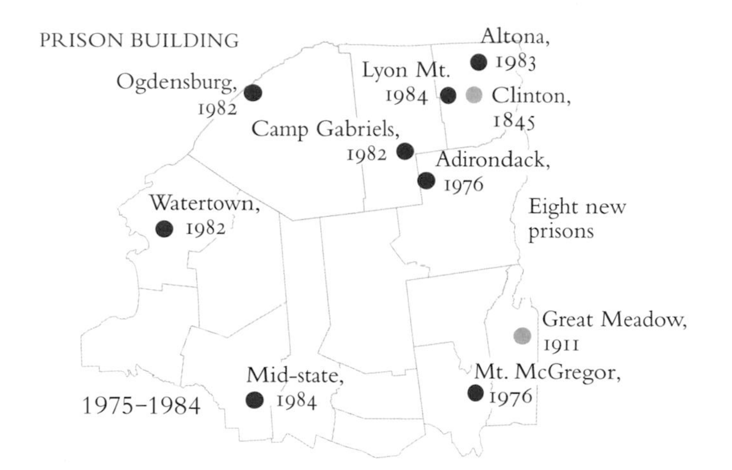

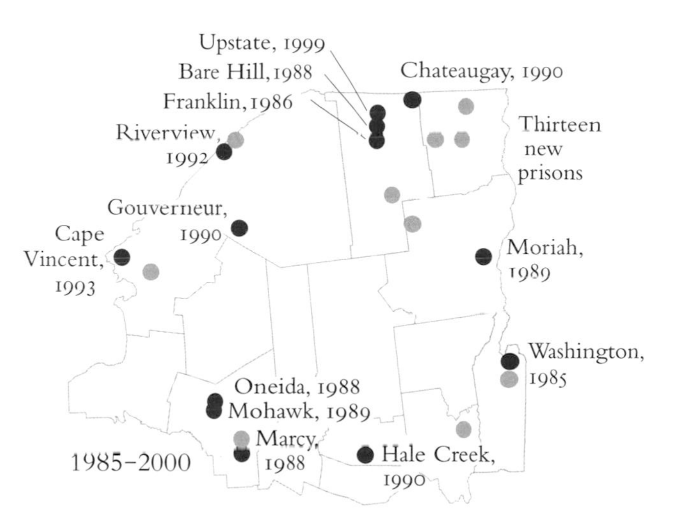

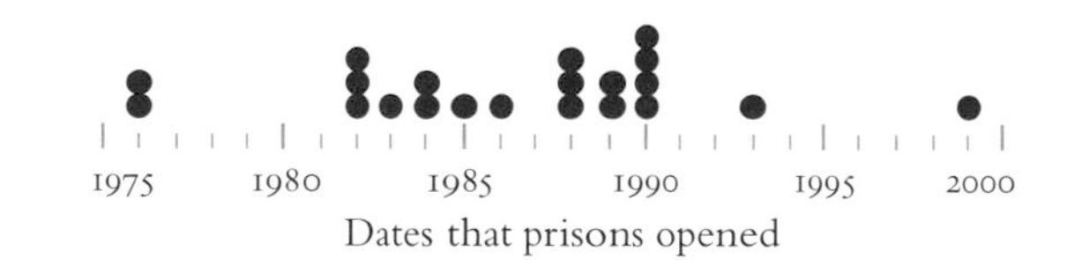

SECONDARY SCHOOLS

The Adirondack Park contains 61 school districts. Twenty-five are completely within the park and 36 partly inside and partly outside. These districts operate 135 schools, 65 within the park and another 70 outside. There are also ten private secondary schools within the park and another seven private schools within the Adirondack school districts but outside the park.

The public school system represents a massive investment by Adirondack communities. In 1999 the Adirondack school districts enrolled about 62,000 children from a total population of about 350,000. To educate them for the legally required 180 days per year, the districts spent over 620 million dollars. Forty-one percent of this came from local taxes and 59% came from state and federal aid. The sixty-one districts employed about 4,500 full-time and part-time teachers plus 400 other supporting professionals. The average full-time teaching salary, before benefits, was about $50,000. Fully 90% of their students completed twelve years of school. Seventy percent of their graduates went to two-year and four-year colleges, but only 45% of the graduates completed the academic requirements for a Regents diploma.

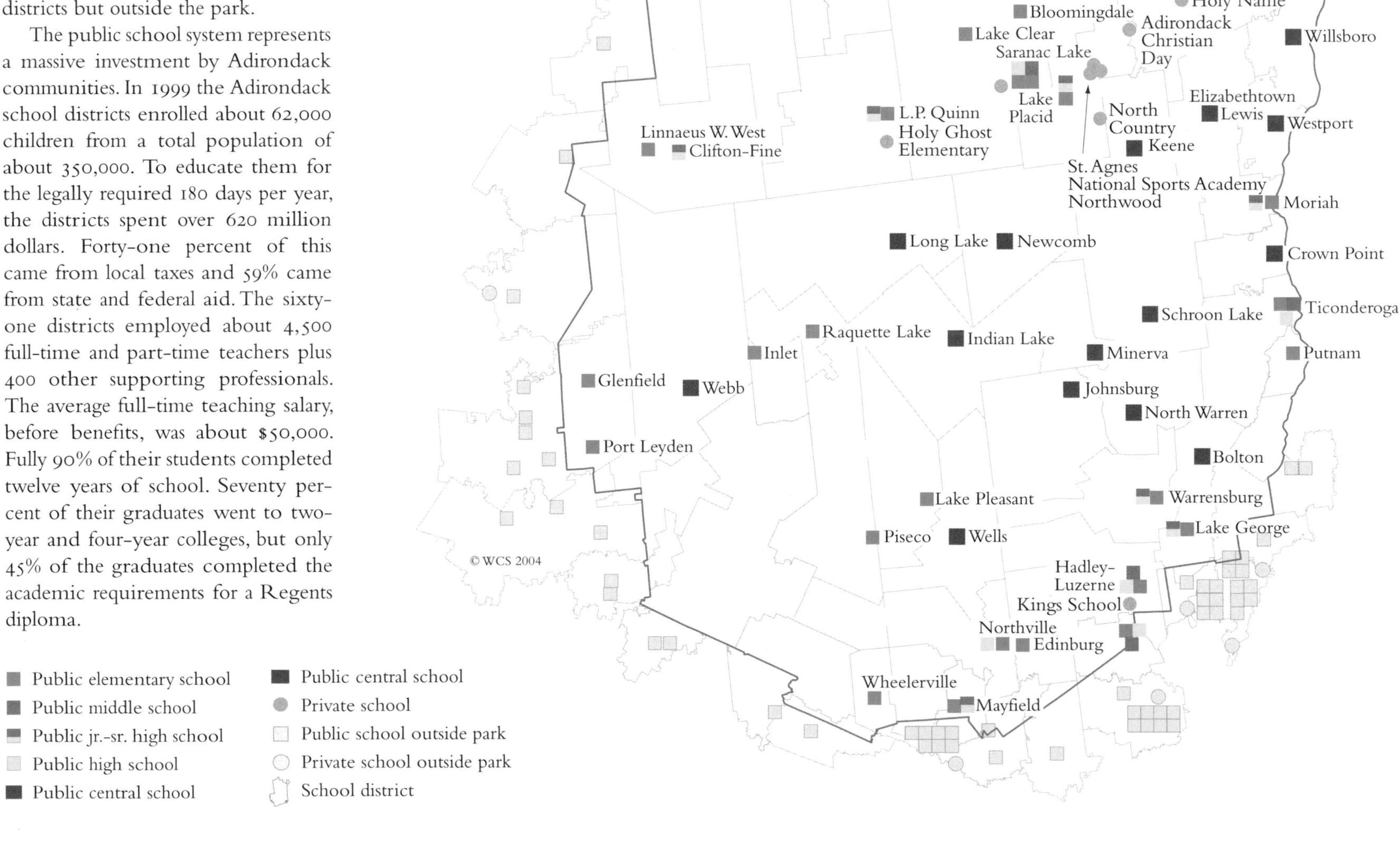

10 Schools & Colleges

In the Adirondacks, as in much of rural New York, good public schools are widely available and over seventy percent of the adult population has graduated from high school. Colleges and college degrees are rarer. The region is served by three State University of New York campuses and three private four-year colleges just outside the park and by nine two-year community college campuses within or just outside the park. But there is only a single four-year college within the park (*Map 10-3*), and it has a relatively specialized curriculum emphasizing hotel and natural resources management. In consequence, all Adirondack students who want a four-year education, either liberal arts or pre-professional, must go to college outside the park. Once outside, with the career expectations that their education has created, it is often hard for them to return.

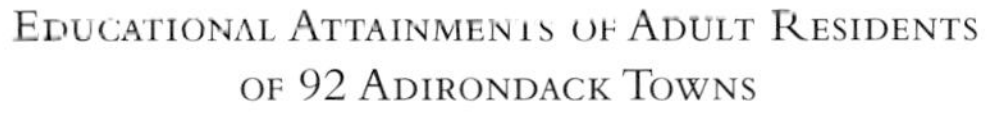

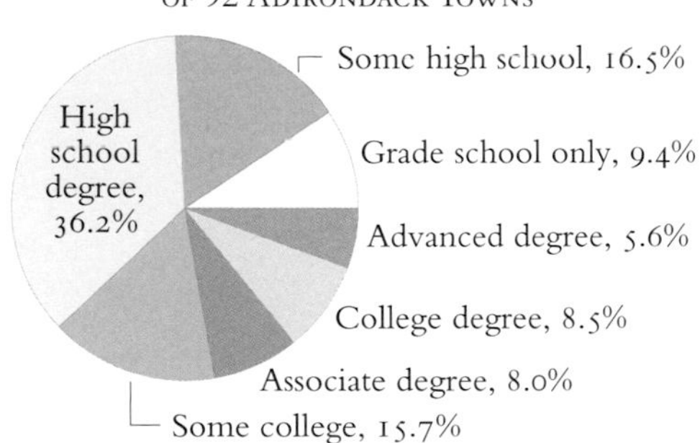

Of an adult population of about 145,000 in the Adirondack towns, 74% (106,000) have at least a high school degree, but only 14% (20,000) have a four-year college degree or an advanced degree.

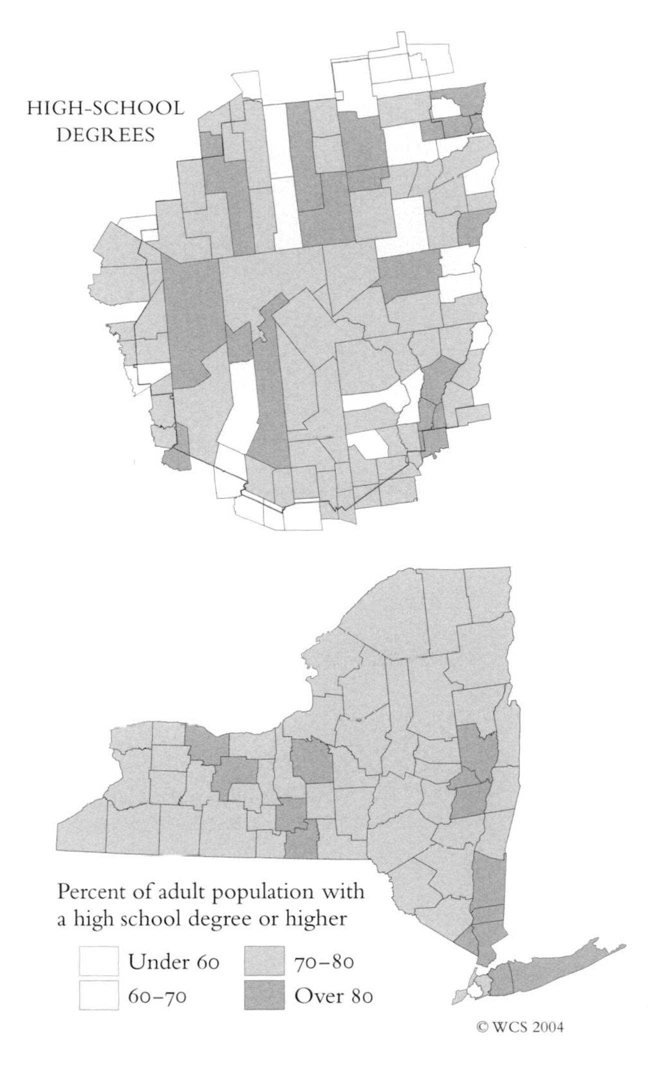

In most counties in New York 70–80% of the adults have high-school degrees. In some suburban counties over 90% have degrees; in the most rural towns and poorest parts of the inner city less than 50% do.

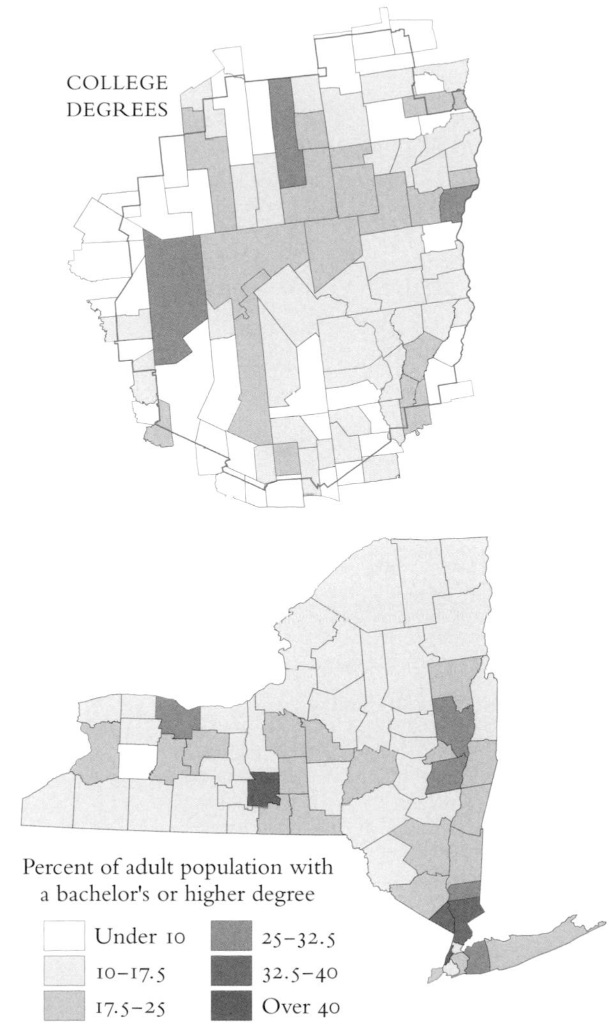

College degrees are commonest in the suburbs, especially in erudite Tompkins County. They are also surprisingly common in the Adirondack interior, though rare near the Adirondack edge.

An average Adirondack school district has a population of about 6,000 persons, sends about 700 children to school, and spends about 8 million dollars per year, or $11,000 per child, to educate them. But there is great variation between districts. In 1999, Saratoga Springs, the largest Adirondack district, had a total population of 41,000 and 6,800 children in its nine schools. Its total budget was 19 million dollars, equivalent to $9,300 per child per year; it had a 13.9:1 student-to-teacher ratio, graduated a very creditable 60% of its students with Regents degrees, and sent 80% of them to college. The same year Raquette Lake, the smallest district, had a year-round population of 139 and thirteen students in its school. It spent $22,000 for each of them or about $300,000 total and, with two teachers covering five grades, had a 6.5:1 student-to-teacher ratio.

Despite the variability, there are clear geographical patterns. *Enrollment* follows population. It is high near the big towns at the edge of the park and also in a few big districts like Saranac Lake where a central town attracts students from a relatively large area. It is low in the childless center of the park regardless of the size of the district. *Spending per child* is the mirror image of enrollment, tending to be high but quite variable in small schools and converging to an average value in the larger ones. *Class sizes*, measured by the number of students per teacher, tend to decrease as spending per student increases.

Taken together, the resulting pattern is one of large, fiscally efficient schools at the edge of the park and smaller, more costly, more heavily staffed ones in the center. If, as many of us believe, spending and staffing levels affect the quality of education, Regents scores and the percentages of students going to college should be higher in the center of the park. Interestingly, they are not.

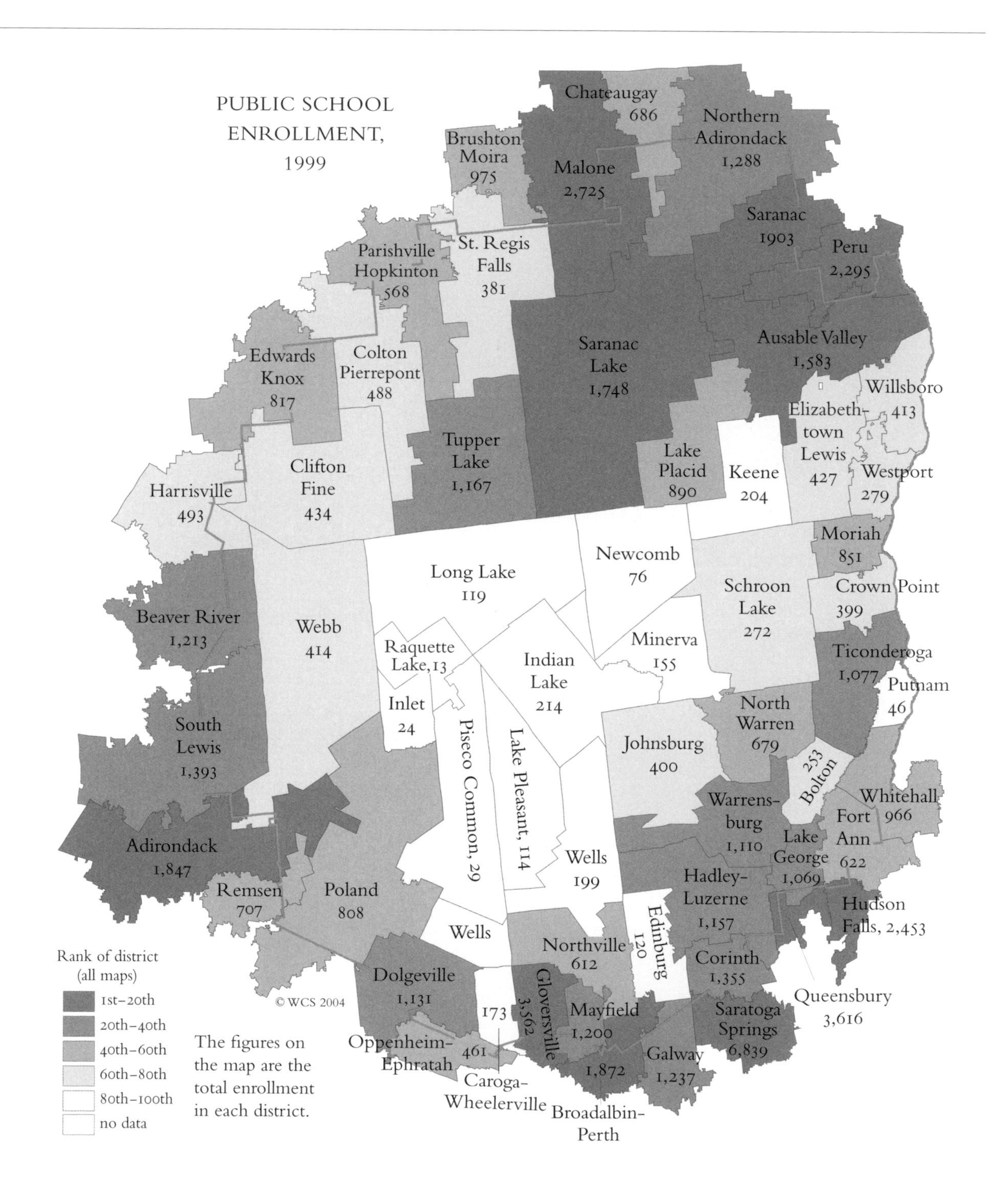

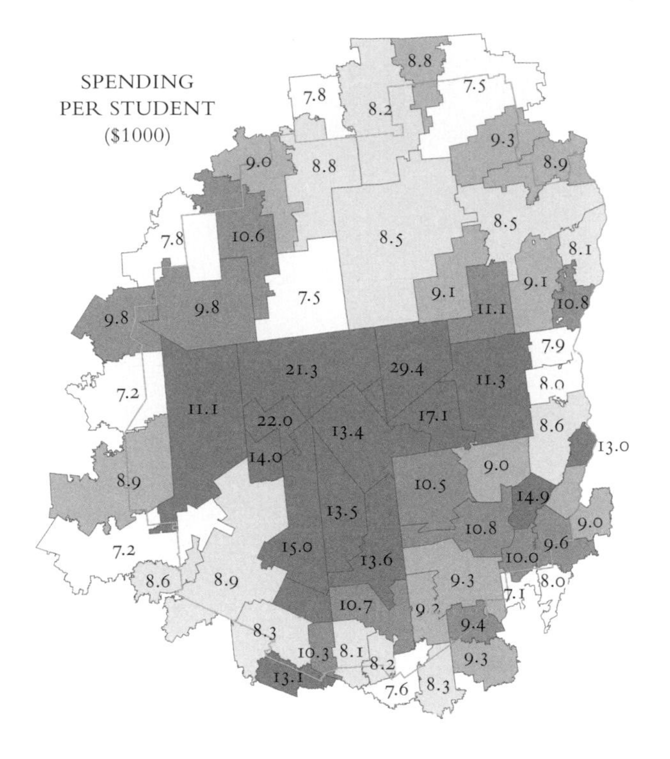

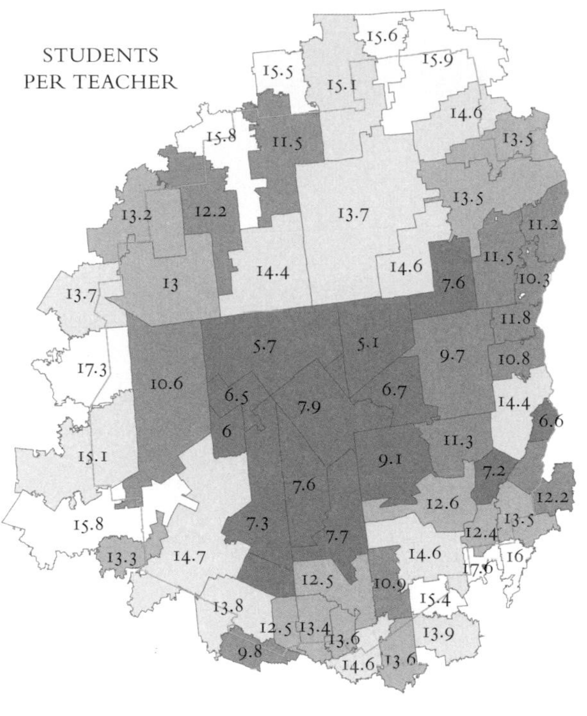

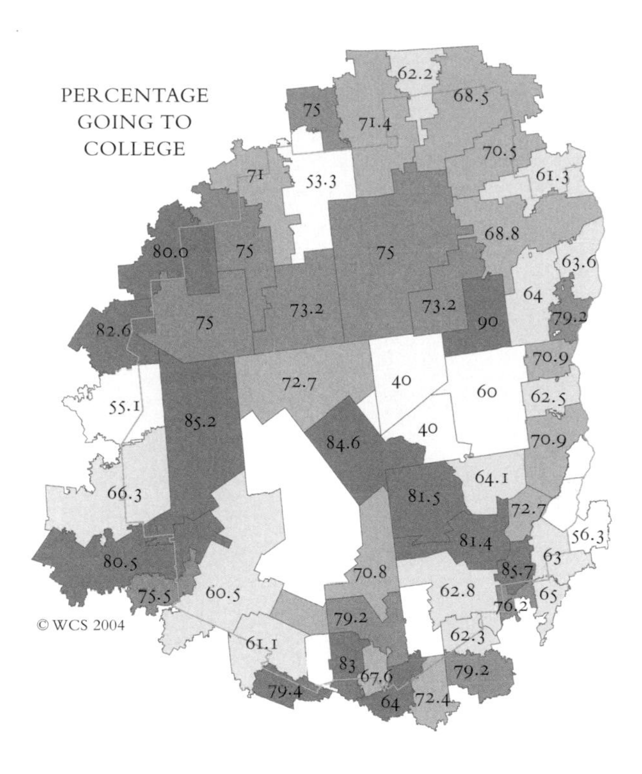

The highest *spending per student* is found in small schools in the central towns. There are at least three reasons for this. First, the larger peripheral schools derive more of their income from state aid than the smaller ones, and this tends to keep their per-child spending near the state average. Second, and probably most important, many of the small schools are in relatively wealthy towns (*Map 10-2*). And third, the small schools may cost less relative to other town expenses than the large ones and so may have more flexible budgets. If you can double your teaching staff for less than the cost of a new plow truck, why not do it?

The number of *students per teacher* (here mapped with dark colors indicating a low student-teacher ratio), is lowest in the small schools. This is partly because these schools are wealthier and have the option of hiring more teachers and partly because they have to cover the same number of grades and roughly the same number of subjects as the larger schools. Because teachers are by far the largest expense in running a school (*Map 10-2*), the cost per student rises steeply as the number of students per teacher decreases. There is no cheap route to small class size.

The percentage of *students going to college* varies greatly. It is strongly correlated with the percentage of students passing Regents exams but not with the number of students, the spending per student, or the students-per-teacher ratio. Long Lake, which spends twice per student what Parishville does, sends the same fraction of students to college. Perhaps aspiration is, after all, a matter of individual students and individual teachers and not of statistics; or perhaps it has its own geography, high in the busy and ambitious towns, low in the peaceful ones where the fishing is good.

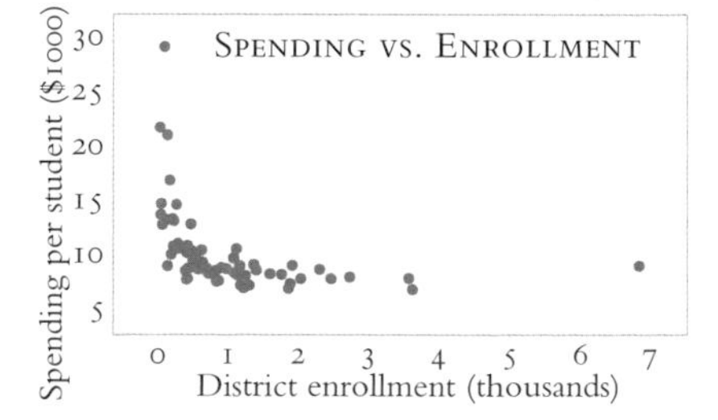

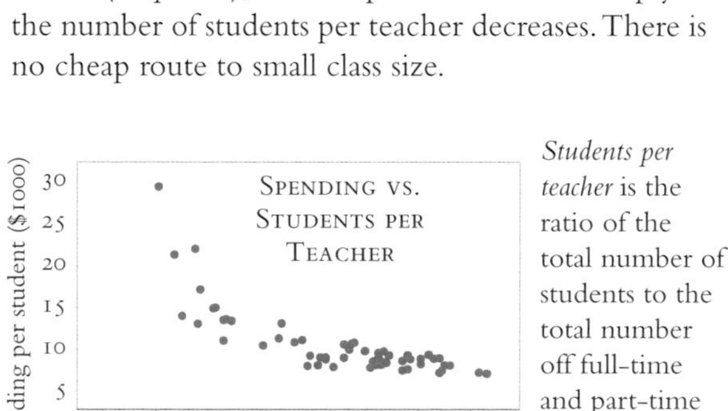

Students per teacher is the ratio of the total number of students to the total number off full-time and part-time teachers.

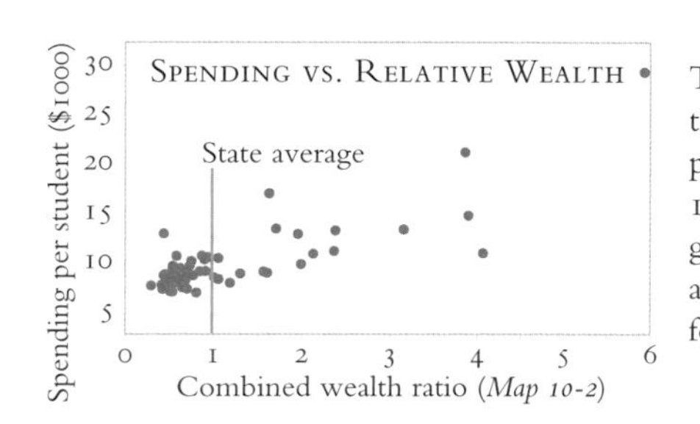

The figures on the map are the percentage of 1999 graduates going to either a two-year or four-year college.

The 600 million dollars required to support the Adirondack school system are derived from a mix of local, state, and federal funding. The mix varies from town to town, depending on what the town chooses to spend and how much aid the state supplies. The town looks at its needs and its inventory of taxable lands (*Map 11-1*) and tries to balance what the parents want with what the taxpayers will stand. The state looks at the number of students in the district and at a *combined wealth ratio,* which compares the total value of the property in the district and the total income earned by the residents of the district to the average for New York State. Towns with wealth ratios below 1.0 (and thus poorer than the state average) receive more aid per student than towns with higher wealth ratios.

By this definition, the districts in the center of the Adirondacks are for the most part wealthier than those at the edge and often wealthier than the state average. The edge districts typically raise about three-quarters of their budget from state aid and one-quarter from taxes. The center districts, especially those with small schools, reverse this and typically raise less than a quarter of their budget from aid.

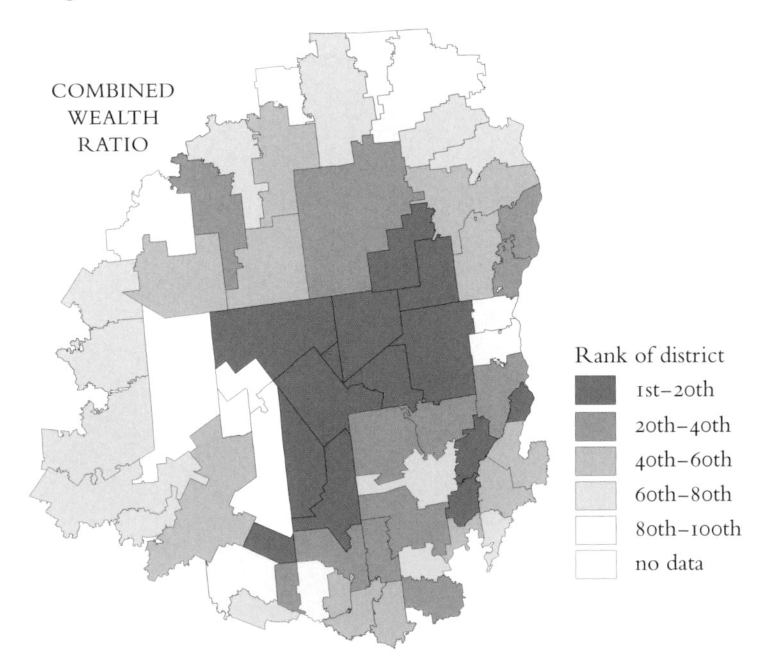

In contrast to the varying sources of funding, all Adirondack school districts seem to spend their money in about the same way. The largest expense, averaging about 50% of the budget, is for the salaries of teachers and other instructional personnel. Transportation, maintenance, operation, and administration together require another 15%, which by corporate or government standards is quite lean. Employee benefits are another 15%, and other miscellaneous expenses, including debt service and tuition paid for students attending schools in other districts, are the remaining 20%. The largest variation is in this last category, probably because the number of students going to other districts varies greatly. Other categories vary less. Instructional salaries are remarkably constant; benefits are in a fairly fixed ratio to salaries; and administration, operation, and transportation costs apparently can't be reduced to much less than 10% or allowed to expand to more than 20%.

This constancy of the noninstructional costs has an interesting consequence. Apparently the percentage overhead of running a school—the portion of the budget consumed by noninstructional costs like buses and school boards—doesn't decrease as a school gets larger. There seems to be no inherent advantage in being a large school or penalty for being a small school. Small schools usually have increased instructional costs because they use more teachers to teach the same number of students. But whether this is because they have to or because they choose to and whether it should be regarded as a penalty or a benefit lies beyond what the numbers can tell us.

Colleges and college-level programs in the Adirondacks conform to the long-established Adirondack social pattern: many visit, several have seasonal homes, but only three are permanent residents.

The seasonal residents are six colleges and universities (Gordon, Colgate, Houghton, SUNY Cortland, SUNY-ESF, St. Lawrence) that maintain Adirondack field stations or camps.

The three permanent residents are North Country Community College, SUNY-ESF, and Paul Smith's College. All are extremely important to the Adirondacks. North Country is a two-year college with about 1,500 students, 38 full-time faculty, and 110 part-time faculty. It has many local students and serves as a stepping stone for many students that go on to other colleges. The SUNY College of Environmental Science and Forestry, besides running a summer field school at Cranberry Lake, has a one-year program in forestry and surveying at Wanakena and a year-round research facility at Newcomb. Paul Smith's began as a two-year trade school, teaching forestry, hotel-management, and cooking. It has recently begun a four-year program, emphasizing environmental science. Such transitions are difficult because two-year and four-year colleges require different faculties with different definitions of competence and scholarship. If Paul Smith's succeeds, it will become the Adirondacks' first four-year college and an important one indeed.

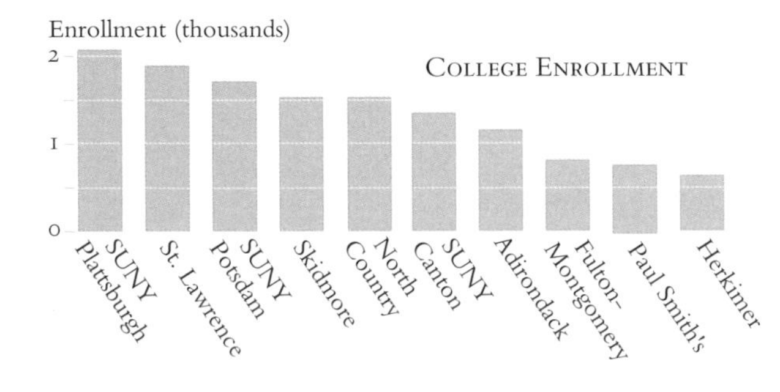

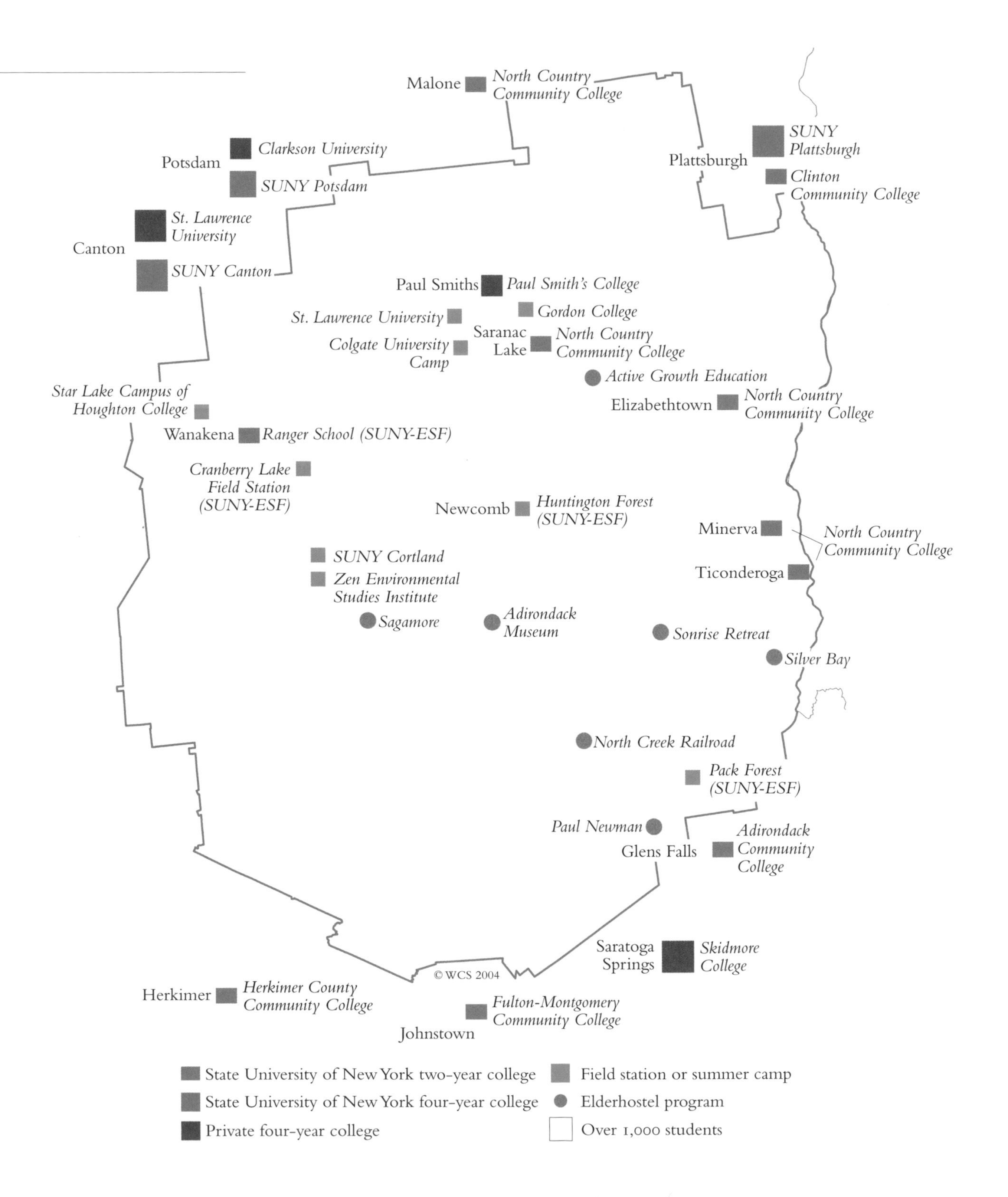

THE EMPLOYEES OF PAUL SMITH'S COLLEGE

PAUL SMITH'S, though a small college, is a large institution by Adirondack standards. With about 750 students, 60 faculty, 130 other employees, and an annual payroll of 6.5 million dollars it is the largest single college and the twelfth largest private employer in the park. Over 80% of its students come from outside the park, making it a major importer of capital and hence of considerable regional importance.

The map shows approximately how the Paul Smith's payroll is redistributed. Most of the staff live within forty miles of the college. Seventy-eight percent of them live within the Adirondack Park, and the majority of these live within fifteen miles of the college. People do not move to the Adirondacks to be commuters. A majority do at least some shopping near their homes, but over half make regular trips of twenty to forty miles to shop at large stores outside the park, where prices are substantially better.

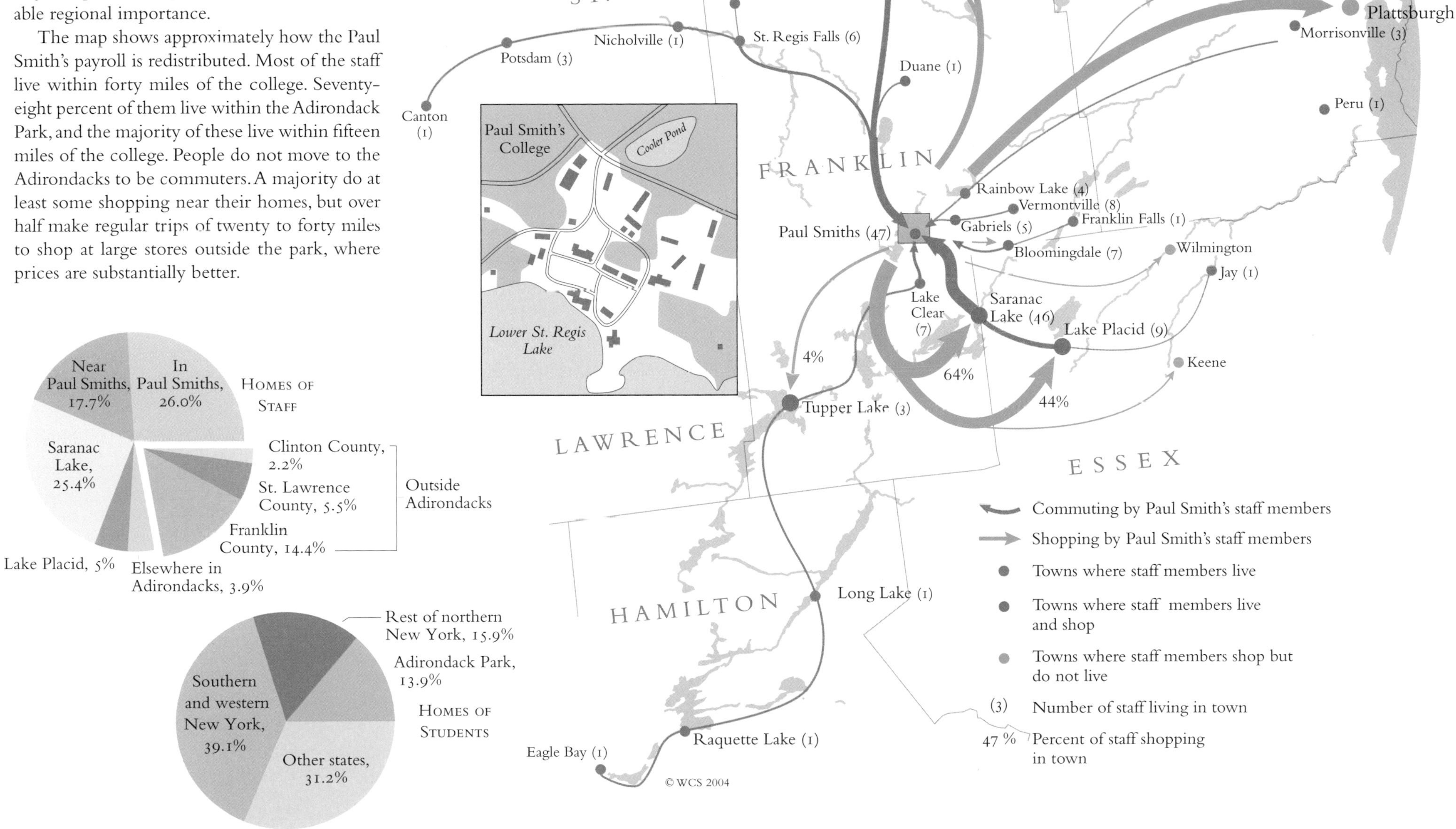

WHERE PROPERTY TAXES GO

The revenue from Adirondack property taxes is divided between 115 village, town, and county governments, 61 school districts, and 92 "special tax districts" that fund water and sewer supplies and some fire departments. On average, slightly over half of the total revenue goes to schools, a fifth to the towns, and another fifth to the counties. The remainder, about six percent, goes to the special districts.

Many towns differ significantly from these averages. Town government takes only one percent of the taxes collected in Queensbury, a third of the taxes in Newcomb, and over half the taxes in Inlet. In the peripheral towns where resident populations are relatively high and fairly young, the schools often receive sixty percent or more of all property taxes. In Greenfield they receive three dollars of taxes out of every four collected. But in the central and western towns where the resident populations are smaller and older, the school taxes are less than half of the total property tax. In Inlet the schools receive only one dollar out of every three collected.

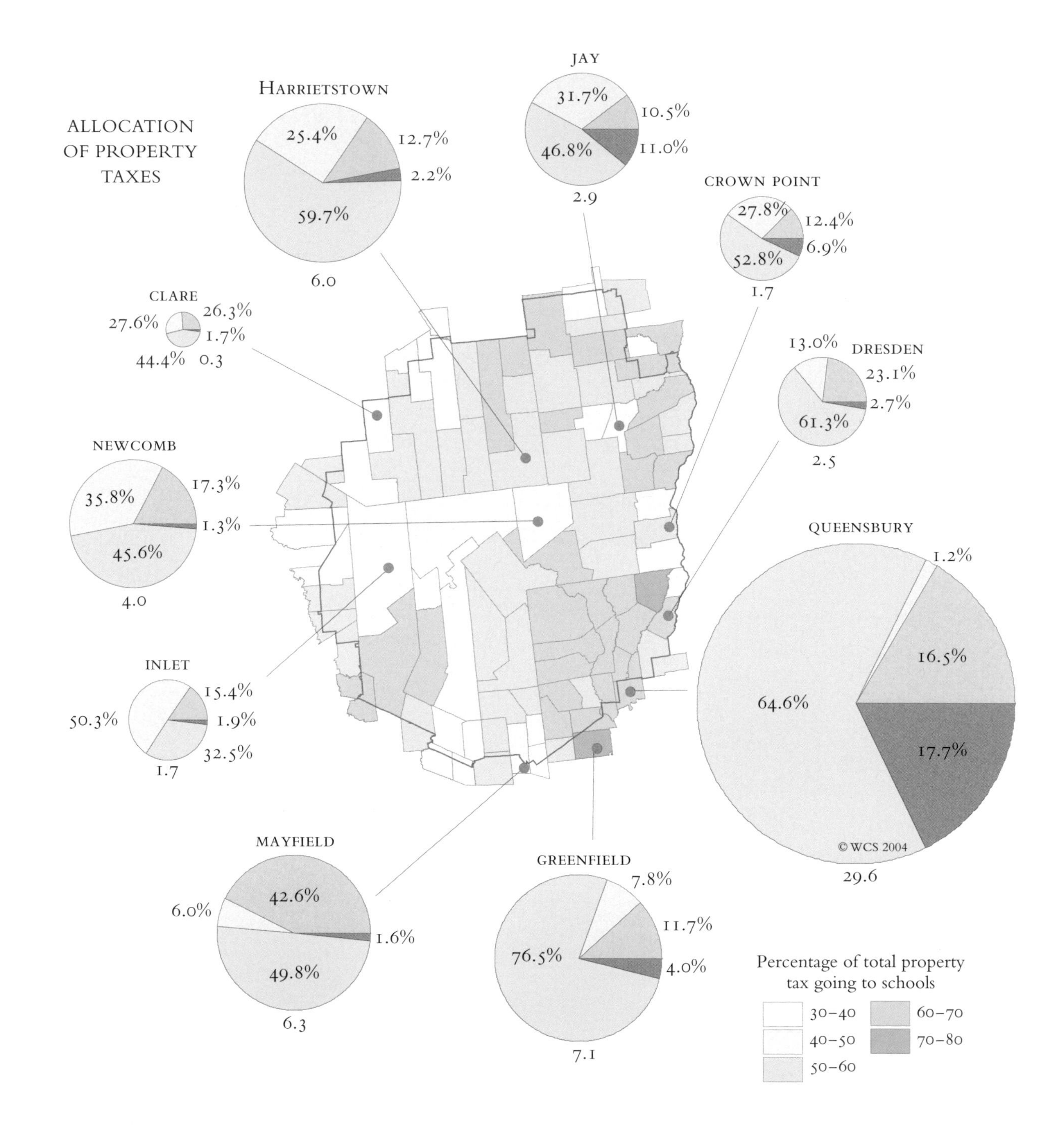

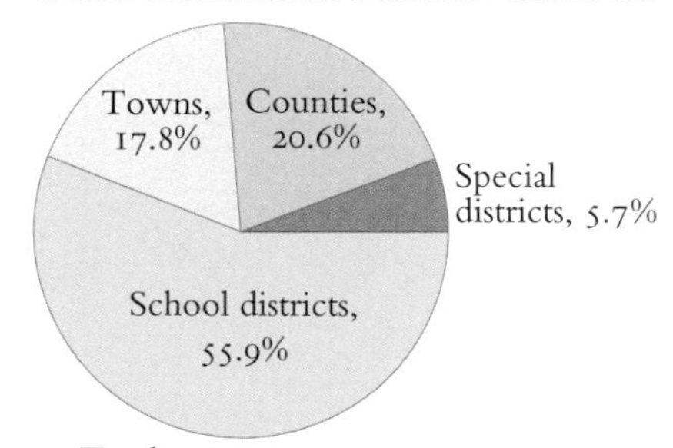

11 Town Budgets & Local Taxes

In 1997, landowners in the ninety-two Adirondack towns paid about 306 million dollars in property taxes. About 175 million dollars from these taxes went to the schools, 65 million dollars to the counties, and 48 million dollars to the towns.

While property taxes are the largest single source of town revenues, they are not the only source. Adirondack towns receive about 26 million dollars in state and federal aid every year, collect 18 million dollars a year in sales taxes, earn 3 million dollars a year in interest on money they are waiting to spend, and receive an intriguing 28 million dollars a year in "other revenues," suggesting that dog licenses, dump stickers, and raffle tickets may be an unexpectedly active economic sector.

Revenue Sources of Adirondack Towns & Villages

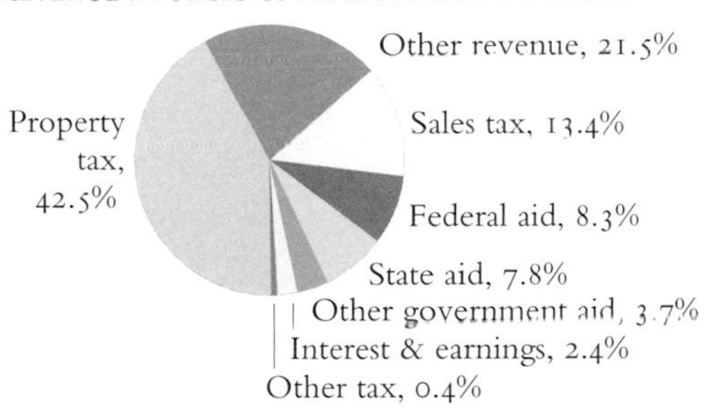

The extent to which towns depend on property taxes varies with the overall level of economic development. The towns with moderate or large populations and significant commercial development have a variety of revenues available and qualify for a fairly wide range of government grants. Most raise less than sixty percent of their revenues from property taxes. Twelve of the largest or most enterprising get less than 30% of their revenue from property taxes. The less developed towns with small populations and few businesses have fewer revenue sources and often raise over three-quarters of their revenue from property taxes.

In 1997, the ninety-two Adirondack towns spent a total of about 135 million dollars on government and services. About twenty percent of this was administration, which is to say the basic cost of being a town. The remaining eighty percent went to the services—local roads, fire and police, beaches and building inspections and ball fields—that we rely on the towns to provide. The budgets ranged from under a hundred thousand dollars to nearly fifteen million dollars. Roads and people are notoriously expensive, and large, well-roaded, populous towns invariably have high budgets. Small empty towns are much cheaper to run but, at least for plow drivers and select-people, probably much less fun.

Expenditures of Adirondack Towns & Villages

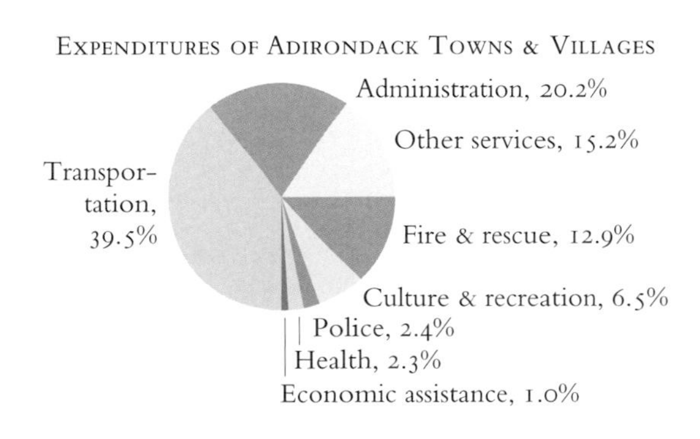

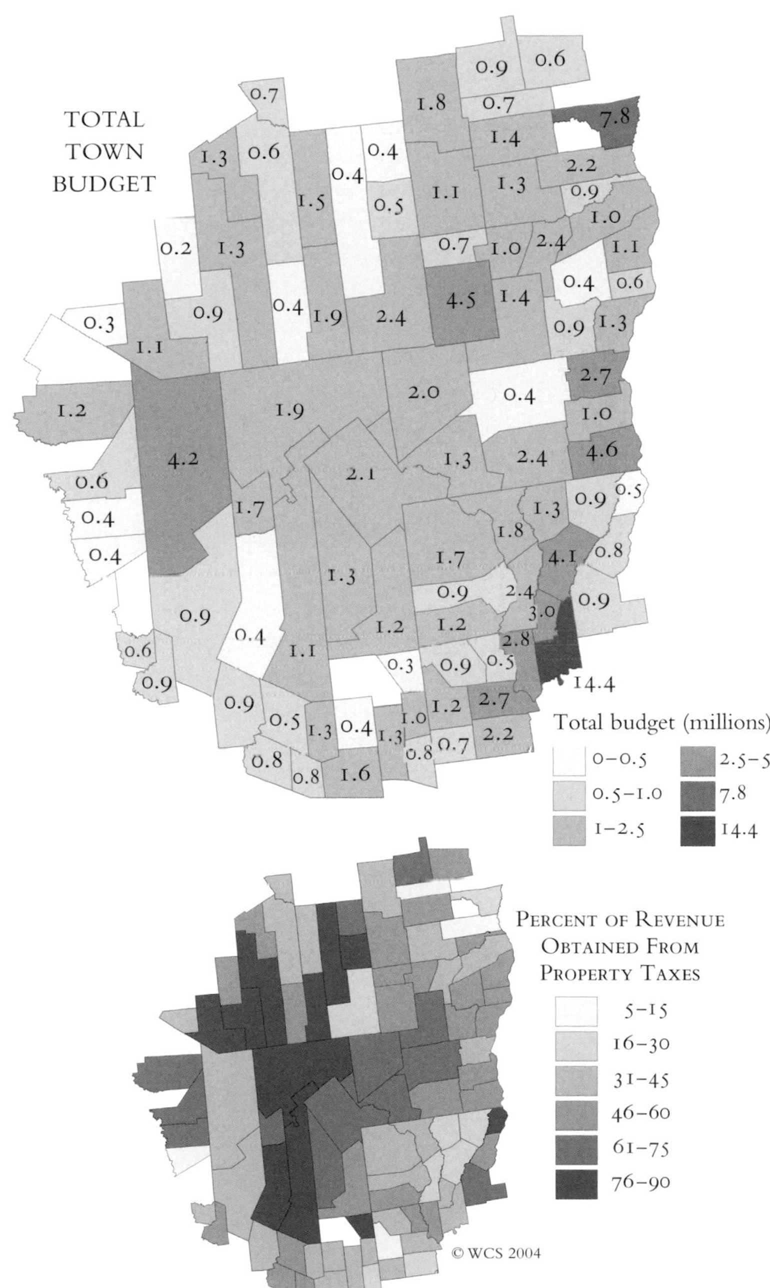

PROPERTY taxes, in one form or another, have been nearly universal in state-run societies for three millennia. In their bad form they are a way that a governing elite lives off the people it governs. In their good form they are a way that people combine to provide services that they can't provide individually. Fortunately, we have the good form.

Property taxes are determined by two numbers: an *assessment* which determines the taxable value of the property, and a tax *rate*, which determines how much has to be paid for every thousand dollars of assessed value. The ability to adjust both assessments and rates gives the towns a flexible way of getting the money they need from the landowners they have. They also can vary the way the total tax burden is divided between residents and nonresidents and, through the use of kind of two-tiered appraisal called the *homestead tax exemption*, increase the share of the town budget paid by commercial landowners and New York State. Income redistribution, an old-fashioned idea that can no longer be mentioned in national politics, is the daily business of every town clerk's office.

If flexibility is the great advantage of property-based taxes, inequity is their great disadvantage. The price of allowing towns to choose their own tax rates is unfairness. Both the tax that will be paid on a property of a given value and the services that the owner of the property will receive vary greatly from town to town. In some cases the differences are minor; in others they are large enough to raise constitutional questions about fairness and equal protection.

The maps illustrate the way that taxable wealth and tax rates vary from town to town. We have called this the tax base and tax burden, but remember that taxes are not only a measure of the burden that taxpayers bear but also of the benefits that they receive. Whether a given tax rate will be a burden or a benefit to you will depend on how much money you have and how badly you need the services that the town provides. Every taxpayer wants low rates, good schools, and well-maintained roads. None, the maps suggest, will get all three.

The property values shown in the maps are the *equalized values* computed by the Office of Real Property Services. These correct for town-to-town differences in the dates when properties were assessed and in the ratio of assessed value to market value used by the assessors.

The ninety-two Adirondack towns are estimated to contain real property—land and buildings—worth 18.2 billion dollars. About 3.7 billion of this is tax exempt (*Map 11-2*); the remaining 15.5 billion is the *total taxable value*, whose distribution is shown in the maps on the right.

Because even a small building is usually worth more than the land it stands on, both the total value of the town and the average value per acre are high in developed towns and low in undeveloped ones. Hopkinton, which is mostly forest, has a total taxable value of 50 million dollars and an average value of $417 per acre. North Elba, which contains the commercial and resort areas around Lake Placid, is worth 627 million dollars total, or $6,319 per acre. Queensbury, the most urban of the Adirondack towns, is worth 1.5 billion dollars total, thirty times that of Hopkinton, and a pricey $34,465 per acre. Trees and rocks, however charming, are worth very little at tax time. Concrete and lumber and sewers, every assessor knows, are what pay the town bills.

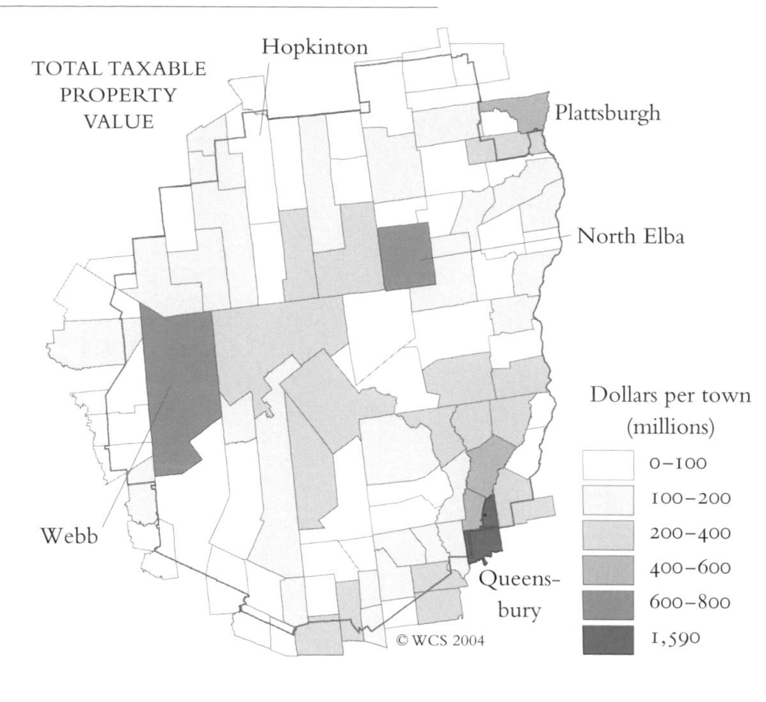

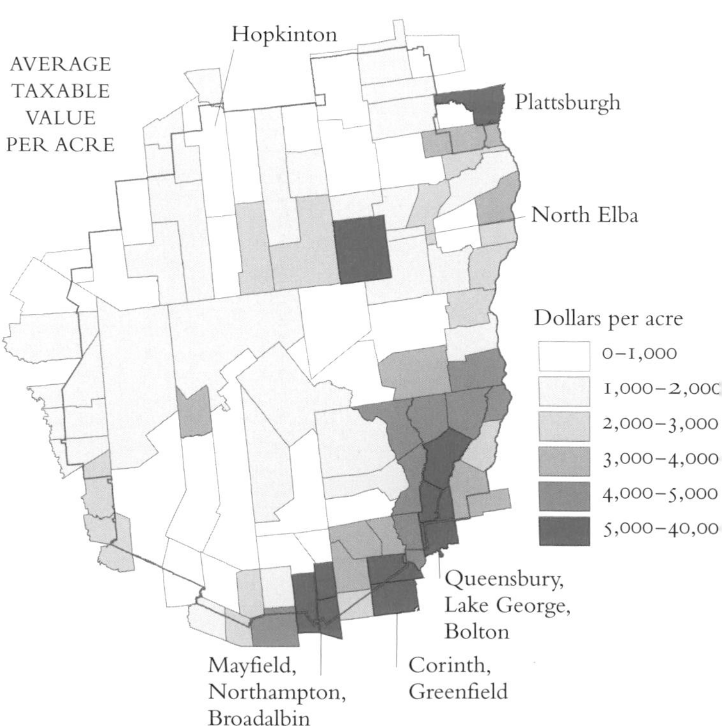

EVERYONE wants to know where the taxes are low and the services high. Like many other simple questions, this has no simple answer. The maps show four different ways of looking at the problem, each of which gives a different answer.

Total taxes. The total tax collected by a town is a measure of the total services that the town provides but not of what the individual taxpayers pay for these services. It is high where populations are large and property values are high. Towns that have people must provide for them; towns with wealth invariably tax it.

Tax per acre is a measure of how concentrated the tax base is and is highest in the most uniformly developed towns. A town like North Elba, with much woodland and a few highly developed areas, ranks high in total tax but lower in tax per acre.

Tax per thousand dollars of property value is the most familiar measure of taxation and the most useful for comparing towns. The results are interesting and a bit unexpected: tax rates tend to be low in the poorest towns (which can't afford high rates) and also in the larger or more developed towns, which have a large tax base and don't need high rates. They are highest in towns that are not particularly wealthy but choose to provide high levels of service and have at least a few wealthy taxpayers who can, willingly or otherwise, pay for them. And they tend to be extremely variable: in Newcomb, you will pay $1,410 on a $30,000 house and lot. Next door, in Long Lake, you will pay $391.

If only residents paid taxes, the *tax per resident* would tell us the average tax paid by individuals. Instead, because many businesses and nonresidents pay taxes, it tells us the average tax *collected* per resident, which is quite different. Not surprisingly, it averages much higher in the empty center of the park than at the populous edges.

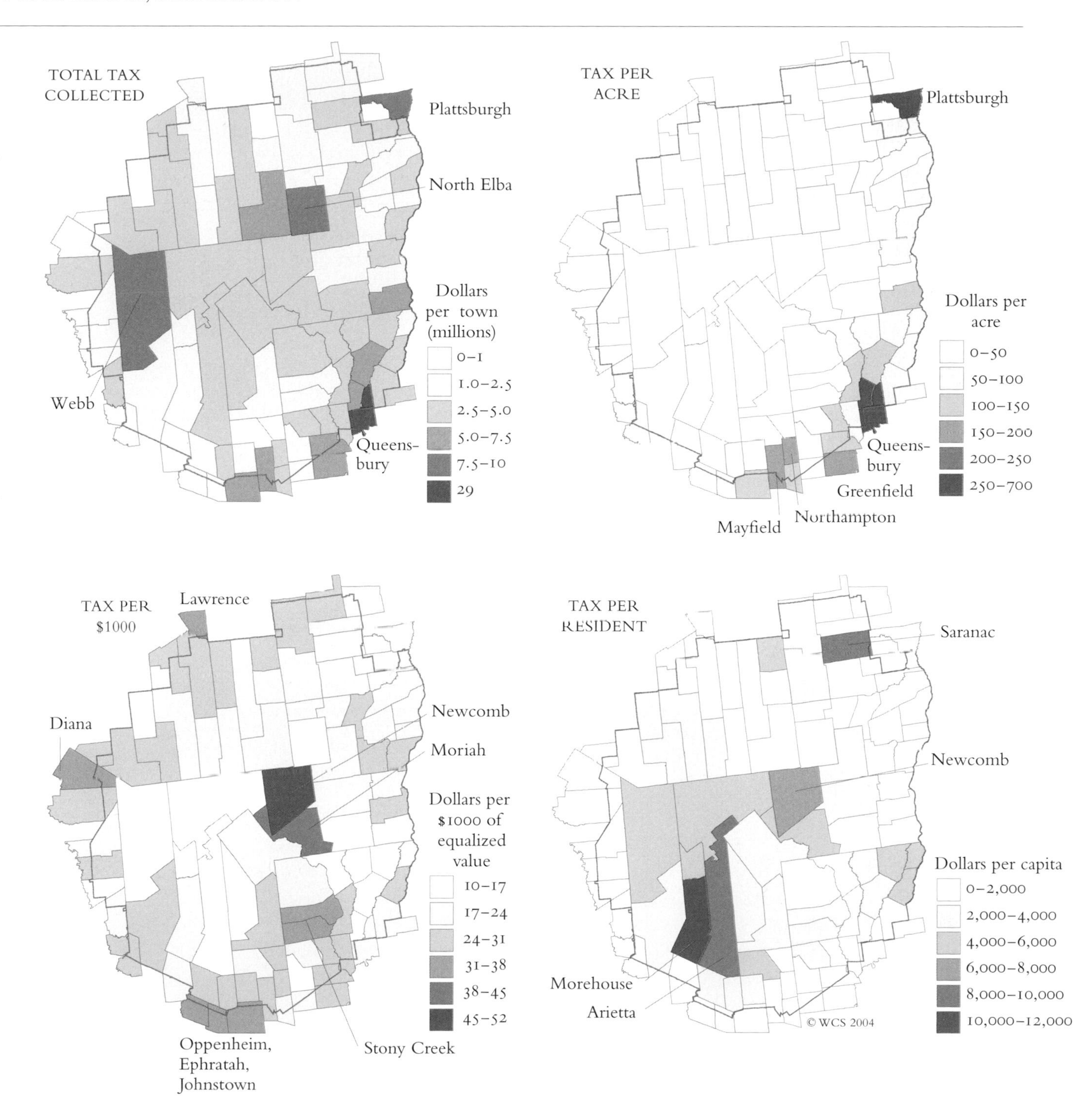

It is widely and incorrectly believed in the Adirondacks that the state does not pay taxes on state land. Every town official knows otherwise. New York State is in fact the largest taxpayer in the Adirondack region. In 1994 it paid a total of 41 million dollars, roughly 4% of all property taxes in the twelve Adirondack counties.

Direct tax payments from a state to the towns in it are legally unusual, which is one reason many people believe they do not exist. By a famous U.S. Supreme Court decision of 1819, a lower jurisdiction may not tax a higher one. Thus the towns, which are the lower taxing unit, may not tax the state unless the state consents to be taxed.

New York State first consented to be taxed in 1886, the year after the Forest Preserve was created. The legislature said that because the Forest Preserve holdings imposed burdens on Adirondack and Catskill towns while conferring benefits on the entire state, the state should recompense those towns. Accordingly, the towns were allowed to assess and tax Forest Preserve lands exactly as if they were any other local property. They did so, gladly, and do so to this day.

Since 1960, the state has paid an additional *Adirondack aggregate assessment* of some 17 million dollars per year, which is basically a carefully disguised subsidy to property-rich towns. The aggregate assessment was instituted to compensate for changing property values. Normally, as a town developed, the value of the developed lands would increase relative to the value of the Forest Preserve lands, and the proportion of taxes paid by the state would decrease. The aggregate assessment prevents this from happening; essentially it guarantees that the state will never pay a smaller share of each town's total tax revenue than it did in 1960. Thus, somewhat unfairly, the towns that have recently developed and so are collecting more money from private landowners can collect more money from the state as well. Deals like that don't come along every day.

Total state payments—taxes and the aggregate assessments—are strongly concentrated in the Adirondack counties which have large amounts of Forest Preserve lands. In these counties they average between 4 and 9 million dollars a years and are a very important source of local revenues. In Essex County, despite a number of developed towns, the state pays 16% of all property taxes. In Hamilton County, with much state land and only a few small towns, the state pays a whopping 39% of all the property taxes. In the remaining counties state payments are less important but still appreciated. No county supervisor ever born, even in prosperous Oneida County, ever turned up her or his nose at a free million or two of unrestricted state funds.

Because the taxes the state pays are based on local assessments, and because the state pays no tax on much state land outside the Blue Line, the amount the state pays per acre differs greatly from place to place. An average acre of state land in developed and moderately prosperous Washington County brings the town that contains it five times as much tax revenue as one in woodsy Hamilton County and—in an even more extreme unfairness—twenty-five times as much as one in struggling Jefferson county.

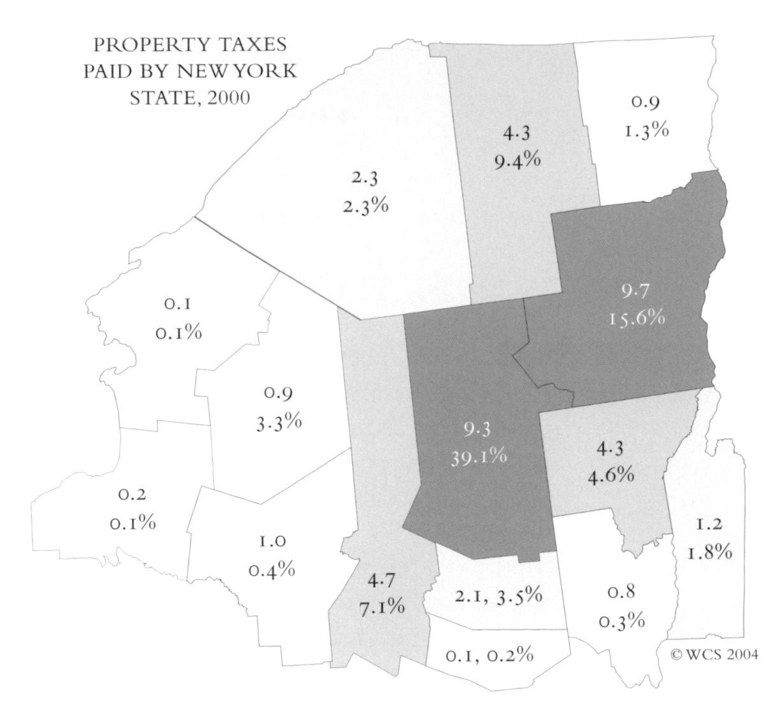

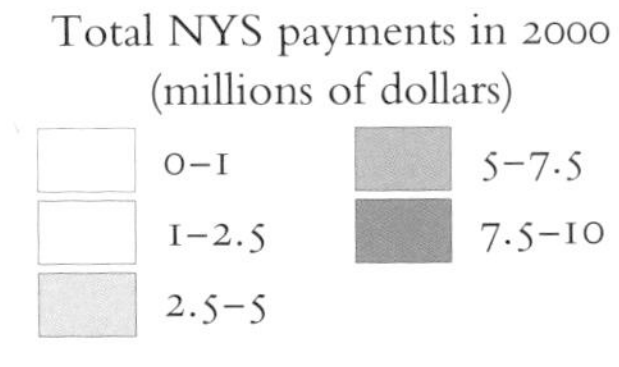

The top number in each county is the total NYS payment in millions of dollars. The lower one is the percent of all property taxes in the county paid by New York State.

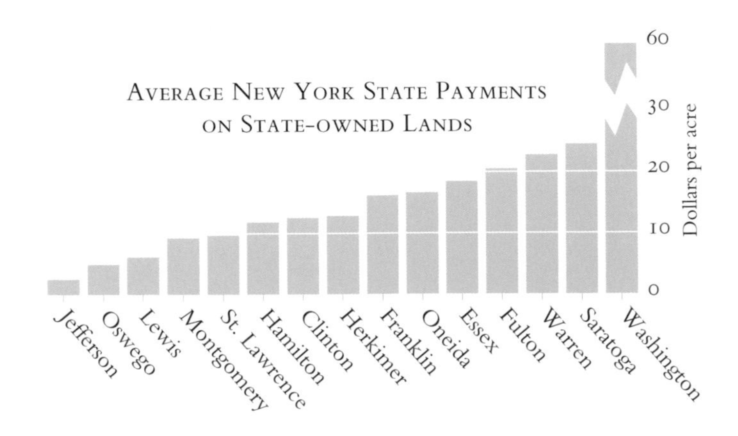

WHILE many landowners pay property taxes, many others do not. Many charities, churches, disabled people, dead people, farmers, foresters, fire departments, governments, hospitals, nonprofits, old people, soldiers, schools, and veterans are wholly or partially exempt from paying taxes on the property they own, use, or lie in. The total is consequential: 19% of the property value in the ninety-two Adirondack towns, 3.5 billion dollars of a total of 18.2 billion, is exempt from taxation. But the Adirondacks are not unique in their level of exemptions, and the majority of their exempt value is not forest land.

In the majority of Adirondack towns, less than 20% of the total assessed property value is tax-exempt. Towns with substantial tax-exempt facilities like prisons, government offices, and public utilities can have larger values. Dannemora, with a large prison, is 54% exempt; Croghan with large power plants is 40% exempt; North Elba with a prison and extensive state-owned Olympic facilities is 48% exempt.

The Adirondack counties average about 30% exempt, fairly typical of the state. St. Lawrence and Clinton Counties, with large amounts of federal land, have among the highest exemption rates in the state. Hamilton County, with little developed land and no major government facilities, has the lowest percentage of exempt value in the state.

Except in the towns with large federal or state facilities, the majority of the tax-exempt value in the Adirondack counties is on private lands where there is at least some development. Although commercial forest owners in the Adirondack counties receive 160 million dollars in tax exemptions, this is only 1% of the total exempt value and only 0.3% of the total property value.

If the forest taxation programs, which now only enroll about 16% of the eligible forest lands, were more attractive to landowners, the total value of the exempt forest lands might be higher. But because trees are cheap compared to the buildings we replace them with, no conceivable forest tax exemption will ever amount to more than a few percent of the total property value.

PERCENT OF PROPERTY VALUE THAT IS TAX-EXEMPT

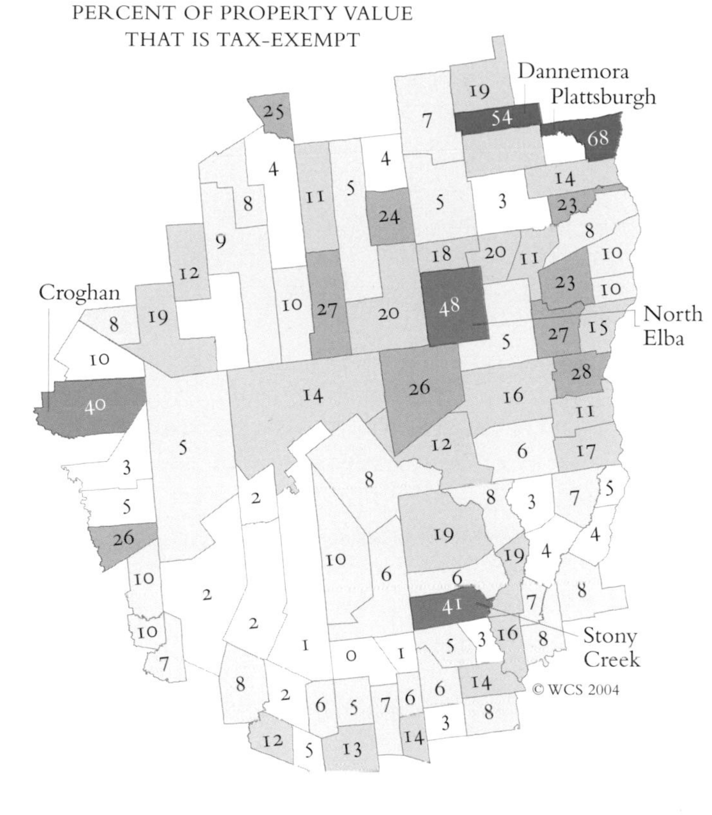

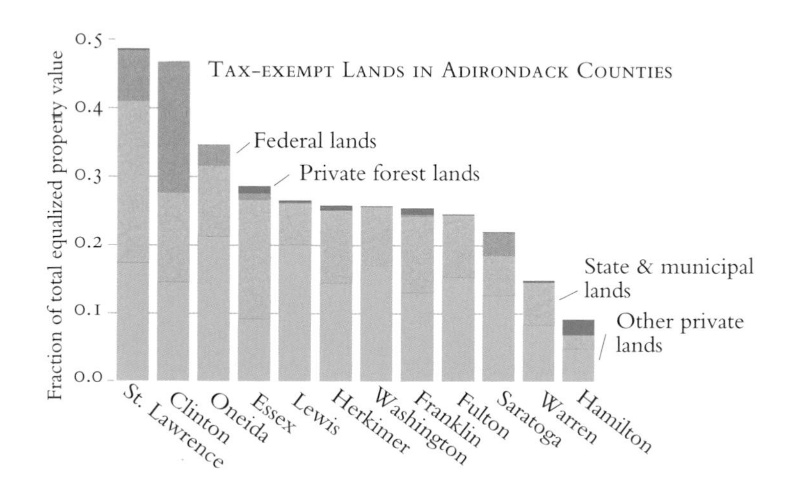

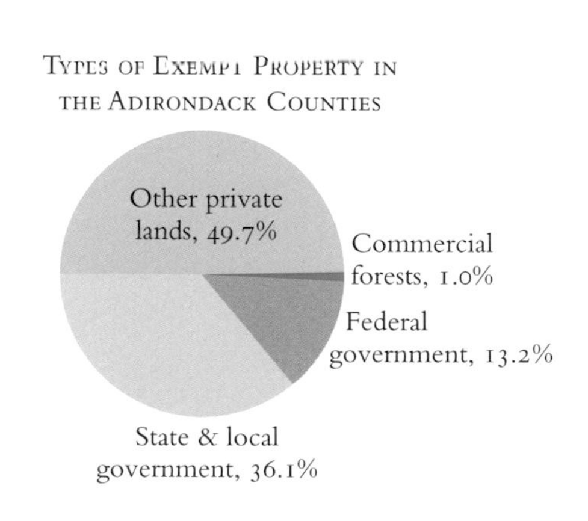

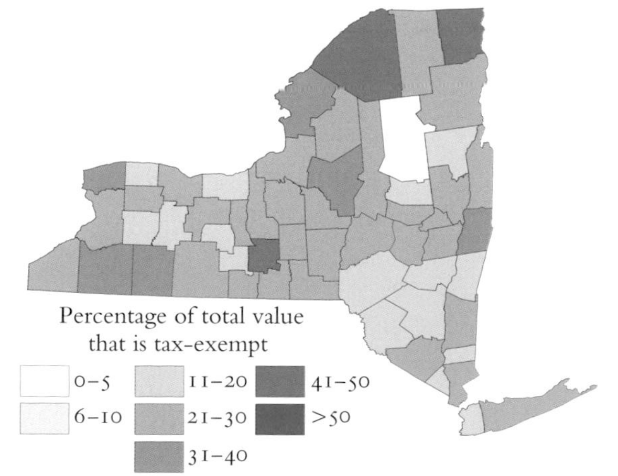

FIRE DEPARTMENTS AND RESCUE SQUADS

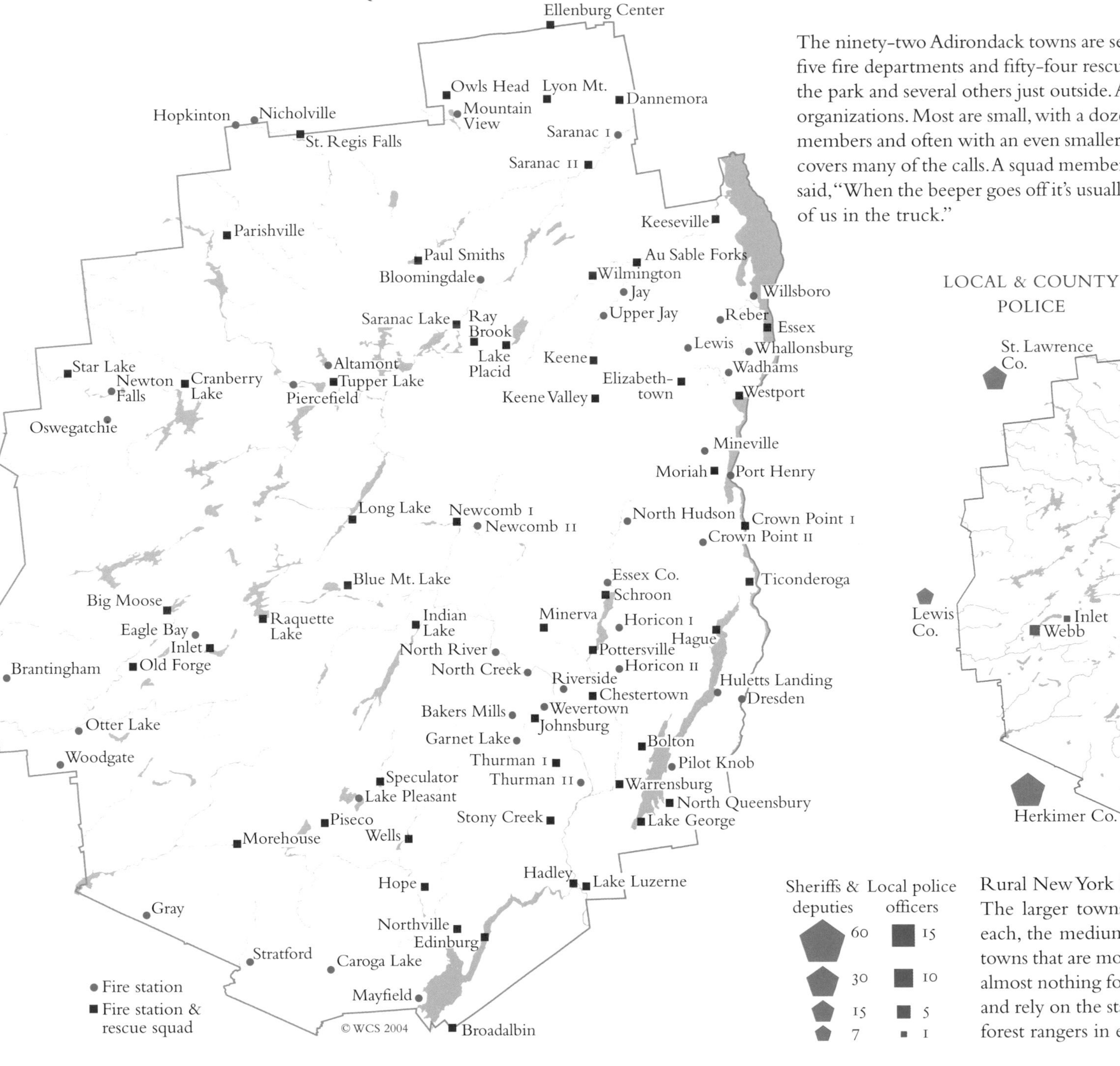

The ninety-two Adirondack towns are served by ninety-five fire departments and fifty-four rescue squads within the park and several others just outside. All are volunteer organizations. Most are small, with a dozen or two dozen members and often with an even smaller core group that covers many of the calls. A squad member in a small town said, "When the beeper goes off it's usually the same three of us in the truck."

LOCAL & COUNTY POLICE

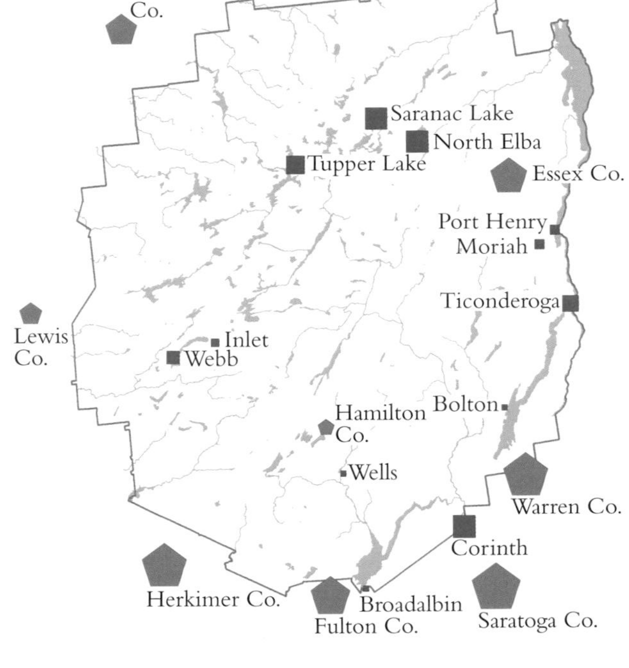

Rural New York has relatively few local police. The larger towns have ten to fifteen officers each, the medium-sized ones one or two. The towns that are mostly woods and highways have almost nothing for a local police force to police and rely on the state police, county sheriffs, and forest rangers in emergencies.

12 *Vital Services*

BECAUSE the Adirondacks are a settled wilderness, they have the same needs for protection, care, and emergency assistance as other settled places. But because they are only sparsely settled, with lots of woods between the towns, the service providers have to stretch their resources and people to cover a large area. They do this impressively, with thrift and competence and very few compromises. But still there are limits to what they can do: a twenty-bed hospital will not have an obstetrician and a delivery room; a hospice with only a few patients at a time will not be able to afford a round-the-clock professional staff.

These limits have led, inevitably, to an emergency service system that is strong in mobile units and first-reponse personnel but weaker in medical specialties and long-term care. The Adirondacks are a very good place in which to break a leg in the woods or get stuck on the third pitch of an ice climb and a better place than many in which to have your chimney catch fire. But they can be a poor place in which to have a heart attack, to try to provide for an elderly or chronically ill family member, or even in which to be pregnant.

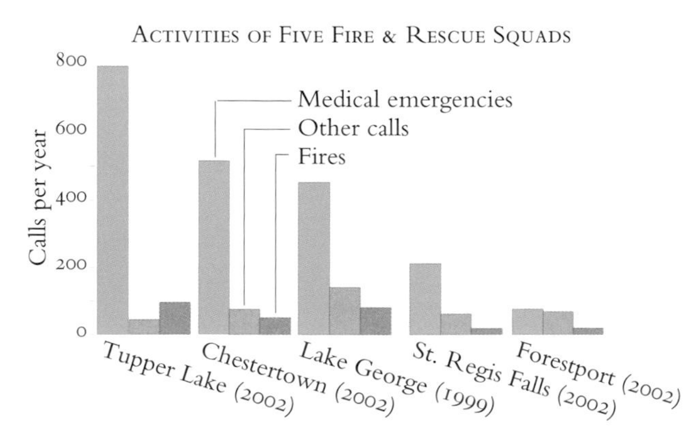

Protecting and caring for the residents and visitors of an area the size of a small state is a substantial undertaking. The park has at least 240 separate emergency and service organizations, including 38 police forces and uniformed services, 95 fire and rescue squads, over a hundred medical facilities and practices, and at least a dozen organizations that provide nursing and home care. These organizations respond to over a thousand fires, about a thousand serious crimes, five hundred to a thousand deaths, and perhaps ten thousand medical emergencies each year. They also answer tens of thousands of nonemergency calls and provide routine nursing and medical care to several hundred thousand people.

FIRE CALLS ANSWERED BY SARANAC LAKE VOLUNTEER FIRE DEPARTMENT, 2001

False alarms, 40.4%
Service calls 23.5%
Brush fires, 1.8%
Chimney fires, 2.4%
Storms, 2.4%
Vehicle fires, 4.8%
Mutual aid calls, 5.4%
Building fires, 6.0%
Hazardous materials, 6.6%
Good intent calls, 6.6%

To supply these services requires the full-time effort of at least eight hundred professionals and part-time work from perhaps twice that many volunteers. The professionals include 46 forest rangers, 53 environmental conservation officers, 46 sheriffs and deputies, 76 local police, at least

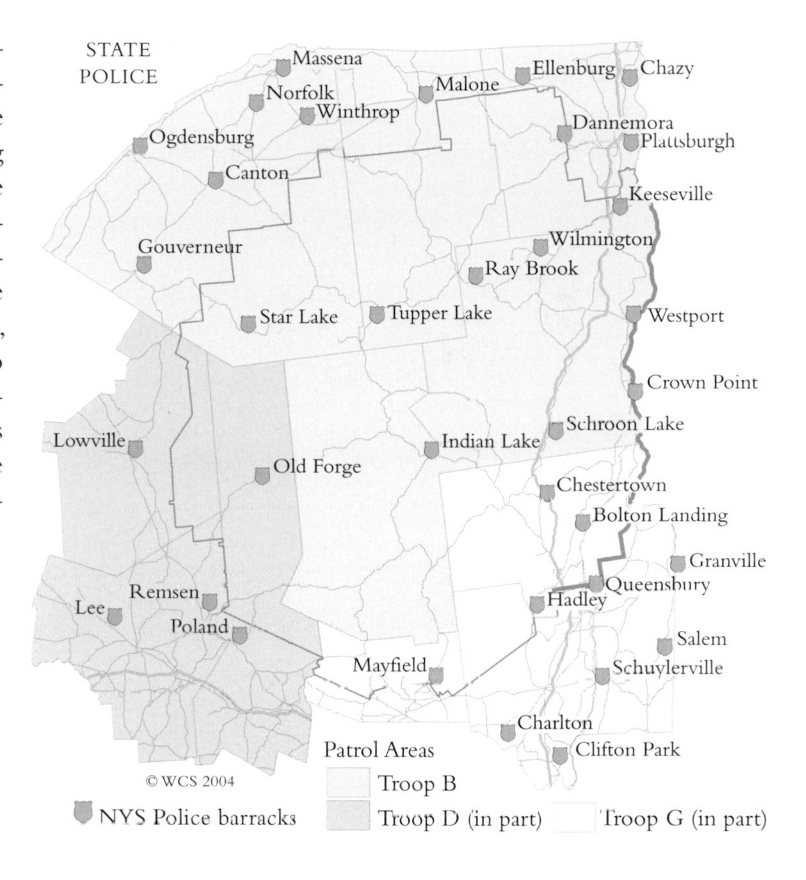

250 state troopers, 190 doctors and dentists, and several hundred nurses and other medical staff. The volunteers include the members of the ninety-five fire and rescue squads, several search and rescue organizations, and hundreds of other volunteers that assist at hospitals, clinics, hospices, and nursing homes. Service—people helping people—is a powerful idea everywhere; in a big place with few people, it is an essential one.

IN competence, toughness, and versatility, the New York forest rangers are the preeminent emergency response organization in the park. They have, by law, a triple duty: they are responsible for the "care, custody and control" of people on public lands. In practice this extends from routine patrol work and police work in campgrounds to general public assistance and education and, perhaps most important, to search, rescue, and wildland fire fighting. Rescue and fire fighting are their technical specialties, for which they are highly trained and at which they excel.

The ranger corps is surprisingly small, both in comparison to their responsibilities and to other uniformed forces. Currently, there are only 105 field rangers and 12 lieutenants in all New York, down from 140 rangers in the 1960s. About 60 of these have districts that include the Adirondack Park, and about 50 live within the park. Each of these rangers is responsible around 50,000 acres of public lands and is also available to cover backcountry emergencies on an equal or larger amount of private land. For each ranger in northern New York there are at least four state policemen, seven sheriffs and deputies, and two village policemen.

FOREST RANGERS, 2003

Jamie Smith
Dan Fox
Jeffery Balerno
Lt. Fred LaRow
Jay Terry
Lt. Bob Barstow
Joseph Rupp
Gary Friedrich
Wayne LaBaff
Keith Bassage
Chris Kostoss
Julie Harjung
Kevin Burns
Lt. Robert Marrone
Howard Graham
Scott Murphy
Joseph Kennedy
Joe LaPierre
Charlie Platt
Will Giraud
Bernard Siskavich
Joel Nowalk
Will Benzel
Jim Giglinto
Scott vanLaer
Robert Zurek
James Waters
John Scanlon
Edwin Russell
Greg George
Paul Clickner
Mark St. Claire
Pete Evans
Lt. Brian Dubay
John Chambers
Douglas Riedman
Bruce Lomnitzer
Steve Ovitt
Werner Schwab
Bob Coscomb
Rick Schroeder
Thomas Eakin
Jamie Laczko
John Seifts
Lt. David Brooks
Steven Guenther
Steve Bazan
Mark Kralovic
Rick Requa
Tony Goetke
John Ploss
John Solan
Rob Praczkajlo

● Field ranger
■ Lieutenant ranger
○ Vacant post
— DEC ranger district

Search & Rescue. Forest rangers describe a mission as a *search* when they attempt to locate someone who is missing, a *rescue* when they assist someone out of the woods, an *attempt to locate* when they carry an urgent message to someone in the woods, and a *recovery* when they bring out a body. In 2000 and 2001 the rangers of Region 5 conducted 196 searches, 107 rescues, 19 attempts to locate, and 10 recoveries. Seventeen of these missions involved specialized back-country medical teams. The success rate of these missions was quite high; in 2000 and 2001 there were no unsuccessful searches and only four cases in which a severely ill or injured person died before they could be rescued.

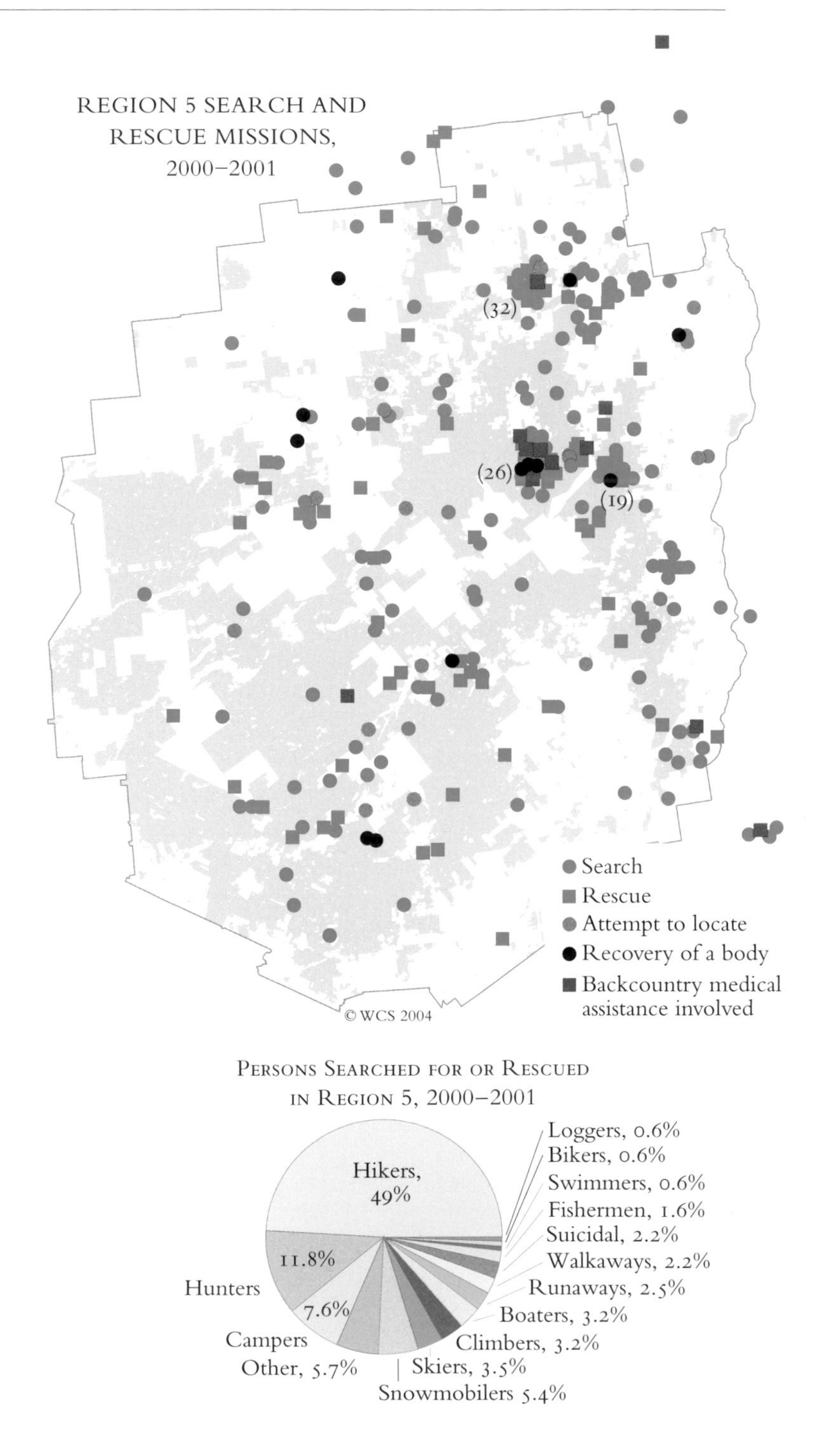

IN 2001, an average bad year, both the winter and the spring were quite dry. The fire season began in mid-April, not long after snowmelt. About half of the early fires, as always, were fires set to burn brush or grass that got out of hand. Most of these "debris fires" were small but three burned twenty acres or more. Spring grass fires don't burn especially hot, but they can burn surprisingly far.

Other human activities and agents—smoking, logging, burn barrels, a lawn mower, and an electric fence—accounted for most of the other spring fires. Arson was suspected in three of them. One, somewhat mysteriously, was blamed on a manure pile. Railroads, perennial fire-starters, started fires that burned 435 acres, a little over a third of the total acreage burned that year.

Almost all of the early fires were in suburbs and rural areas rather than in forests. The first woodland fires, all starting from campfires, were in May. In the next seven months a total of forty-one campfires escaped to the woods. The total area burned was only forty acres, less than an acre per fire. The only naturally ignited fires were thirteen lightning fires in August, which burned a total of 80 acres and included the year's third-largest fire.

By December, the rangers had fought 166 wildfires. A total of 1,153 acres burned, insignificant compared to the great logging and railroad fires of the early 1900s (*Map 6-2*) but a busy season by modern standards. The geography and causes were modern too. Most of the fires were in settled areas or along the second-growth edges of the park and not in the central woods. The central woods, which are wetter and older, can burn too, but getting them started is much harder. And unlike the great fires of the past, most of the 2001 fires were not only caused by people but *deliberately made* by people—they were useful fires that had grown, through carelessness or nature's old perversity, larger than their makers intended.

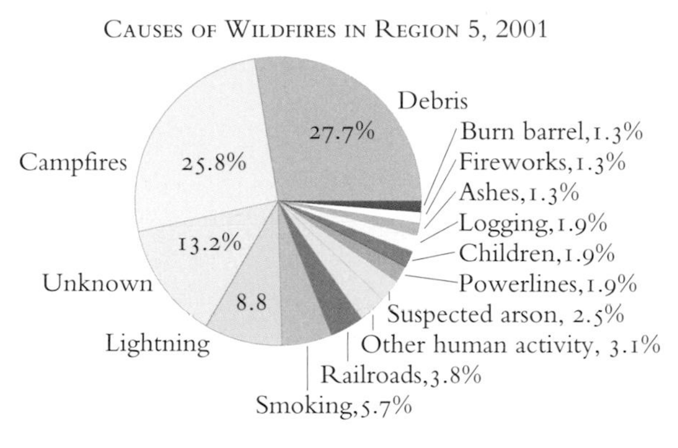

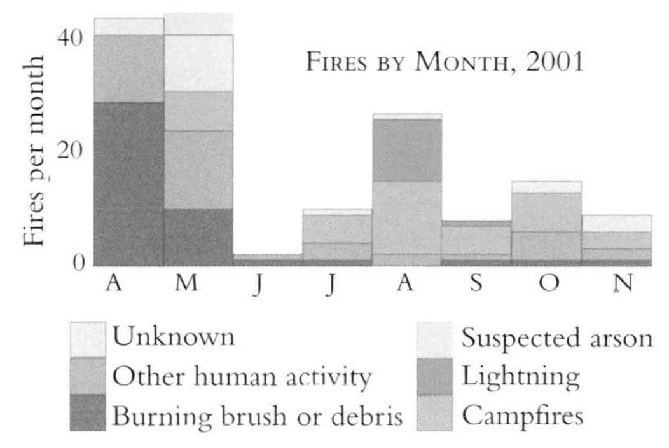

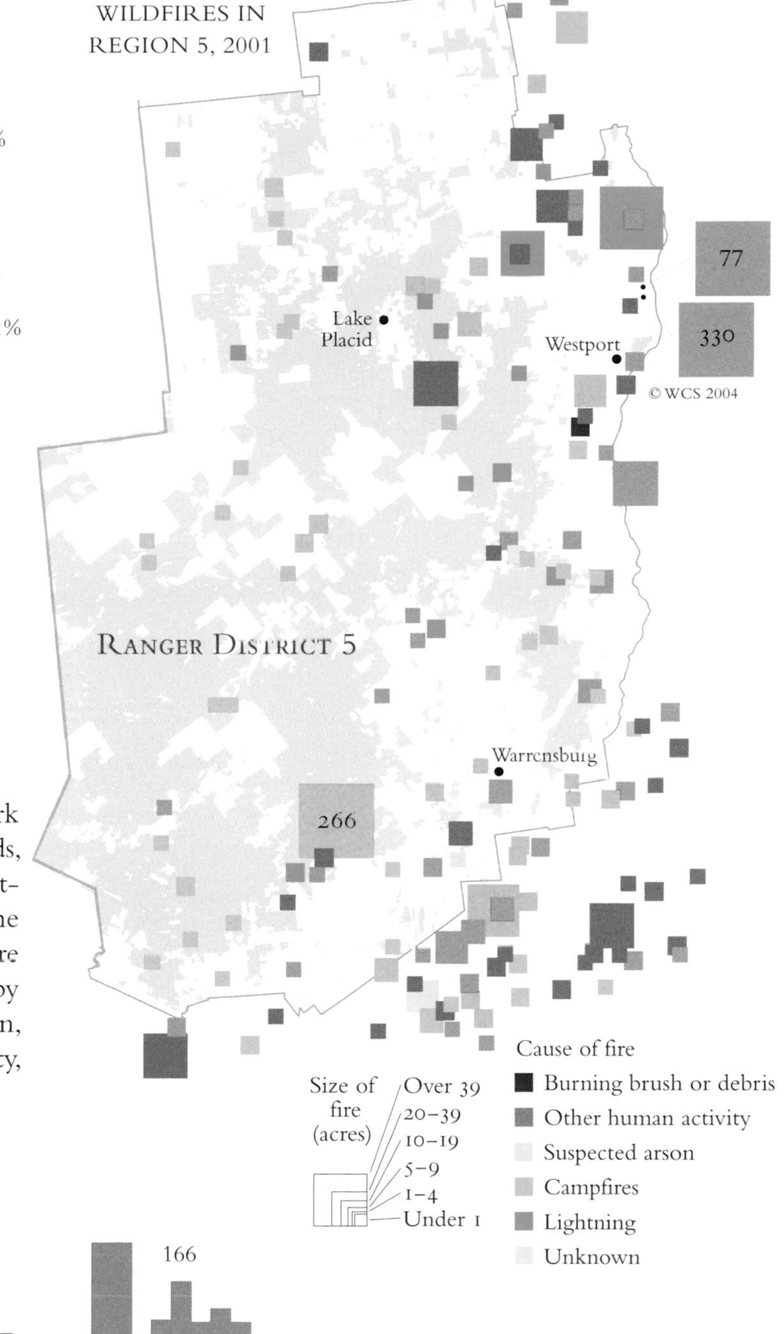

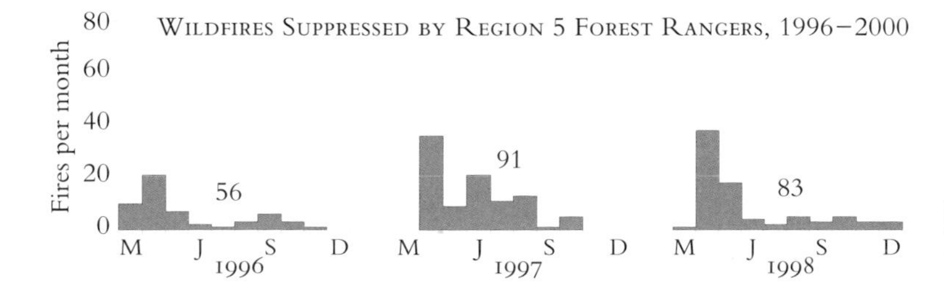
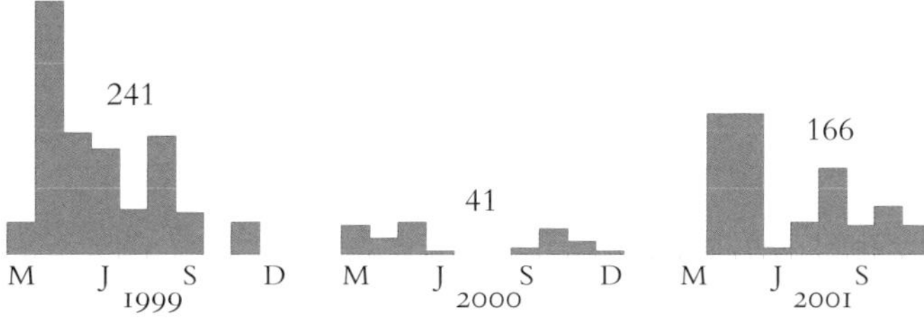

BECAUSE medical practices are expensive to equip and run, how big towns always have more medical services than small ones. A family practice may need to see four to six thousand patients a year to meet its expenses and survive. A specialty practice, which can charge more but may have higher equipment and insurance costs, may need to see three thousand patients or more a year. A town with two thousand people—and two-thirds of the Adirondack towns have that number or less—will not have enough patients to support a specialty practice and will only support a family practice if everyone makes several visits to the doctor every year.

In 1970 the Adirondacks were served by a sparse and to some extent troubled health-care system with one large hospital, four small ones, and perhaps twenty scattered clinics and general practitioners. The small hospitals were struggling, the costs of individual practice were rising, and older doctors who retired were not being replaced.

Thirty years later, thanks to local initiative and federal aid, primary care is probably more available in the Adirondacks than it ever has been. Key elements in the change have been the creation of networks of clinics that share personnel and facilities and the use of para-professionals in clinics too small to support a full-time doctor. The largest primary-care network is the Hudson Headwaters Health Network, which was started by Dr. John Rugge in 1974 and now has eleven clinics and a medical staff of over seventy.

Hospital and specialty care, however, remain limited. No new emergency medical facilities have been built, the specialists remain concentrated near the large hospitals, and small hospitals are still struggling to stay open. And because the park is large and hospitals widely separated, getting critically ill patients to them is not easy.

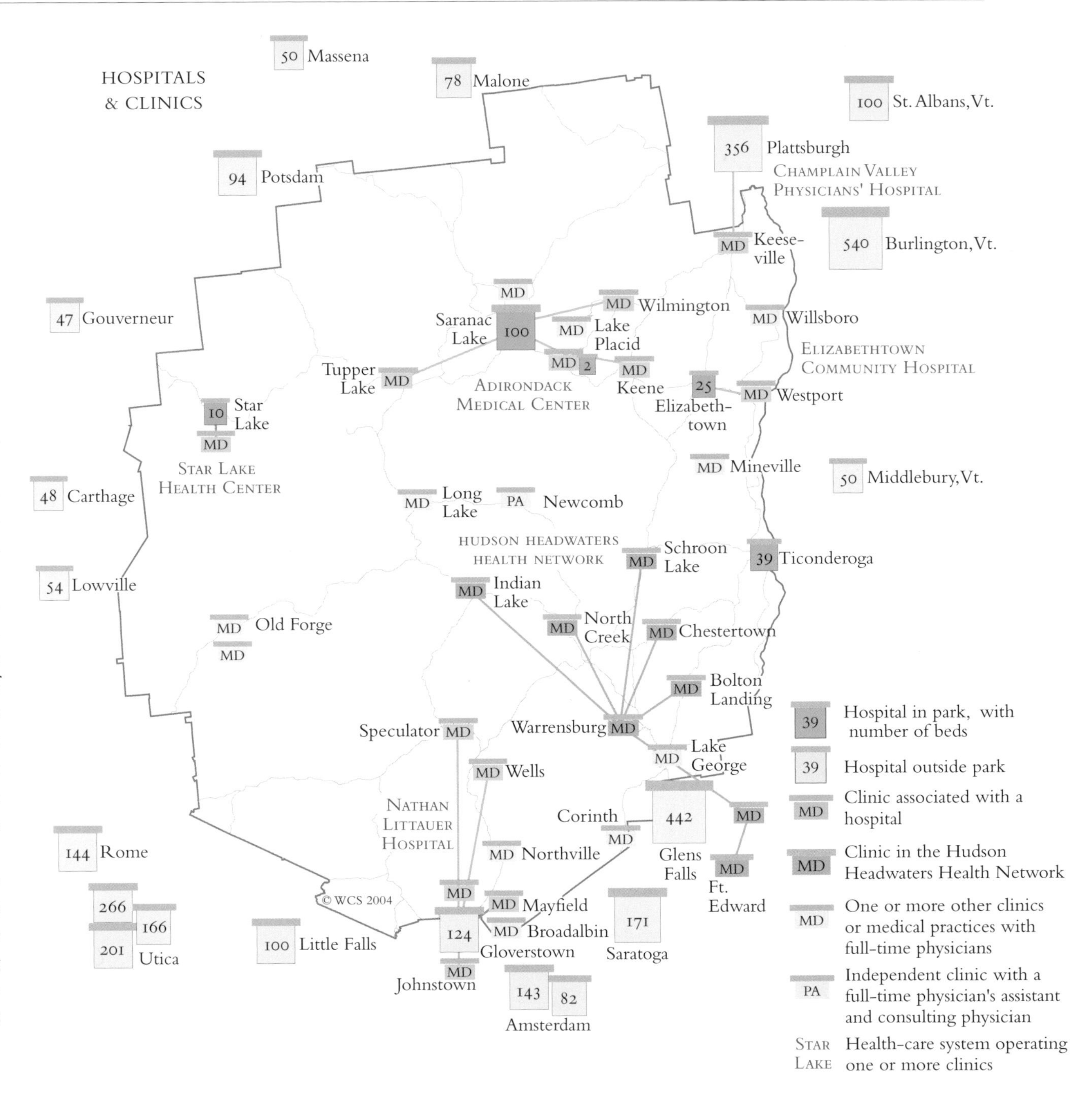

In most medical emergencies in the park, the first responder is one of the fifty-four local rescue squads. They are fascinating organizations: autonomous, highly trained, all-volunteer, locally funded, free to their users, usually run by elected officers, and usually dependent on a dedicated cadre of long-term members. They are perhaps our most remarkable example of just how well communities can organize to provide for their own needs and should make us wonder what other essential needs could be met in the same way.

Because the rescue squads can now do many things that used to require an emergency room, in most emergencies what matters is how fast the rescue squad gets to the patient and not how fast the patient gets to the hospital. But in a few cases where a critical hospital procedure must to be started as quickly as possible, travel time matters a great deal. In heart attacks, for example, the survival rates of the patients that reach a hospital have greatly improved. But as a result, half of all heart attack fatalities now occur before the patient reaches the hospital. Many of these might have been prevented if the travel time had been less.

It is in time-critical emergencies like strokes and heart attacks that a basic fact of the park's geography, the eighty-mile wide central circle without emergency medical facilities, may have a great effect on the outcome. Under good conditions the state police helicopter from Saranac Lake may be able to make a rapid evacuation. But when conditions are not good—and in the mountains they are often not good—the responsibility belongs to the local rescue squad. And no matter how well drilled they are or how daring their driver is, if the patient is within the inner blue circle on the map, it is likely to be an hour or more from the time the call arrives to the time the patient enters the hospital.

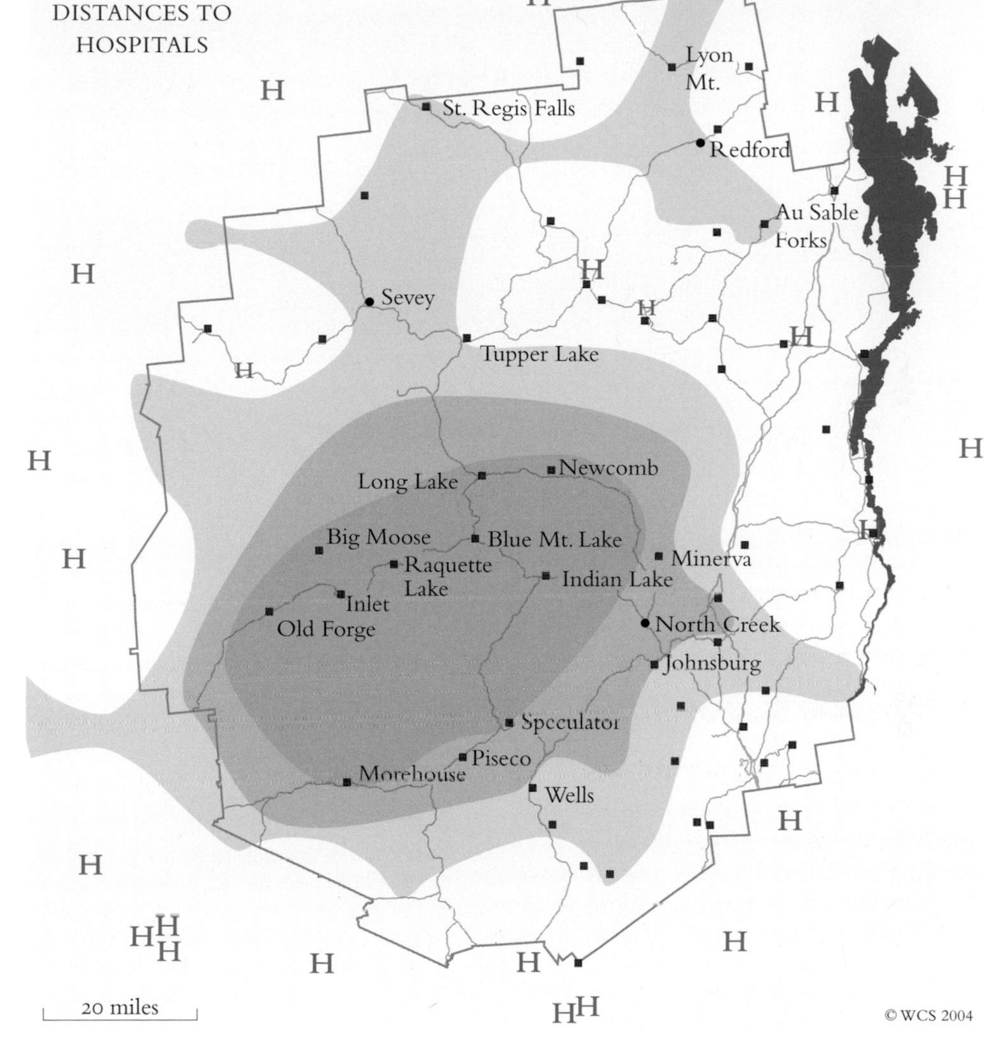

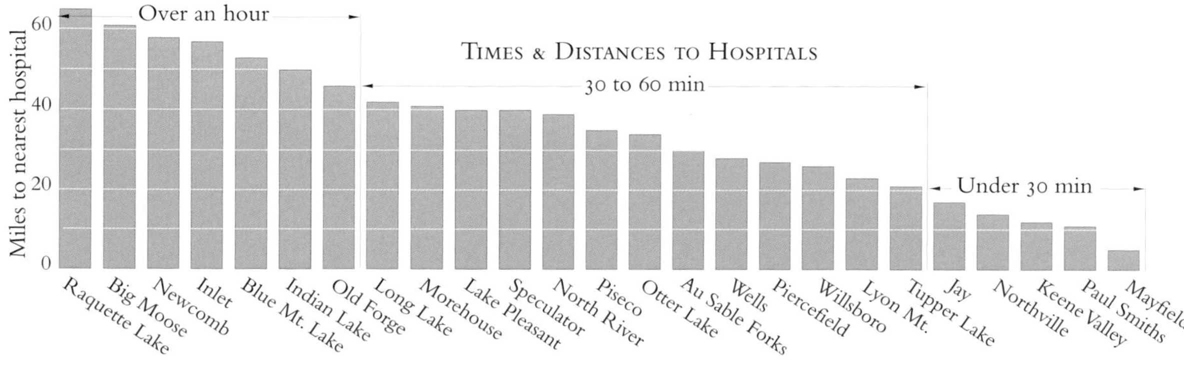

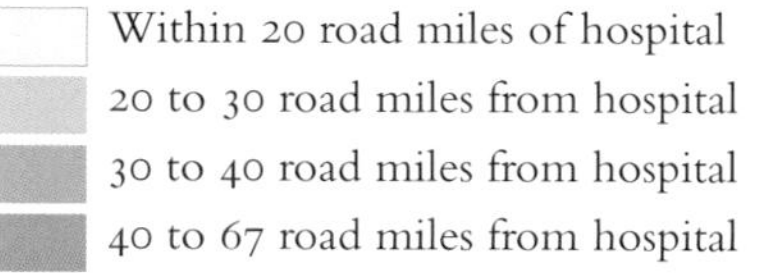

MEDICAL care is more than general practices and emergency rooms. Chronically ill patients need skilled nursing, access to specialists, and often specialized treatments like dialysis and chemotherapy. Elderly patients need nursing homes and hospice care. Women need gynecological care, midwives and obstetricians, and delivery rooms. Everyone needs dentistry, and surprisingly often, the everyday specialties like ophthalmology, radiology, and surgery.

It is in the provision of supportive and specialized care that the Adirondacks, like most rural areas in the U.S., are most limited. Specialty practices require diagnostic facilities, the support of other specialists, and a reliable supply of well-insured patients. An oncologist needs a CAT scanner or an MRI machine, and surgeons and pathologists and radiologists to collaborate with. An obstetrician needs a delivery room and a surgical team. A hospital with a delivery room needs to equip it and insure it and, because babies can't be scheduled, to have at least two obstetricians on their staff. Small hospitals tend to have only a few specialists and limited diagnostic facilities and have to send patients—and the income those patients generate—to larger hospitals. This loss of income is alarming because the small hospitals are critically important Adirondack institutions. All provide essential care to their communities, and the loss of any of them would create a serious gap in the park's emergency care system.

The maps shows the current system and its gaps. Specialized care is present at all the big hospitals and, through a rotating staff of out-of-town specialists, at the main Hudson Headwaters clinic in Warrensburg. It is limited at the smaller hospitals and absent otherwise. Hamilton County has none of any kind. Counseling services are available in nine Adirondack towns, gynecological services in four, but a delivery room only in one. Even ordinary supporting businesses, like pharmacies, are limited: eleven towns have clinics but no pharmacies.

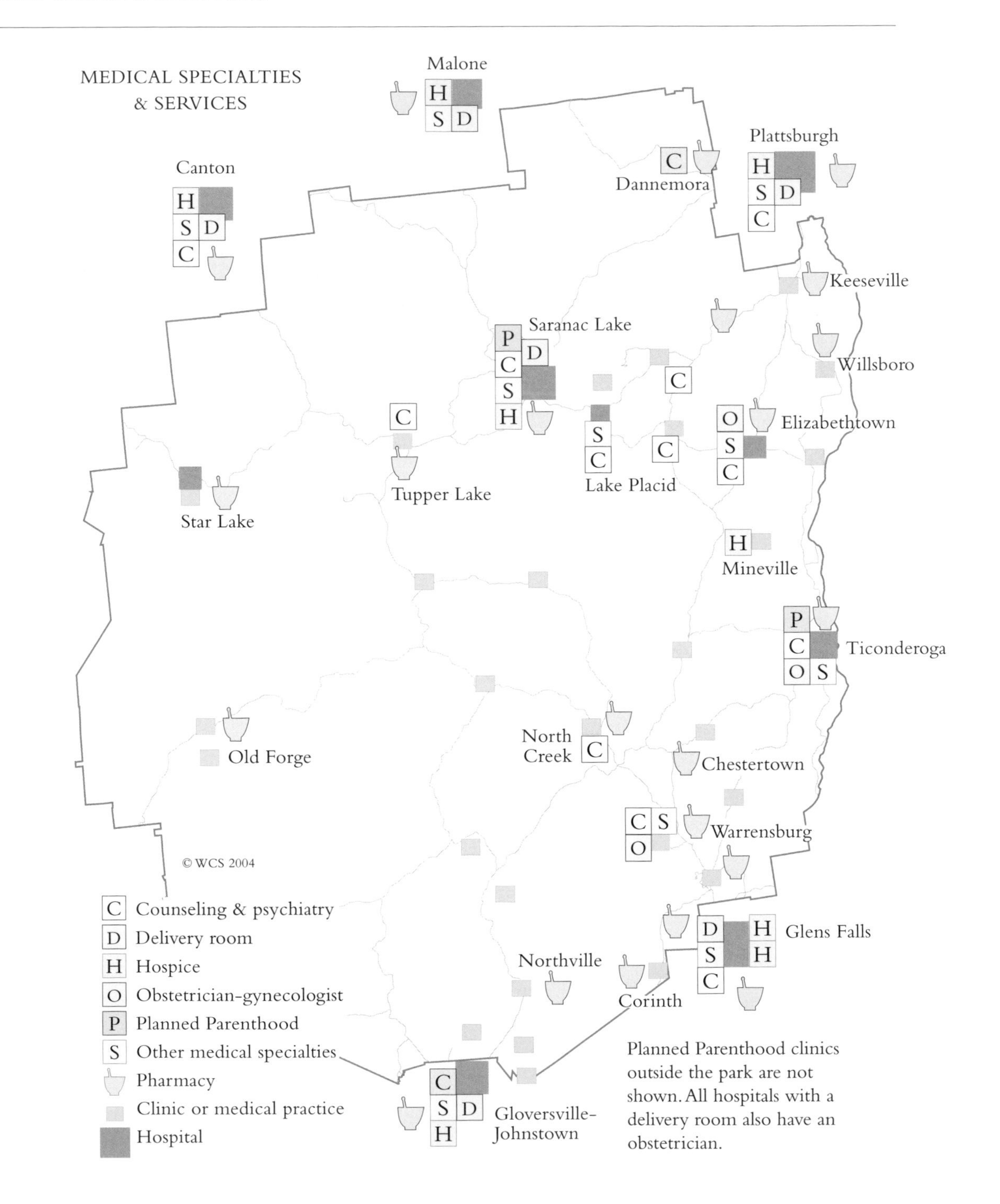

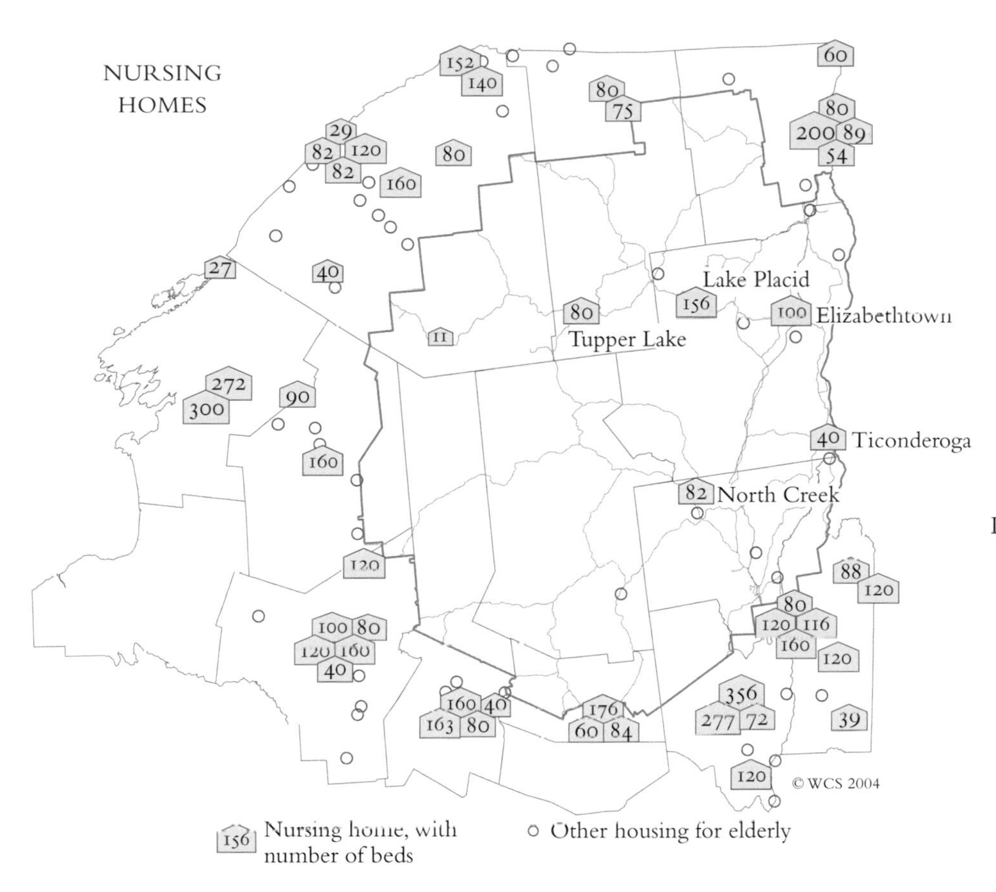

Nursing homes, which are relatively common outside the park, are among the scarcest medical facilities within it. At present there are only six within the blue line, all but one located in the north and east. There are none in the Adirondack portions of Clinton and Herkimer Counties and none at all in Hamilton County, despite its relatively large elderly population.

The five Adirondack nursing homes have a total of 469 beds, or roughly about one for every 330 residents. Since about nine people out of every 330 in the park are eighty years old or over, it is very likely that the need for nursing home beds is greater than the supply.

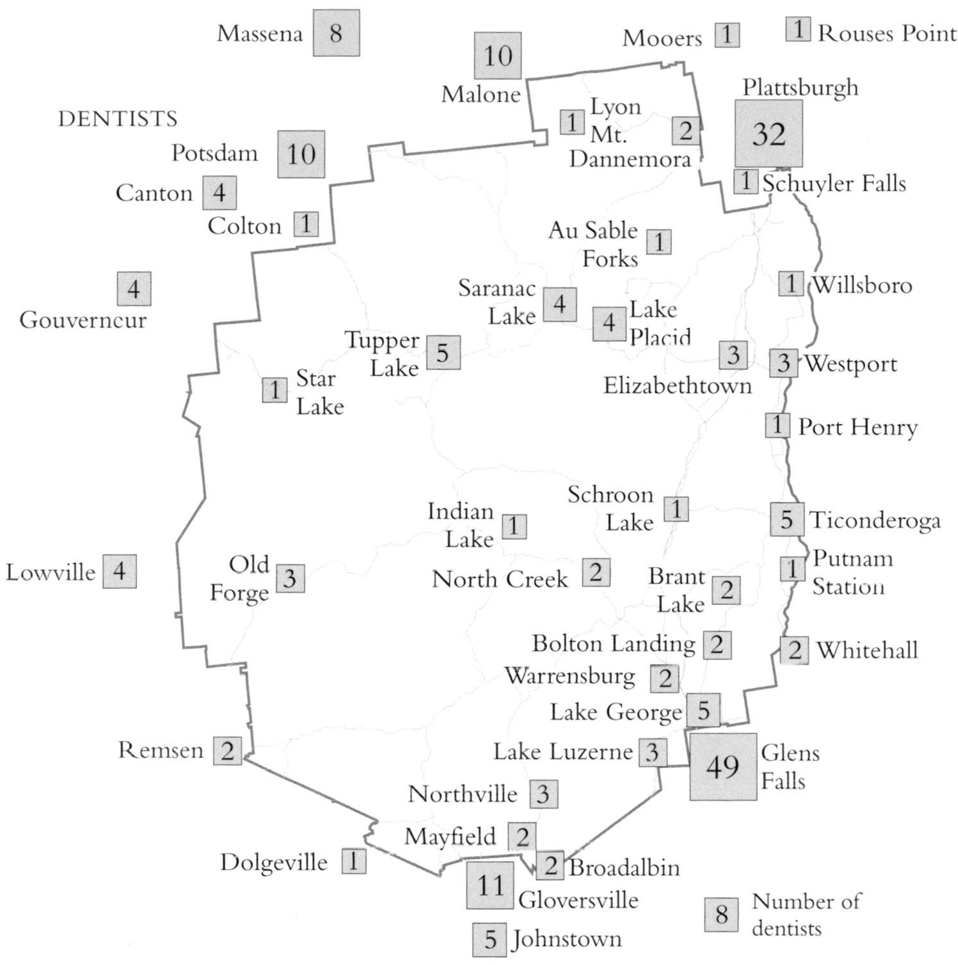

About sixty dentists practice within the park, or, very roughly, one for every 2,500 people. This is about what we would expect for a rural population. Dentists spend more time with their patients, both per visit and per year, than doctors and require fewer patients to support them. In theory a population of 1,000 receiving regular dental care could keep a single dentist busy and prosperous. In practice, because many people neglect or can't afford regular care, it takes about 2,000 to 3,000 people to support each dentist.

The park's sixty dentists practice in twenty-six different towns, making them the most well-dispersed medical providers in the Adirondacks. Only the smallest towns like Jay and Newcomb, or the most seasonal like Long Lake and Inlet, do not have a dentist.

THE BANKING SYSTEM

The banking system of northern New York consists of approximately thirty-six banks with about three hundred offices and total deposits of 10 billion dollars per year. Within the Adirondacks there are currently fifteen banks with about thirty-seven offices. In the ten Adirondack counties for which we have good records, two-thirds of the money deposited in 2000 went into local and regional banks and a third into large national chains.

Northern New York Bank Deposits, 2000

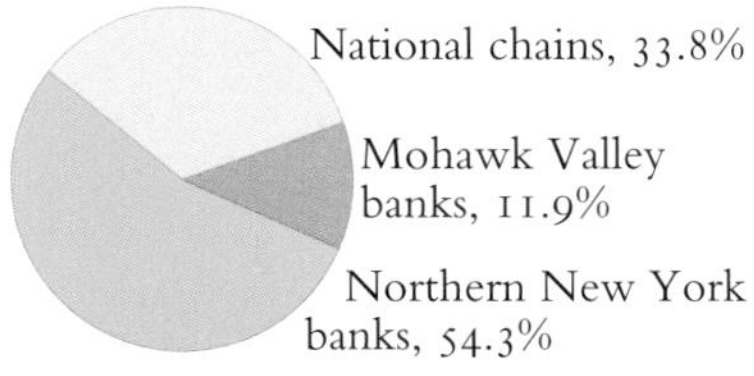

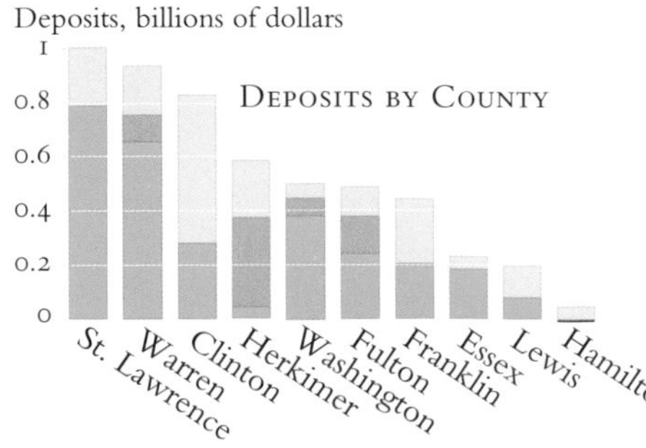

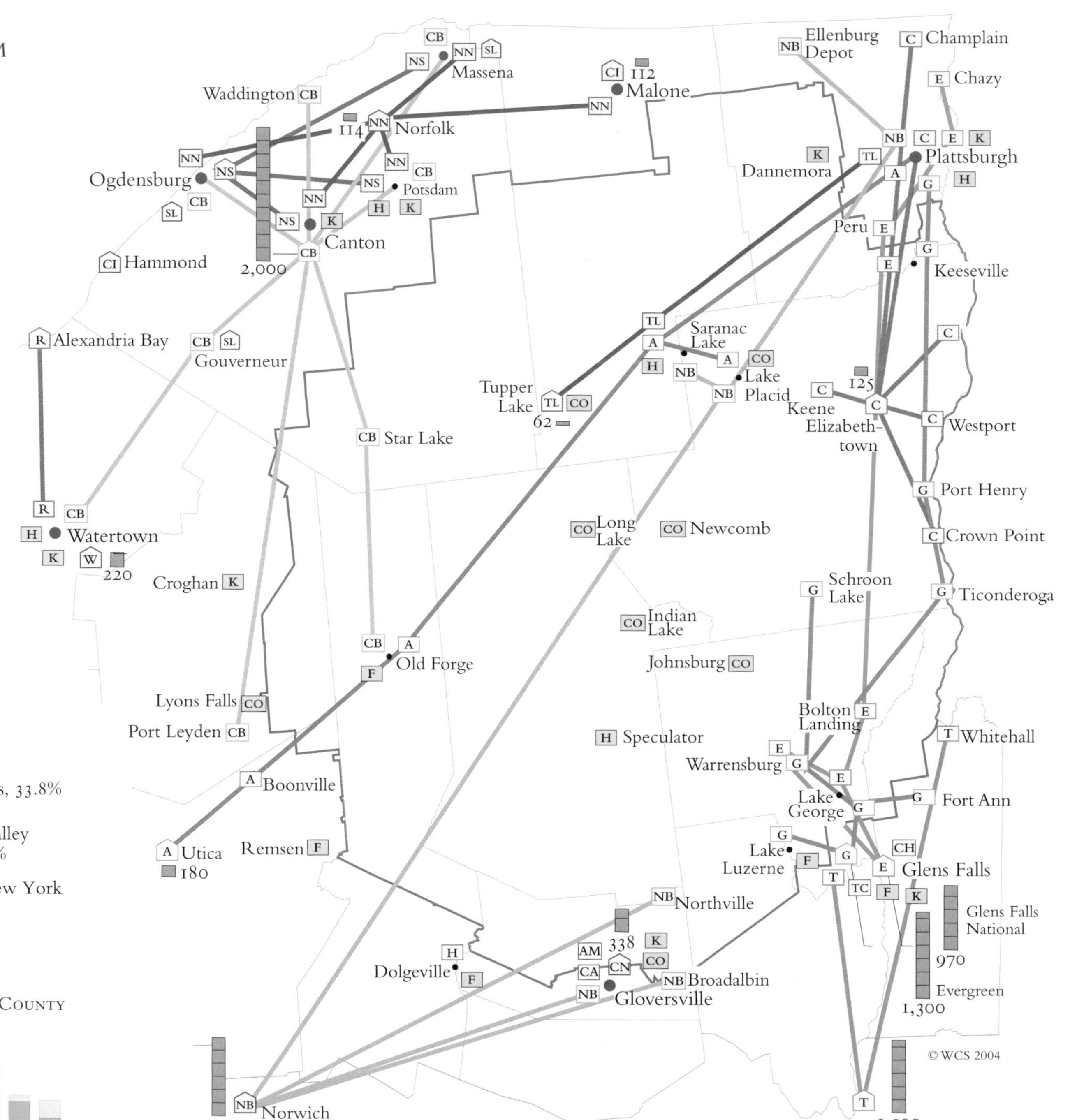

[A] Home office
[ ] Branch office

Banks with home offices in northern New York

- A Adirondack Bank
- C Champlain National
- CB Community Bank
- CI Citizens National
- CN City National
- E Evergreen
- G Glens Falls National
- NN First Nat. of North N.Y.
- NS Northcountry Savings
- R Redwood
- SL Savings & Loan Assoc.
- TL Tupper Lake National
- W Watertown Savings

Branch offices of Mohawk Valley banks

- AM Amsterdam
- CA Canajoharie
- CH Cohoes
- H Herkimer Trust
- NB NBT
- T Troy
- TC Trustco

Branch offices of large banks from outside the region

- CO Charter One
- F Fleet
- H HSBC
- K Key Corp

Total assets of bank

338 Total assets (billions of dollars)

13 *Business & Industry*

THE Adirondacks have at least three main economies: a resource-based, industrial economy; a tourist economy centered around lodging, food, and services; and a public service economy, largely nonprofit, centered on schools, local governments, and prisons. Chapter 8 described these economies, somewhat abstractly, with pie charts and graphs of occupations and sectors. This chapter is more geographic, showing where the businesses are and what they do.

The maps suggest that the old economy of mines, mills, woods-work, and farming, while neither extinct nor unimportant, has to a great extent been replaced by a new economy of hospitality, professional work, and service. The new economy is still resource-dependent and, to the extent that it requires a supply of developable land, still consumes and depletes resources. But it is no longer, and probably never will be again, primarily based on extraction and manufacturing.

The main elements in the new Adirondack economy are:

- Many small, locally owned businesses, especially in retail sales, accommodations, small manufacturing, maintenance, and repair.
- A few branches of large chains like Stewart's, Wal-Mart, Rite Aid, and Key Bank.
- A substantial number of big public-sector employers, especially public schools, government agencies, and prisons.
- A few large private for-profit companies in manufacturing and service.
- A substantial number of private nonprofits, in health, education, and services.

The map to the left shows the distribution of commercial property and hence where most economic activity happens. Almost all commercial lands, whether part of the old economy or the new, are in narrow corridors along lakes and rivers. In medieval Europe, money changed hands most freely at fairgrounds and in harbor towns. In the Adirondacks, eight centuries later, topography still controls trade; Adirondack money flows most freely near quiet, pretty water.

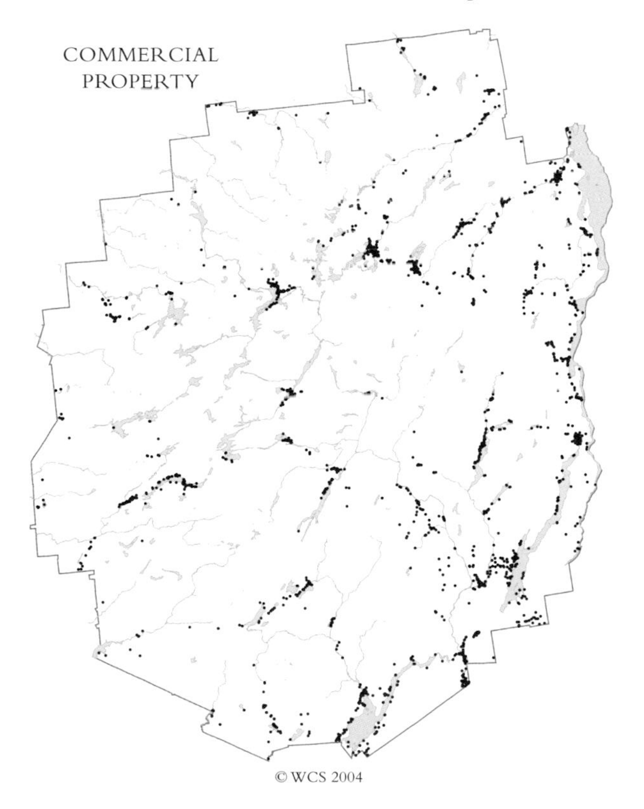

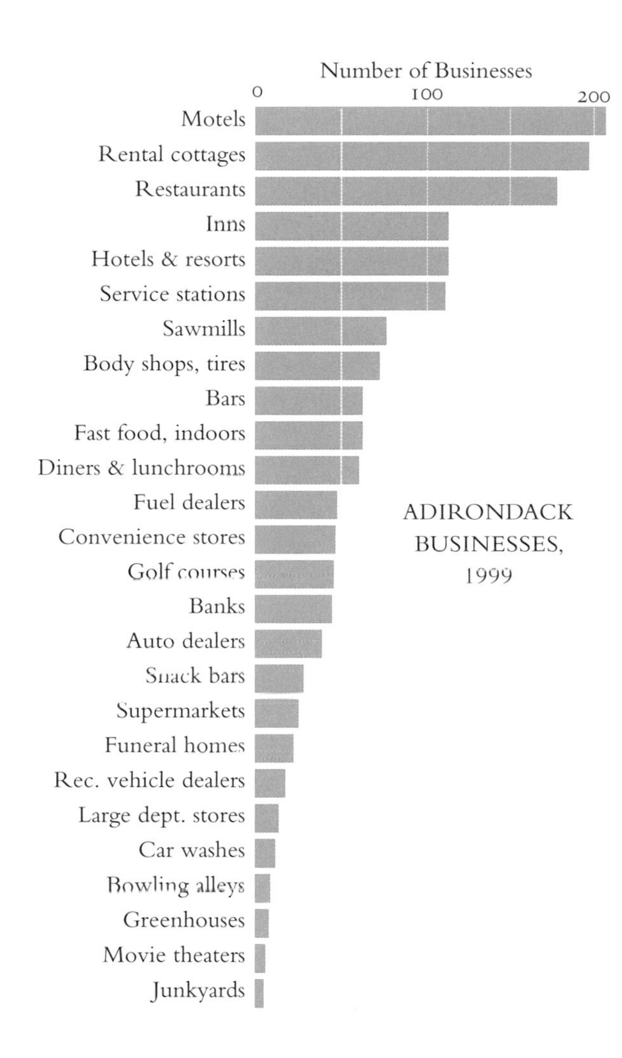

ADIRONDACK BUSINESSES, 1999

Motels are the commonest retail businesses in the Adirondacks, followed closely by inns, restaurants, hotels, and resorts. Year-round fast food joints are twice as common as the more seasonal snack bars. Car washing and bowling, never Adirondack priorities, are near the bottom. Movie theatres, which have lost much ground to home video, are almost gone.

HERE are the thirty-two largest Adirondack employers, conveniently dividing into sixteen private firms and sixteen governments, authorities, and agencies. Together they provide about 11,000 jobs, or about 30% of the 36,600 jobs in the park in 1997.

They are an interesting and very modern group of organizations, heavily weighted towards government and services. Only 44% of the jobs provided by the large employers are in the private sector; only 23% are in for-profit businesses, only 12% in manufacturing, and only 8% natural-resource related. Only one of their workers in ten actually produces a physical product. The rest provide services: they govern, teach, nurse, jail, administer, inspire, and care for each other and the rest of us. If the percentage of people employed in serving one another's needs is a measure of civilization, then the Adirondacks appear to be a very civilized place.

LARGE PRIVATE EMPLOYERS

Georgia Pacific (500)
Paul Smith's College (190)
NYCO Minerals (145)
Commonwealth Home Fashions (200)
Uihlein Mercy Center (235)
Lewis Concrete (100)
Oval Wood Dish (230)
Lake Placid Vacation Corporation (200)
Elizabethtown Community Hospital (100)
Franklin Co. ARC (210)
Holiday Inn (150)
Lake Placid Hilton (145)
Essex Co. ARC (485)
North Country Home Services (300)
W. Alton Jones Cell Science (82)
Trudeau Institute (144)
Adk. Medical Center (490)
International Paper (920)
Adk. Daily Enterprise (38)
Word of Life Fellowship (310)
Moses Ludington Hospital (110)
Adk. Bank (85)
Barton Mines (130)
Lincoln Log Homes (55)
Green Island Associates (840)
© WCS 2004
Finch Pruyn (600)

For-profits
- Mine
- Mfr. plastic utensils
- Bank
- Concrete & asphalt
- Paper mill
- Newspaper
- Curtains
- Hotel
- Log homes

Nonprofits
- Assoc. for Retarded Citizens
- Cell biology research
- Nursing or skilled care
- Hospital
- Ministry & camp
- Paul Smith's College

Companies shown in red are listed in the Dun & Bradstreet *Million Dollar Directory* and have over 180 employees or nine million dollars in sales in sales.

TYPE OF JOBS PROVIDED BY THE LARGEST ADIRONDACK EMPLOYERS

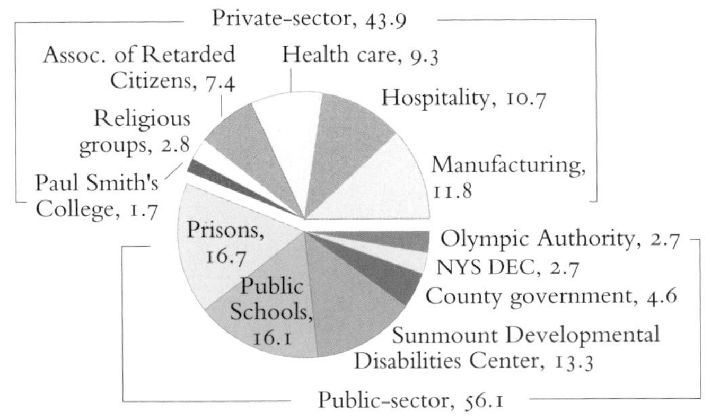

The numbers give the percentage of the 11,000 jobs provided by the park's thirty-two largest employers that are provided by each group of employers.

THE LARGEST PRIVATE EMPLOYERS • For-profit firm • Nonprofit firm

- 1 International Paper (920)
- 2 Green Island Associates (840)
- 3 Adk. Medical Center (490)
- 4 Essex Co. ARC (485)
- 5 Amer. Management Assoc. (370)
- 6 Word of Life Fellowship (310)
- 7 N. Country Home Services (300)
- 8 Uihlein Mercy Center (235)
- 9 Oval Wood Dish (230)
- 9 Franklin Co. ARC (210)
- 11 Lake Placid Vacation Corp. (200)
- 12 Com. Home Fashions (200)
- 13 Paul Smith's College (190)
- 14 Holiday Inn (160)
- 15 Lake Placid Hilton (150)
- 16 NYCO Minerals (145)
- 16 Trudeau Institute (145)

Collectively, governments are the largest employers in northern New York, providing at least 11,000 jobs within the park, and another 12,000–15,000 just outside it. Governments provide just under a third of all the jobs within the park, more than any other economic sector, public or private.

Within the park, prisons and schools, with roughly two thousand employees each, are the most important types of public employer. Sunmount Developmental Disabilities Center, with 1465 employees, is the largest single employer of any kind, public or private, in the park. Clinton Prison, with 1200 employees is next, and Essex County, with 500, is third. The town governments, which typically employ twenty to fifty people, are too small to show on the map but collectively represent at least another thousand jobs.

Immediately outside the park, the counties and schools are the largest public employers with roughly 5,000 employees each, followed by state prisons with 1,800 and state colleges with 1,600. Farther away, and large because a global organization must be large, is Fort Drum, the home base of the 10th Mountain Division of the U.S. Army, with some 11,000 military personnel and another 2,500 civilians.

The Largest Public Adirondack Employers

1 Sunmount Developmental Center (1,465)
2 Clinton Prison (1218)
3 Essex County (500)
4 Saranac Lake Schools (356)
5 Adirondack Prison (320)
6 Ray Brook Prison (300)
6 Olympic Development Authority (300)
6 Saranac School System (300)
6 Dept. of Environmental Conservation (300)
10 Warrensburg School System 230)
11 Corinth School System (200)
12 Lake George School System (190)
12 Hadley-Luzerne School System (190)
14 Ticonderoga School System (187)
15 Mayfield School System (152)
16 Tupper Lake School System (144)

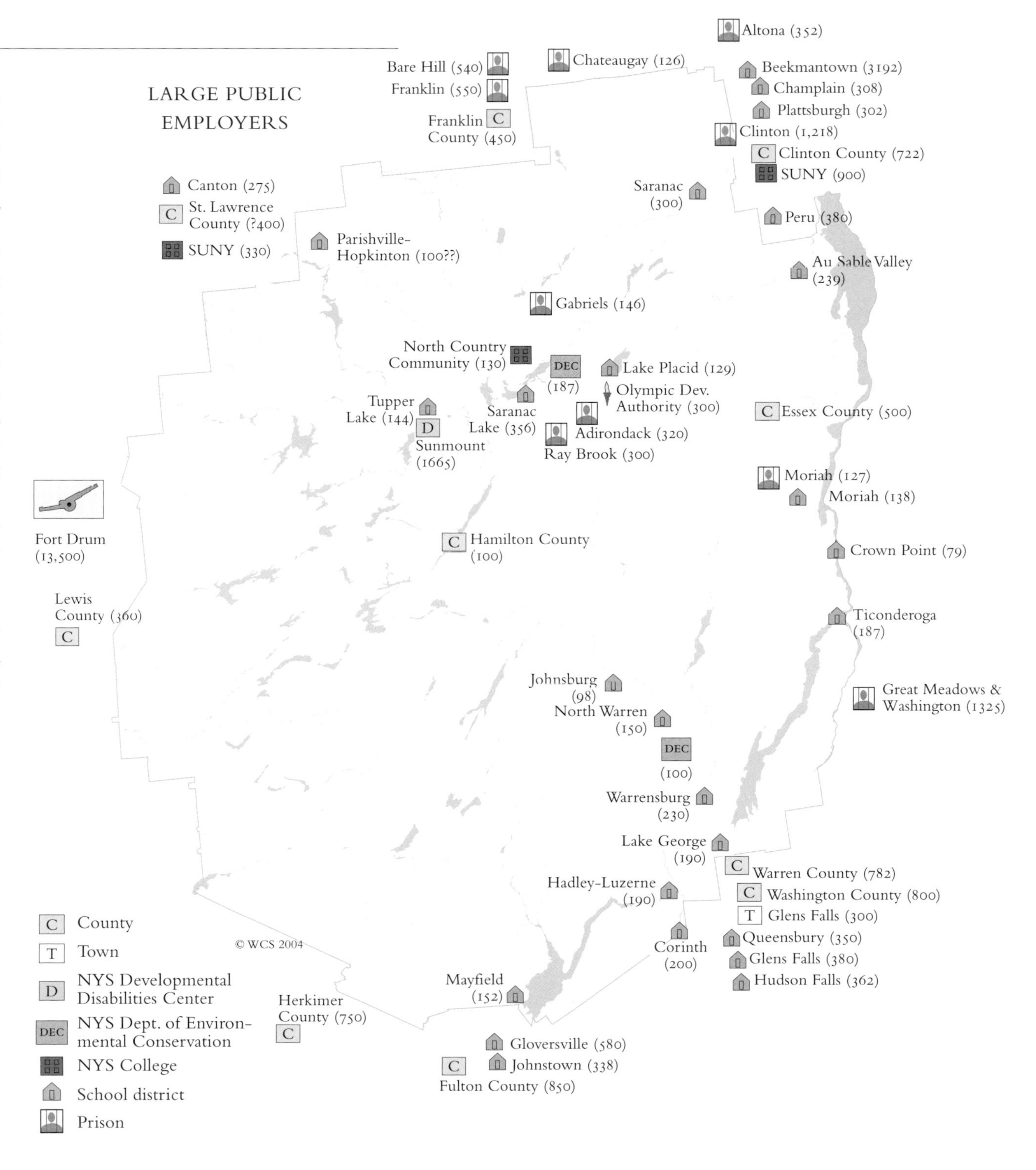

MANUFACTURING is almost any activity that modifies raw materials and produces a saleable product. Formerly, it was mostly heavy, dangerous, resource-based work, dominated by a few big firms. In 1916, for example, Witherbee & Sherman employed 1,454 men to produce 1.2 million tons of iron ore. Seven of them were killed doing it, and many others injured in the process.

Contemporary manufacturing is lighter, smaller in scale, less dangerous, and less resource-dependent. The symbols on the map represent about 130 manufacturers employing about 3,500 persons, or roughly 25 per firm. The product sand activities of these manufacturers are quite varied. Some, like logs, minerals, lumber, furniture, and paper, are traditional regional specialties. Others, like printing, pallet-making and welding, reflect the almost universal tendency of civilized people to write, move boxes, and break things made of metal. And still others, like curtains and brake pads, are things that could be made anywhere but just happen, by circumstance or luck or because the owner is also fisherman or mountaineer, to have an Adirondack label.

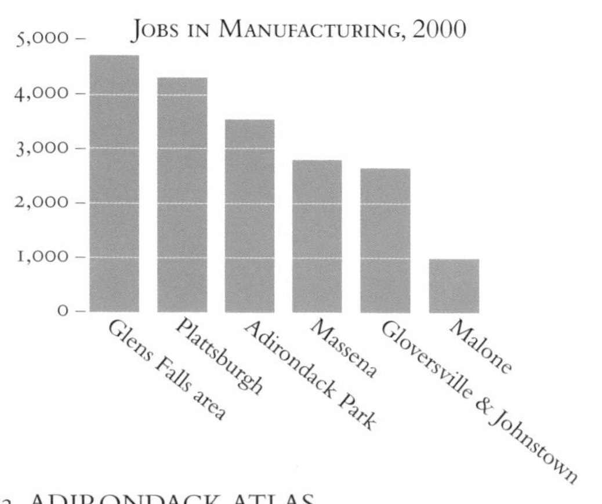

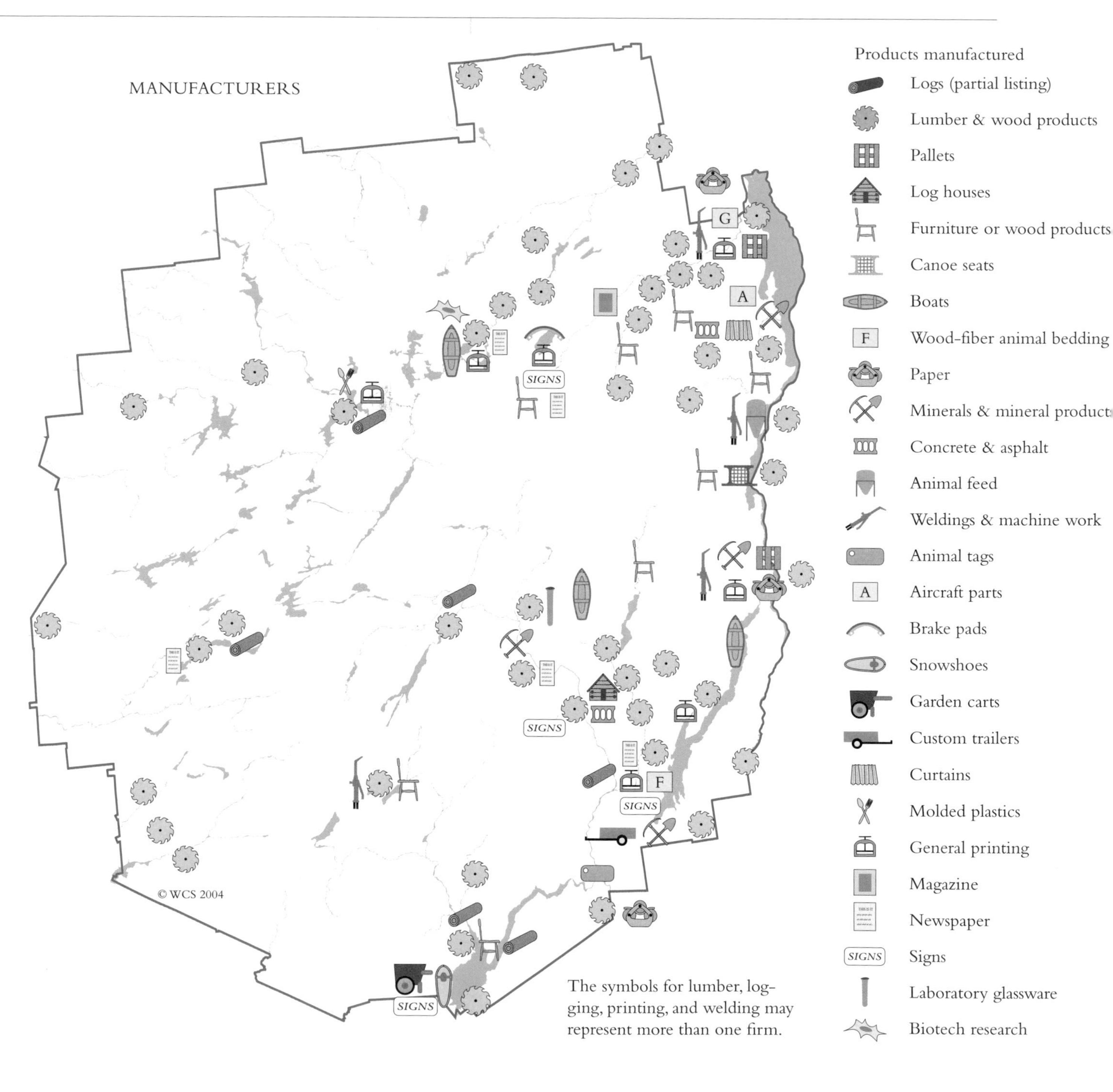

The symbols for lumber, logging, printing, and welding may represent more than one firm.

In forestry, as in farming, the profits lie with the processors and not the producers. Relatively little money is made by growing trees or cutting them. Substantially more is made by turning them into lumber or furniture. Hence the local wood-using industries matter twice: once to the region because they are the sector of the forest industry that generates jobs and profits and once to the timber-owners and loggers to whom they transfer a badly-needed portion of these profits.

Forests grow two sorts of commercial logs, sawlogs and pulp logs. Saw logs, the good ones, become lumber and furniture; pulp logs, the rest, become paper and laminates. To have healthy commercial forests you must harvest both. To have a healthy forest economy, you must have local processors who will buy both.

By this standard, the Adirondack forest economy is not very healthy. Essentially, we are growing trees and shipping them off for other places to make profits on. Within the park there are many small mills but only a few big ones. Regionally, there are many mills buying hardwood saw logs but only a few buying softwood saw logs or pulp. Five paper mills have closed recently, and two others only buy preprocessed pulp and not logs. Currently, though we have buyers fighting for every stick of good hardwood available, we have only a weak market for softwood saw logs and the worst market for pulp in anyone's memory.

The results are predictable. Too many bad trees are staying in the forests, where the foresters don't want them. Too many good ones are not wanted by local mills and are going instead, by concentration yards and roving dealers and big trucks, across the border into Canada.

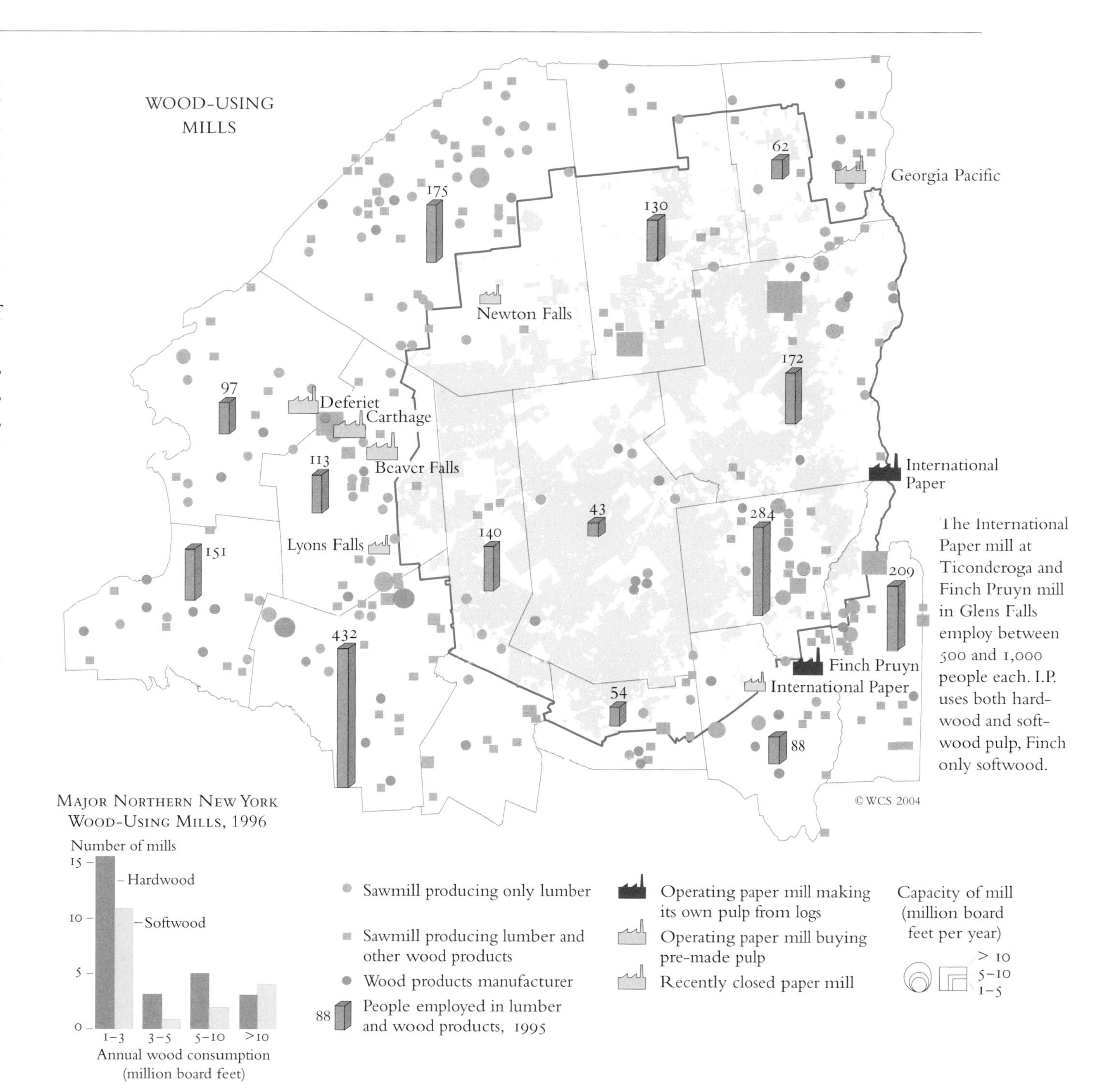

NORTHVILLE is a village of about a hundred and fifty houses. It is an old Adirondack-edge settlement located at the end of an agricultural valley, about where the serious woods once began. It began as a tannery town, lost its tanneries, acquired lake-front when the river was dammed in 1930 and is now both a large summer colony and a small commuter town.

Northampton, the town containing Northville, is a small town of about 35 square miles. It has 2,800 permanent residents and 770 year-round houses, giving it a density of 80 persons per square mile. In addition there are 780 weekend or seasonal houses along the river, plus another 1,355 next door in Edinburgh. The total seasonal and weekend population of the two towns is probably two to three times as large as the permanent population.

The village of Northville serves both the residents and the visitors. Its facilities are typical of small Adirondack towns that expect to feed and profit from their guests but don't need to house, equip, or entertain them. It has a central school of six hundred students, a recently built library, several churches, a single grocery, a few service stations and two minimarts. There are three marinas, two banks, one motel, and one inn. And, outnumbering all other commercial establishments, there are about fifteen antique shops, boutiques, galleries, and other visitor-oriented retailers. All of the speciality shops and service providers are locally owned. But the grocery, both banks, one minimart, and the drug store are owned by out-of-town chains. This, as we will see, is also typical.

CHAIN stores are owned by a parent company that manages many branch stores. Chain stores are different from *local franchises*, locally-owned stores affiliated with a national distributor, in that they are run by the parent company and send their profits back to it. McDonald's and Charter One and Rite Aid are chains; True-Value and NAPA are local franchises.

Chain stores are common in the larger towns, but almost absent from the smaller ones. The maps show 74: one Wal-Mart, seven McDonald's, 23 Stewart's, 10 chain banks, 15 chain pharmacies and 18 chain groceries. The extent to which they dominate local markets varies with the type of business. Local snack bars, pizza shops and minimarts still compete very successfully with the chains, especially in the smaller towns. Local and regional banks are firmly established in the older commercial towns, but almost all the new branch banks in the resort towns belong to national chains. Chain pharmacies, often highly profitable because drugs are highly profitable, are established in all the large towns and have put many local pharmacies out of business. Chain department stores and supermarkets are impossible for all but the most enterprising local stores to compete with and monopolize the food and dry-goods business in all towns large enough to support them.

Chain stores are neither all good nor all bad. They hire people, pay taxes, and sell things people need at pretty good prices. But they also bankrupt local stores and export profits—often a lot of profits—to investors elsewhere. And, whether predatory or benign, they are inevitably responsible for the spread of the uniform commercial architecture so common in America and so foreign to Adirondack sensibilities and traditions.

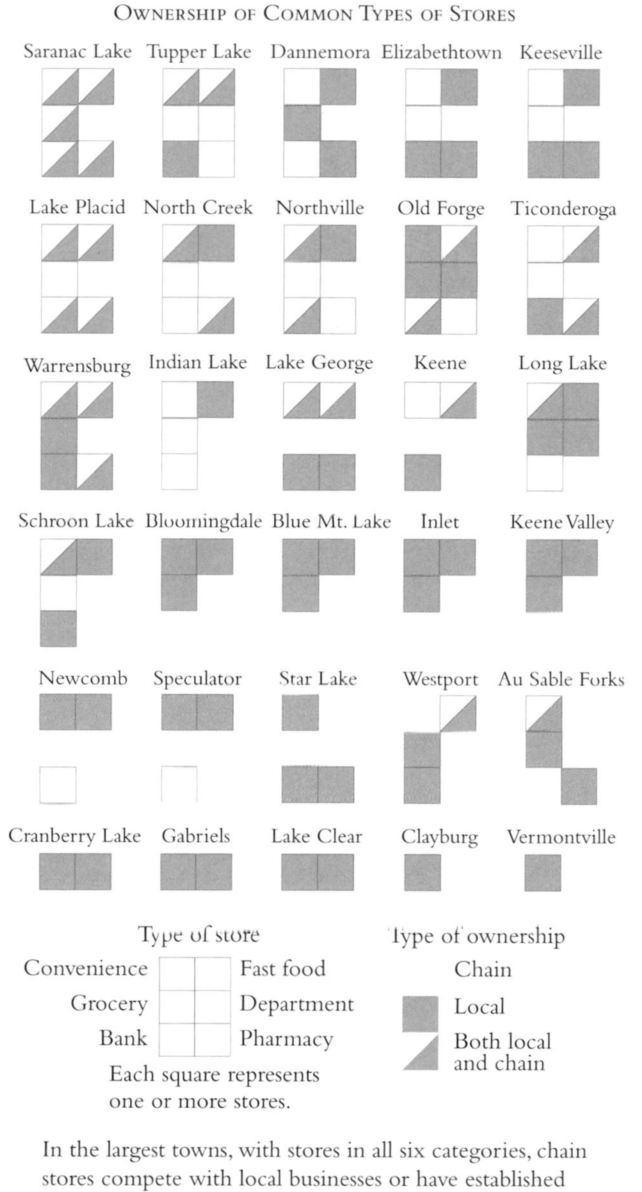

In the largest towns, with stores in all six categories, chain stores compete with local businesses or have established monopolies in all six categories. As the town gets smaller and has fewer types of stores, the number of chain stores drops. The one-store towns, like Clayburg and Vermontville and many others, never have chain stores.

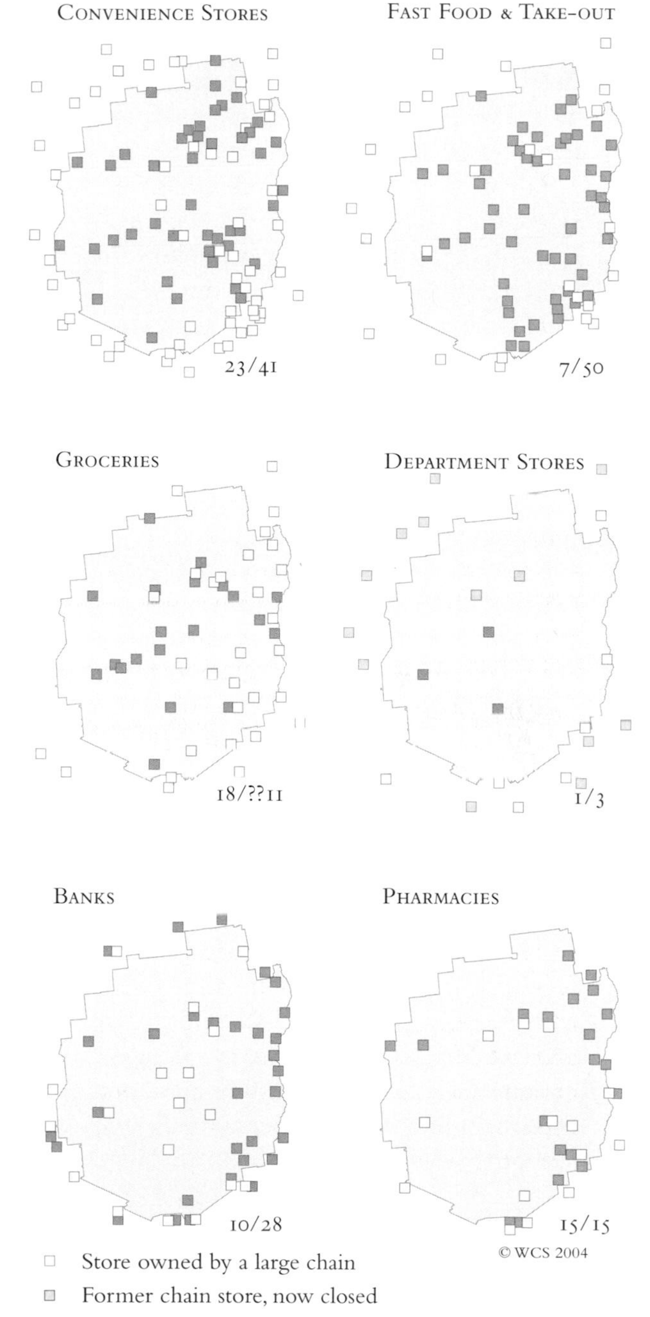

MANY people come to the Adirondacks not for exercise or adventure but simply to be near woods and water and pretty places. For every visitor that camps somewhere in the backcountry, many others stay at camps, campgrounds, cabins, and summer houses.

The Adirondacks have a large supply of these. The map shows 75 summer camps, 147 campgrounds with at total of 11,700 campsites, and 20,028 houses classified as seasonal residences. There are in addition some 28,800 other "unoccupied" houses that are used seasonally and 200 landlords who rent at least a thousand summer cottages (*Map 13-5*).

Many of the users of these houses, camps, and cottages are a group of somewhere between 100,000 and 200,000 regular visitors whom we call the other Adirondack population. They are neither residents nor tourists. Like residents they are tied to particular communities, and often make substantial economic contributions to those communities. Like tourists they live and vote elsewhere. What they are, in many cases, are people who live in one place but whose real home, the Adirondacks, is someplace else. In this they are not particularly unusual; the exigencies of life and work separate many of us from our real homes.

Because the other Adirondack population does not vote here, it is often ignored politically. This may be unfortunate because it disenfranchises many people who matter to the Adirondacks and to whom the condition and future of the park are very important. If this was not the case—if the Adirondack constituency was expanded to include everyone who supports the park and cares about it—the terms and outcomes of many Adirondack debates might be very different.

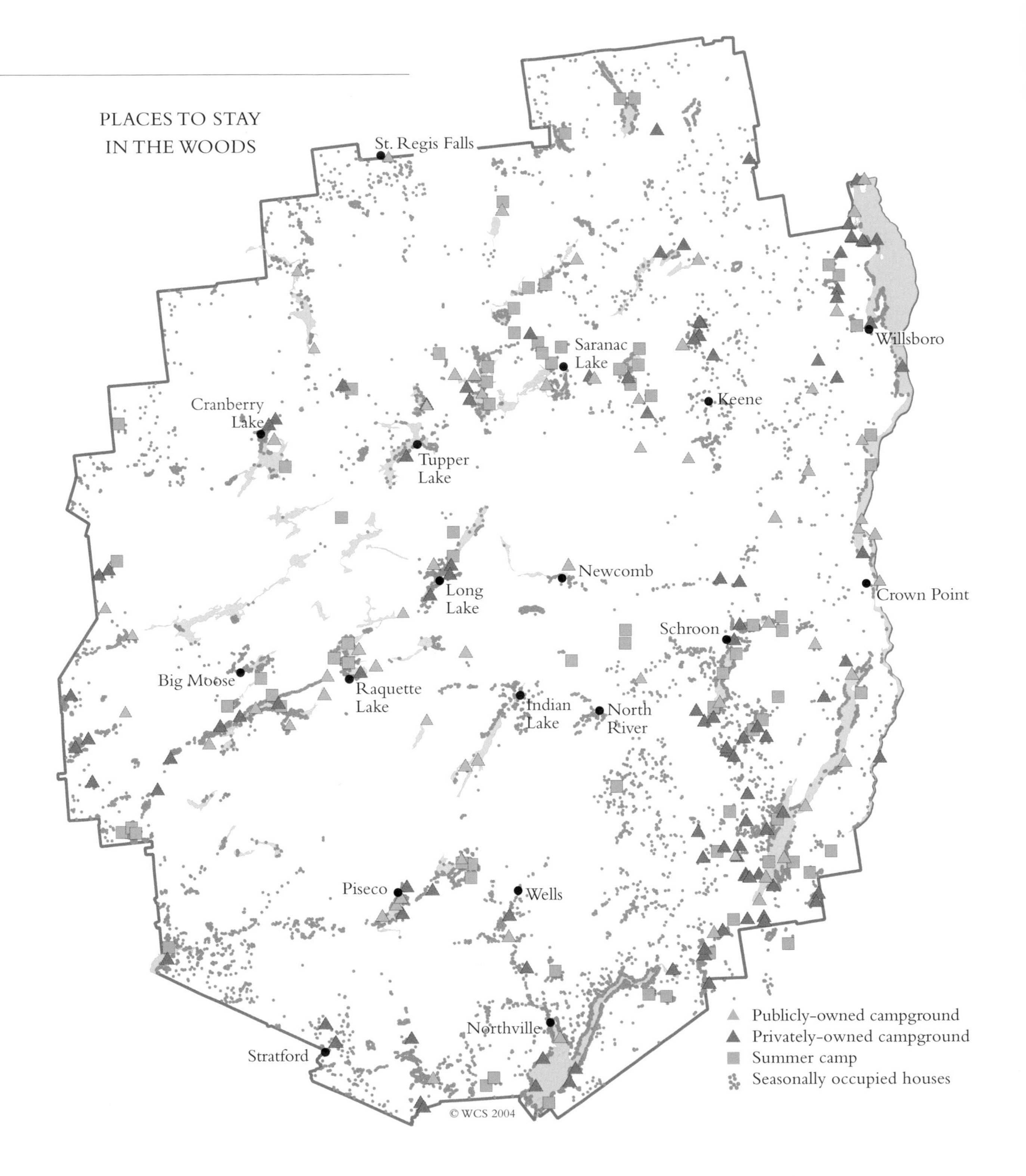

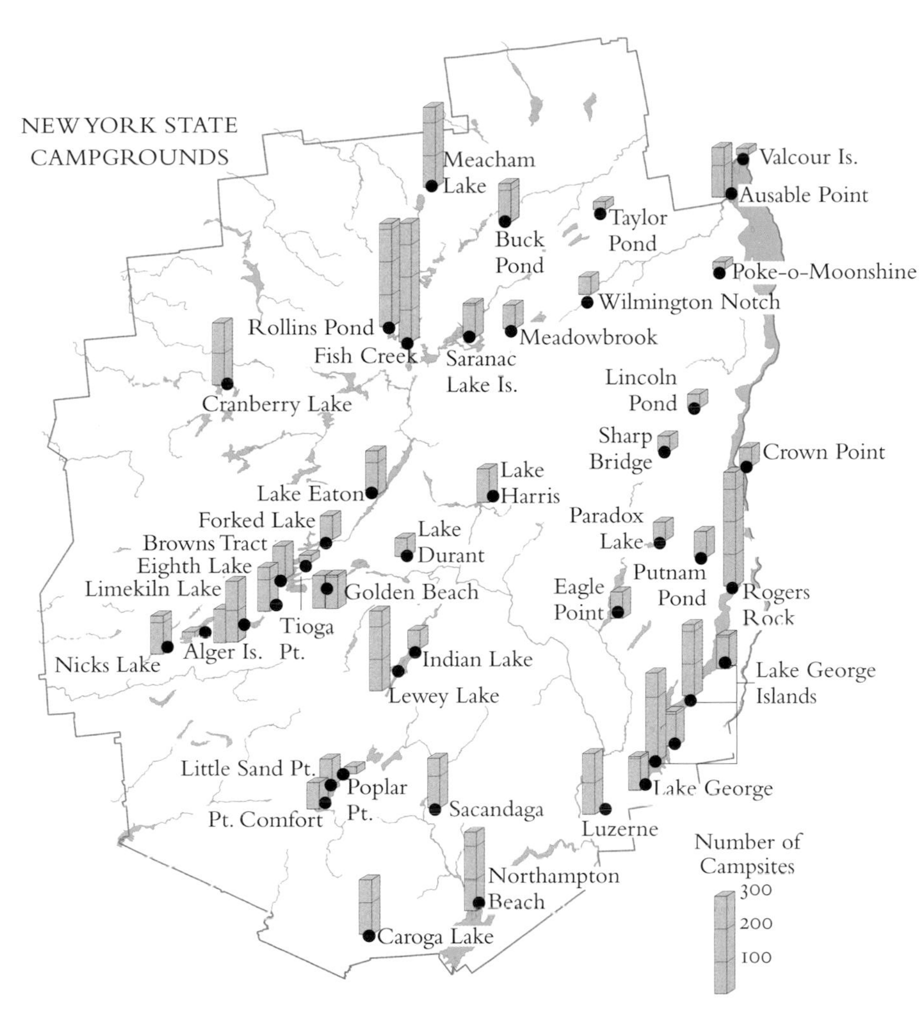

New York State operates 45 campgrounds with about 5,500 camping sites, a little over a third of all the campsites in the park. About 1.3 million people use them each year. The first ones were built in the 1920s and 1930s when automobile camping first became popular and many of the remainder in the 1950s and 1960s as the highway system expanded. Taylor Pond, the last one built, was completed in 1974.

Because of constitutional issues, it is possible that any future campgrounds will be on leased land rather than on the Forest Preserve. A campground is, after all, a small outdoor village with roads, buildings, and cleared campsites. It is quite possible the courts, which held in 1930 that the clearing of 4.5 acres of state land for a bobsled run was unconstitutional, might say the same thing about a new Forest Preserve campground of a hundred or so acres. And while no one seems particularly interested in challenging the constitutionality of the existing campgrounds, it is quite possible that no one would risk raising the issue by proposing a new one.

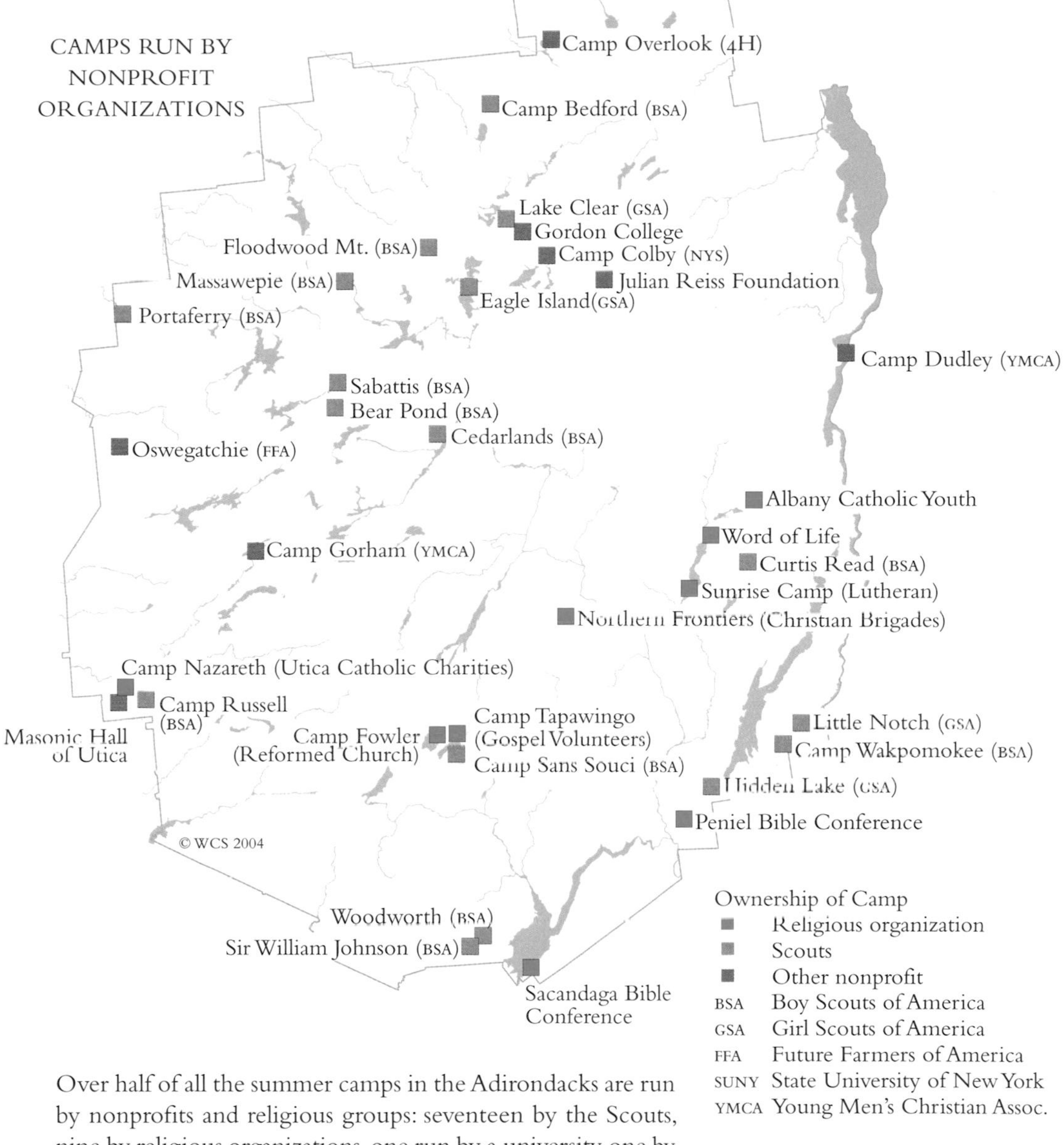

Over half of all the summer camps in the Adirondacks are run by nonprofits and religious groups: seventeen by the Scouts, nine by religious organizations, one run by a university, one by the state, and eight by other organizations. Almost every educational and inspirational organization believes in the healthiness of fresh air, camp songs, and group hiking. Many sneak in a bit of their favorite message around the campfire too.

THE Adirondacks, which have about 11,000 rental rooms in hotels, motels, inns, and cottages, can accommodate at least 20,000 short-term visitors. Unlike the campgrounds and summer houses, which are spread over every part of the Adirondacks that has public roads, the rental rooms are concentrated in the resort towns and along lake shores. Some are private and isolated and cater to people who come for peace and quiet; others form dense shoreline communities and accommodate people who can tolerate, or who want, a good deal of resort-town activity going on around them.

Rental cottages, the dark green bands on the map, are most abundant in the old lake towns—Indian Lake, Schroon, Saranac Lake, Inlet—that have been accommodating visitors for many years. Motels and hotels are more concentrated in the main resort towns like Lake Placid and Lake George. Cottage guests, a self-contained bunch, cook for themselves and sit on their own porches. Motel and hotel guests, more sociable and perhaps less sufficient, like to go out. Wise hostelers site their establishments carefully to make sure they have someplace to go.

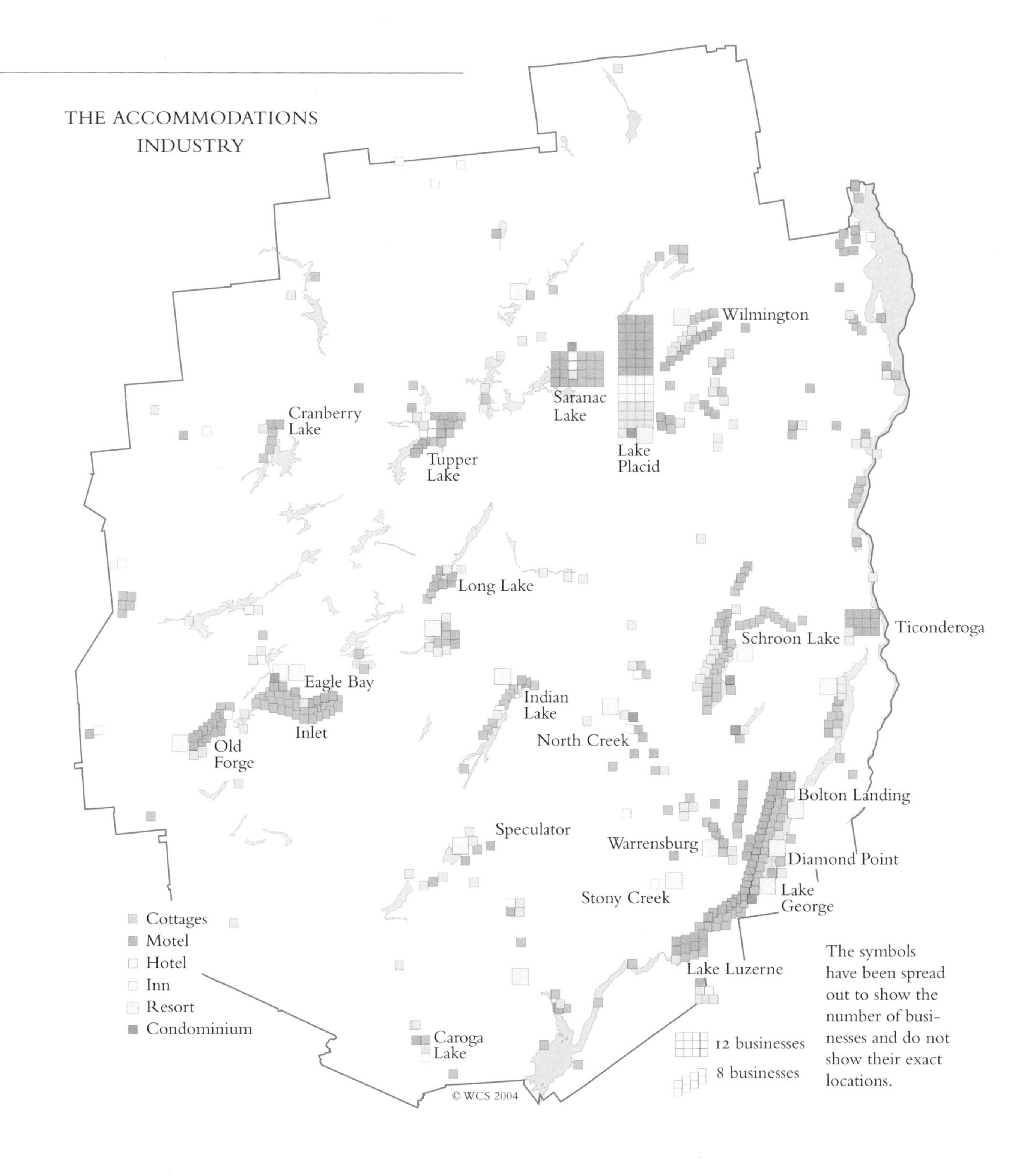

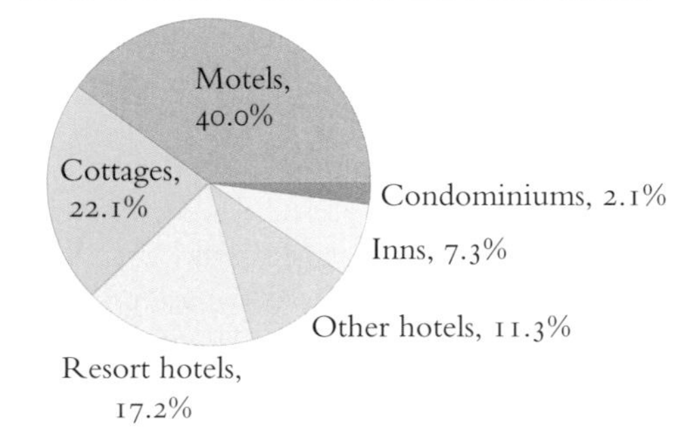

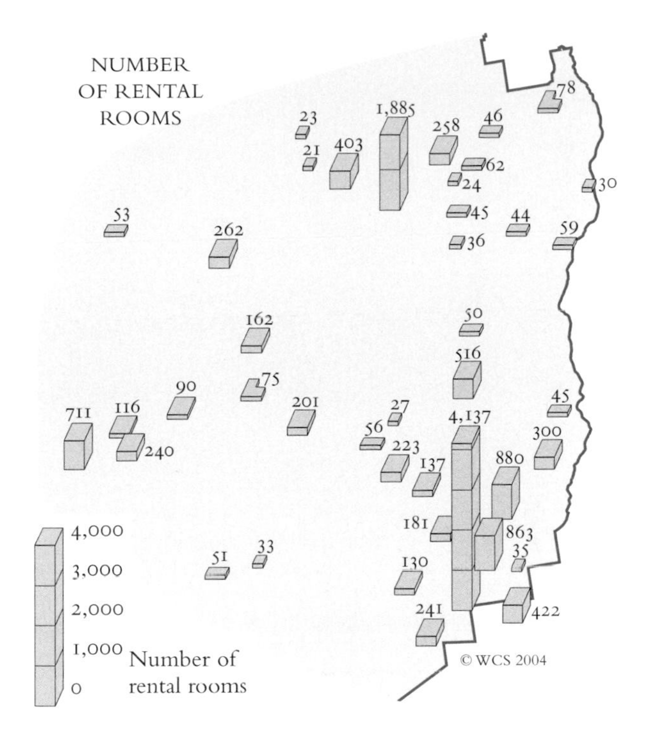

Businesses Offering Camping Lodging & Food

	Number	Rooms	Campsites
Motels	207	5,538	
Cottages	197	2,886	
Inns	118	959	
Resort hotels	77	2,254	
Other hotels	39	1,473	
Condominiums	7	274	
Total lodging	645	13,384	
Private campgrounds	90		6,411
Public campgrounds	57		5,257
Summer camps	75		
Total camping	222		11,668
Restaurants	179		
Snackbars & fast food	55		
Diners & lunchrooms	62		
Total food service	296		
Total businesses	1163		

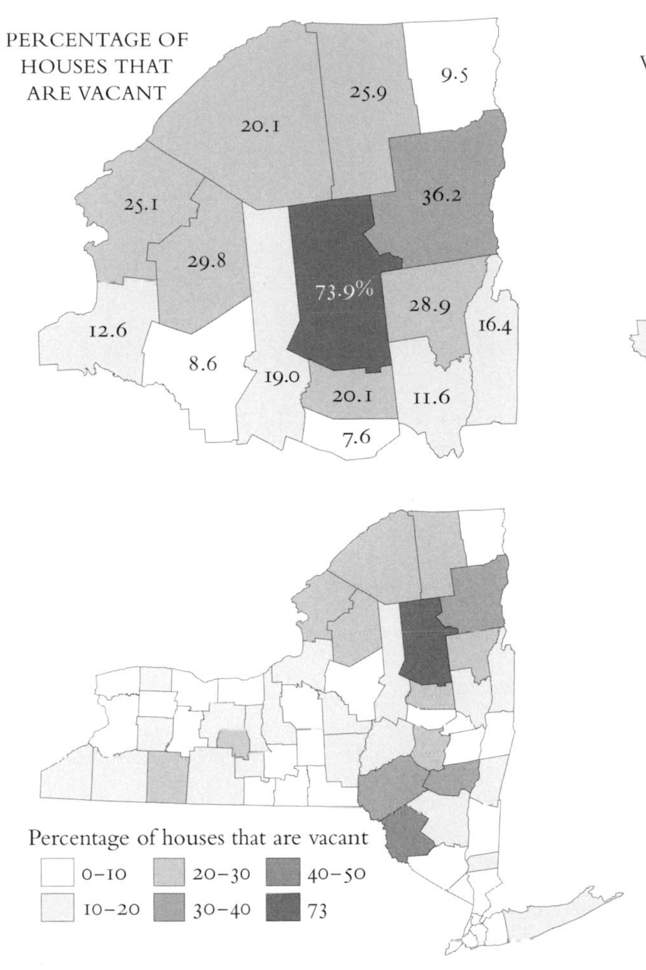

Although the hospitality industry is undoubtedly one of the largest and most specialized sectors in the park's economy, no one seems to have computed its size, employment, or earnings. The figures given here, which are only approximate, are based on the property-use codes in the Office of Real Property Services database of commercial property and an inventory of rental rooms by Nadia Korths. They suggest between 1,100 and 1,200 businesses, providing about 25,000 rooms and campsites and at least 300 places to eat. This would suggest that the park can accommodate well over 50,000 overnight visitors at full capacity. If, as seems reasonable, half or more of the 10,000 people in the park who work in service industries are involved in accommodations and hospitality, then the hospitality industry as a whole may be one of the park's largest employers, bigger than schools and prisons combined.

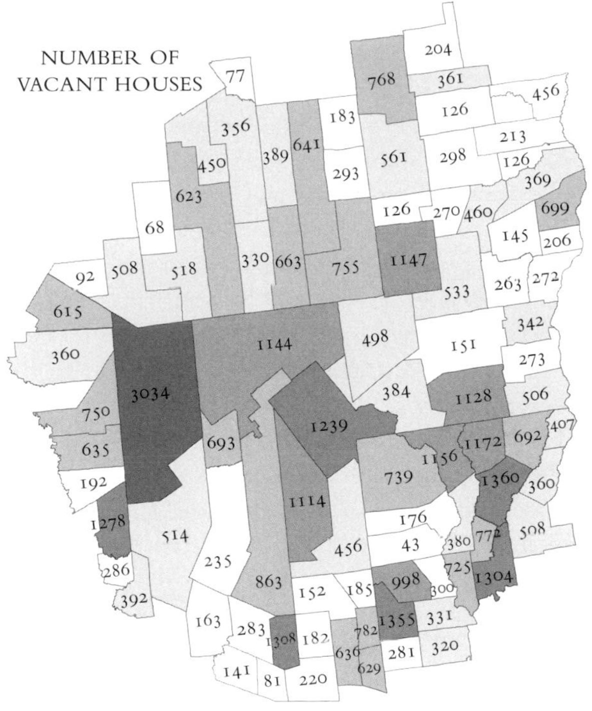

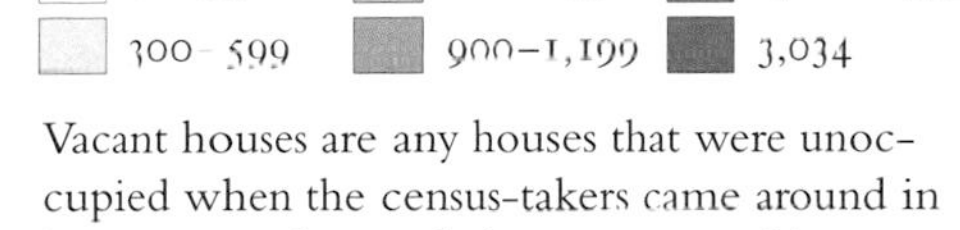

Vacant houses are any houses that were unoccupied when the census-takers came around in late winter. They include many second homes and summer camps and a few houses waiting to be rented or sold.

WHILE making a living by housing and feeding visitors is fairly straightforward, making one by equipping or entertaining them is substantially harder. The problems are structural. Many visitors need no entertainment or equipment at all. Those that do need equipment may not buy it in the park—very few people set off on a hunting trip planning to buy a gun and dog along the way. And those that want entertainment will usually spend less on it than on food and lodging.

In consequence there are, if anything, fewer attractions and outfitters than might be imagined. An approximate survey suggests less than two hundred in the park, compared to some eleven hundred businesses offering food and lodging.

These two hundred businesses solve the structural problems of being what might be called secondary tourist businesses in a variety of ways. Many survive by staying small. One of the park's most successful attractions is an ice cream stand with a twenty-foot building and a staff of three. Many others sell services—guiding, equipment rentals, instruction—that can't be purchased ahead of time. A particularly interesting group, exemplified by the theme parks, concentrate in the gateway towns and offer pay-to-view attractions that in some sense compete with the free attractions of the park itself.

That this last group should be here at all is somewhat odd. Why, after all, should visitors to the Adirondacks, a naturally wet and wild place, patronize water slides and wild-west villages? But patronize them they do. There are fourteen theme parks, none of them Adirondack-related, in the park or at its edge. If what they tell us is that people enjoy fantasy landscapes surrounded by real woods, this is perhaps no surprise: they are neither the first incongruous entertainments to please our guests nor the first fantasies to be enacted in these woods.

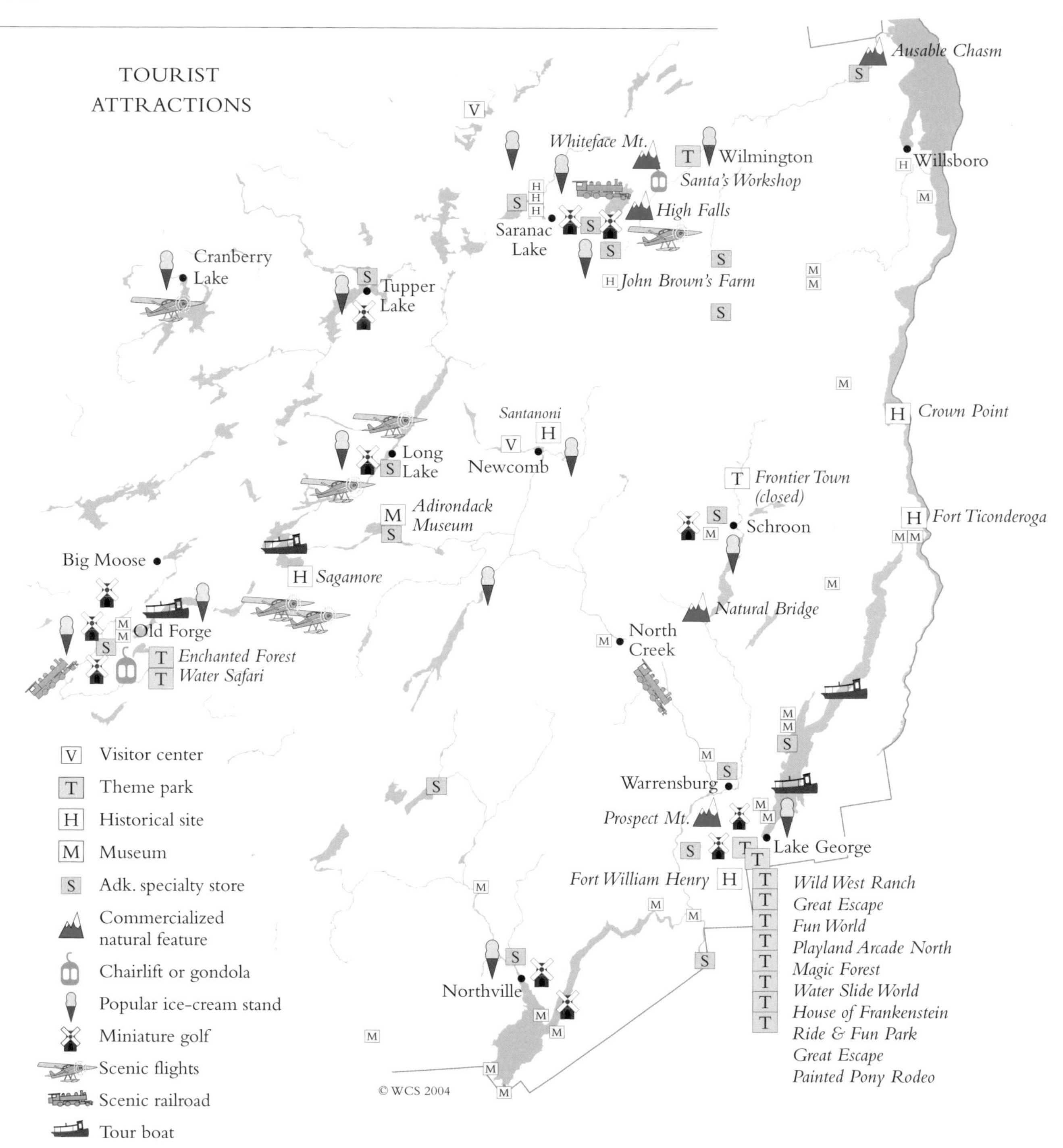

SOMEWHERE between 150 and 170 businesses and over six hundred guides and instructors (many of whom are employed by or run these businesses) serve recreationists in the park. These include at least seventy-five marinas, twenty-four powerboat, snowmobile, and ATV dealers, twenty-two hunting and fishing suppliers, fourteen ski stores, thirteen rafting companies, eleven canoe and kayak dealers, and eleven general camping outfitters, two of which also sell rock-climbing and mountaineering equipment. Interestingly, despite the increasing numbers of hikers, paddlers, climbers, and campers, about three-quarters of these businesses serve more traditional (or perhaps more affluent) recreationists: hunters, fishermen, skiers, and users of powered recreational vehicles.

While serving recreationists is a traditional Adirondack business, many individual services are traditionally controversial. Guides and outfitters, after all, make money by getting their clients to places that other people would like to have to themselves. A hundred years ago the local fall hunters disliked the guides who got the big bucks for their summer clients. Today the wading fishermen dislike the guides who bring clients in drift-boats; the drift-boat guides, in turn, say the raft releases are ruining the fishing. Neither the drifters nor the rafters have much good to say about guys in bass-boats who, perhaps coincidently, don't like guides or rafters all that much either.

Some, perhaps many, of these quarrels are selfish and curmudgeonly. Others raise important questions. What, fundamentally, is the distinction between a public preserve and a public playground? Just how consistently do we want to see that distinction observed? And, if consistently, then at what cost and who profits and who pays?

ADIRONDACK LIBRARIES

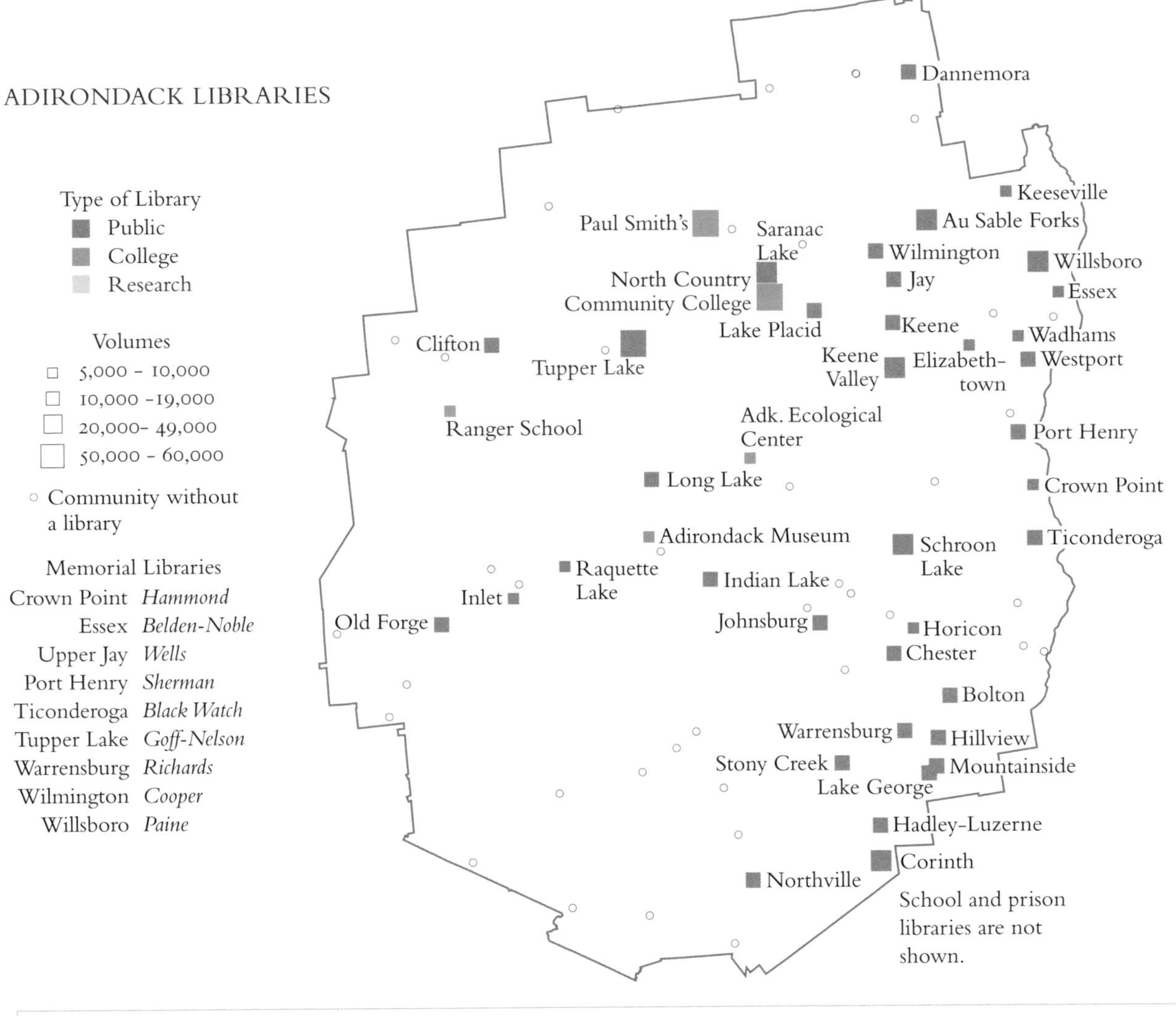

The Adirondacks have thirty-seven public libraries, two college libraries, one major museum library, and about sixty school libraries. There are also libraries at the six Adirondack prisons and at the larger hospitals but not at either the Park Agency or the Department of Environmental Conservation.

The public libraries have about 64,000 members, hold about 500,000 volumes, and circulate some 600,000 books annually. How much it costs to house and circulate these books has never been tabulated. The ninety-two Adirondack towns spend about nine million dollars annually on recreation and culture, which includes everything from books and concerts to ball fields and fireworks. The libraries, a civic bargain at any price, are probably only a small fraction of this.

The first public libraries, in Lake Placid, Elizabethtown, and Westport, were built in 1884. The last, in Inlet, was built in 1999. The largest, and one of the most beautiful, is the Goff-Nelson Library in Tupper Lake, with 52,000 volumes. The smallest are the village libraries in Essex and Wadhams, each with about 5,000 volumes. Five thousand books, as anyone who reads or builds shelves knows, is still a large number of books.

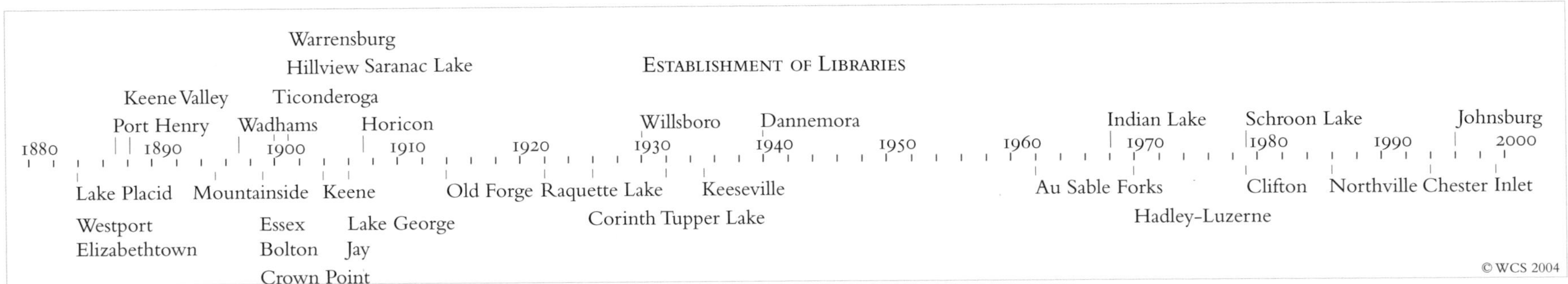

14 *Media & Culture*

THIS chapter is about the channels through which the public knowledge is maintained and transmitted. These channels include the both the institutions—the libraries, newspapers, broadcast media, museums, galleries—that store and transmit artifacts and information, and also the formal and informal exchanges—the meetings, performances, shows, classes, contests, conversations—in which that information comes alive.

Obviously no map will capture all of this. The chapter includes six maps of Adirondack cultural institutions and in addition, as sample of the enormous layer of noninstitutional culture not shown in these maps, two calendars of local events.

Our particular concern is with Adirondack culture. We assume that the larger American culture is everywhere and that any news, personality, or entertainment known nationally will also be known everywhere in the park. What interests us is whether Adirondack culture is similarly pervasive and if Adirondack art and news also reach every part of the Adirondacks.

The maps suggest that while in general they do there are some important limitations of scale and geography. In the largest or most lively towns, local news is well reported and cultural events plentiful and varied. Outside these towns culture is more private and news less reported. Though musicians, for example, play all over the park, in some towns they will be in the library or on the town green, with posters and announcements; in others they will be at private gatherings in houses or in yards.

If local culture is strong, the culture of the park as a whole is weaker. Two important institutions, the Adirondack Museum and North Country Public Radio, and two equally important publications, *Adirondack Life* and the *Adirondack Explorer*, deal in very substantial ways with the history, culture, and politics of the park. But besides these four there is very little. No daily newspaper circulates throughout the park. No regional radio or television station has live broadcasts of Adirondack events. No museum holds a major collection of contemporary Adirondack art, and no college has a major collection of Adirondack books, maps, or documents.

The Adirondack cultural landscape is then an archipelago: many vigorous, fascinating islands, with empty spaces around and between them, and no center or peak from which the whole can be viewed.

The distribution of libraries illustrates this pattern. The towns with libraries fall into two groups: older towns where the libraries were often the gifts of local patrons and actively growing towns where the libraries are the result of a community decision that life is better with a library than without one. As a result, while there are libraries in all Lake Champlain towns and in the main resort towns of the interior but almost none in the small towns away from the main tourist routes. Neither Lyon Mountain, Wells, Caroga Lake, nor any of the old railroad towns between Bloomingdale and Saranac have libraries.

As with other cultural institutions, it is when we ask where the *regional* libraries are that we encounter the real gap. Only one library—the research library at the Adirondack Museum— deals with the whole park. Nowhere is there any comprehensive collection of Adirondack science or literature, or any archive of Adirondack politics and administration, or any comprehensive archive of regional news. In consequence, the Adirondacks, though notably mindful and literate at the local level, are in some important sense forgetful and illiterate as a whole.

LARGE RADIO STATIONS

ALMOST a century into the electronic age, printed works on paper, printed in two-thousand-year-old characters with a thousand-year-old printing technology, are still our best source of local and regional information. The broadcast media have only limited local content and no patience with complexities or details. The Internet, though currently of great importance as a source of state-wide information, still has little Adirondack information. At present, if you want to know where the Brooklyn Cooperage Railroad went, or how many deer were shot on opening day, or what the board decided about the controversial K-6 lunch program, you have to go to a printed source.

Since northern New York has only two regional book publishers and a single magazine, its fifty newspapers are of central importance. Like newspapers everywhere they have important limitations. Physically, they are short-lived, resource-intensive, and inflexible in format. Think of how much paper and fuel it takes each day to deliver the real-estate ads to people who are not buying houses. Editorially, they are limited by staff and resources and, all too often, by the absence of competition and their fear of offending advertisers and subscribers.

But limitations are one thing and intrinsic value is another. Today, and likely for a long time to come, our newspapers will be what they have been since their ancestors the broadsides appeared three hundred years ago: fascinating, relevant, detailed, alive, and essential.

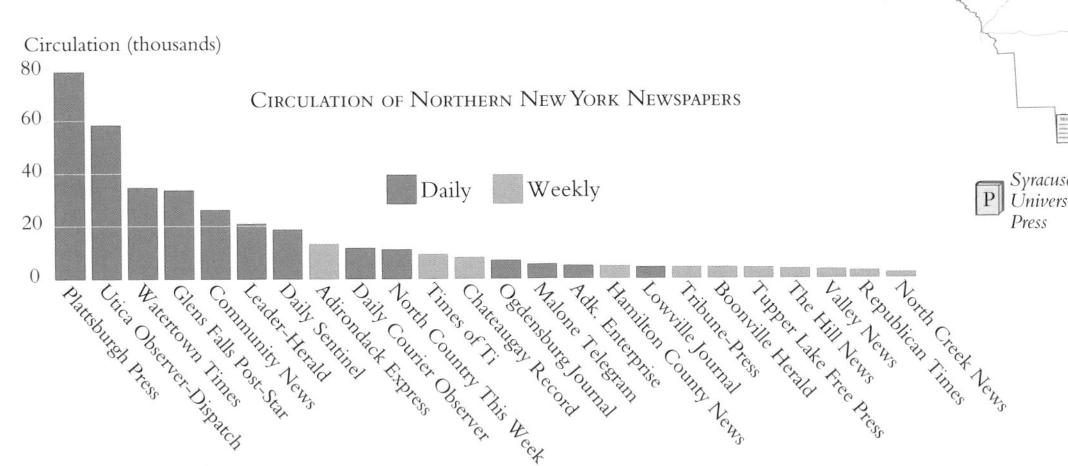

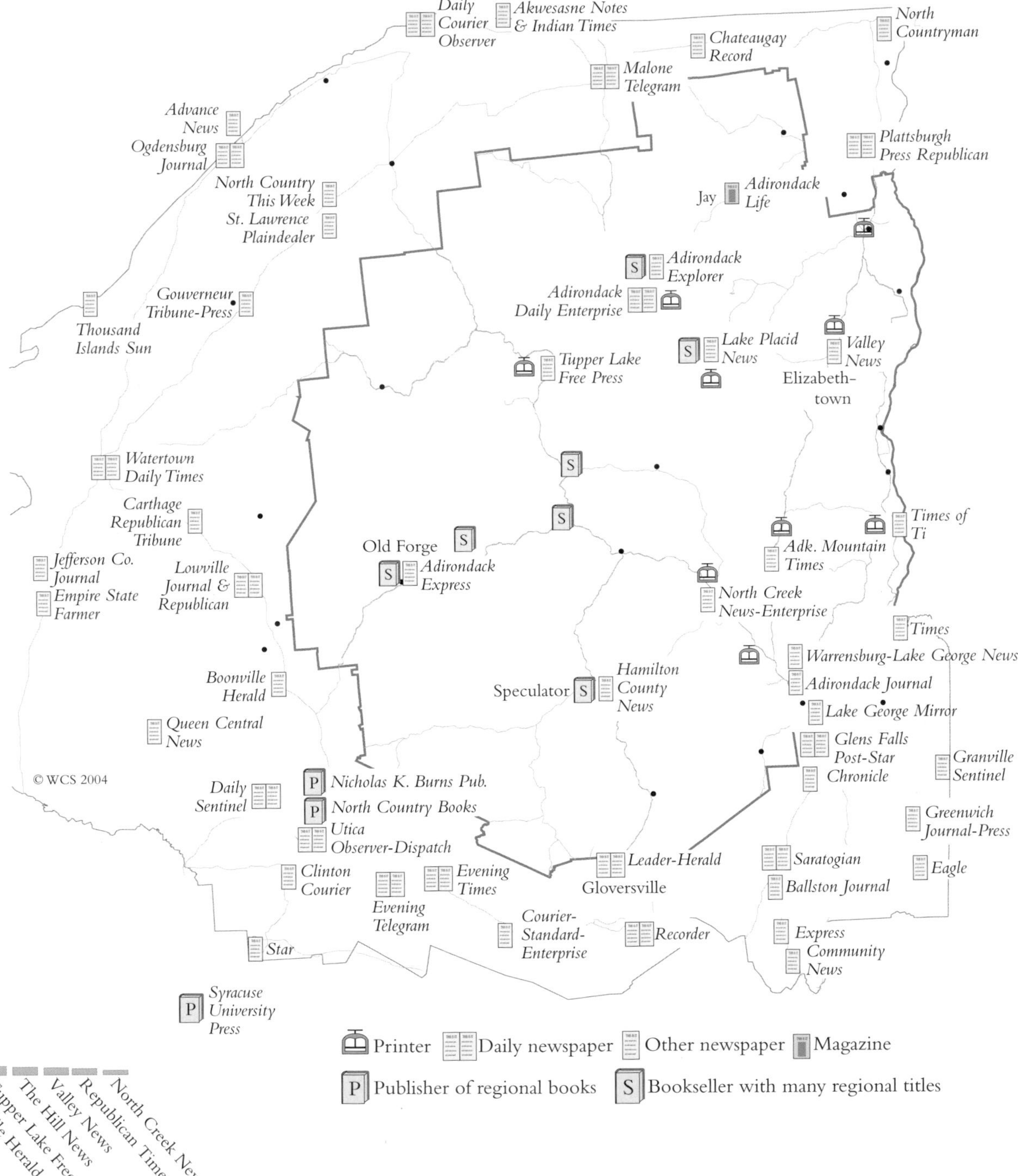

OVER 300,000 copies of twelve daily papers are distributed every day in northern New York. One or more of them reaches every town in the Adirondacks except, their circulation managers swear, the stubbornly out-of-touch town of Morehouse.

While some paper reaches every town, none of them reaches all the towns, and only few towns get more than one. No paper's regular circulation covers more than a quarter of the park, and a close look at the maps shows that, except for the bravely competitive *Daily Enterprise*, most of their circulation areas meet like puzzle pieces but barely overlap.

Paradoxically, some of the best coverage of the Adirondacks as a whole is in large papers published outside the park. This is largely a question of resources. The two largest dailies, the *Utica Observer-Dispatch* and the *Plattsburgh Press Republican*, employ between thirty and fifty people each in their newsrooms and have the resources to cover regional news and to do investigative reporting and feature stories. The three smallest dailies, the *Malone Telegram*, the *Lowville Journal*, and the Saranac Lake *Daily Enterprise* have circulations of about 5,000 each and are near the minimum size at which a daily paper can exist. They employ only a few (hardworking) reporters each, focus on local news, and have only limited resources for investigation and features.

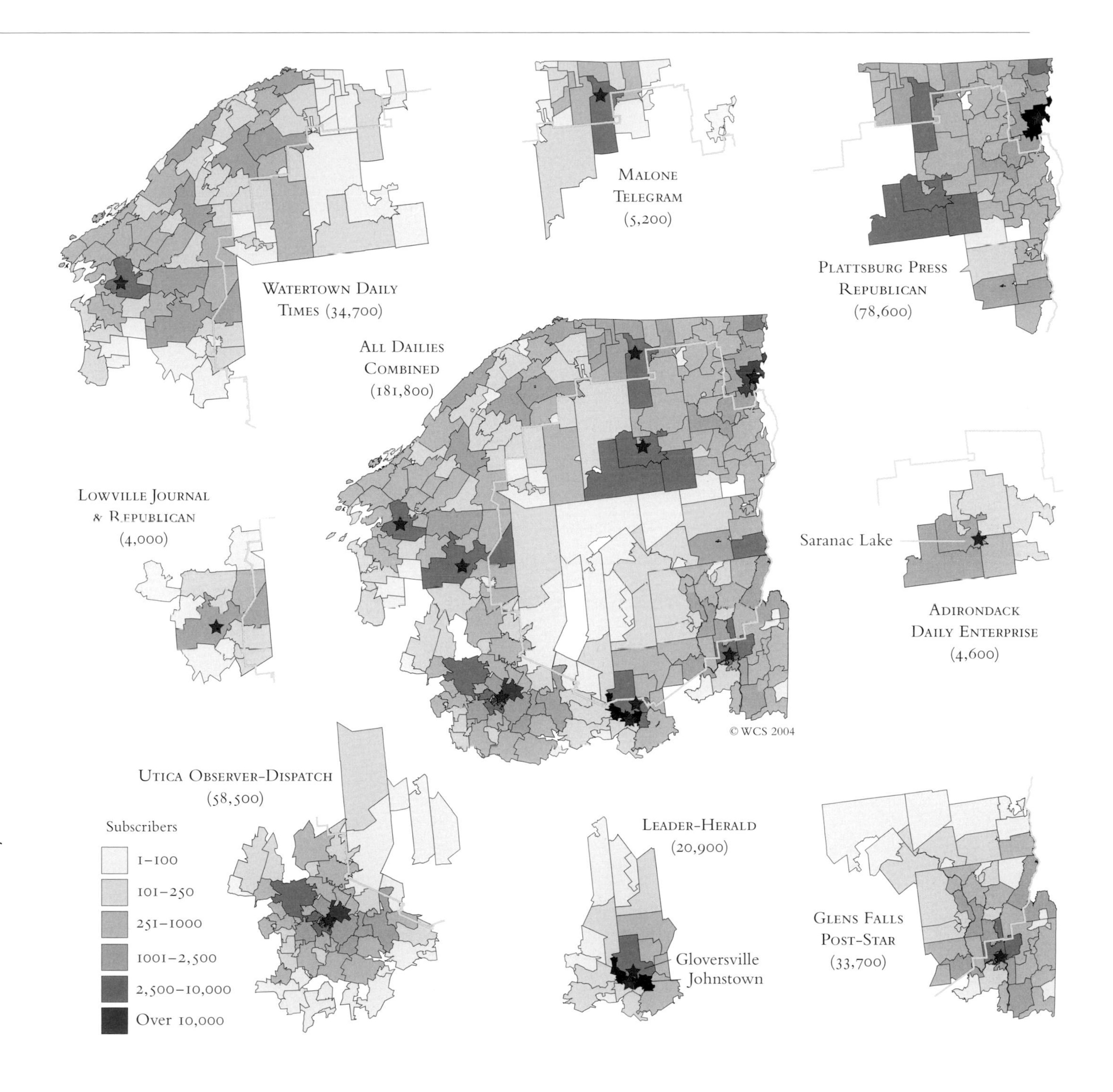

THE broadcast media are, simultaneously, numerous, pervasive, formulaic, and limited. Some of the limitations are technical; the high-frequency signals necessary to carry complex programming propagate poorly in hill areas and must be limited in power to avoid interfering with each other. Other limitations are economic and cultural. Many broadcast media are free to listeners and thus highly dependent on their sponsors. Because their revenues are limited they have very small staffs. In order to deliver a predictable number of listeners to their sponsors, many specialize in a single, easily identifiable, and cheaply generated kind of programming—talk, news, sports, or any of several dozen flavors of music.

The combination of limited resources and a uniform format means that much of gets aired either originates in the studio or is pre-packaged programming from a network. This in turn means that local content is very limited. Local news, with the notable exception of North Country Public Radio, is at best inadequate. Live comedy, drama, music, sports, and local events—all things that the broadcast media used to do extremely well—are now almost gone.

ESTIMATED FRACTION OF MID-ATLANTIC RADIO AUDIENCE LISTENING TO DIFFERENT TYPES OF PROGRAMS

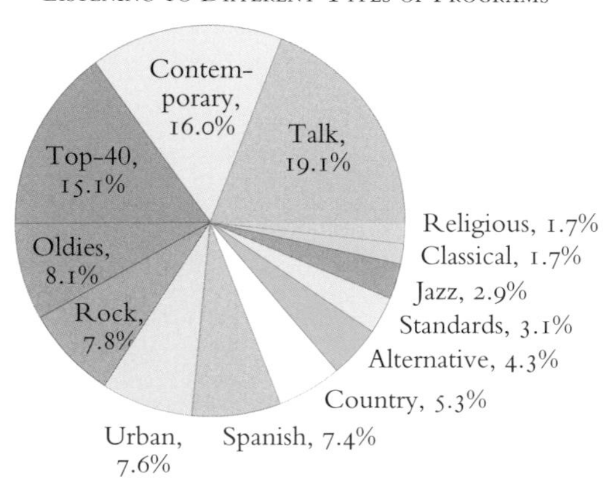

The map shows 149 radio stations, 27 in the park and 122 outside. Most are low-powered. High power isn't really needed if your tower is high enough and isn't all that useful if it isn't. Only one of the nineteen stations that broadcast 10 kilowatts or more is within the blue line, and none are in the park interior.

Ninety-three of the radio stations are commercial stations. Most are independents, and all but one are single-format stations. The talk stations have a substantial amount of local programming, though not necessarily local news or information. The remaining eighty broadcast largely recorded music and national news and sports.

The other 56 stations, including 24 public radio stations, 21 religious stations, and 14 college stations, are to some extent subsidized by their listeners or by parent organizations and so are freer to offer noncommercial programming. The college stations are all independents. The religious stations are about half independents and half affiliates or parts of networks. The public radio stations are all members of regional networks; they carry important local programming and occasionally rebroadcast local concerts but almost never broadcast a live event from outside their studio. Considering that most cultural life does not take place inside broadcast studios, the failure to broadcast live events seems very sad. Any self-respecting 1930s station director would be shocked.

The limitations of contemporary radio stations, and particularly of the commercial stations, become clear when they are compared to newspapers. The fifty northern New York newspapers employ somewhere between a hundred and two hundred reporters. Most of radio stations in northern New York employ no reporters at all.

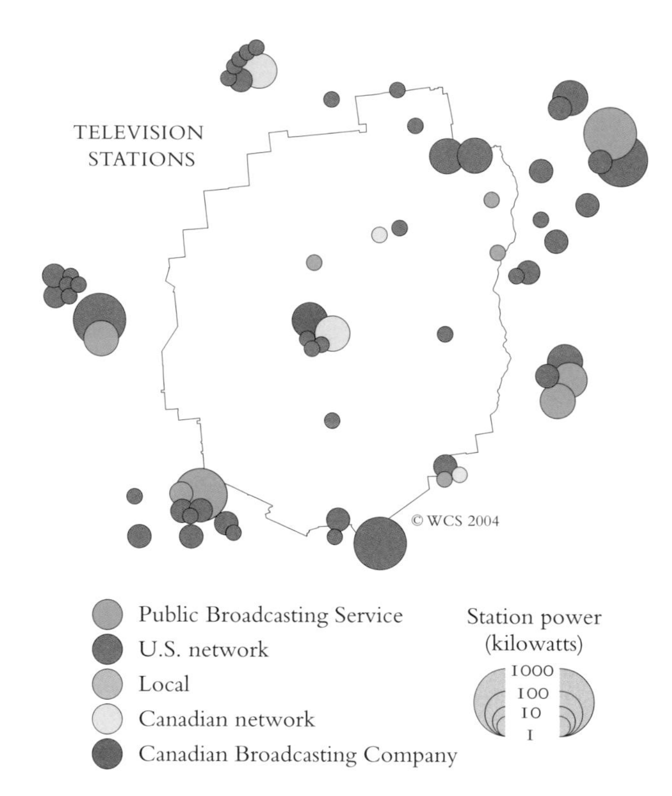

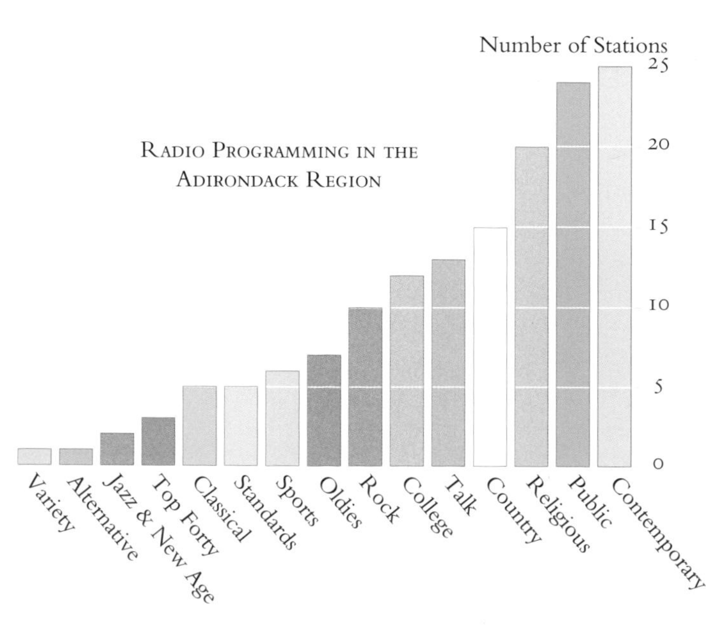

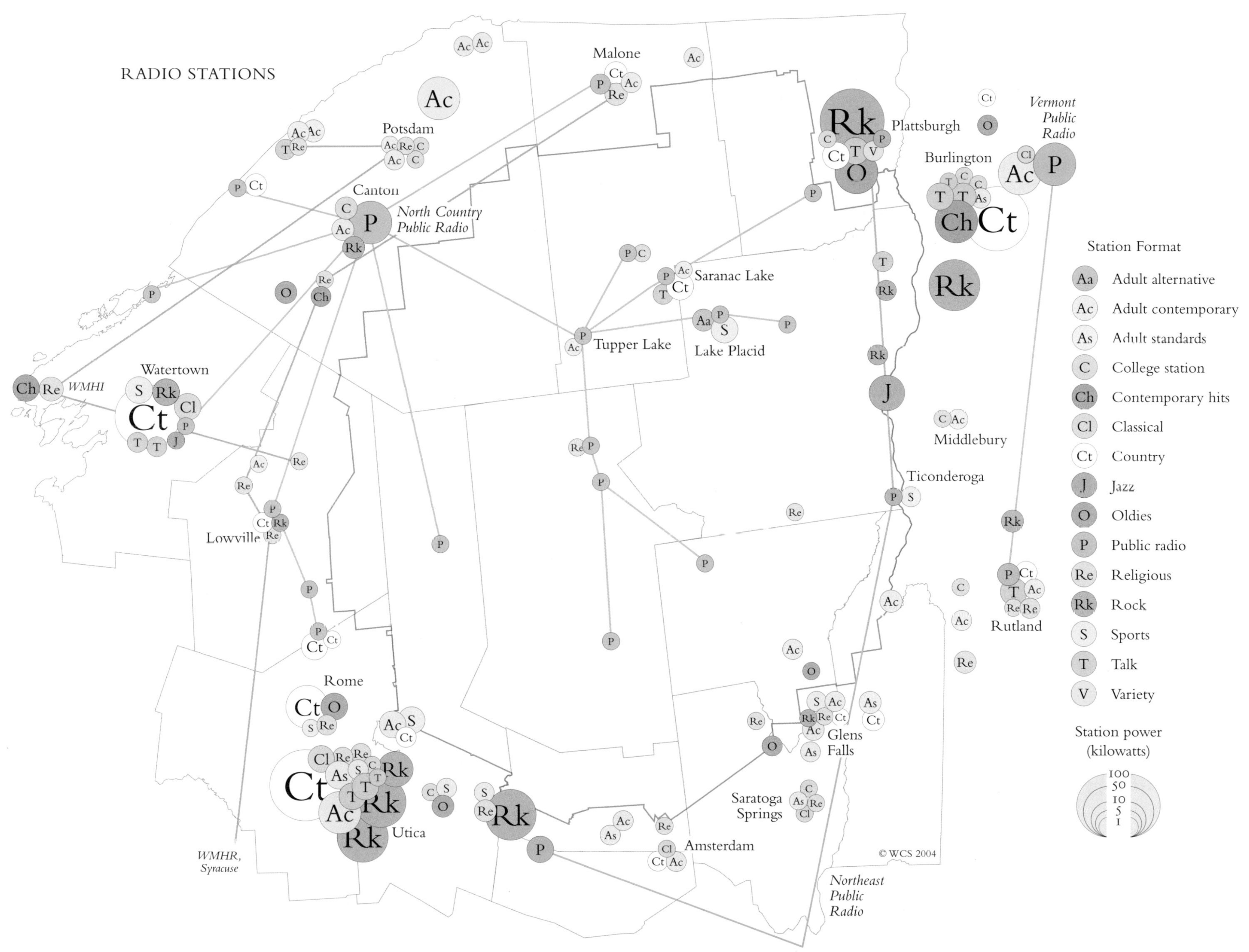
RADIO STATIONS
Malone
Potsdam
Canton
North Country Public Radio
Plattsburgh
Vermont Public Radio
Burlington
Saranac Lake
Lake Placid
Tupper Lake
Watertown
WMHI
Lowville
Middlebury
Ticonderoga
Rutland
Rome
Utica
Glens Falls
Saratoga Springs
Amsterdam
WMHR, Syracuse
Northeast Public Radio
© WCS 2004
Station Format
Aa Adult alternative
Ac Adult contemporary
As Adult standards
C College station
Ch Contemporary hits
Cl Classical
Ct Country
J Jazz
O Oldies
P Public radio
Re Religious
Rk Rock
S Sports
T Talk
V Variety
Station power (kilowatts)
100
50
10
5
1

If you want to find a place where sponsorship does not control content and where live and local are the norm rather than the exception, you have only to look at the arts.

The map shows a gratifyingly lively cultural scene: galleries in twenty-five towns, lectures in twenty-six, live music in fourteen, and live theater, which is commoner in the Adirondacks than movies, in another fourteen. What it hints at but doesn't show is the community of artists that makes the cultural scene possible. There must be several hundred artists and craftspersons to supply the galleries and an equal number of actors, musicians, and writers giving the performances and lectures. Adirondack artists, when finally counted, may turn out to be as numerous as Adirondack guides or innkeepers and just as characteristic of the landscape.

The usual Adirondack exceptions apply. Cultural events, except in the largest towns, draw heavily on summer visitors and are mostly limited to the vacation towns along the main tourist routes. Many of the galleries and theaters close after Labor Day, though, interestingly, fully half are now open all year. And in roughly a third of the park—west of Tupper and Speculator, north of Jay and Saranac Lake—there are no theaters or galleries to close.

Also interestingly, although the audiences go and some of the halls close, many of the artists stay. Some rest, some practice, many visit schools and give performances to small groups in libraries and churches. It must not be easy being an artist in an economy with such a short season, but then it must not be easy being a guide or a store owner either. If your life or art is in some way dependent on the Adirondacks, you may not really have a choice. If you use Adirondack subjects or materials, or if the serenity in your cello's lines derives in some way from the mountains and the trees, you would probably be poorer if you left than if you stayed.

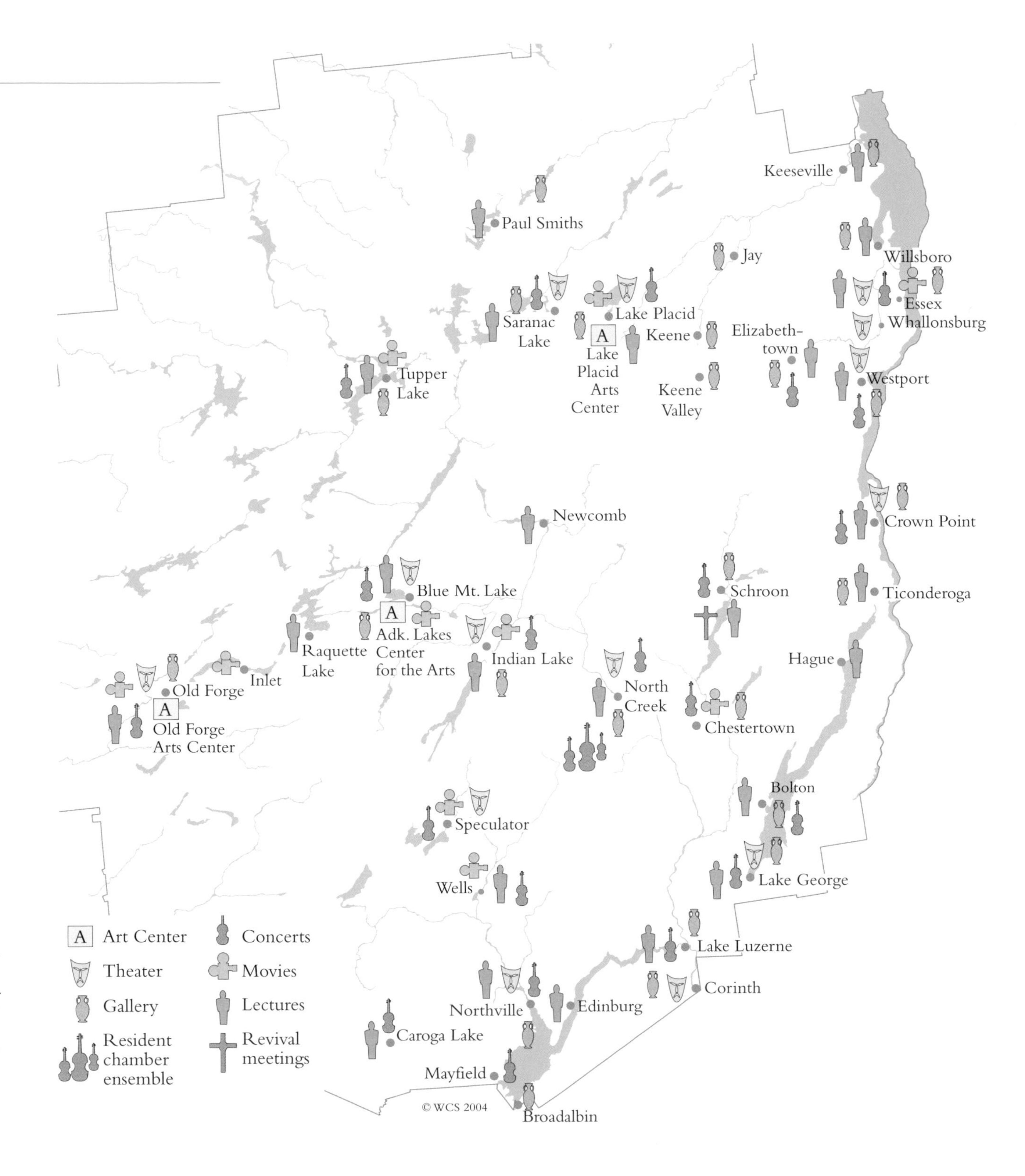

ADIRONDACK history is well preserved and displayed. The map shows five major historical sites and museums, forty-five smaller historical sites and museums, five "submerged heritage" preserves containing shipwrecks, and about 220 buildings on the National Historic Register. There are, in addition, 4,700 unregistered early buildings, many equal in beauty and dignity to those on the register. Many of the historical sites are small but, as with John Brown's farm, storied and evocative. Others are large and complex and, as with Fort Ticonderoga and the Adirondack Museum, nationally famous for content and interpretation.

It is important to notice what is and is not on the map. The ever-popular history of wars and recreation and wealth is very well represented. The history of the working Adirondacks—mines and immigrants and mills and forests—is much less represented. The fascinating and ambiguous history of the park itself—what the Adirondacks are and how and why we have a park at all—is not displayed anywhere.

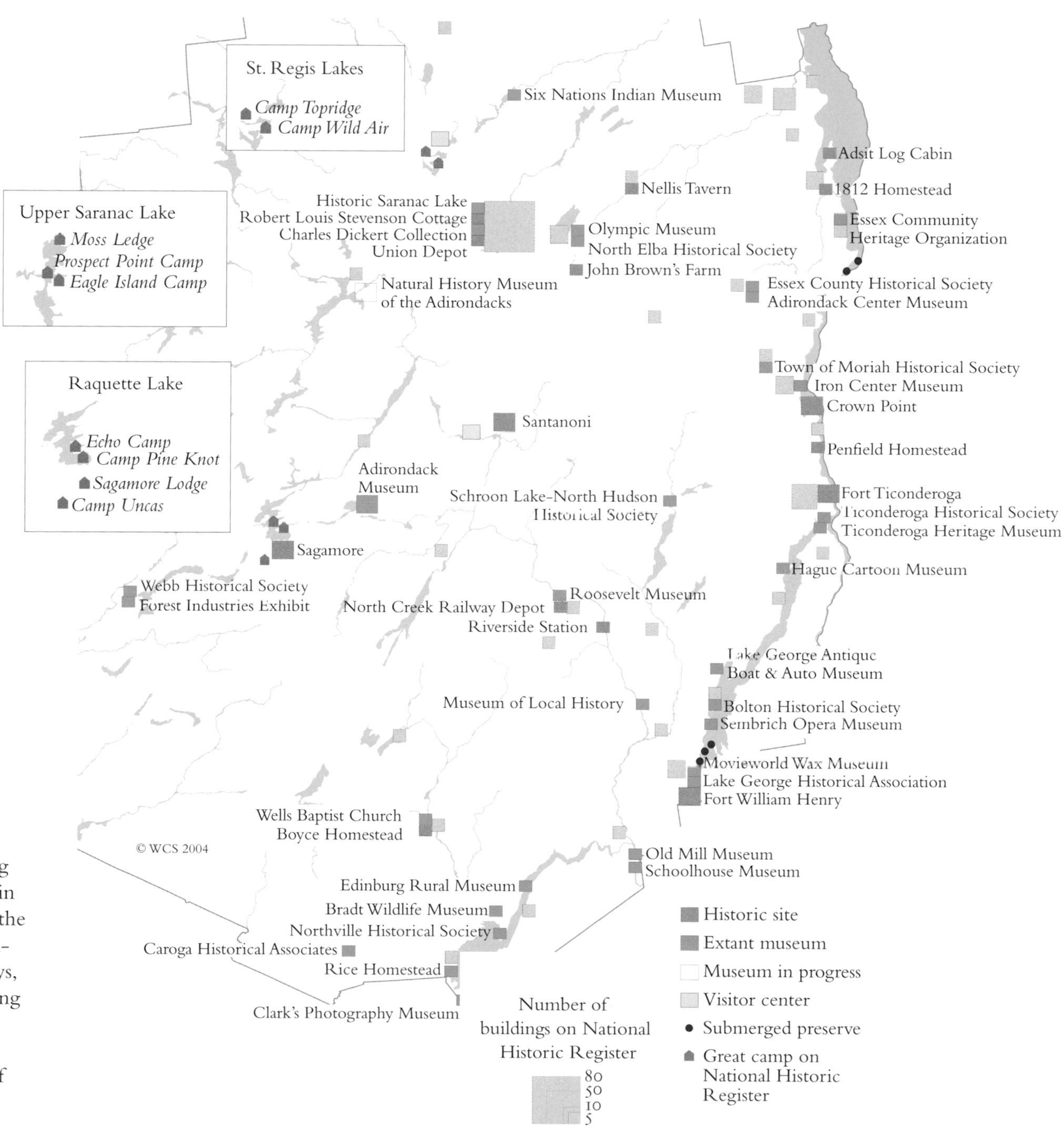

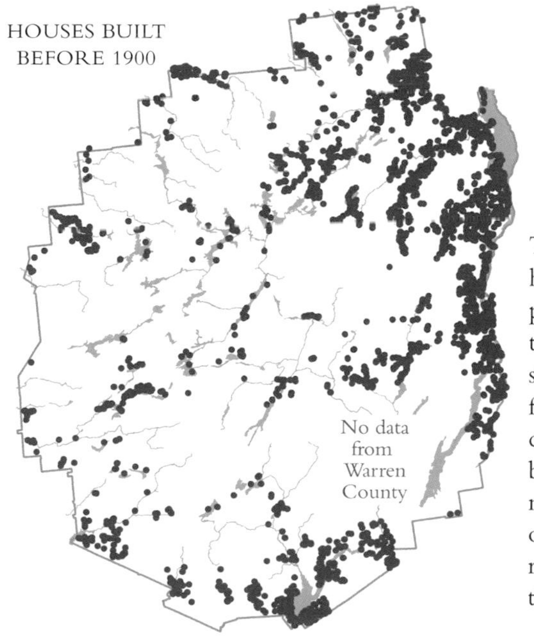

The oldest houses in the park are along the Champlain shore and in the formerly agricultural valleys, but a surprising number also occur on the resort lakes of the interior.

"ALL thought," Fernand Braudel wrote, "draws life from contacts and exchanges." These two calenders chronicle some of the public exchanges—auctions, bashes, challenges, classes, clinics, concerts, dinners, expos, festivals, meetings, performances, rallies, and races—that make two lake towns lively.

Old Forge is both a resort town and a gateway town. Like other resorts it needs to attract and amuse its visitors, and like other gateways it tries to delay some of those who are headed further into the park. To do this it relies on a mixture of attractions (*Map 13-6*) and special events, concentrated in the summer months.

Because the tourist season in the Adirondacks starts late and weekends before fly time are at a premium, spring in Old Forge is a bit hasty. The Paddlefest participants have to share a weekend with the Powerboat Expo, and, for good measure, the Thunder in the Forge motorcycle rally. How willingly they do this, and how often they will be willing to do this, we do not know. Summer is longer and very full, though between postcard shows, gangster reenactments, and the climactic Library Bash, perhaps a bit hokey. Fall comes quickly in the mountains: the Octoberfest is in early September, and, save for a few last paddlers and quilt-makers, the season ends when the leaves fall a few weeks later.

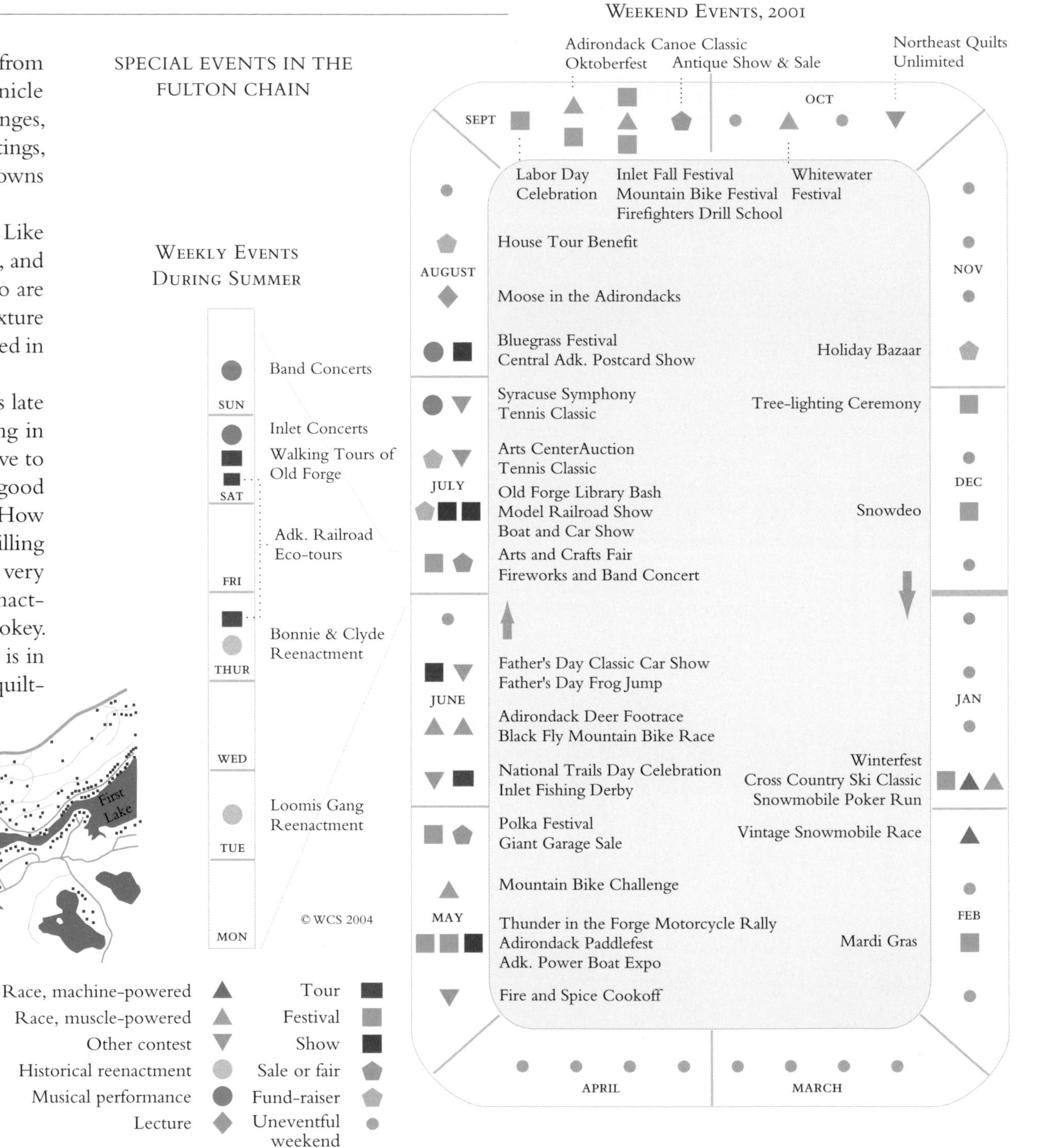

SARANAC LAKE is a tourist town in the summer and a ski town in the winter. In between it belongs to the residents. The calender shows what the residents did, with and for each other, in one week in early spring and another in mid fall.

Clearly they did a lot. These two unspecial weeks have some forty-five public events, not counting school events, church services, and sports. There were fourteen meetings, ten lectures and classes, seven performances, three clinics, four fund-raisers, and a half-dozen breakfasts, dinners, services, dedications, and distributions.

Fall is apparently the busier season: funds to raise, flu and weight to control, things to organize and dedicate and celebrate. Early spring is more subdued: travelogues, a few improving lectures and necessary meetings and, perhaps on a fine thaw weekend with the first blackbirds, chain-saw classes and sugar on snow and the high school musical.

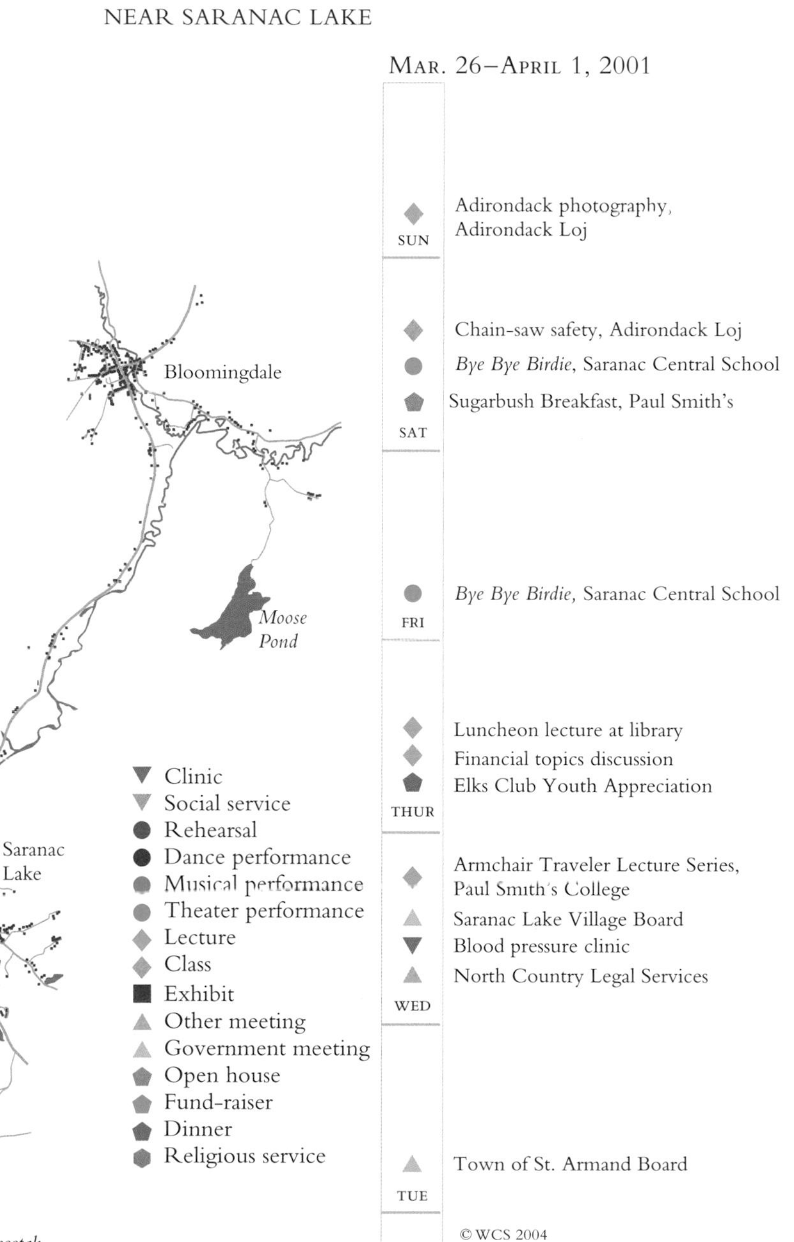

MAR. 26–APRIL 1, 2001

SUN
- Adirondack photography, Adirondack Loj

SAT
- Chain-saw safety, Adirondack Loj
- *Bye Bye Birdie*, Saranac Central School
- Sugarbush Breakfast, Paul Smith's

FRI
- *Bye Bye Birdie*, Saranac Central School

THUR
- Luncheon lecture at library
- Financial topics discussion
- Elks Club Youth Appreciation

WED
- Armchair Traveler Lecture Series, Paul Smith's College
- Saranac Lake Village Board
- Blood pressure clinic
- North Country Legal Services

TUE
- Town of St. Armand Board

MON

NOV. 12–18, 2001

SUN

SAT
- Dance Performance at Lake Placid
- Spaghetti dinner for St. Bernard's fifth grade
- Dedication of New Adventist Church
- Musical celebration, Presbyterian Church
- Critters in the Cold, at VIC
- Flea market, St. Paul Church, Bloomingdale
- First Aid Class

FRI
- Dance Performance at Lake Placid
- Fall Classic at Pendragon Theater
- Opening for Art Show at VIC

THUR
- Moose Lodge 457
- Franklin Co. Highway Dept., Vermontville
- Knights of Columbus Bingo
- Fall Classic at Pendragon Theater
- Board of Cooperative Education Services
- Luncheon lecture at library
- Youth mentoring program, Lake Placid
- Flu Clinic, Bloomingdale
- Food Distribution, Vermontville

WED
- Winter Carnival Committee
- Adk. Park Agency Public Meeting
- Nature for the Very Young, at VIC
- Fall Classic at Pendragon Theater
- Adk. Genealogical & Historical Society
- Memorial Mass for Catholic Daughters

TUE
- Library Board of Trustees
- Tri-Lakes Amateur Radio Club
- Weight-control support group
- Harrietstown special work session
- Town of St. Armand Board
- Adirondack Singers

MON
- Saranac Lake Village Board of Trustees

OUTDOOR RECREATIONAL FACILITIES

The map shows about three hundred of the several thousand outdoor recreational facilities in the Adirondacks. Besides those shown here there are several thousand miles of forest roads and trails and their associated parking areas and trailheads (*Maps 15-1,4,5,6*), several hundred individual campsites on state land, over a hundred private campgrounds and their associated beaches and boat launches (*Map 13-4*), 77 marinas, 187 lean-tos, and about 35 dams creating recreational lakes.

The creation of some of the facilities on New York State lands raised constitutional questions. The mountain roads and large ski areas were clearly inconsistent with the requirement that the Forest Preserve be kept wild and required constitutional amendments. The first to be built, in 1927, was the Memorial Highway to the top of Whiteface. It was followed by the Whiteface ski area in 1941, the Gore Mountain ski area in 1947, and the Prospect Mountain scenic highway in 1954. The legal status of the smaller facilities like campgrounds is uncertain; it has neither been established by amendment nor tested in the courts.

The forty-five state campgrounds, three ski areas, two scenic highways, and assorted Olympic stadia, jumps, tracks, and courses make New York State the largest commercial provider of recreation in the park. Had it not been for the voters it might have been even bigger: at least ten other proposals for major recreational developments, including ski developments for the Blue Ridge and McIntyre Range and eight constitutional amendments that would have authorized the building of closed cabins and resort complexes have been defeated.

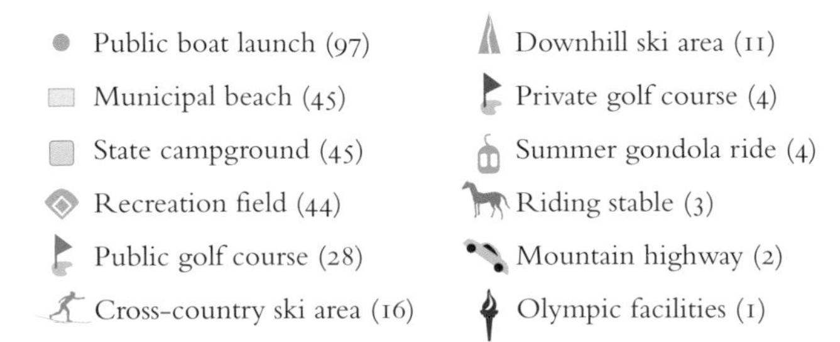

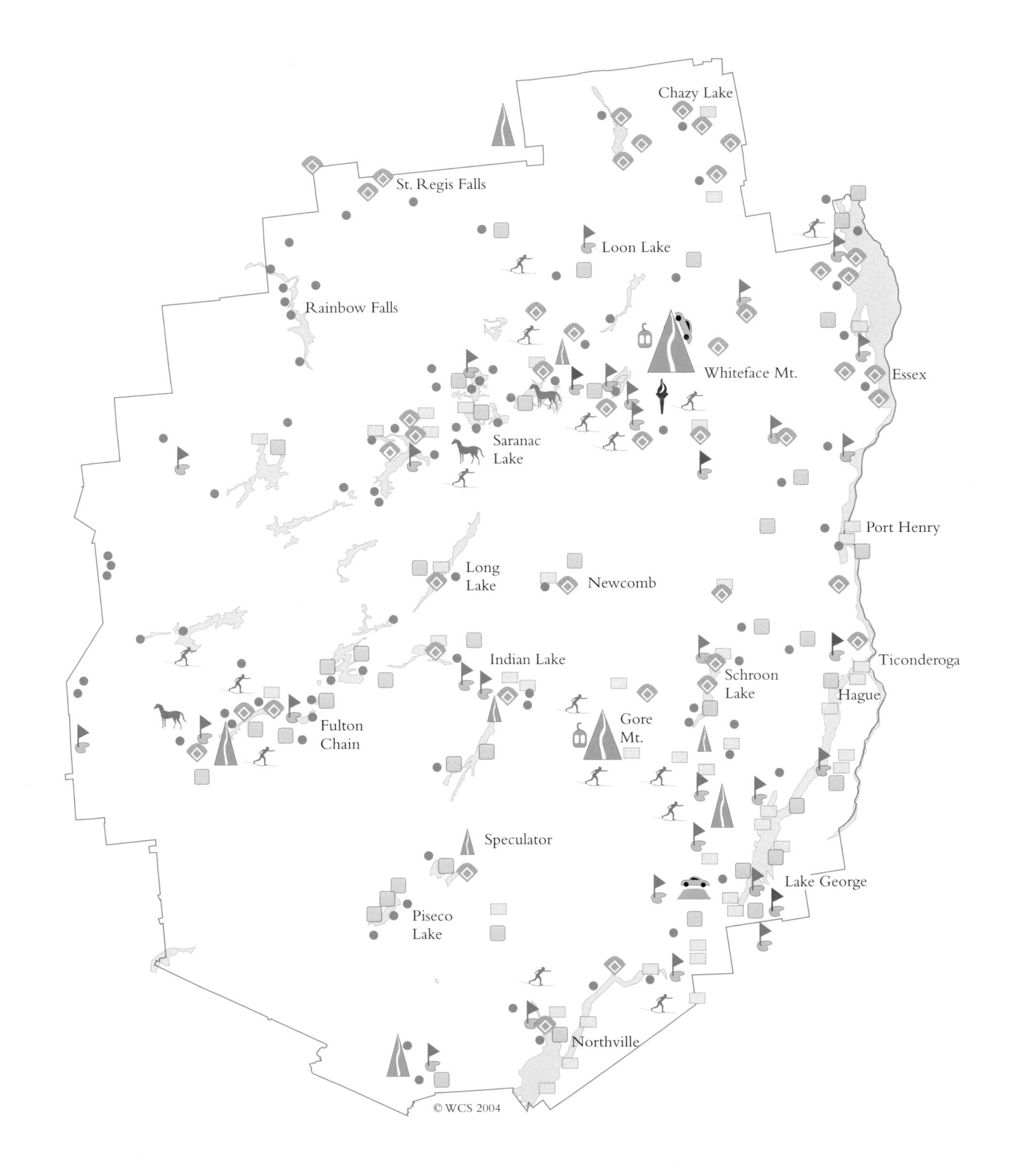

15 *Outdoor Recreation*

ONE of the ironies of Adirondack history is that the recreationists, for whom the founders had hoped to preserve the park, have been responsible for some of the greatest changes in it.

The forests and lakes that the founders of the Adirondack Park knew and that, in 1894, they obligated us to keep forever wild were far from completely wild even then. The founders traveled by train and steamer and stayed in fancy hotels on lake shores. But when they left the hotels and went into the woods, they still saw a very wild park. The woods through which their guides led or rowed them were roadless, machineless, uncrowded, and almost trailless. This was what wild meant to them, and what they hoped could be preserved so that other people could enjoy it as they had.

What they did not know was that their way of enjoying the woods was about to vanish. Thirty years later the big hotels and their guests and guides were almost gone. A new generation of recreationists, much less interested in being cosseted than their predecessors, were arriving by automobile, using modest accommodations, and setting out to hike and camp on their own.

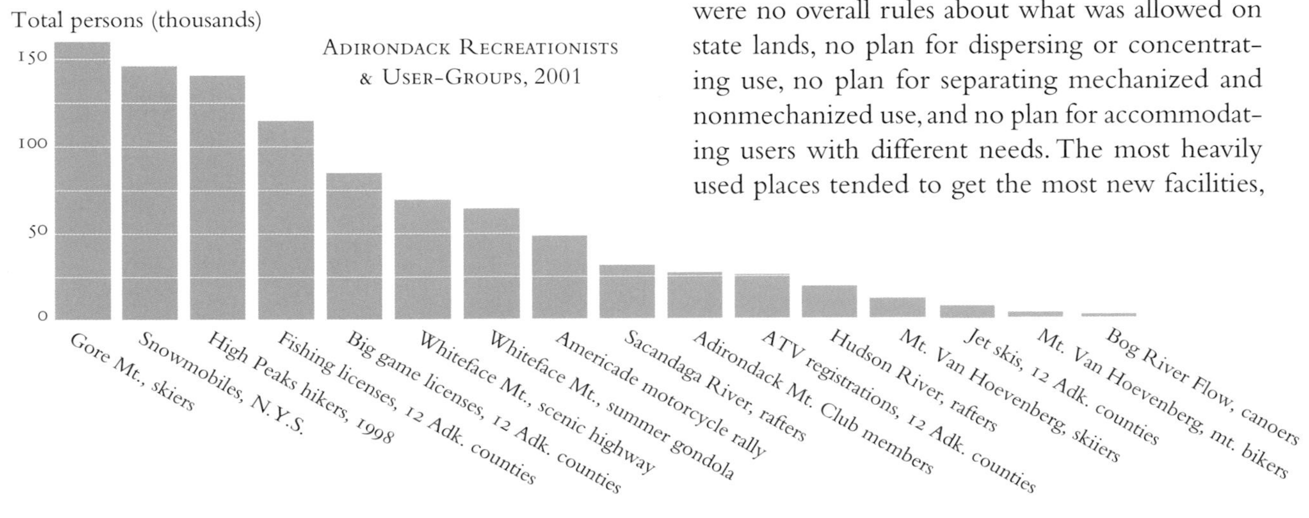

Very few of these new visitors were interested in a trailless wilderness. What they wanted instead was a developed recreation area, whose trails, shelters, motels, and cabins, would, in a sense, do for them what the hotels and guides had done for their predecessors.

By the early 1920s both the state and several hiking clubs were working hard to provide the conveniences the hikers wanted. Soon, inevitably, the providers built additional facilities for themselves. Quite rapidly, and with the best of intentions, the wilderness acquired cabins, access roads, telephone lines, fire towers, and ranger stations.

By 1950 both the woods and the private lands adjoining them had been very thoroughly modified. The map on the facing page shows some of the more conspicuous modifications and suggests how thoroughly the park had changed.

The modification of the park, though extensive, was haphazard. There never was, and is not now, a master plan for Adirondack recreation. There were no overall rules about what was allowed on state lands, no plan for dispersing or concentrating use, no plan for separating mechanized and nonmechanized use, and no plan for accommodating users with different needs. The most heavily used places tended to get the most new facilities, increasing their use even more. It was eighty years before the park had official wilderness areas. After a hundred and ten years there are still no trails specifically designed for disabled users.

Every decade since the 1950s has added new groups of users. The 1960s brought snowmobilers and large numbers of hikers. The 1970s brought an influx of climbers and rafters. The 1980s brought the first jet skis and ATVs. The 1990s brought large recreational vehicles with toilets and lights and generators. The 2000s may well be the decade of the kayakers and snowshoers.

Each group of users claimed a space for their own and requested that the park be modified to accommodate their needs. Some, like the climbers, used spaces that no one else wanted and were extremely self-conscious about the modifications they made. Some, like the rafters, needed few facilities and were able to share spaces that were already in use. And some, like the ATVs, tended to damage the spaces that they used and displaced or angered other users.

The extent to which the park has been able to accommodate a wide variety of users has been one of its great successes. But the competition between these users, and especially between their visions of what the park is for, has been the source of much conflict. Today, a hundred and ten years after the park was created, that conflict is not only still present but thoroughly institutionalized. It has been built, the maps in this chapter suggest, into the laws governing the park, into the plans that implement those laws, and even, surprisingly deeply, into the geography of the park itself.

The Adirondacks contain 1,500 miles of designated hiking trails plus an equal or greater mileage of old roads and undesignated trails. The system developed informally, in part by the construction of new trails and in part by the conversion of forest roads to trails. Only gradually did the present system of officially designated trails develop.

The original trails were all in the mountains. The first were built by the Adirondack Trail Improvement Society in the late 1890s. Many others were added by the Adirondack Mountain Club, which was founded in 1922, and the Camp and Trail Club, founded in 1923.

The clubs worked hard, and most of the current trail system in the High Peaks was in place by the time the second Adirondack Mountain Club guidebook appeared in 1934. Despite the enormous increases in use since then, only a few additional trails have been built in the High Peaks and surprisingly few in the park as a whole. In the High Peaks several connecting trails have been added, ten high-elevation shelters removed, and nine new shelters built in the valleys. The sixteen 4,000-foot peaks that were trailless in 1934 remain officially trailless today, though unofficially well-tracked and thoroughly traveled.

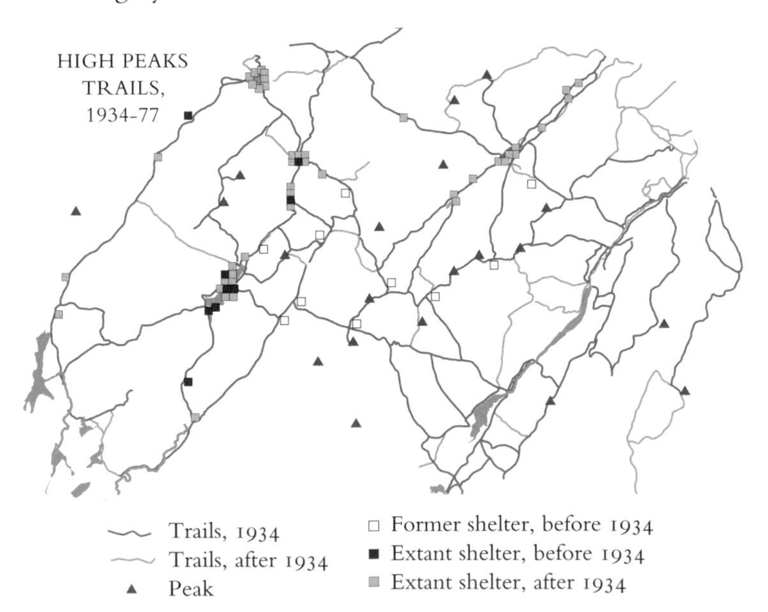

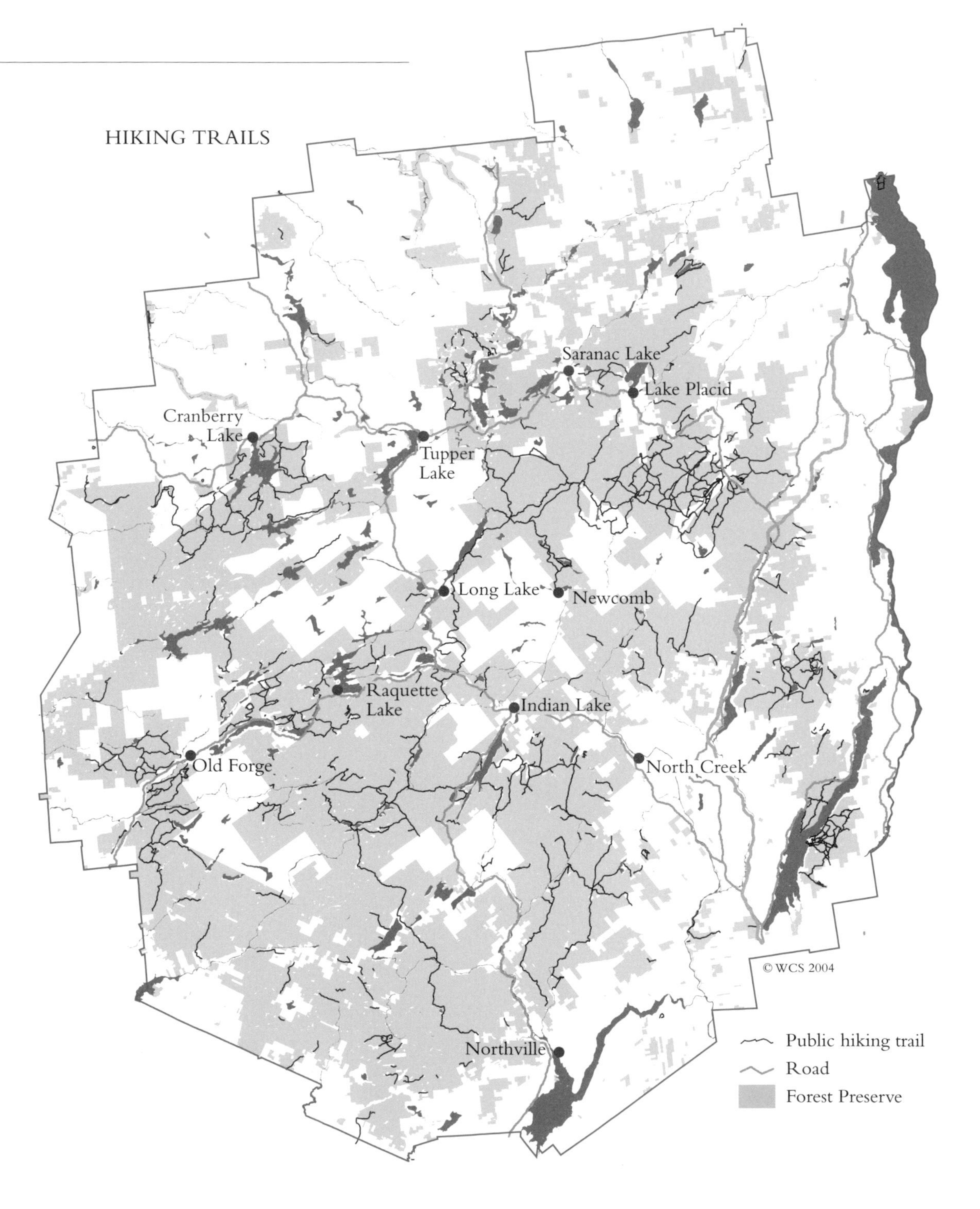

THE High Peaks are by far the most intensely used wilderness in the park. About 100,000 hikers register at the trailheads shown on the map each year and another 40,000 at other trailheads to the north and west. Registration studies suggest that many users don't register and that the actual use may be between 200,000 and 300,000.

To accommodate these visitors a substantial infrastructure is required. In 1998 the High Peaks management unit contained 303 miles of trails, 300 signs, 107 pit privies, 72 lean-tos, 49 bridges, four ranger cabins, four dams and artificial lakes, and, on inholdings, a lodge and two rental cabins operated by the Adirondack Mountain Club.

At least in part because this infrastructure makes hiking safe and convenient, the use of the High Peaks is both concentrated and increasing. Registered use more than doubled between 1983 and 1998. Currently, on an August weekend, between one thousand and two thousand people will hike in the High Peaks. Most of them will use the two or three most popular trails, and many will camp at the same half-dozen popular places.

This much use, and the crowding and trampling associated with it, raises questions about what is best for mountains and their users and what if anything can be done to reduce or disperse use. Probably few hikers really like sharing a trail with several hundred others, and certainly no one wants to see a lovely camping area turn into a backcountry slum. But to decrease the use of the High Peaks you will either have to regulate the users or send them somewhere else: one restricts personal freedom and the other compromises other wildernesses. Adirondackers value freedom and wilderness highly and may well prefer continuing the customary overuse of the High Peaks to instituting the new rules and increased use elsewhere that might be required to fix it.

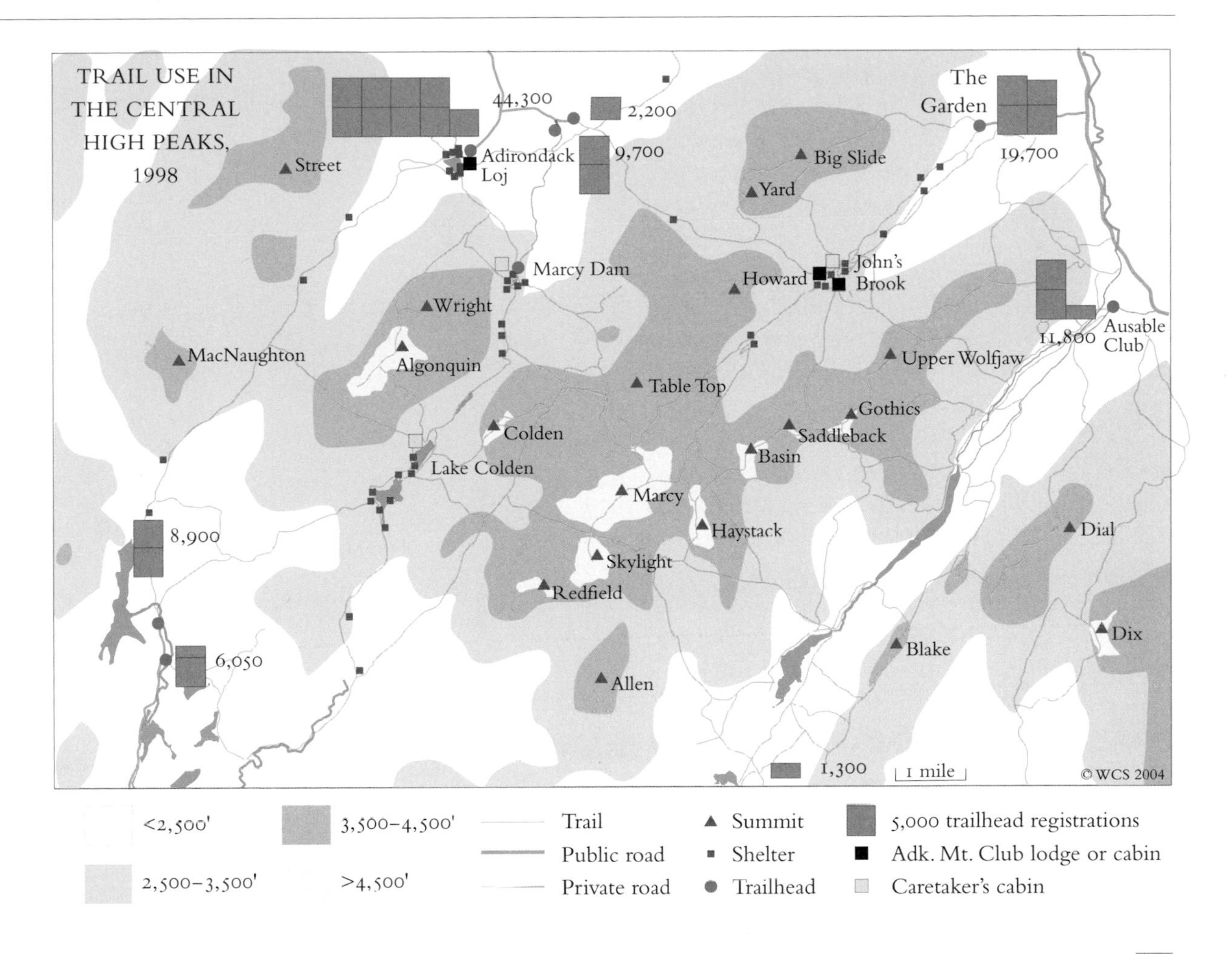

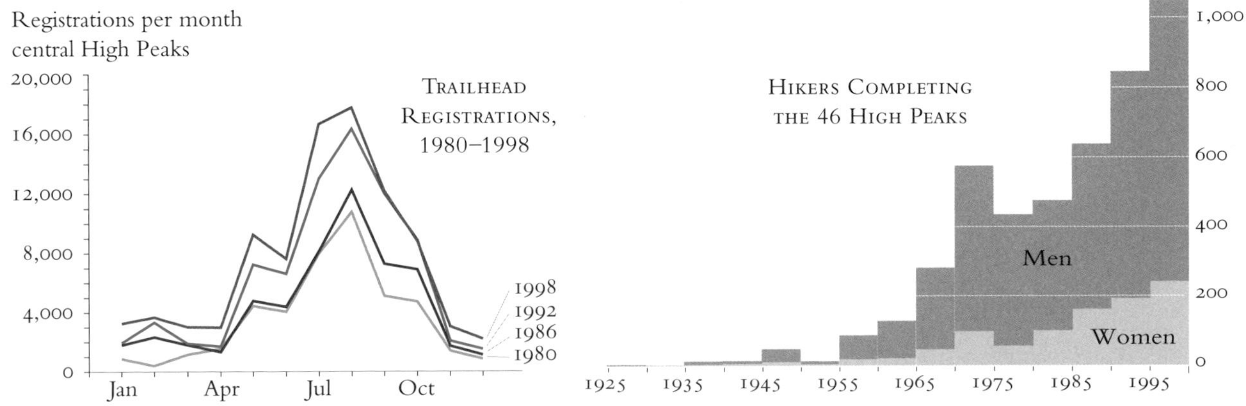

Climbing, perhaps the most dangerous and certainly among the most difficult of outdoor sports, is also one of the youngest. Before 1960 only a few Adirondack cliffs had been climbed and these only by a few pioneering climbers. By the mid-1970s, nylon ropes and chrome-molly nuts had made climbing safer and more accessible: Thomas Rosecrans' 1976 *Adirondack Rock and Ice Climbs* listed 103 named climbs, ten of these graded 5.8 and six 5.9. By the 1990s, sticky rubber shoes, new protection systems, and the efforts of a generation of dedicated local climbers and a few visiting stars like Henry Barber and Martin Berzins had again redefined what could be climbed. The 1995 edition of Don Mellors' *Climbing in the Adirondacks* listed over eight hundred climbs, two hundred fifty a humbling 5.10 or higher. The first of the Adirondacks' twenty-eight 5.12 routes was put in 1988; its first and only 5.13, the one-hundred-foot *Salad Days* on the awesome main face of Poke-O-Moonshine, was put in the same year. Because many Adirondack cliffs remain unexplored and because there are increasing numbers of fine climbers exploring them, it is very likely that it will not be the last.

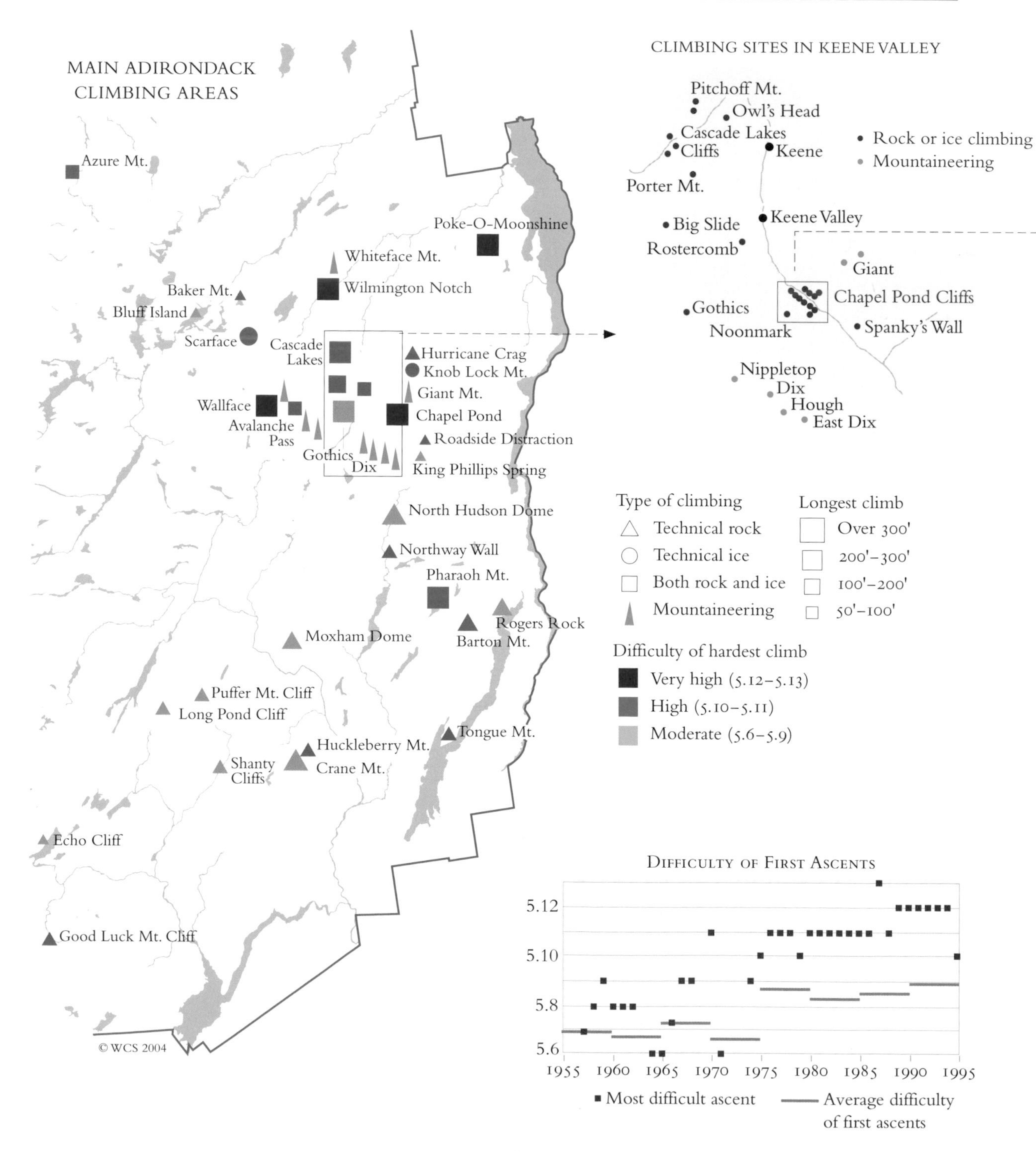

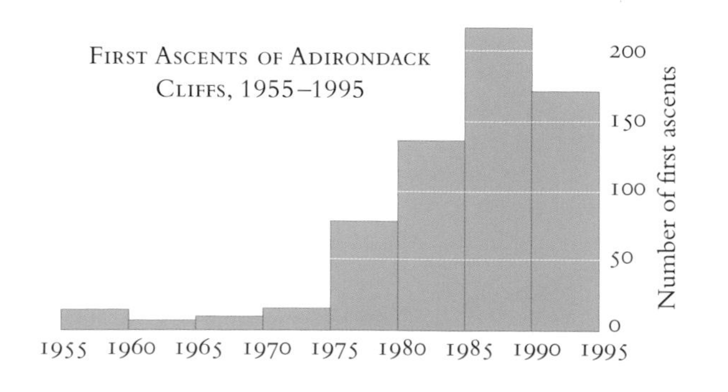

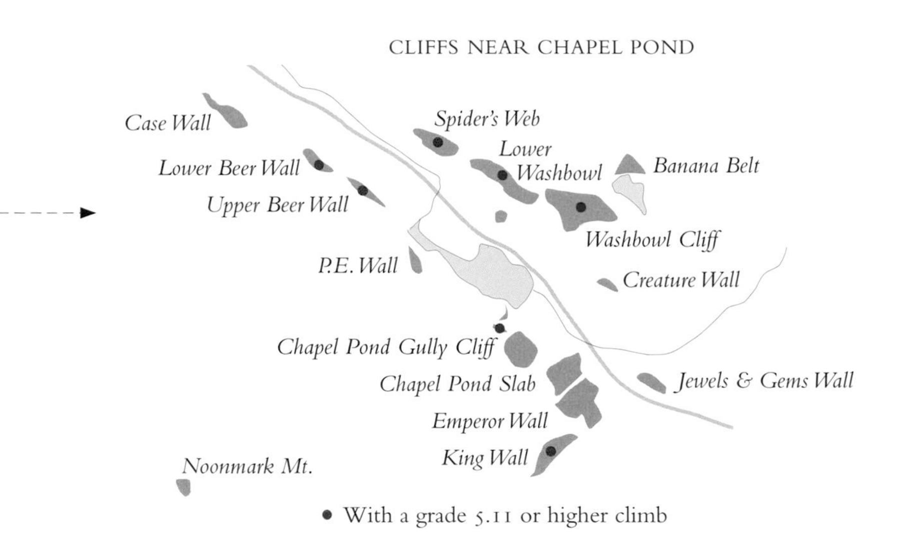

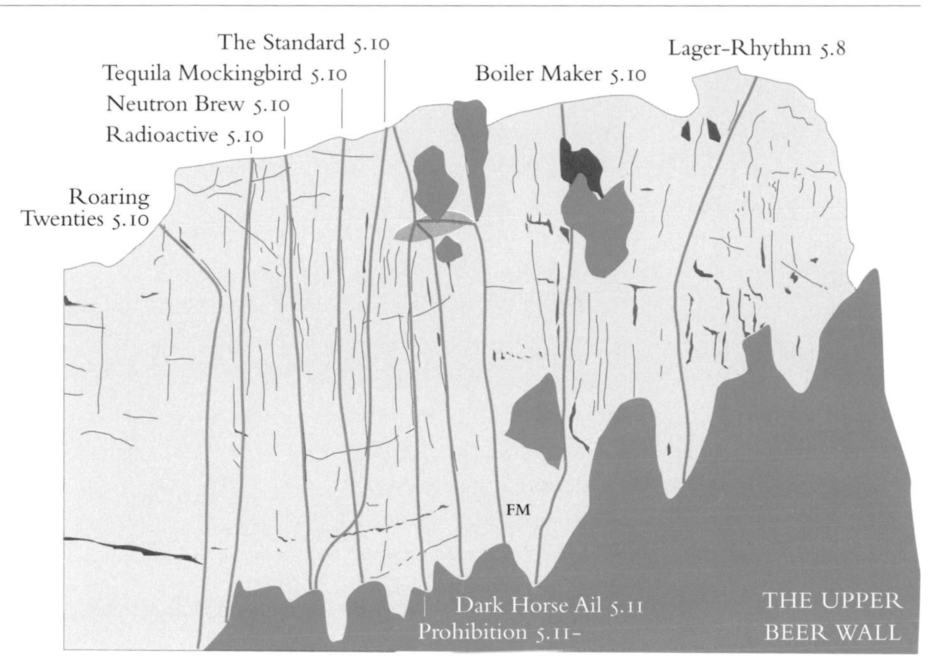

The crack marked FM has four routes: *Frosted Mug*, 5.9; *Frosty Edge*, 5.10+; *Flying & Drinking and Drinking & Driving*, 5.10; and *Labatt-Ami*, 5.7.

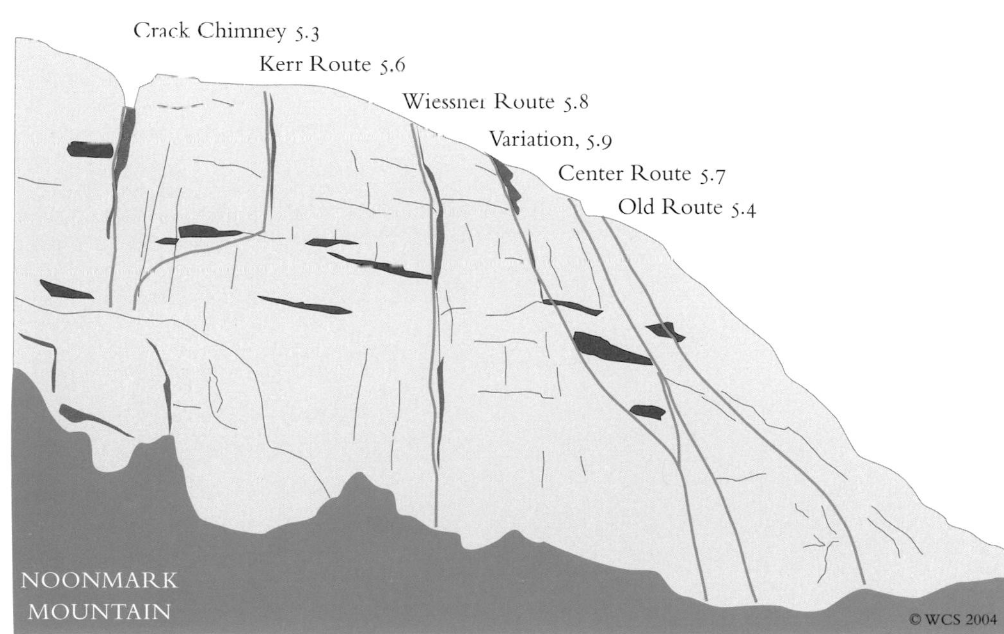

The Beer Walls are newly explored cliffs near Route 73 that have become very popular with mid-level and advanced climbers. The Upper Beer Wall was first climbed in 1982. Its sheer 140-foot middle face contains no less than twelve routes rated 5.10 to 5.12, put in by eleven different climbers between 1983 and 1988. Not only had the level of difficulty increased by three grades since Fritz Wiessner's first Adirondack ascents but by 1990 almost every cliff had people climbing at that level.

The ninety-foot-high, south-facing cliff on Noonmark Mountain is a classic Adirondack anorthosite face with a wilderness location and spectacular views. It was probably first climbed in the 1930s. The three center routes were discovered by Fritz Wiessner and his climbing partners James Kerr and Garfield Jones. Wiessner, a remarkably daring and enduring climber who pioneered difficult routes all over the world, did his first climbing in the Austrian Alps in 1911 and was still climbing difficult routes in the Shawangunks in the 1980s. His routes on Noonmark are considered only moderately difficult today but were among the most difficult yet done in this country when he put them in.

AMONG the most prized of the park's natural assets are the rivers whose headwaters meet near the center of the Adirondack dome and make a three-season waterway, navigable in both directions, across the park. A hundred and fifty years ago this waterway was where the loggers got their best spruce. A little later, as the spruce was running out, it was where the guides from the big hotels took their sporting and eco-touring guests for week-long cruises. Today it is one of the most popular canoe-touring areas in the Northeast. Its longest tour is the hundred-mile diagonal route on the Moose and Raquette Rivers to the Saranac Lakes and St. Regis Canoe Area. Shorter tours on the Oswegatchie, Bog, and Beaver Rivers are also extremely popular. Long tours on the big eastern and southern lakes are possible and have some fine segments, but, because of shoreline development and heavy powerboat traffic, they are much less popular.

The rapids of the upper Hudson are equally prized but much more demanding to use. Between Tahawus and Warrensburg there is an invigorating thirty miles of whitewater in eighty-five miles of river. The whole segment is exceptionally wild, lovely, and rich in biological and historical detail. It is considered, rightly, to be one of the park's and the country's great treasures.

It was a treasure that we almost lost. In the 1960s, two major dams that would have flooded thirty miles of the upper Hudson were defeated after great effort by citizen groups. That the dams could have been proposed at all is frightening. That it was the citizens and not the government that ultimately rejected them is inspiring but cautionary. Democracy, it seems, does work, though neither automatically nor easily.

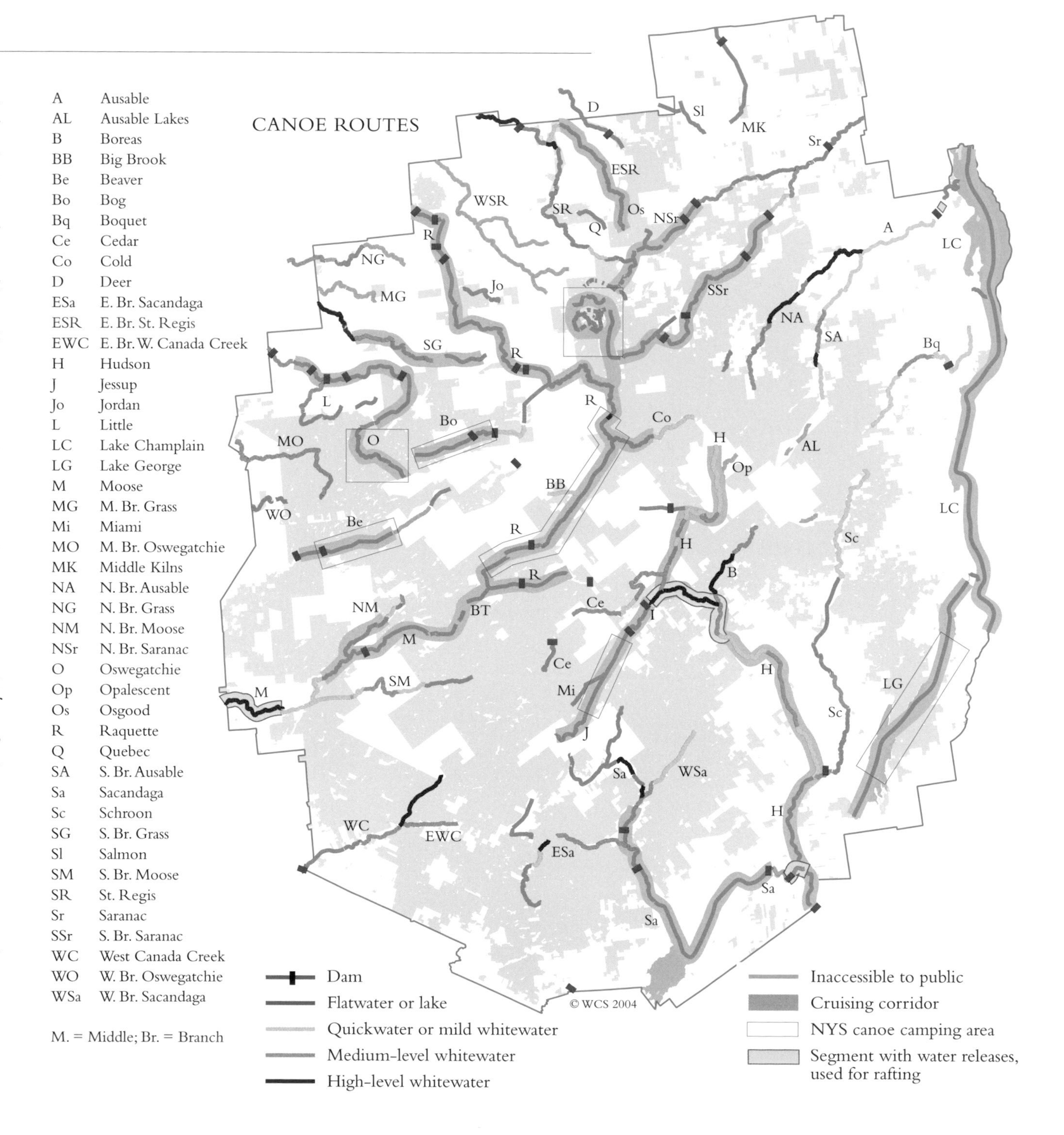

FIVE Adirondack rivers, the Ausable, Moose, Sacandaga, Indian, and Hudson, are used for commercial rafting. The Ausable and Sacandaga are short, relatively mild, very popular rivers. The Moose is a longer run, awesome at high water, fairly gentle in the summer. The Indian and Hudson offer a full-day wilderness run, with heavy straight-ahead whitewater in the spring and long complex technical rapids during summer releases. Their centerpiece is the magnificent Hudson Gorge.

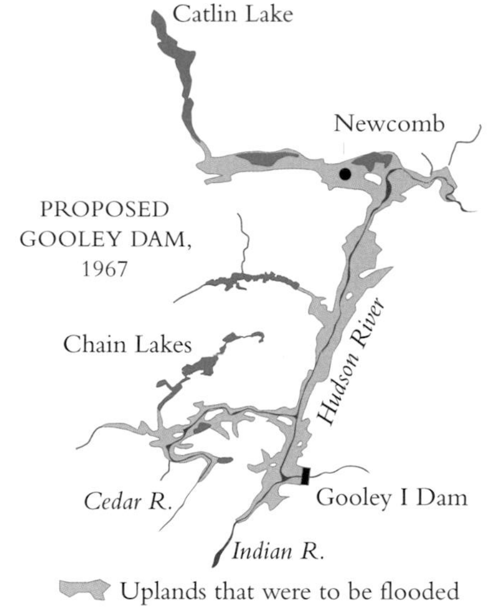

The proposed Gooley Dam. In 1967, when New York City was suffering from a severe drought, the leading proposal for increasing its water supply was for a two-hundred-foot high earth dam in the Hudson Gorge. The dam would have flooded 16,000 acres of land, including fifteen miles of the Hudson River, eight miles of the Cedar River, and much of the settled part of the town of Newcomb. It was finally defeated by the passage of a landmark Hudson River Protection Bill in 1969.

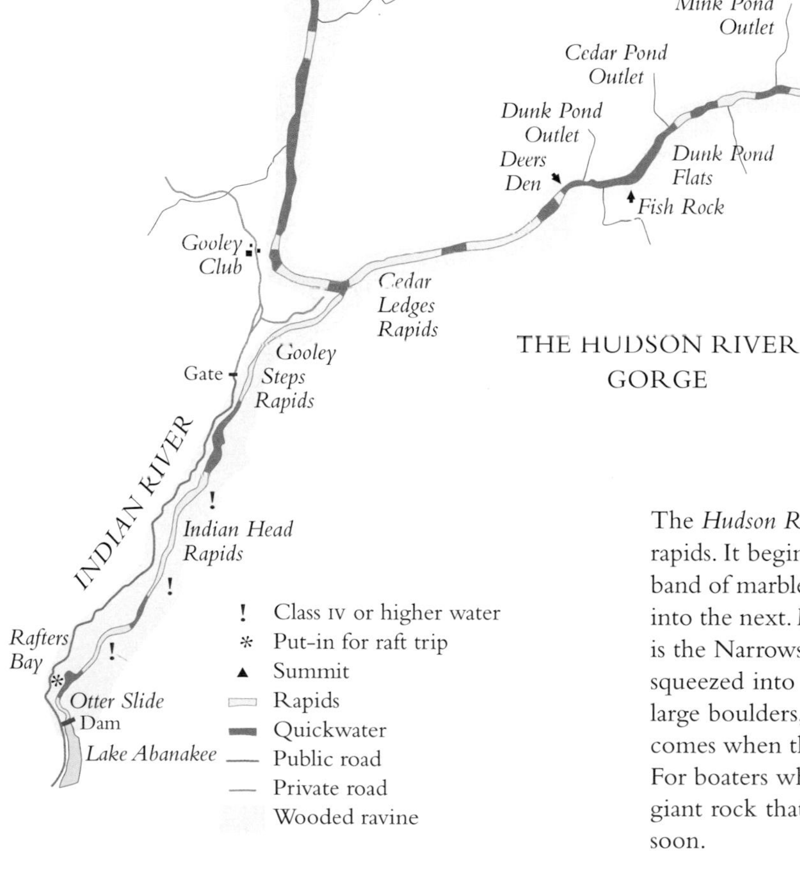

The *Hudson River Gorge* is a steep-walled ravine filled with ledges, boulders, and rapids. It begins at the mouth of the Indian River, where the Hudson follows a band of marble sideways out of one rift valley and then drops 250 feet eastwards into the next. Most of that drop occurs in eleven major rapids. The most intense is the Narrows. Here the river, after a steep drop out of the Blue Ledge Pool, is squeezed into a ninety-foot slot between rocks. The most complex rapid, with large boulders, multiple channels, and oblique breaking waves, is Harris Rift. It comes when the end of the run is in sight, after most of the heavy water is over. For boaters who approach it with diffidence or who miss the ferry just above a giant rock that blocks the channel, the end of the run can come unexpectedly soon.

ADIRONDACK skiing takes place in two very different venues. *Backcountry* skiing uses any patch of snow, open or wooded, flat or steep, that someone is good enough or crazy enough to think they can ski. Backcountry skiers ski trails, railroads, woods, roads, lakes, rivers, and even, in the spirit of Geoff Smith's Ski-to-Die Club of the mid-seventies, some gullies and landslides that to an average skier's eye look more like mild ice-climbs than ski slopes. Backcountry skiing is largely on ungroomed trails and entirely dependent on natural snow. It is currently flourishing in the Adirondacks, especially in the mountains and in the south and west where the lake-generated snow is heaviest and most reliable.

Ski area skiing, with lodges, lifts, groomed trails, and snowmaking, is, unlike backcountry skiing, a highly-organized, capital- and labor-intensive business that requires predictable conditions. Ski areas can survive bad snow with snow-making or, in the case of cross-country areas, simply because bad skiing beats no skiing. But they cannot survive rain and extended thaws. The last ten years have had more poor winters than good, and many small areas throughout the East have been driven out of business.

In contrast to Vermont, most Adirondack downhill skiing is community-based or wilderness-based rather than resort-based. Seven of the ten downhill areas are publicly owned, and none of the private ones are associated with the kind of condominium and ski-village development that is common around commercial ski areas elsewhere in the Northeast. Gore and Whiteface, the largest areas, are operated by the New York Olympic Development Authority. Whiteface, with a 3,400-foot vertical drop and formidable upper slopes, was used for the alpine events in the third Winter

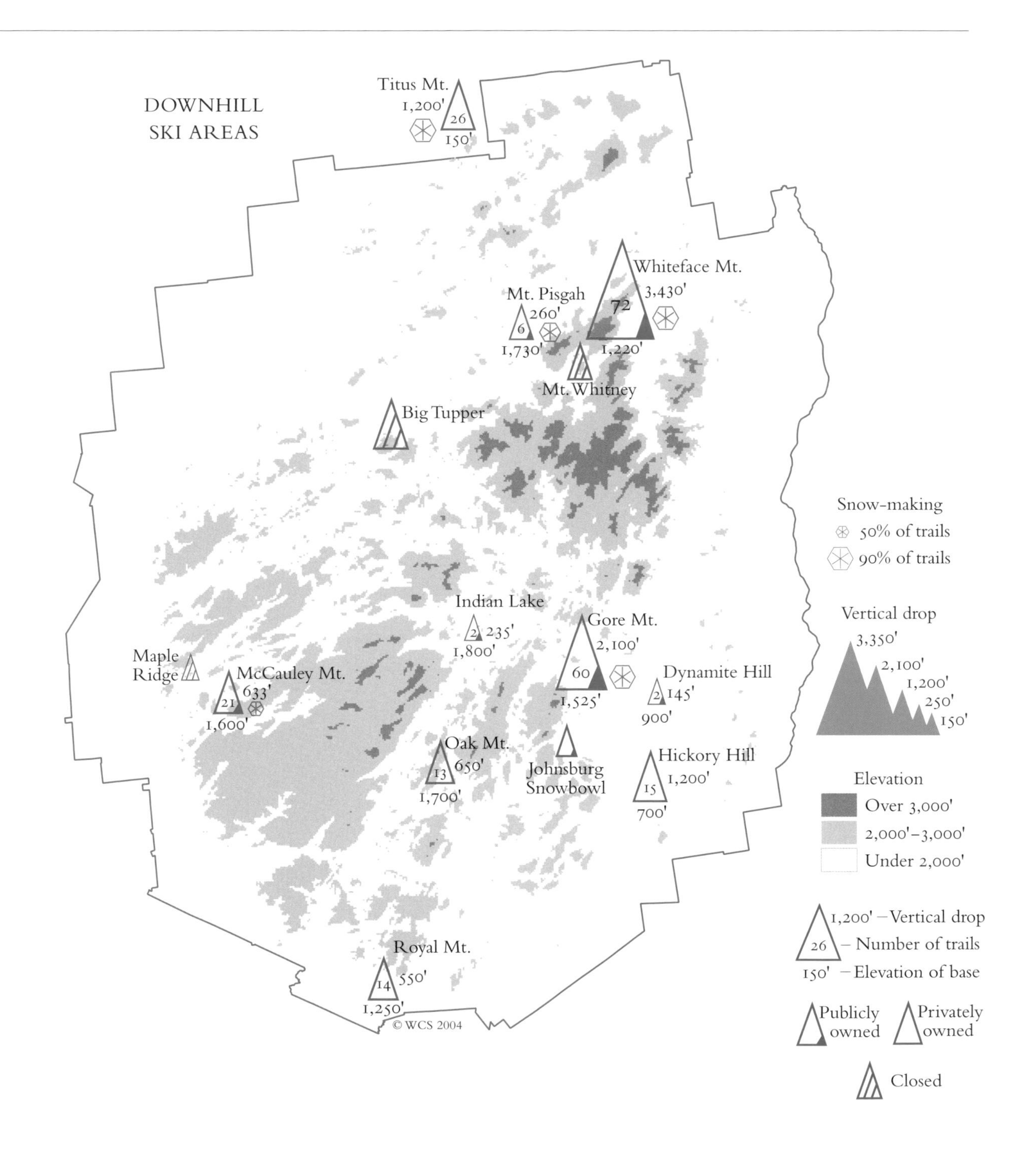

Olympics in 1932 and the fifteenth Winter Olympics in 1980. Gore Mountain, though somewhat smaller, is still a major and demanding ski area by any standard. Both have snow-making on ninety-five percent of their trails; contemporary winters, even in the Adirondacks, cannot supply enough snow for large-scale commercial skiing.

Four of the eight other downhill areas are small, town-owned areas used by kids and families. Only the four midsize areas, Titus, Oak, Royal, and Hickory Hill, are private. They have to be big enough to intercept day skiers who would otherwise go to Gore and Whiteface but small enough to keep management costs and ticket prices down. Because even a medium-sized ski area is very expensive to run, this may be difficult to do. Big Tupper, a midsize Adirondack ski area appraised at 2.5 million dollars, is now closed, as are a number of other similar areas in Massachusetts and Vermont.

The fourteen cross-country areas divide in a similar way: the smallest and largest are public, the private ones in between. Mt. Van Hoevenberg is a large state-run area. Lapland Lake and Garnet Hill are large private areas with extensive trail networks. Cascade, Friends Lake, and several others are midsize areas associated with inns. The remainder are local areas with relatively small trail systems. The big areas get most of the dedicated skiers and especially the skate-skiers and racers; the small areas, several of which provide free skiing, get the kids and locals. The inns and the midsize areas, which have to compete both with the ungroomed trails in the Forest Preserve and with an Olympic facility which offers a day's skiing for $15 and private lessons for $25, position themselves uneasily in the middle and make their money on meals and accommodations.

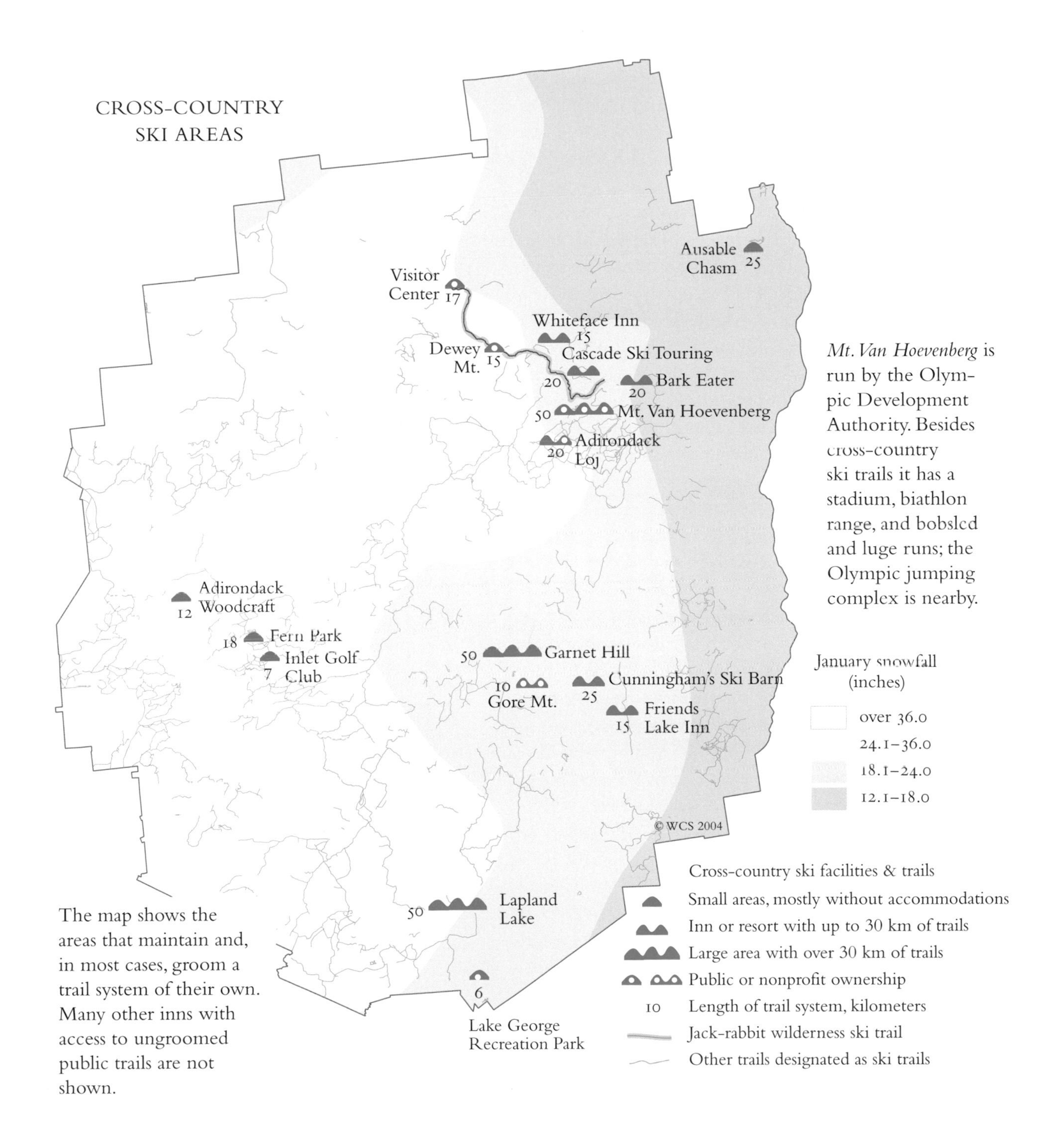

Mt. Van Hoevenberg is run by the Olympic Development Authority. Besides cross-country ski trails it has a stadium, biathlon range, and bobsled and luge runs; the Olympic jumping complex is nearby.

The map shows the areas that maintain and, in most cases, groom a trail system of their own. Many other inns with access to ungroomed public trails are not shown.

MOTORBOATS are allowed on all public lakes in the park that are not completely surrounded by state land and on all private lakes unless forbidden by the owners. In practice this means that almost every lake large enough to be motorized is motorized.

The result is a *de facto* zoning by which the motorboats get all the big lakes and the canoeists and kayakers must either contend with motorboats or settle for the small. On highly developed lakes like Schroon, this is clearly consistent with the traditional use of the lake. But on lakes whose shores are largely state land, and especially those, like Cranberry Lake, that border wildernesses, there is an obvious contradiction between the current motorized use of the lake and the use of its wilderness shores. Though no motor vehicles are allowed on the shores, you may still carry your gear to a wilderness campsite by motorboat or float plane. To motorized users this is an appropriate public use. To the nonmotorized users who go to the wilderness to escape from noise and machines, it is illogical and invasive.

Jet skis, while legally just small, wet motorboats, are exceptionally fast and noisy and have made an exceptional number of enemies in the few years they have been around. A 1998 law allows towns to regulate them independently of other watercraft. Thus far, seven towns have banned jet skis on fifteen lakes and rivers; another ten towns are considering bans.

Because much of the park is already awash with machines and mechanical noise, regulating jet skis involves some fine distinctions. North Elba, for example, permits water-ski boats on Lake Placid but prohibits jet skis. That this is discriminative goes without saying. Whether it is a reasonable or a baseless discrimination will, in all likelihood, be argued for some time.

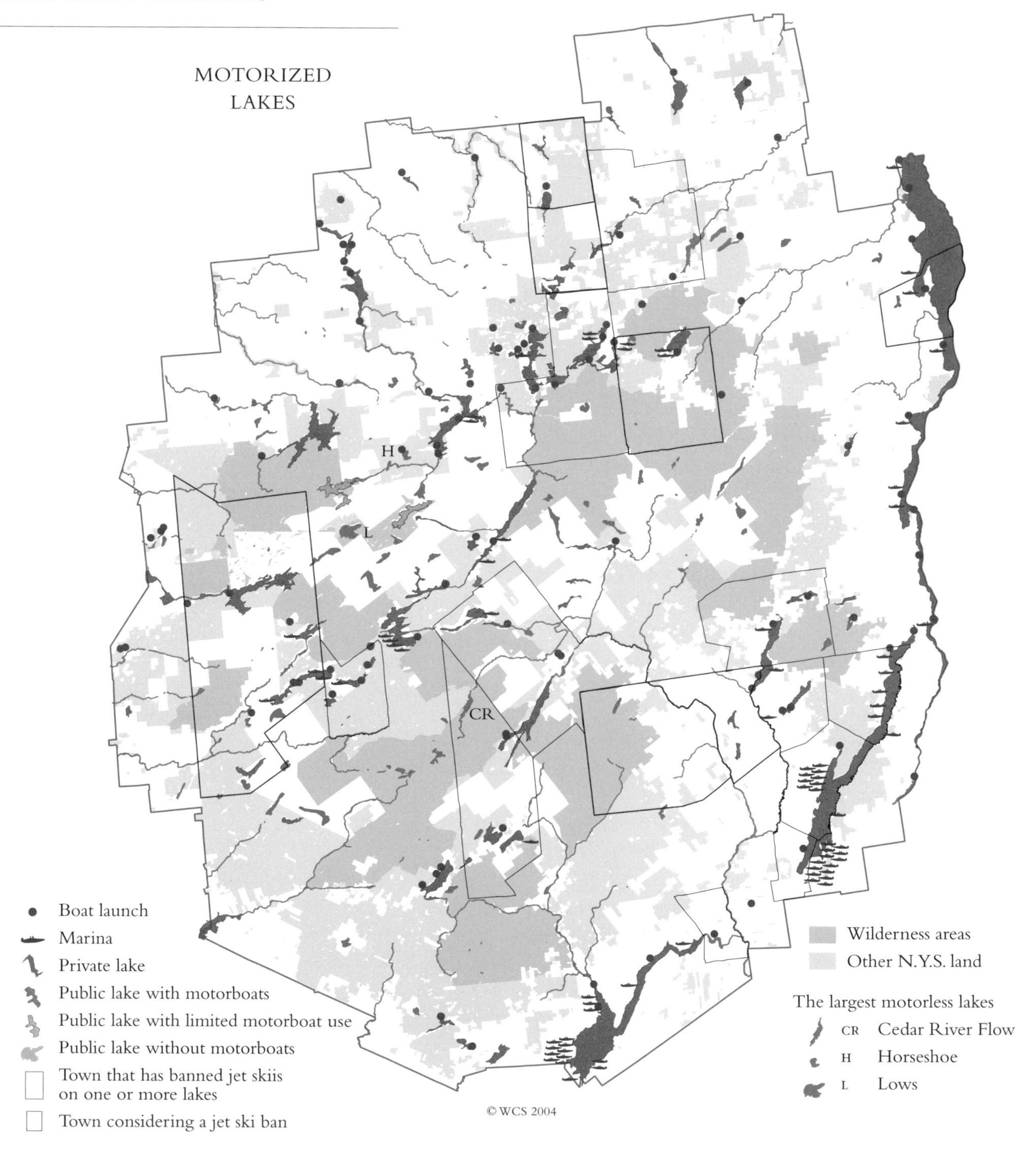

FIFTY years after the first snowmobiles arrived in the Adirondacks, snowmobiling is accepted but still controversial. Proponents see it as a popular sport, comparable to downhill skiing, which is an economic lifeline for some small towns. Opponents see it as a threat to peacefulness and wilderness.

Both, it turns out, are right. Snowmobiling is an extremely popular sport (there are six snowmobiles in New York for every member of the Adirondack Mountain Club) and an important part of the winter economy of several Adirondack towns. But snowmobiles are also noisy and polluting, and few nonsnowmobilers want to travel in, much less live next to, the popular snowmobile corridors. And all machines, however quiet or clean, still make the woods busier and less private.

At present there is no complete map of snowmobile trails. The map shows about 1,150 miles of trails funded or maintained by the state. Their layout, with many short loop trails and lake crossings, reflects the way snowmobiles were used thirty years ago when machines were used mostly for short trips on unpacked trails and the frozen lakes were safe travel routes. Today the machines are wider and faster and their riders more interested in long trips on groomed trails. The warming climate and unreliable snow elsewhere are bringing more riders into the Adirondacks but also shortening the season and making many of the traditional lake routes unusable or unsafe.

Clearly, if snowmobiling is to remain an important Adirondack sport, new long-distance trails will have to be created. Where these trails might go is currently the subject of an interesting and surprisingly open debate. But planning trails is one thing and building them another. In the absence of a master recreation plan, any new trails will have to be authorized by the Unit Management Plans of each management area they go through. Since most of these plans are thirty years overdue (*Map 15-6*), the approval process for any new trail may be difficult and protracted indeed.

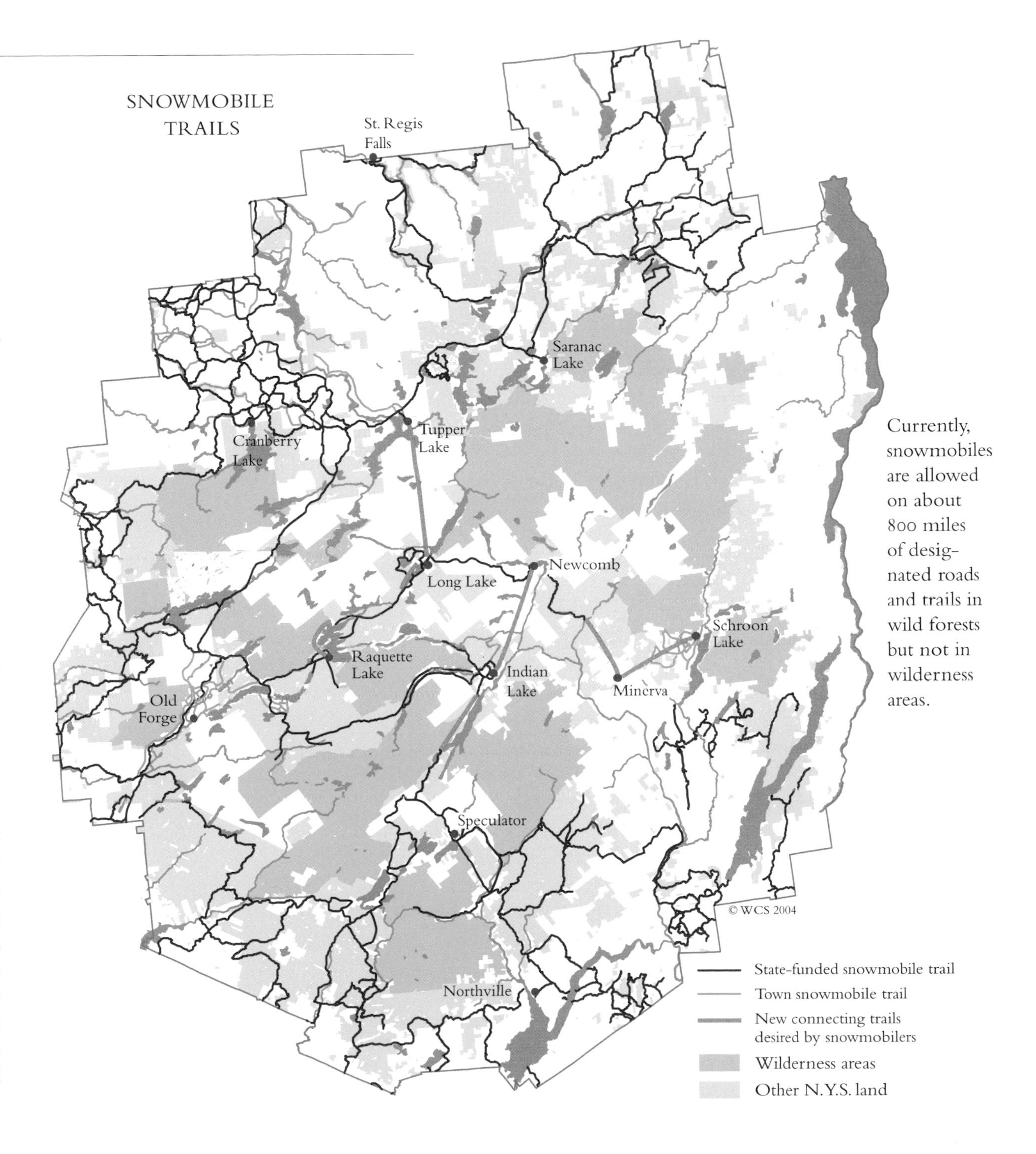

Currently, snowmobiles are allowed on about 800 miles of designated roads and trails in wild forests but not in wilderness areas.

By New York law, cars, trucks, and ATVs are all motor vehicles. Within the Forest Preserve they may be operated on designated forest roads in wild forests and intensive use areas but not on trails or off-the-road. They may not be used in wilderness areas except in an emergency.

All-terrain vehicles, which damage roads more than other vehicles and are very difficult to regulate or police, are further restricted. They may not be used on private lands without the owner's permission; they may only be used on designated segments of public highways when it is otherwise impossible to gain access to a legal riding area adjacent to the highway; and they may only be used on Forest Preserve roads in wild forests that are specifically approved for ATV use.

These restrictions concentrate most of the motor vehicle use of the Forest Preserve, and most of the controversy about that use, on the wild forests. If you wish to use a car or truck off the public highways and do not have access to private lands, you will have to use a Forest Preserve road in a wild forest. If you wish to use an ATV on public lands, which if you don't belong to a hunting club is almost the only legal way you may use an ATV in the park at all, you will have ride on a designated ATV road in a wild forest. And if you wish to argue for more or less motorized use, what you will be arguing about are the roads in wild forests and why some are designated roads and some trails.

To either use or argue about the wild forest roads, you will need to where the roads are and which roads are approved for which uses. Currently this is not public information. The map shows, in red, forty-eight miles of roads that are open for motorized use by the disabled. Another thirty-nine miles of roads in the Aldrich Pond Wild Forest, for which there does not seem to be a public map, are open to both vehicles and ATVs. Significantly, ATVs from Aldrich Pond have turned up deep within the adjacent Five Ponds Wilderness. At present everything else about open and closed roads is a deep and, one might surmise, guilty secret.

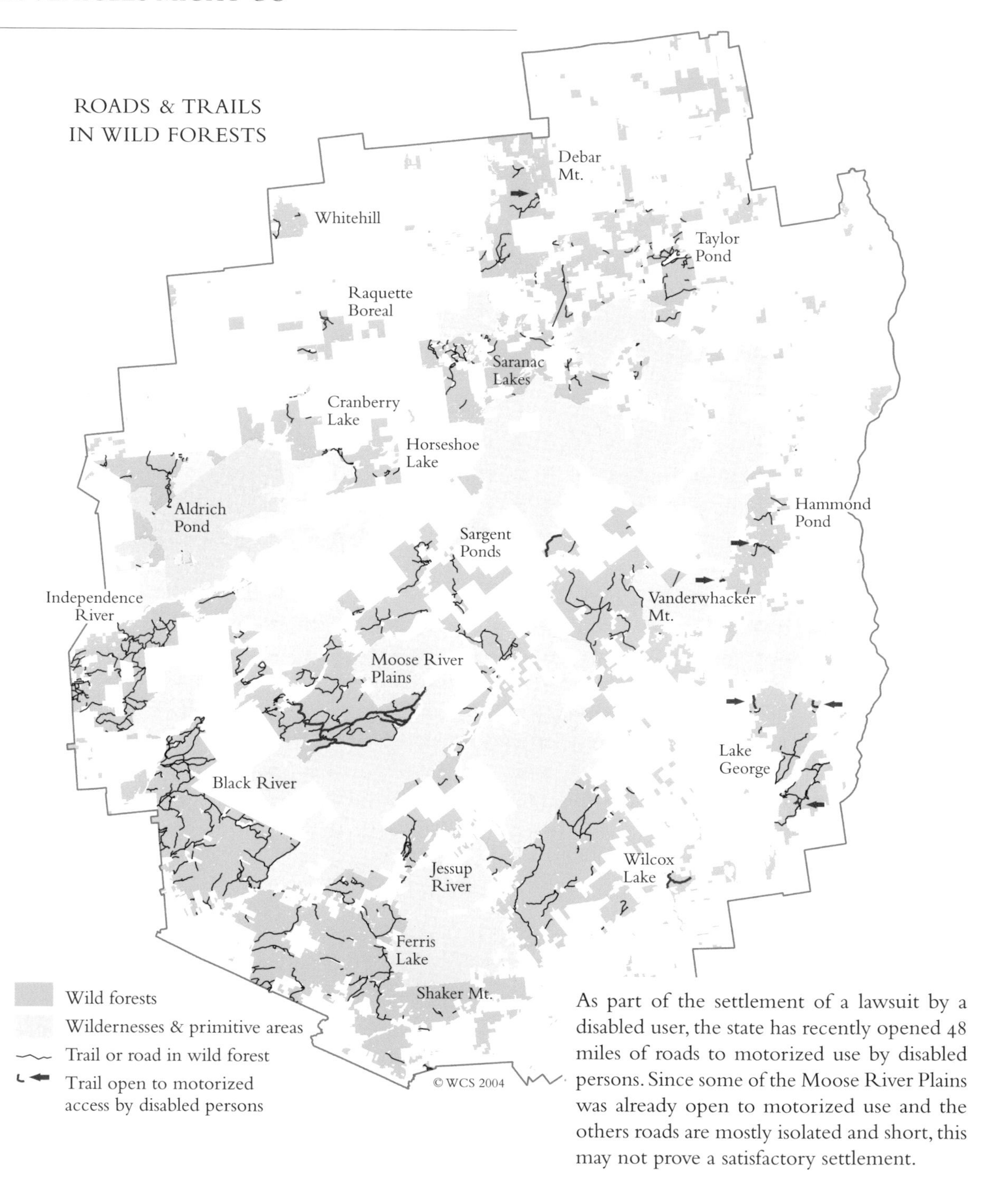

As part of the settlement of a lawsuit by a disabled user, the state has recently opened 48 miles of roads to motorized use by disabled persons. Since some of the Moose River Plains was already open to motorized use and the others roads are mostly isolated and short, this may not prove a satisfactory settlement.

MANY of the current uncertainties surround recreation in the park—where motorized and disabled access would be allowed, how to connect the existing snowmobile trails and bypass the dangerous lakes, what trails and shelters should be removed from wilderness areas, whether new trails and shelters should be built—were supposed to have been resolved by the preparation of a *Unit Management Plan* (UMP) for each of the roughly ninety Adirondack management units. The management plans, which were mandated by the Adirondack Park Agency Act of 1972, were to contain resource inventories, discussions of current use, and recommendations for future facilities and allowable uses.

Progress on the UMPS has been very slow. There are a number of reasons for this, some technical and some political. Management plans are difficult to prepare and invariably controversial. They require information that often isn't available, open an in-house decision-making process to public view, require the termination of inconsistent but long-tolerated uses, and are generally criticized by everyone when they were done. It is no wonder that the DEC and other agencies found it easier to delay than complete the process.

In any event, by 1977 work had begun on only five of the some forty major Adirondack plans required by the Park Agency Act. Twenty-six years later, in January 2004, only eleven have been completed. The first plan for an Adirondack wild forest was not completed until 1986; the first plan for a wilderness area was not completed until 1987.

Considering the importance of recreation planning and the seriousness of some of the issues involved, the delay is very unfortunate. Neither the issues nor the controversies are going to go away; all that is really being delayed are the benefits, to the users and to the park, of their eventual resolution.

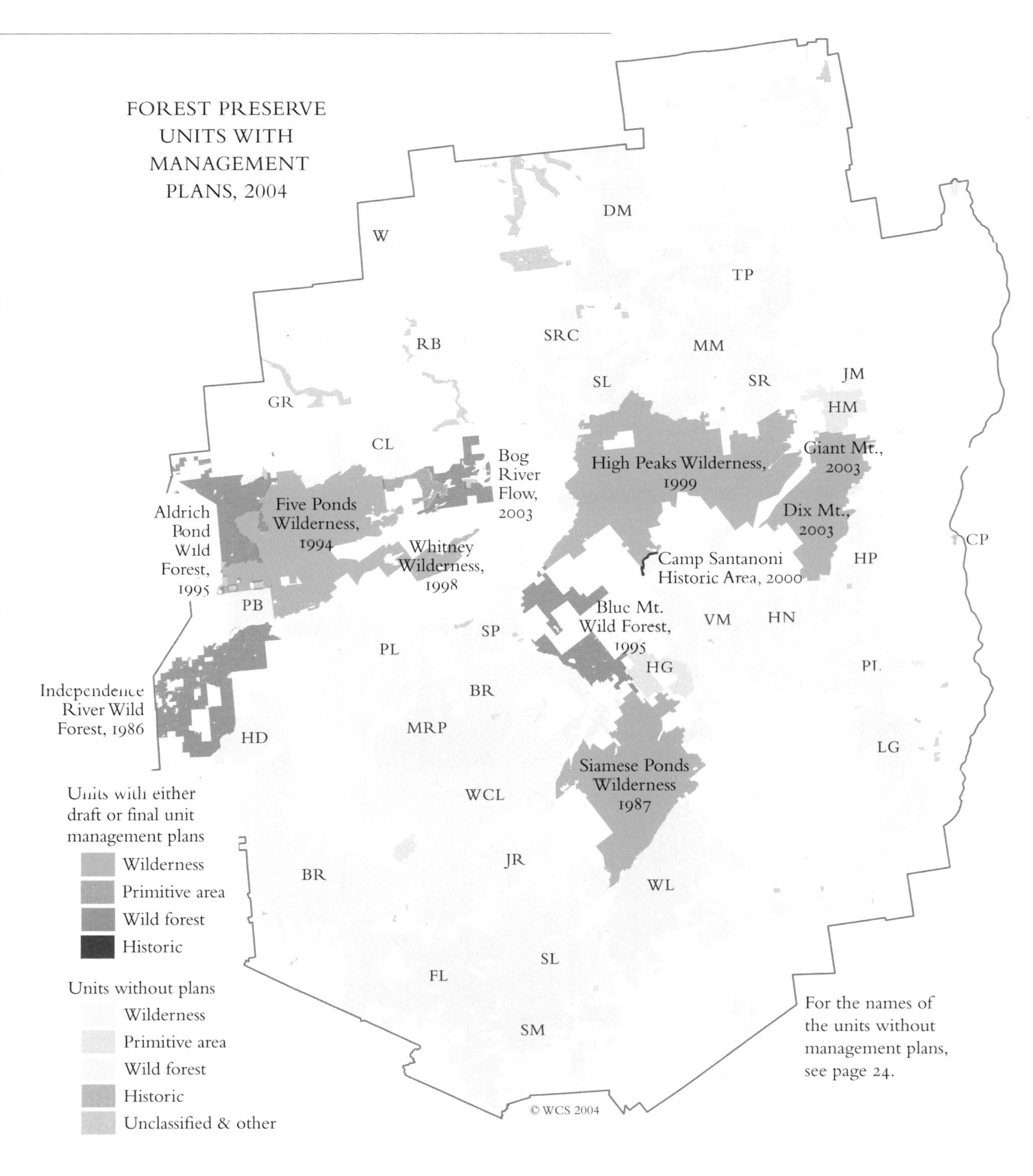

Of all the conflicts between recreationists—between commercial and noncommercial users, between fishermen and rafters, between disorganized individuals and organized groups—none is deeper or more fundamental to the park than that between motorized and nonmotorized users.

To many of the disputants, it seems odd that there should be a conflict at all. To the motorized users it seems obvious, even if nowhere explicitly stated in statute, that public lands should be open to public motorized use. To the nonmotorized users it is equally obvious, though also missing from statute, that the constitutional obligation to keep the park wild entails an obligation to keep it motorless. One group, in consequence, wonders why motor vehicles are allowed anywhere on public lands; the other is angry that they are not allowed everywhere. Both ask how we came to this pass.

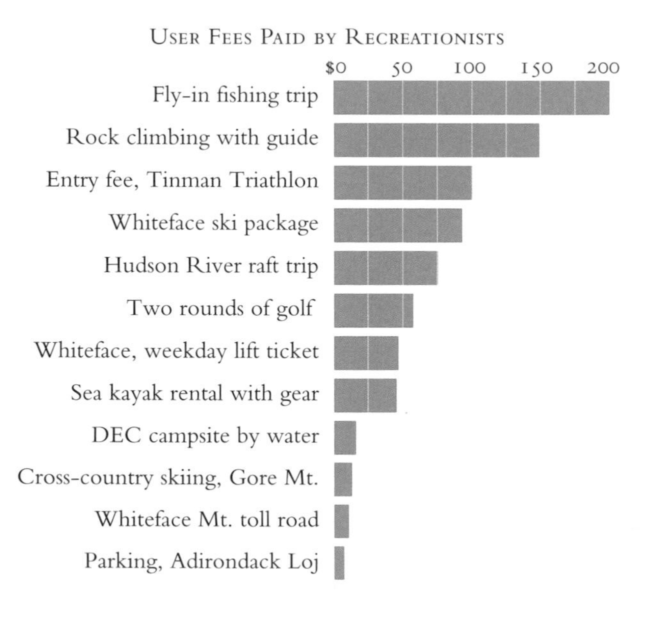

For them, and anyone else interested in how a conflict became a policy, here is a quick history.

For its first eighty years, the park was declared a wilderness but not managed as one. The park was big and relatively poor, and its managers were as much interested in developing it as protecting it. Recreation brought in money, both to the residents and to the managers, and the promotion of recreation was everybody's business. The constitution was read as prohibiting big things like new paved roads and ski areas but not smaller ones like campgrounds and shelters. And even more important, it was not read as regulating *activities* at all: anything that left the deer and the trees alone and was legal off the Forest Preserve was usually legal in the preserve. In those laissez-faire days both the users and the managers cleared trails, built campsites, drove trucks, and landed aircraft pretty much throughout the Forest Preserve.

By the 1960s, things were changing. A growing wilderness movement was alarmed at the proliferation of backwoods roads and cabins and shelters. A growing motor-sport movement was using increasing numbers of light machines—especially outboards and snowmobiles—on public lands. Predictably, the two movements collided.

It was the two-stroke engines in these machines that were particularly controversial. Two-stroke engines are cheap, simple, and have the high power-to-weight ratios desired for light machines. But they also burn a frighteningly dirty fuel-oil mixture and, to increase usable power, often have as little muffling as they can get away with.

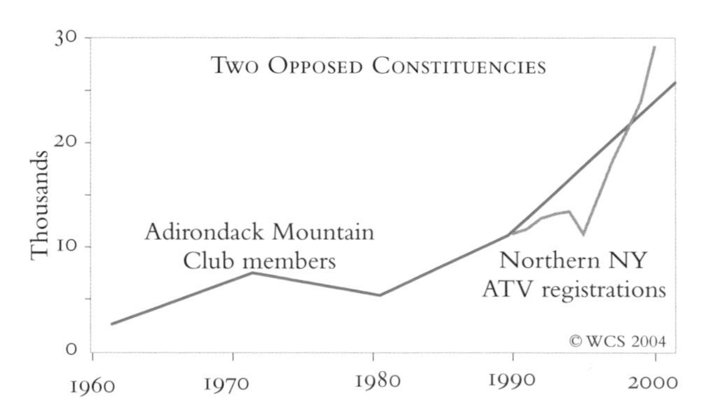

To the wilderness movement, two-stroke engines were particularly objectionable because they were loud and dirty. Snowmobiles at 30 miles per hour were as loud as cars at 60 and emitted perhaps fifty times as much pollution per user mile. Outboards discharged their exhaust and with it large quantities of unburned fuel and oil into whatever water they were used on.

By the late 1960s, there was an all-out fight about wilderness and machines. The question was not whether there would be machines in the park. That had been settled a hundred years earlier when the first locomotives and naphtha-steamers appeared. Rather the question was how far the machines, having gotten people *to* the public lands, would be allowed to continue on *into* them. In other words, were the state lands for people or for machines, or had the two become inseparable?

The ensuing fight was prolonged, bitter, and very political. The fundamental issue for the motorists was not to further penetrate the park—they were

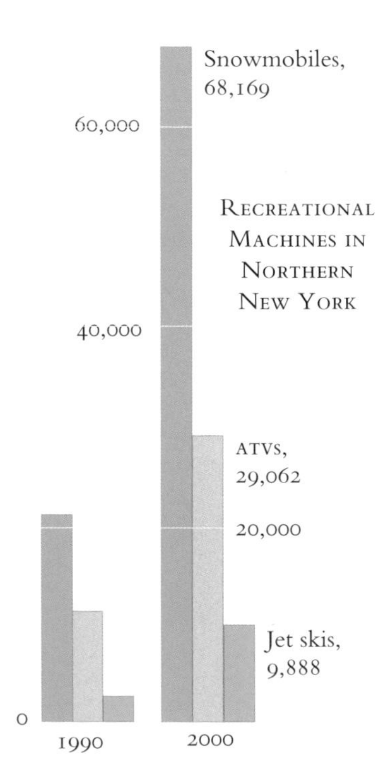

already almost everywhere—but to legitimize the informal access that they already had. The issue for wilderness advocates was not to demotorize the entire Forest Preserve but to create and defend a motorless core. Interestingly, although both sides cited the constitution for support neither was sure enough of its position to appeal to the constitution in court.

The result of the fight was not an outright victory for either side but rather a large-scale compromise, embodied in the State Land Master Plan of 1972, which gave each of the disputants something they wanted badly.

The motorists got legitimacy. Powerboat users, big winners, were given access to all lakes that were not completely surrounded by state land. Since there is private land on almost every lake much larger than a mile long in the park, in practice this meant almost all lakes where a fast boat can go for much more than a minute in any direction were motorized.

The snowmobile users, who for the previous ten years had ridden through a chaotic regulatory maze, got an 800-mile trail system of their own, restricted to wild forests and intensive use areas.

Car and trucks users got an implied promise that designated vehicle roads would be included in the soon-to-be-completed Unit Management Plans for wild forests. Thirty years later only three UMPs for wild forests exist, and motorized users are still waiting to collect on that promise.

The nonmotorists, at last, got motorless areas. Half the public lands were to be designated wildernesses. Roads were to be closed, nonconforming structures removed and all motorized use forbidden, except for emergency vehicles and a few fortunate inholders. The other half of the public lands would be designated wild forests and would have continued, though restricted, motorized use. Thus, in the eyes of the wilderness advocates if not the law, they would not in fact be wild.

This compromise essentially said that the public lands would be divided in two and each half held to a different standard of wildness. It has been the foundation of Forest Preserve management ever since.

The compromise was not popular, either at the time or today. Both sides had wanted something close to complete use of the park, and neither wanted to recognize that the more complete their use was the more someone else's use was excluded.

Predictably, the compromise was least popular with the users who were not considered when it was crafted. Jet skis and ATVs, unlike snowmobiles, do not have any guaranteed access to the Forest Preserve. Disabled users were not provided for by the State Land Master Plan; touring canoeists and kayakers have only a fragmented system of long-distance routes and must contend for sea room with powerboats on all the large lakes.

Currently, many of the new and neglected users are asking for recreational domains of their own. When, like the jet skiers, they are unorganized and widely disliked, they are easy to ignore. But when they are polite and insistent and legally empowered—as the disabled are and as the ATVs and jet skis may someday become—they will be very hard to refuse.

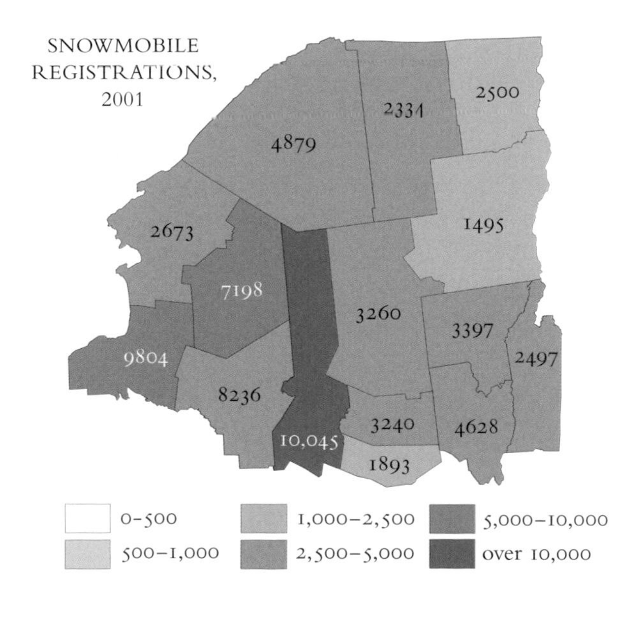

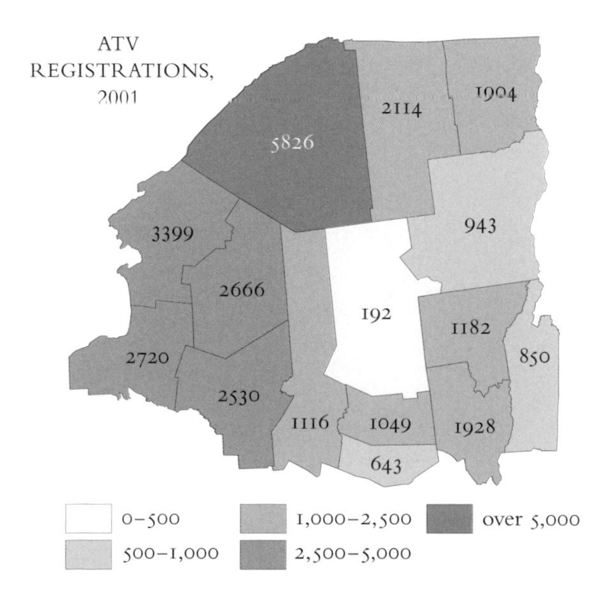

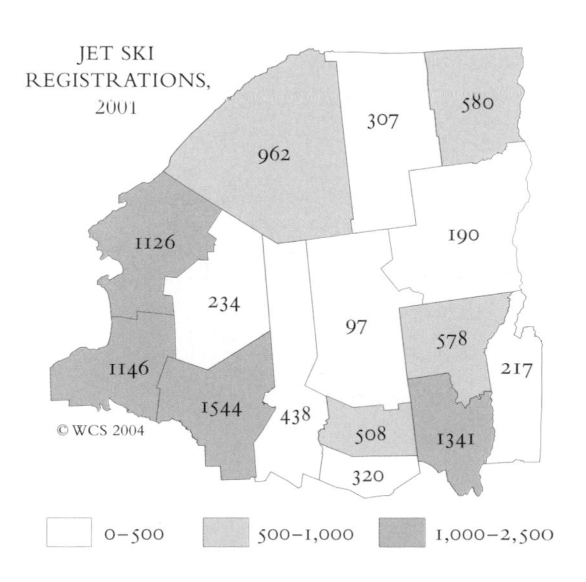

The map shows the sixty-five most central Adirondack towns. Webb (W), the largest, is 483 square miles; Northampton (NH), one of the smallest, 35 square miles. Benson (B), is 90% Forest Preserve; Ausable (A), Putnam (P) and several others are under 1% Forest Preserve.

16 Changing Towns

ADIRONDACK towns and villages differ greatly in character and activities. A few are industrial, many are service-based or hospitality-based, and a growing number are commuter-based. Some are highly seasonal, some occupied and busy year-round. Some, like Lake George, have elaborate facilities for feeding and entertaining guests and are full-fledged *resorts.* Others, like one lake town where the pizza parlor was out of cheese at 7 P.M. on a Saturday in July, are really *summer colonies*, where visitors are expected to fend for themselves.

What a town was a hundred years ago has much to do with what it is today. In 1900, the major business of the interior towns was accommodations. Though many logs were cut in these towns, the money from logging was made elsewhere. In the forest towns it was the boarding houses and hotels that made money for the residents. When the log drives ended and the big hotels closed, many of the interior towns built cottages to accommodate the new visitors in automobiles. Some went to become major resorts, others remained summer colonies; almost all are still hospitality-centered today.

The former industrial towns have had a much harder time. Almost all have lost their original industries. Some vanished with their industries. A few have found new industries; many others have struggled to survive. Seemingly, the bigger the former industry and the longer it lasted, the harder the struggle has been and the more the town has been marked by it.

A number of the ex-industrial towns, particularly those on lakes, have become summer colonies. None seems to have become a resort. Being a resort requires a reputation and perhaps a history as well. All the current resort towns were already resorts in stagecoach days; no automoble-era upstart, whatever its aspirations, has thus far joined their ranks.

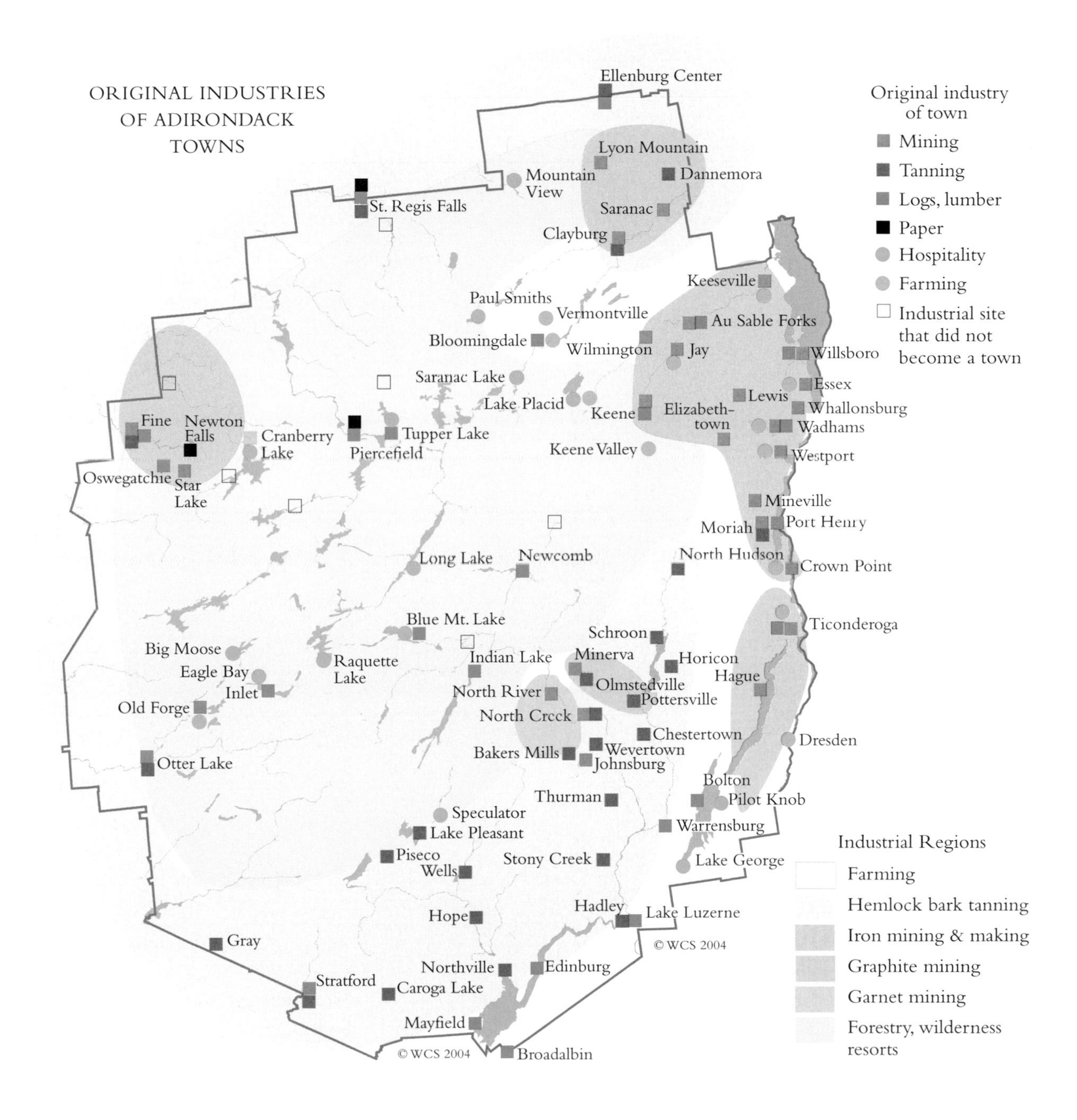

TODAY, the lake towns are the economic backbone of the park. What iron ore and trees were to the early Adirondacks, rentable or developable lakeshore is today. Lakeshore is the thing that the Adirondacks have that other places lack, and it is what brings the people that the service economy serves. At least sixty-eighty Adirondack settlements, including a large majority of the settled areas within the park, are on or near a developed lakeshore. But lakeshore is also a limited and nonrenewable commodity, and the maps suggest that it is surprisingly close to being used up.

Most of the northern lake towns are one-season towns, with relatively few year-round businesses or residents. Some no longer have a grocery store or post office, and a surprising number have no stores at all. Saranac Lake and Lake Placid, which are year-round resort towns, and Tupper Lake which has a mixture of industry, tourism, and snowmobiling, are the main exceptions.

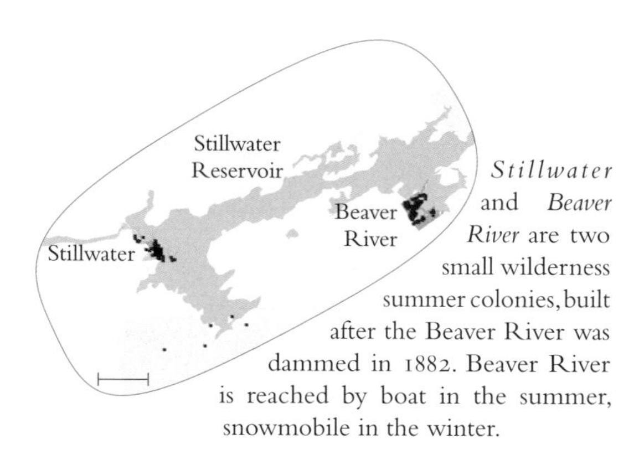

Stillwater and *Beaver River* are two small wilderness summer colonies, built after the Beaver River was dammed in 1882. Beaver River is reached by boat in the summer, snowmobile in the winter.

Rainbow Falls Reservoir
Stark Falls Reservoir
Carry Falls Reservoir

Carry, Stark, and *Rainbow Falls* are hydroelectric reservoirs on the lower Raquette. They were built in the 1950s, in an area of commercial forests where there were no villages and few state lands. They remain only sparsely settled, without stores or commercial recreational facilities.

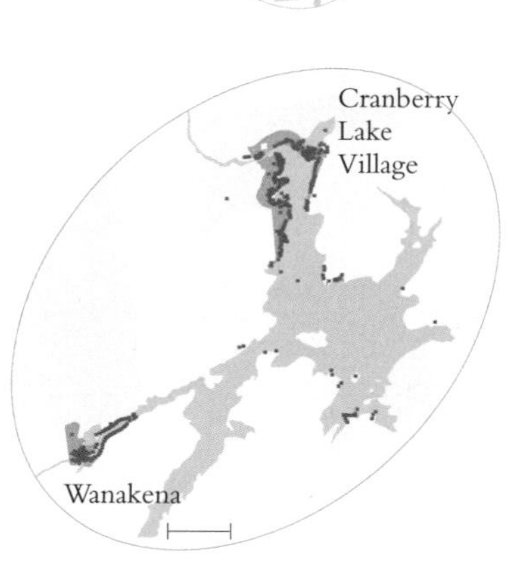

Cranberry Lake and *Wanakena* were major railroad-based industrial towns from about 1900 to 1940 (*Map 16-6*). The railroads and the sawmills they served are now gone and the lake has become a summer colony with a large state campground and about five hundred houses along the shores. Cranberry Lake village has three small stores, a lunchroom, and a restaurant; most close in the winter.

Wolf Pond
Raquette Pond
Gull Pond
Tupper Lake

Tupper Lake, the largest town in the western half of the Park, is an industrial town with a substantial year-round population that has added a moderate-sized summer colony along the lake and river. It has about 20 hotels and motels with 250 rooms to rent, many of which stay open in the winter to serve snowmobilers.

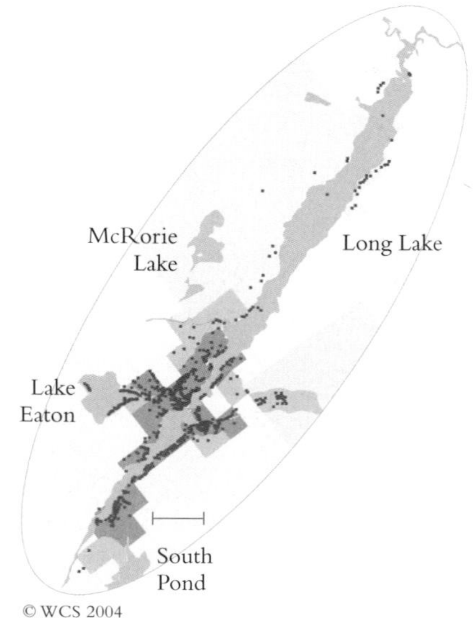

Long Lake is a former lakeshore resort that has become one of the larger summer colonies in the park. It has about 500 houses along the lakeshore and 160 rooms to rent. It also lies on important canoeing and snowmobiling routes and serves to some extent as a gateway to the backcountry.

- Private land, zoned for moderate to high density settlement, 500 houses per square mile or more.
- Private land, zoned for low density settlement, 75-200 houses per square mile.
- Houses
- Private land, zoned for very low density settlement, 15 houses per square mile or less.
- New York State Forest Preserve, no settlement
- One mile

The *far northern lakes*, from St. Regis to Loon Lake, were originally wilderness resorts, far from any villages, served by railroad and stagecoach. Now they are an extended summer colony of some six hundred houses. Paul Smith's College, on the site of the largest 19th-century hotel in the northern Adirondacks, is the only major year-round business.

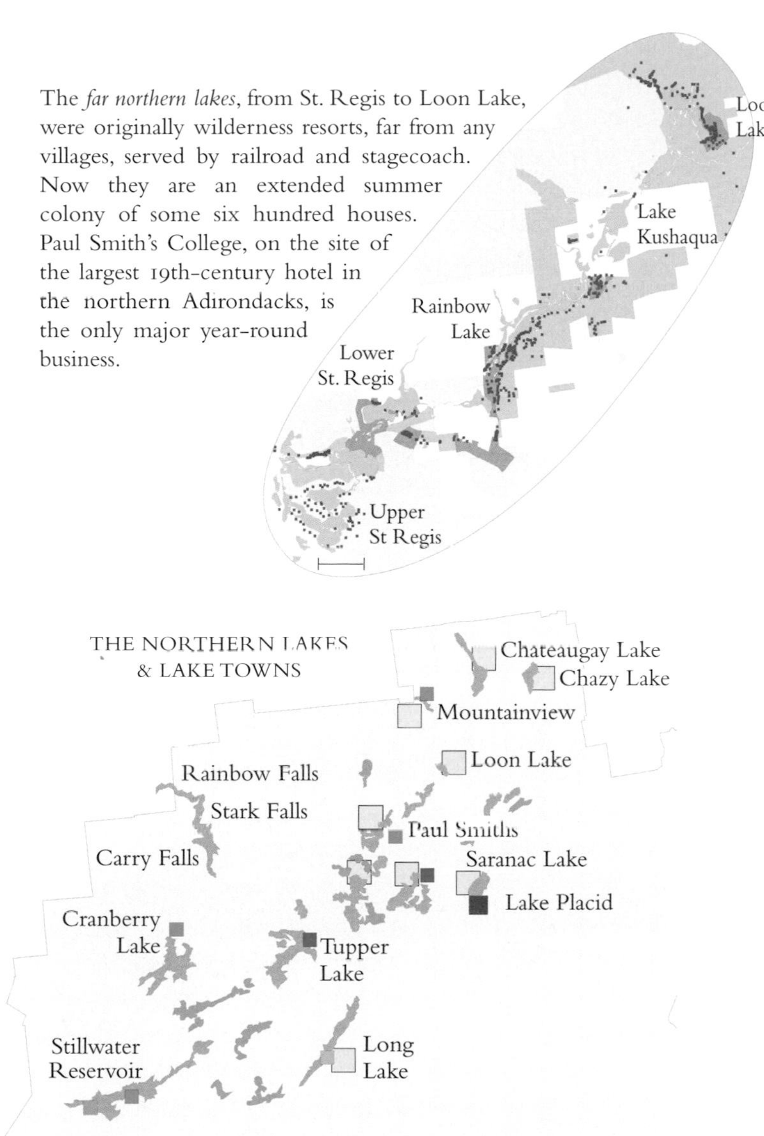

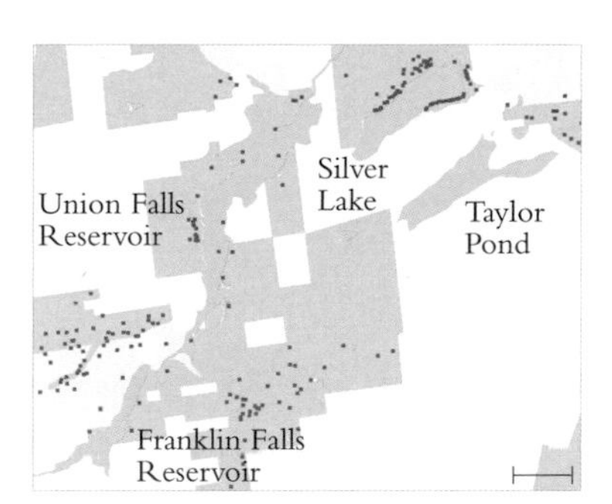

Franklin Falls and *Union Falls* are hydroelectric reservoirs without village centers. At present they have only scattered development.

Chateaugay Lake
Merrill
Chazy Lake

Chazy and Chateaugay Lakes are fairly recent summer colonies near the abandoned iron mines at Lyon Mountain.

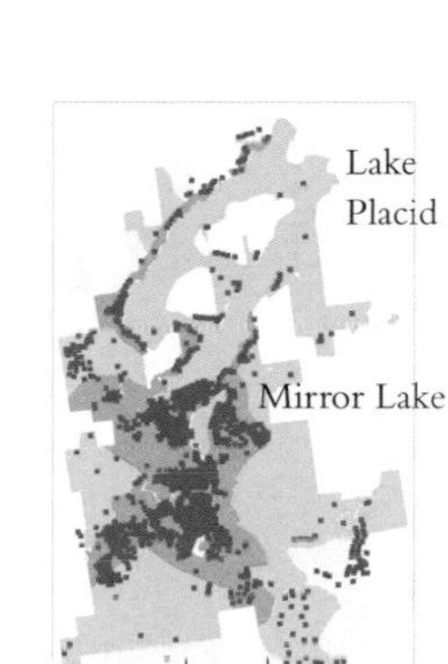

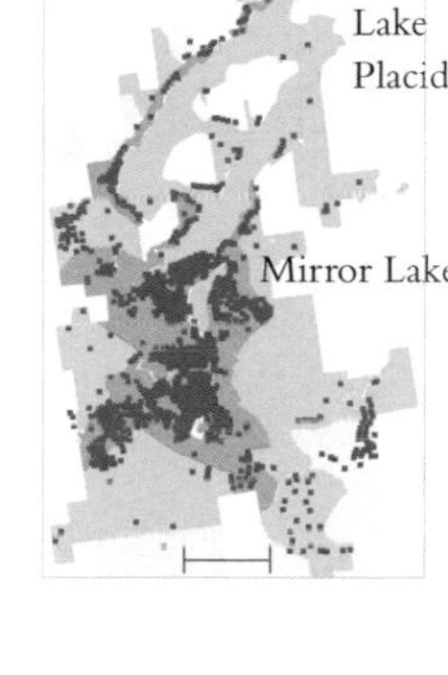

The developed area of *Lake Placid Village* surrounds Mirror Lake and then runs north and east along the shores of Lake Placid itself.

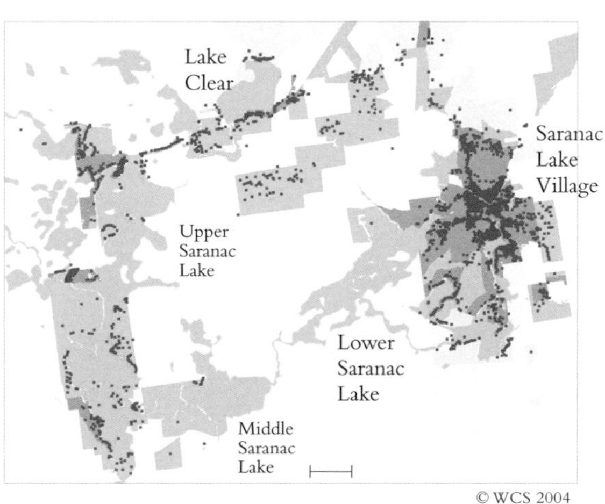

Saranac Lake and *Lake Placid* are 19th-century resort towns; both had luxury hotels, and Saranac Lake had in addition the clinics and cure cottages associated with Dr. Edward Trudeau's tuberculosis sanitarium. Lake Placid (*Map 16-4*), which has the facilities from the 1980 Winter Olympics in town and two major ski areas nearby, has become a major four-season resort. It is also the only upscale shopping destination in the park. Saranac Lake has some ski business and winter tourism and a substantial and fairly prosperous year-round population; but it is still very much a lake and second-home town, and the summer is by far its busiest season. (*Map 14-4*).

THE southern and eastern lakes, which are nearer main roads and population centers and have little or no state land along their shores, are much more continuously developed than the northern ones, with thousands rather than tens or hundreds of houses on their shores. There are about 8,000 houses on and near Sacandaga Lake, about 3,000 on the Fulton Chain, and a city-like 2,000 on little Caroga Lake. Because of the high development pressure, the medium-density land-use zones (shown in light tan) were made wider here than around the northern lakes, and there are far more houses scattered in the woods between the southern lakes than there are in the north.

The amount and pattern of commercial development is different as well. The gateway towns, Lake George and Old Forge, have cluttered edge-of-the-woods road strips of stores, motels, and attractions. The smaller towns are less cluttered but still have, compared to the northern lake towns, surprising amounts of commercial development. Northville, for example, has at least twenty-five businesses that are principally dependent on summer visitors.

The waters of the southern lakes are also busier and much less wild than those of the northern ones. Lake George has thirty marinas and a resident fleet of over 10,000 boats, plus three launching ramps for visitors. Its character changes greatly in different seasons. In late fall the lake is deserted, and the center part, in the Lake George Wild Forest, looks as lovely and wild as when the Mohawks paddled it three hundred years before. In midsummer, when the lake is roiled by hundreds of motorboats and there are gasoline fumes and noise everywhere, it looks about as wild as downtown Glens Falls.

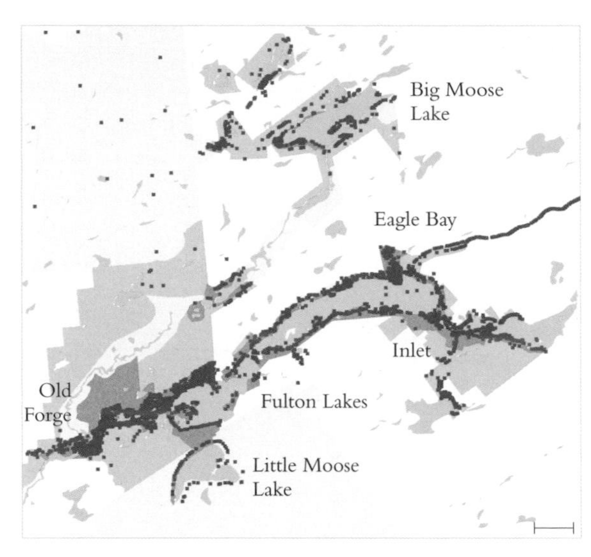

The *Fulton Chain* was created in 1850 when the Moose River was dammed to supply water to the Black River Canal. *Old Forge*, at the foot of the chain, is part summer colony, part amusement park, and part resort (*Map 14-4*).

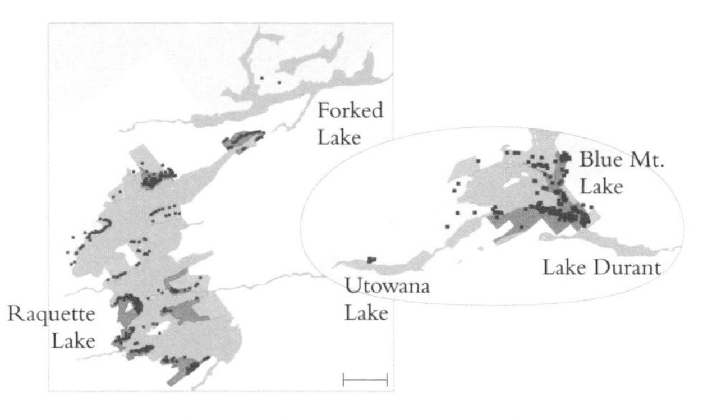

Raquette Lake and *Blue Mountain Lake* are old wilderness resort towns of the stagecoach age (*Map 5-11*). They were famous for big 19th-century hotels, the first great camps, and the steamboats and one-mile-long Marion Carry Railroad that once connected them. Now they are well-to-do vacation towns, filled with summer camps and vacation homes and cottages to rent, relatively busy in the summer and fall, deserted when the leaves are gone.

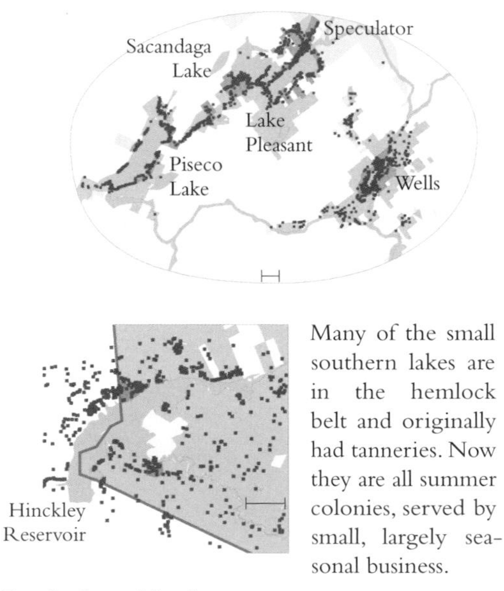

Many of the small southern lakes are in the hemlock belt and originally had tanneries. Now they are all summer colonies, served by small, largely seasonal business.

For the legend for these maps, see page 210.

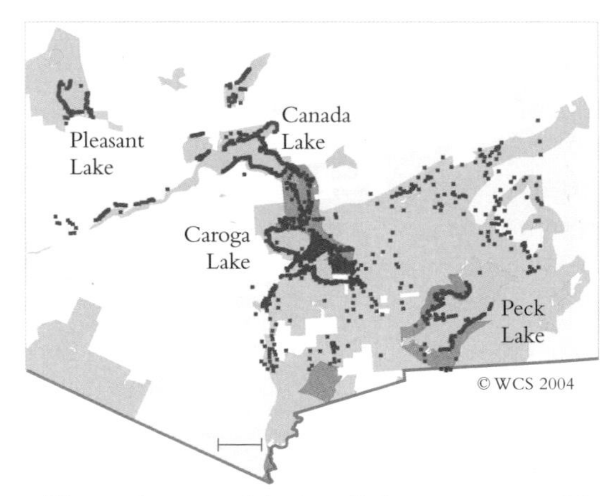

The southernmost lakes have little or no commercial development but are very densely settled. The three largest, *Canada, Caroga,* and *Peck,* lie within a broad corridor of developed lands.

THE SOUTHERN LAKE TOWNS

For legend, see p. 210.

The shoreline of Lake Champlain is almost continuously developed wherever topography permits; in the seventy miles between Valcour and Whitehall shown on the map there are several thousand shoreline houses and less than five miles of public land. The northern towns, particularly Westport, Essex, and Valcour, are active summer colonies; the southern towns are or were industrial. Oddly, no resort town has ever developed on either side of the lake.

Schroon Lake, Lake George, and *Great Sacandaga Lake* are probably the most developed lakes in the park. Sacandaga, built in 1930, is almost entirely developed as a summer colony, close to and serving the Albany-Schenectady area. Lake George is a highly developed resort lake, with restaurants, amusements, luxury hotels, and more rooms to rent than the rest of the Adirondacks put together. Schroon Lake is to some extent a mixture of the two, with several thousand private homes near the lake, lodging for visitors, and a large camp and conference center run by the Word of Life Ministry.

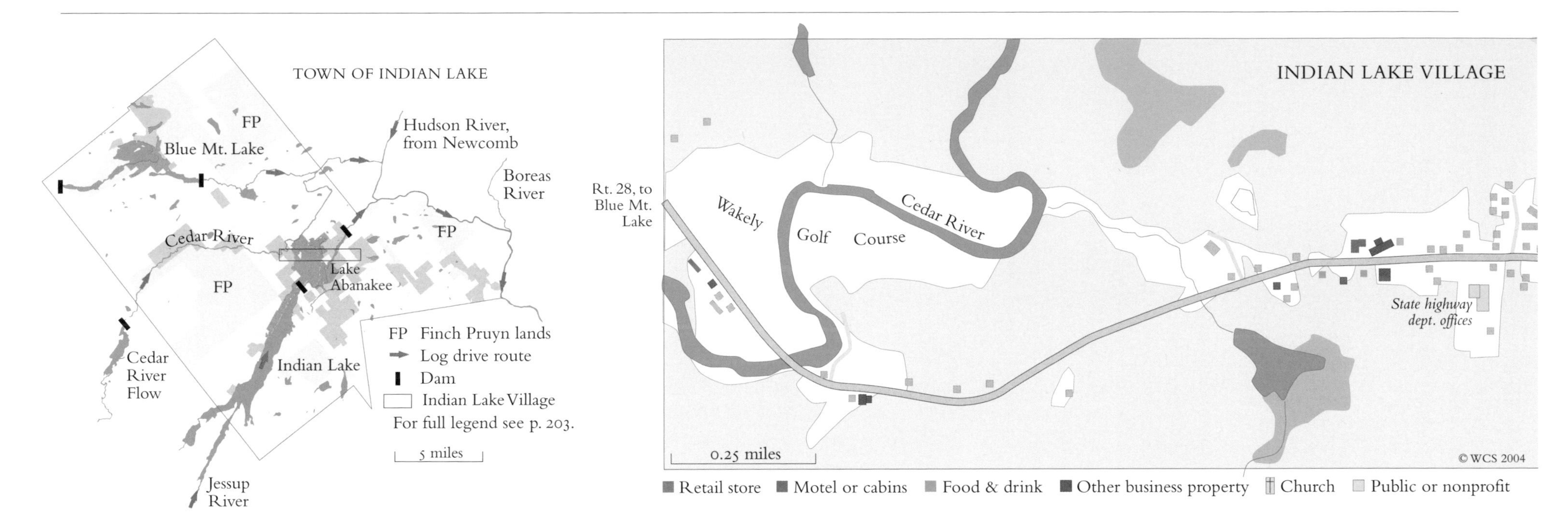

INDIAN LAKE originated as a logging town and stagecoach stop. For overland travellers it was the halfway point where passengers and horses rested on the stage road from North River to the big hotels at Blue Mountain Lake. For rivermen it was the meeting place of the drives from the north and west and the last access to the Hudson River before the formidable Hudson Gorge (*Map 15-3*). Its business has always been hospitality. Although by accommodating loggers and river-drivers it facilitated much industry elsewhere, it has never had a major industry itself.

Currently, its major assets are its two lakes and the water releases on the Hudson they make possible. The lakes are the artificial descendents of a much smaller natural lake. They were created when the Indian River was first dammed for log driving in 1845 and enlarged when the current dams were built in 1898.

Perhaps because it lacked a resort tradition, the town grew very slowly in the early part of this century. Growth accelerated after the war and has been steady for the last fifty years. Because of steep slopes and large woodland ownerships, almost all the building has been concentrated in the town center and along the lake shores; the developed area of the town is almost exactly the same as it was fifty years ago, and there has been little or no sprawl.

At present, Indian Lake is typical of a number of summer colonies that are edging toward being year-round towns. It has a central school, a clinic, a fire station, a post office, and two churches. The largest employers are the school district, the state Department of Transportation, and the town itself. There are a surprising number of small businesses as well. In the area shown by the map there are about ten stores, six restaurants, cafés and delis, five businesses offering accommodations, three rafting companies, two snowmobile and ATV dealers, and at least ten other businesses. Also surprisingly, most of them are open year-round. The rafting companies, movie theater, and tourist cabins close; at least twenty-five other business are open through the winter. Compared to Glens Falls or Saranac Lake this is a very small commercial core; but by central Adirondack standards, and especially compared to central Adirondacks of thirty years ago, it is a significant and promising one.

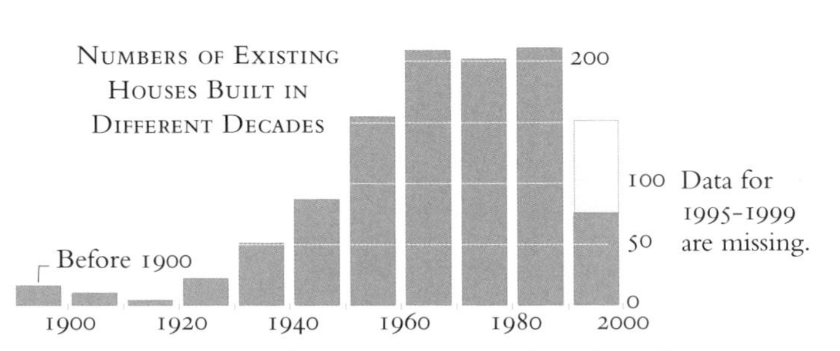

This graph includes only existing houses and not those that have burned or been torn down.

Chain Lakes Rd.,
to Rafters Bay
& Gooley Club

ADIRONDACK LAKE

LAKE ABANAKEE

r.s.
Legion hall
Health center
p.o.
Central
school
c.c.
f.d.

Rt. 28, to
North River

c.c. Chamber of Commerce

Rt. 30,
to Lake
Pleasant

f.d. Fire dept. p.o. Post office r.s Rescue squad

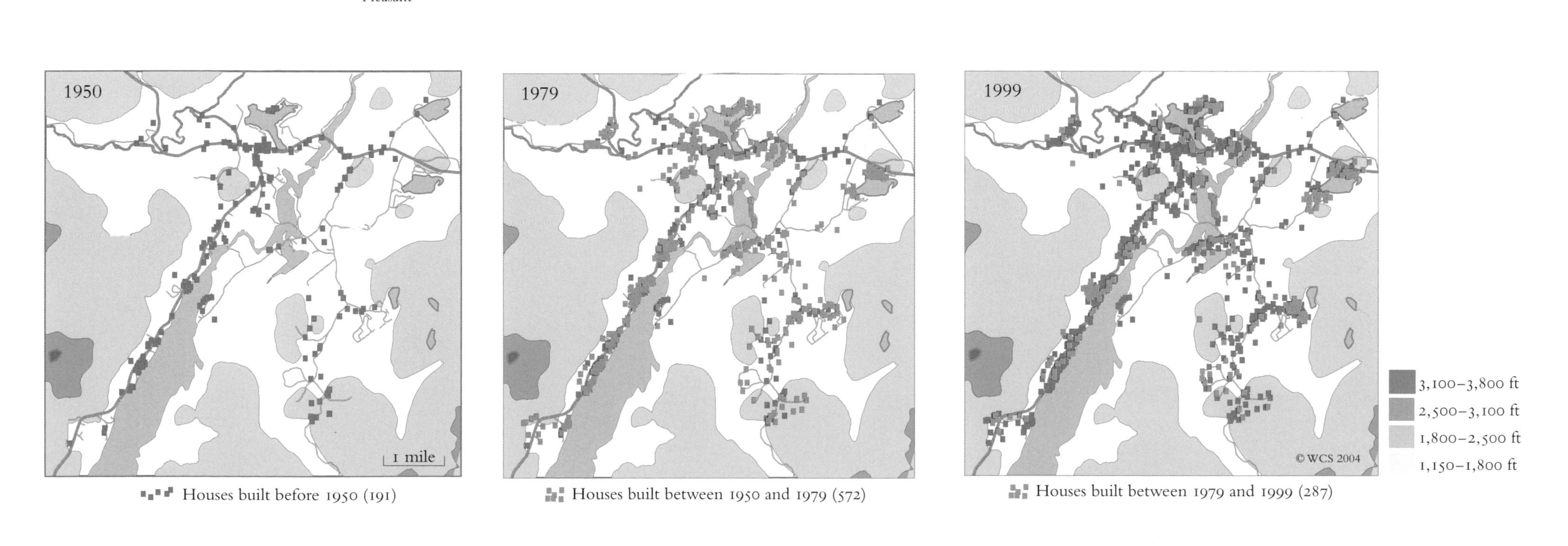

Houses built before 1950 (191)

Houses built between 1950 and 1979 (572)

Houses built between 1979 and 1999 (287)

LAKE PLACID is a village in the town of North Elba. North Elba had agricultural land and good stage connections to the Champlain Valley and so was settled early. Archibald McIntyre and David Henderson were making iron unsuccessfully here in 1809, thirty years before they began making iron unsuccessfully in Tahawus. The abolitionist Gerrit Smith, whose father had made a fortune in the fur trade, began giving land in the town to would-be black farmers and ex-slaves in 1846. To a black family from Brooklyn, North Elba must have seemed farther away than the moon. The fiery John Brown settled here briefly, also in the 1840s, before starting on the road that led him and his sons to Osawatomie, Harpers Ferry, and finally to the scaffold. His body was returned here in 1859; the bodies of eight of his sons were brought secretly forty years later.

Gradually, others less dedicated or fanatic settled here as well, and industry and abolition gave way to fancy camps and wealth. A hotel was built on Mirror Lake in 1871, and a large and highly exclusive club was founded by Melville Dewey (of the decimal system) in 1895. By 1900 there were about a hundred camps on the lake. By 1950 Lake Placid had hosted a Winter Olympics, gained a major ski area, and was the Adirondacks' most successful winter resort. By 1990 it had hosted its second Olympics and had grown to be a village of about 2,000 houses with over 50 stores, 60 motels, hotels and inns, and accommodations for about 1,900 guests.

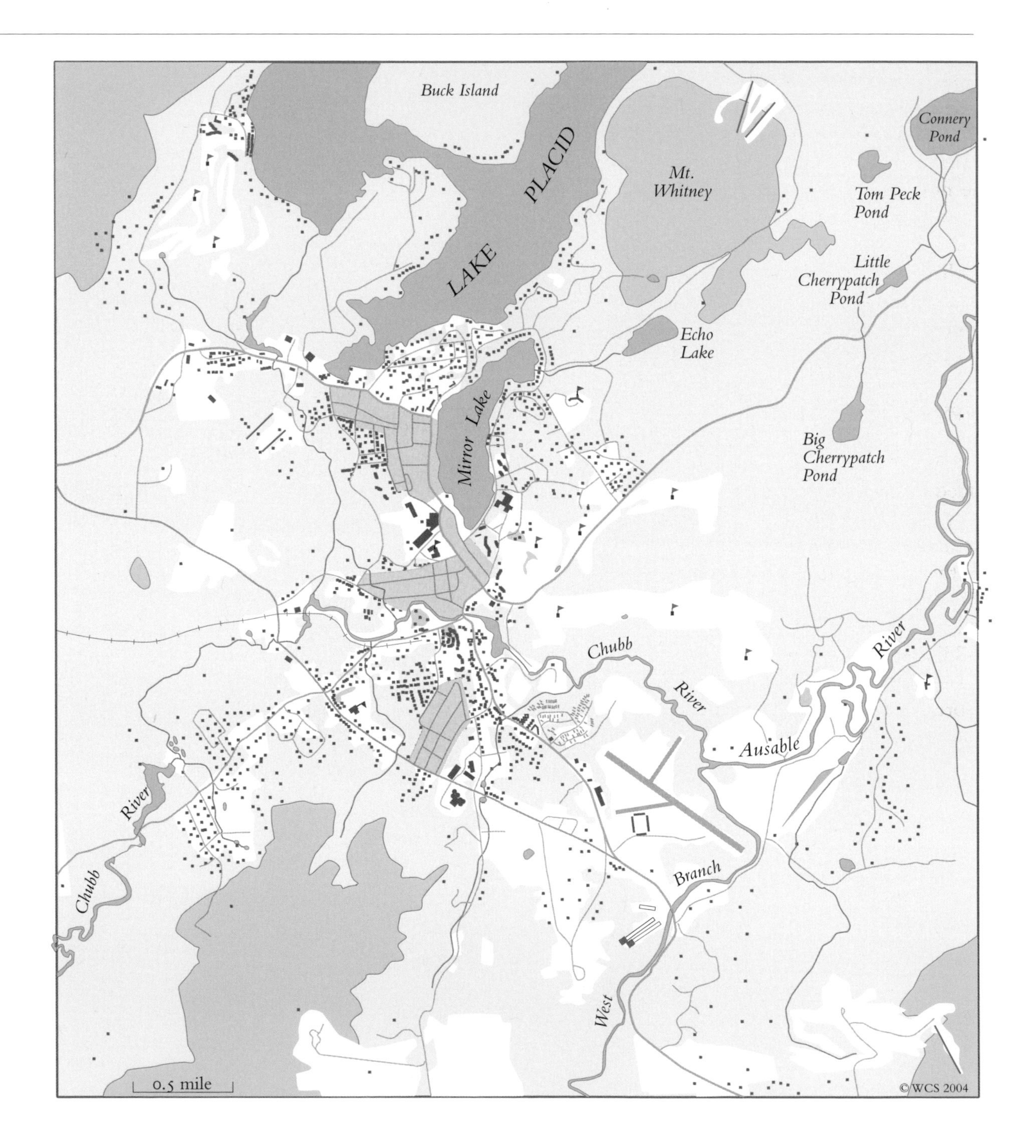

IN 1900 Lake Placid was, by Adirondack standards, a medium-size village. It had one hamlet by the lake and another by the river, farms scattered in the intervale, and clubs and camps by the lake. For the next six decades it grew steadily, though not spectacularly, at a rate of ten to fifteen houses per year.

Until 1950 the building was largely in the two hamlets and along the lake shores. After 1950 several new subdivisions appeared outside town, and growth spread outward along the main roads. Much commercial development, not shown in the maps, also occurred.

In the 1970s, ski and Olympic development greatly accelerated the building rate. The town added nearly a thousand houses and much additional commercial development, changing from a small vacation town to a serious destination resort. By this time the lakeshore and town center were fully developed, and much of building was in new developments on the edges of town. The rate of building slowed abruptly with sharply rising land prices in the 1990s, but the growth of popular resorts is always cyclical, and it could easily accelerate again.

Both the graph and the maps include only existing houses and not those that have burned or been torn down.

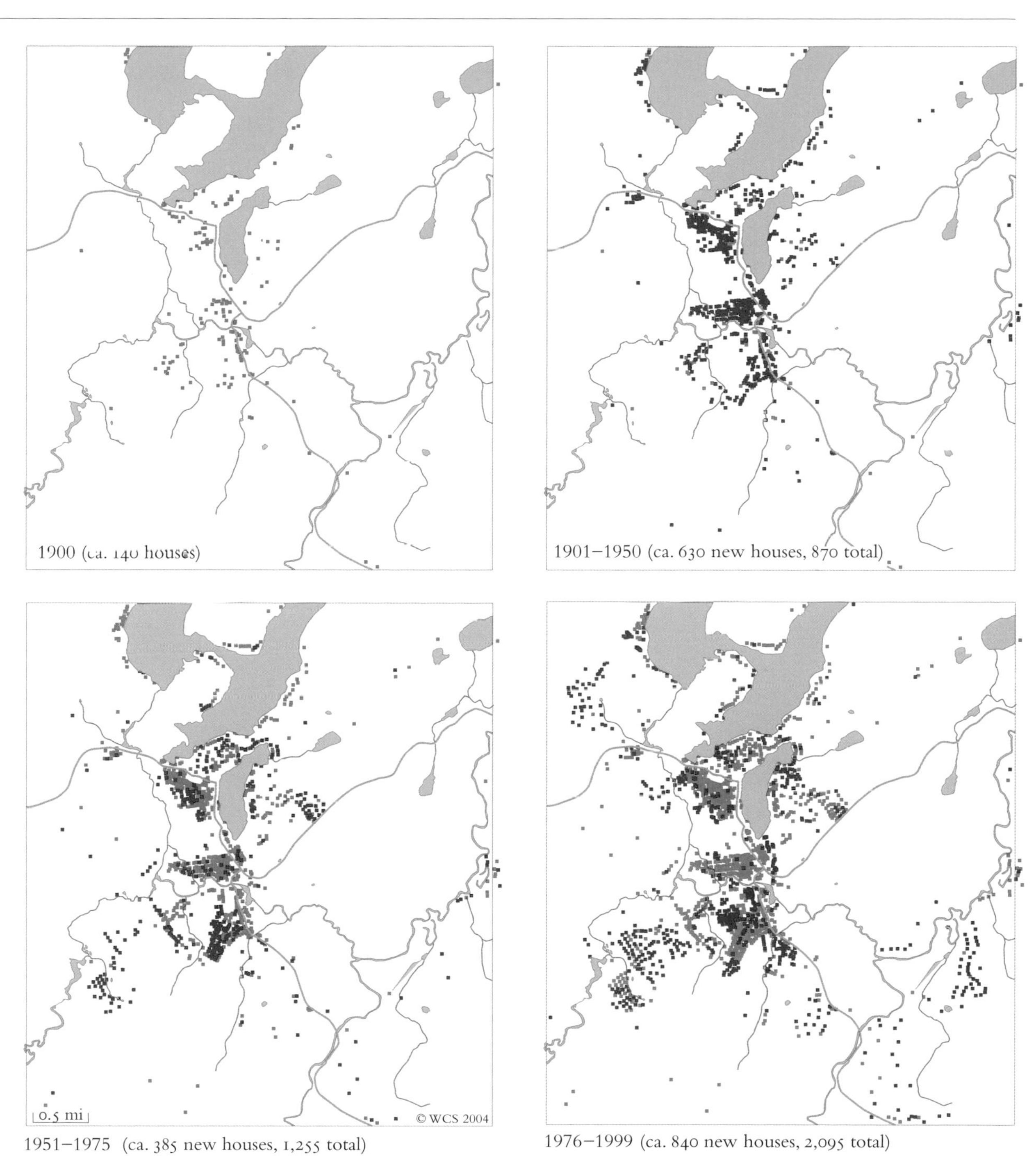

1951–1975 (ca. 385 new houses, 1,255 total)

1976–1999 (ca. 840 new houses, 2,095 total)

HERE are a dozen towns that were remarkable for their vitality and energy fifty or seventy years ago and might have been expected to be prosperous and thriving today. Most, sadly, are neither. The few that have prospered have had to fight for their prosperity and have found it mostly by abandoning, rather than preserving, their original occupations.

Their stories are fascinating and cautionary. Tahawus and Benson, the Adirondack's biggest surface mines, are now wastelands of spoil piles and abandoned rusting buildings. Hitchens Pond, where Augustus Low had his timber and maple sugar empire and where he was wiped out in a week of wildfires in 1911, is now a favorite camping area surrounded by a handsome young forest. Lyon Mountain, an iron town for a hundred and forty years, now has a prison in the center of town and fancy houses along the lake. Hadley and Corinth, once paper-mill towns, have become commuter towns at the edge of suburbia. Willsboro has found other industries, and Port Henry is on the edge of becoming a summer colony. Mineville, Star Lake, and Crown Point, once among the most prosperous industrial towns in the park, have been badly damaged by deindustrialization and are still trying to discover a livable future. Only Ticonderoga, where International Paper runs one of the two pulp-and-paper mills left in New York, remains, for a time, an old-fashioned, one-industry town.

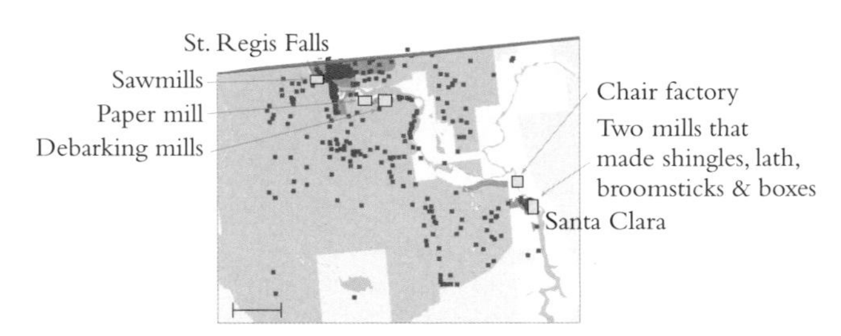

St. Regis Falls was a sawmill and paper town. The river was first dammed for a small mill and tannery in 1860s; the lumber baron John Hurd built a big mill in 1882 and the first railroad in 1883; the St. Regis Paper Company arrived sometime in the 1890s. The industrial heyday was short, and most of the mills closed by 1930.

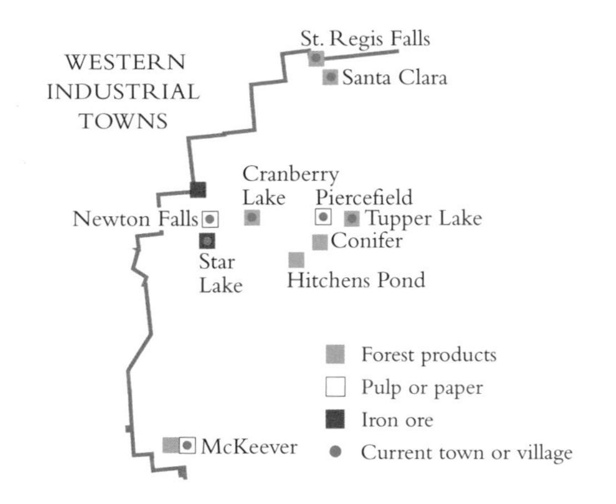

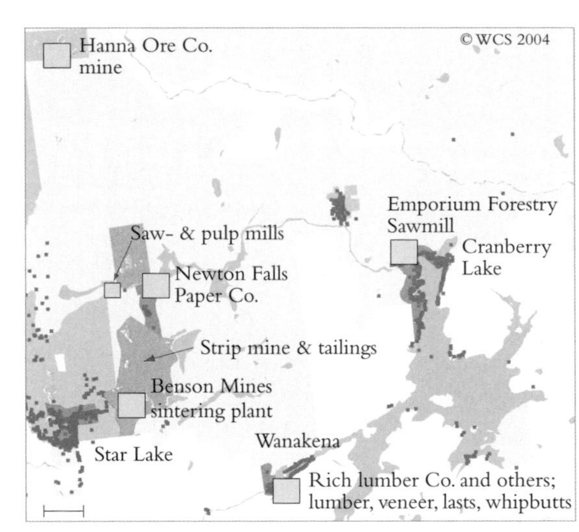

Star Lake and *Newton Falls* were major industrial towns, for much of this century. Currently they are developed, year-round towns, with a central school system and the only hospital in the west of the park, but with few year-round jobs. *Cranberry Lake* and *Wanakena*, quiet summer colonies today, began as saw mill and railroad-towns, and were the central points of two large but short-lived woodland empires in the first go-go decades of hardwood logging (*Map 16-6*).

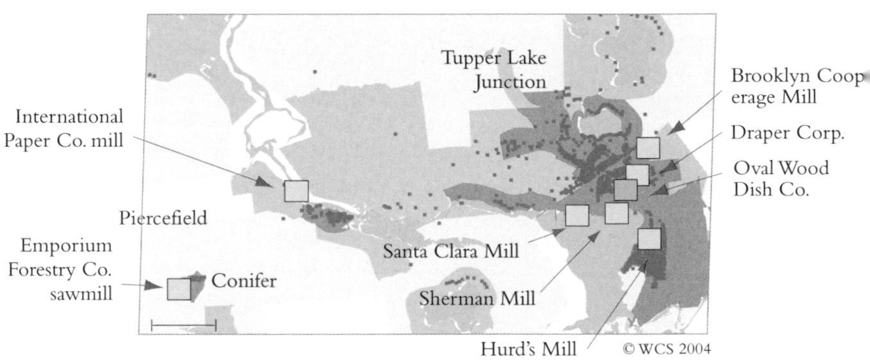

Tupper Lake was created to drive logs down the Raquette, and the town grew up to service the drives. By 1900 it was a railroad junction with roundhouses, wyes, engine shops, freight depots, and a union station. After some years of shipping lumber elsewhere, it acquired big mills of its own and by 1910 had passed Glens Falls as the largest lumber producer in New York. It also developed a substantial wood products industry, making veneer, bobbins, chairs, dishes, mangles, and wood alcohol. Currently, though it produces lumber and plastics and likes to think of itself as an old-time lumber town, its major industries are schools, therapeutic facilities, lodging, and vacation homes.

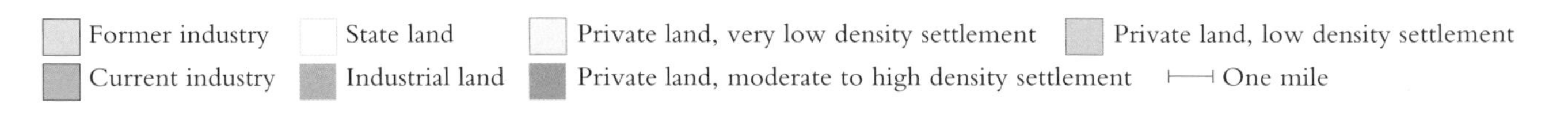

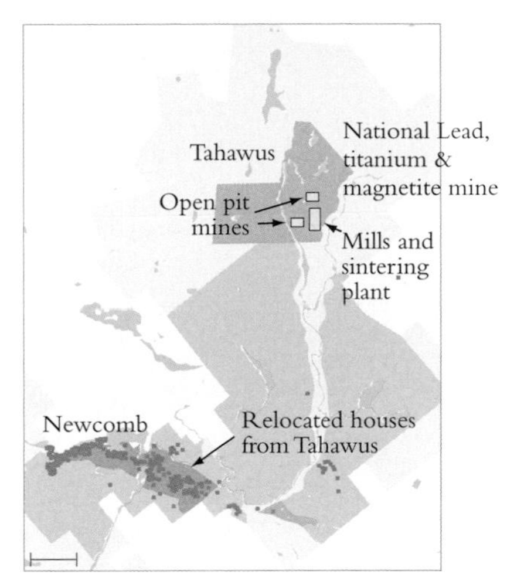

Tahawus was a wilderness iron mine in the middle nineteenth-century. Despite some radical experiments in furnace technology and a brave early attempt to produce steel, it had no water or rail transport and was an economic failure. The ore was rich in titanium, bad for making steel but good for making white paint, and a major titanium-mining operation was developed during World War II. The original village, which sat on a particularly desirable ore body, was relocated ten miles to the south in 1955. Tahawus is now deserted except for lumber trucks and a small two-man sawmill in one of the mill buildings.

Hadley and *Lake Luzerne* are big-river, paper mill towns turned to summer colonies and tri-cities suburbia. *Palmer* had, until 2002, the oldest operating paper mill in New York. The paper industry currently has much excess capacity, and the mill is shut down.

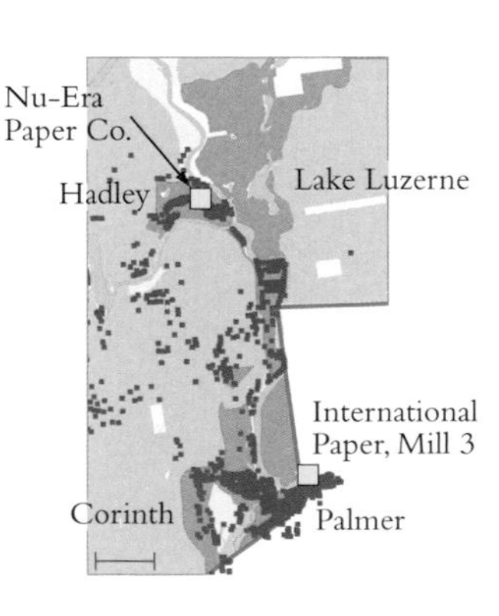

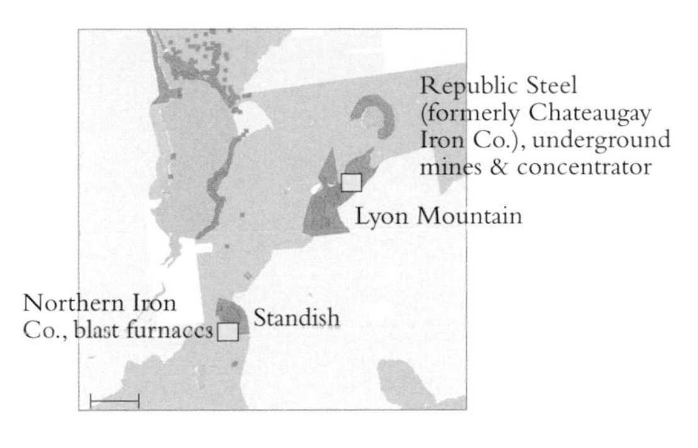

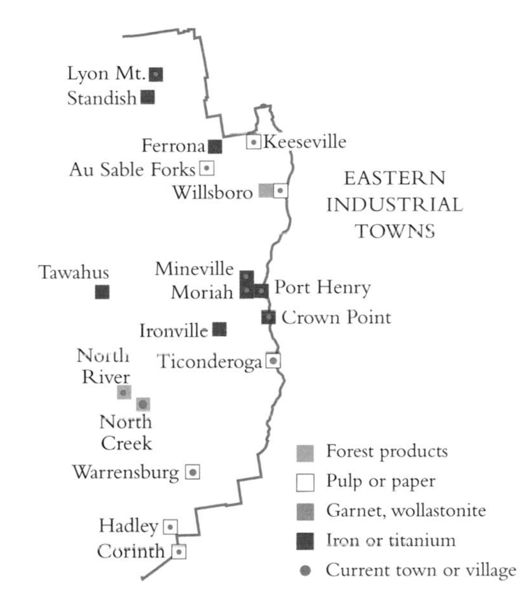

The northeastern Adirondacks have been iron-mining country for over two hundred years. In 1900, with perhaps ten major furnaces running, it was a major iron-making area as well. By 1940 the mines and concentrators were supplying 750,000 tons of iron ore per year to the blast furnaces of Pennsylvania and Ohio and were the major domestic source for most eastern United States furnaces. The last Adirondack blast furnaces, at Port Henry, closed about 1937. The mines at Lyon Mountain closed in 1967, and the mines at Mineville, the last underground mines in the Adirondacks, in 1971.

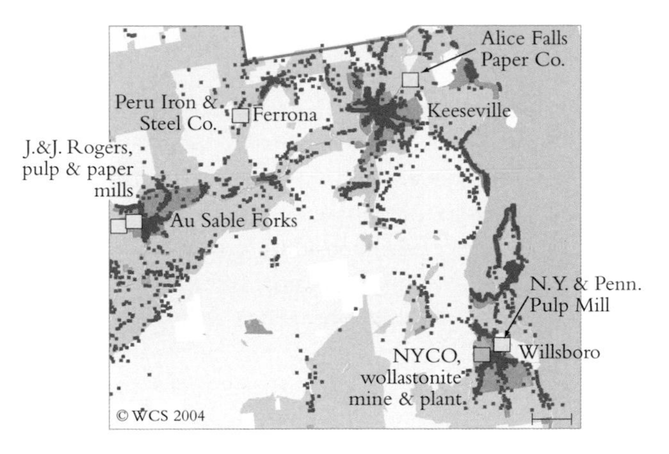

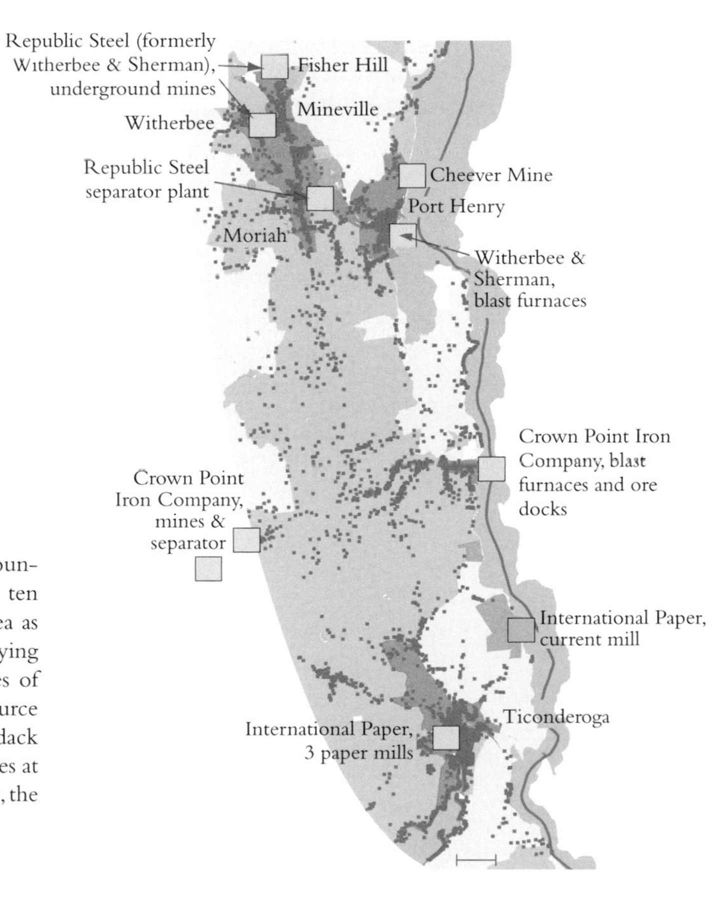

CRANBERRY LAKE, now the third largest lake in the park, was originally a small stillwater on the Oswegatchie River. Log drives began on the upper Oswegatchie about 1854. In 1864, the Tooley Pond Road, the first stage road into the northwest Adirondacks, was built to serve the drives and allow sportsmen to visit Cranberry Lake. The thirty-seven mile trip from Canton took two days with an overnight stop in Degrasse, roughly like flying to Africa today. People will go a long way for logs and fish.

The lake was enlarged by a thirteen-foot-high timber-crib dam in 1867. The dam was raised to eighteen feet by 1910 and replaced by a concrete dam in 1916. The dams greatly enlarged the original lake and provided access to large stands of spruce that bordered the swampy upper river, allowing the development of large-scale logging and lumbering.

The development came quickly. In 1880 Cranberry Lake was a tiny wilderness village with hotel, a school, and a few houses. By 1905 it had become a prosperous mill and resort town, with a railroad, two churches, five hotels, a large sawmill, and a number of private homes and clubs. A few miles to the west there was a major iron mine, a paper mill, and another large sawmill. Fishing and lumber had paid off.

The un-developing that followed was gradual and allowed the town time to adapt. The sawmills were both closed by 1930. By 1950 the railroad was gone and the hotels mostly replaced by private cottages. The Benson Mines, the last iron mines in the park, closed in 1977. The Newton Falls paper mill, the next-to-last papermill in the park, closed in 2001.

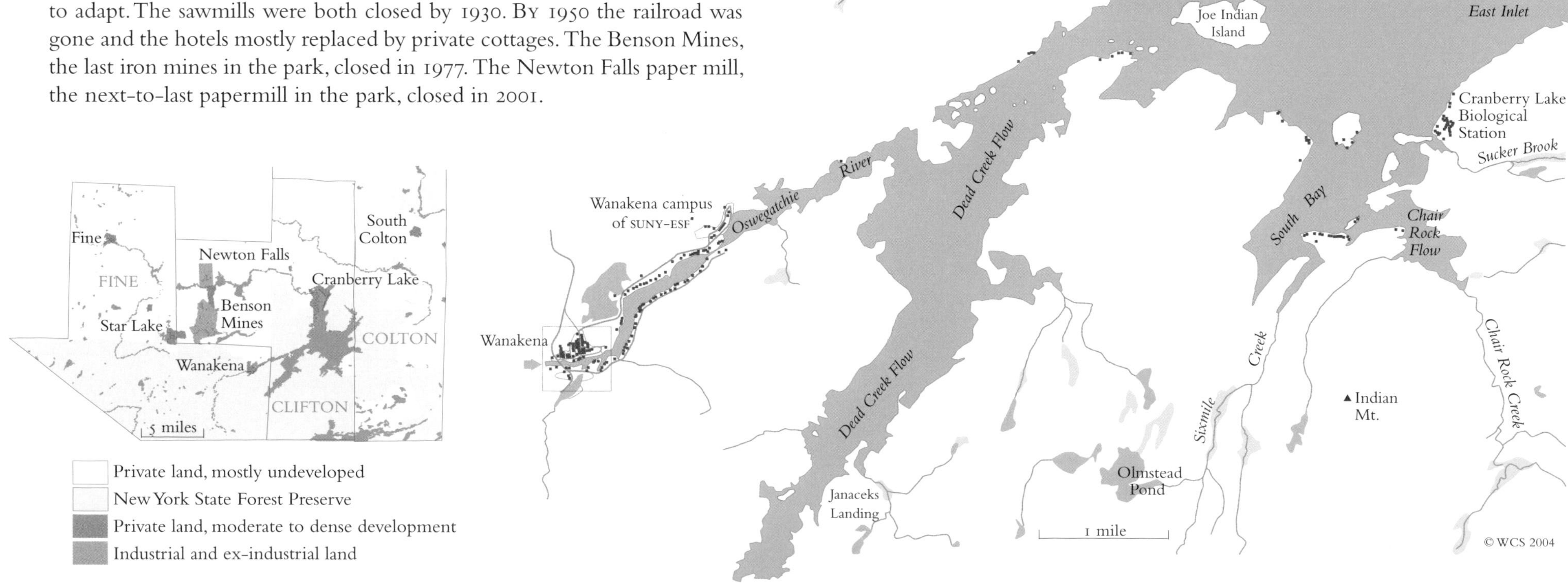

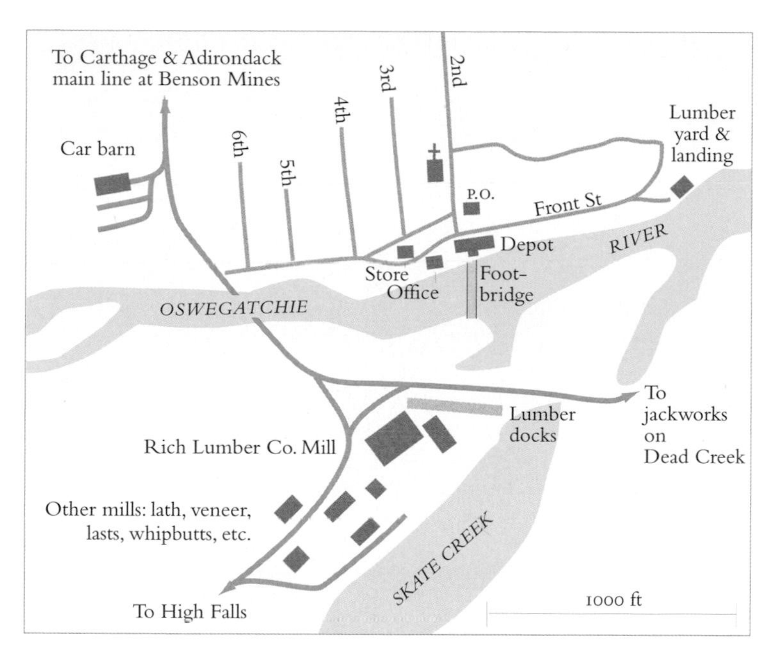

RICH LUMBER COMPANY, WANAKENA, 1905

In 1902, the Rich Lumber Company, a successful company from Pennsylvania that was running out of timber at home, purchased 16,000 acres around Wanakena, then just a campsite on the shore of a newly-made lake. By 1903 they had built a sawmill, laid out a town with a main street and five numbered side streets, and built a railroad connecting with the Carthage & Adirondack main line six miles away at the Benson Mines. By 1905 they had pushed a logging railroad fifteen miles up the Oswegatchie and were cutting everything sawable at a rate of about 2,000 acres a year. By 1910, just about on schedule, they had run out of marketable softwoods. By 1912 they had cut all the good hardwoods and closed their mill. In nine years, fast even by the standards of the time, 16,000 acres of virgin forest had become commercially worthless. They gave some land to Syracuse University, sold the rest to the state, and moved to Arlington, Vermont. There, they bought 15,000 acres of virgin softwoods on the Green Mountain Plateau and built a mill and a railroad. This time, wiser and thriftier, they were able to make the timber last twelve years.

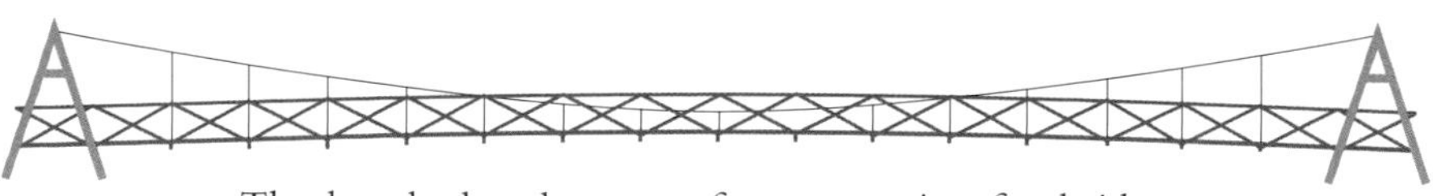

The hundred-and-seventy-foot suspension footbridge at Wanakena, connecting the town and the mill, ca. 1903.

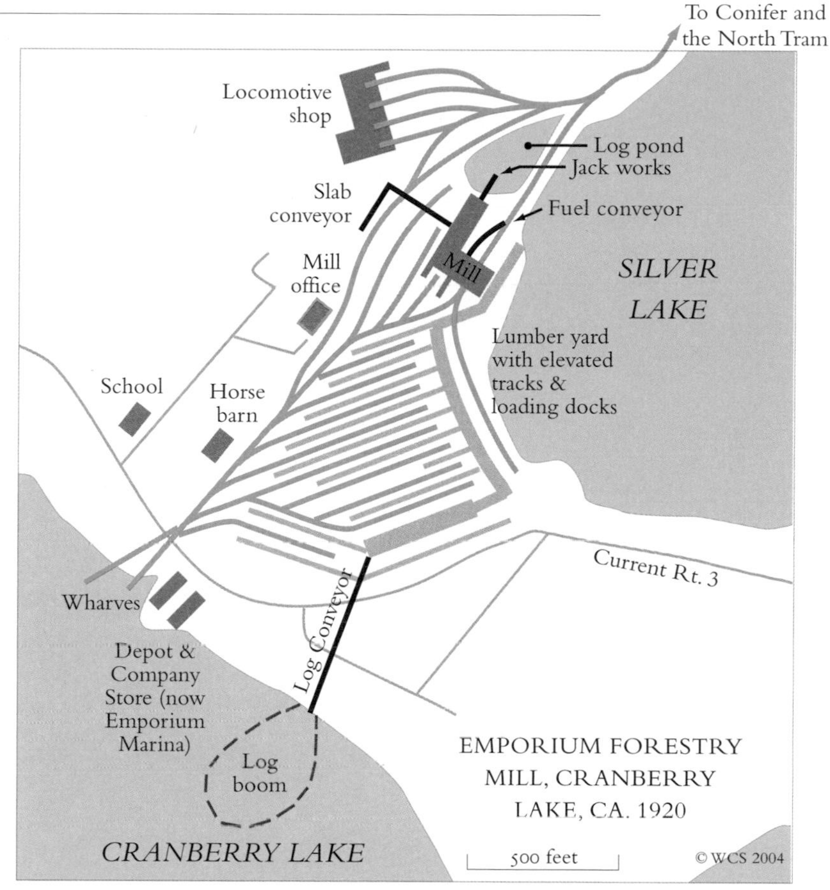

EMPORIUM FORESTRY MILL, CRANBERRY LAKE, CA. 1920

The Emporium Forestry Company, another Pennsylvania company, was the largest and longest-lived Adirondack timber company of the early 20th century. At their peak they owned at least 110,000 acres in the western Adirondacks and had large mills at Cranberry Lake and Conifer. Their logs came from a forty-mile logging railroad, the North Tram, with perhaps a hundred miles of tracks and spurs (*Maps 5-10, 16-7*), stretching northwest from Cranberry Lake down the drainage of the Grass River. They were organized, well managed, and reportedly very efficient. But, like everyone else in those days, when they cut they cut hard.

The Cranberry mill, one of the largest ever built in the Adirondacks, opened in 1917. In the next ten years it sawed an amazing 200 million board feet of virgin hardwoods before closing for lack of timber in 1927. The North Tram continued to serve their mill at Conifer, bringing down about 10 million board feet of logs a year. By the early 1940s it had produced some 1.1 billion board feet of lumber from 2.6 billion logs. Virgin timber was almost completely gone from the western Adirondacks; Emporium began importing logs for the mill at Conifer, dismantled the North Tram, and began disposing of its lands.

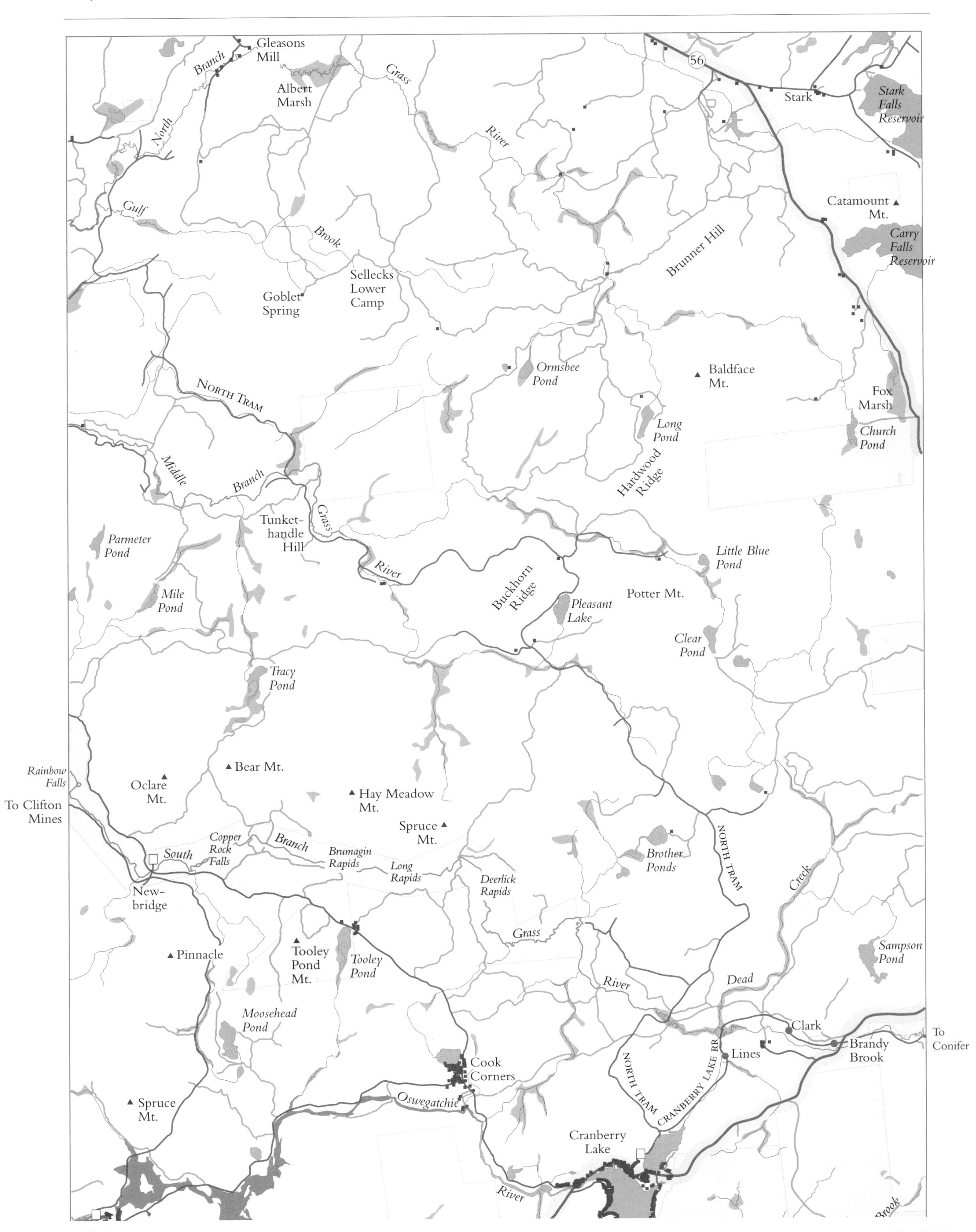

Gleasons Mill
Branch
North
Gulf
Albert Marsh
Grass
River
56
Stark
Stark Falls Reservoir
Catamount Mt.
Carry Falls Reservoir
Brunner Hill
Brook
Sellecks Lower Camp
Goblet Spring
Ormsbee Pond
Baldface Mt.
Fox Marsh
Church Pond
Long Pond
Hardwood Ridge
North Tram
Middle
Branch
Tunket-handle Hill
Grass
River
Parmeter Pond
Mile Pond
Buckhorn Ridge
Pleasant Lake
Little Blue Pond
Potter Mt.
Clear Pond
Tracy Pond
Rainbow Falls
To Clifton Mines
Oclare Mt.
Bear Mt.
Hay Meadow Mt.
Spruce Mt.
Copper Rock Falls
South
Branch
Brumagin Rapids
Long Rapids
Deerlick Rapids
Brother Ponds
North Tram
Creek
New-bridge
Grass
Tooley Pond Mt.
Tooley Pond
Pinnacle
Sampson Pond
Dead
River
Moosehead Pond
Clark
To Conifer
Brandy Brook
Lines
Cook Corners
North Tram
Cranberry Lake RR
Oswegatchie
Spruce Mt.
Cranberry Lake
River
Brook

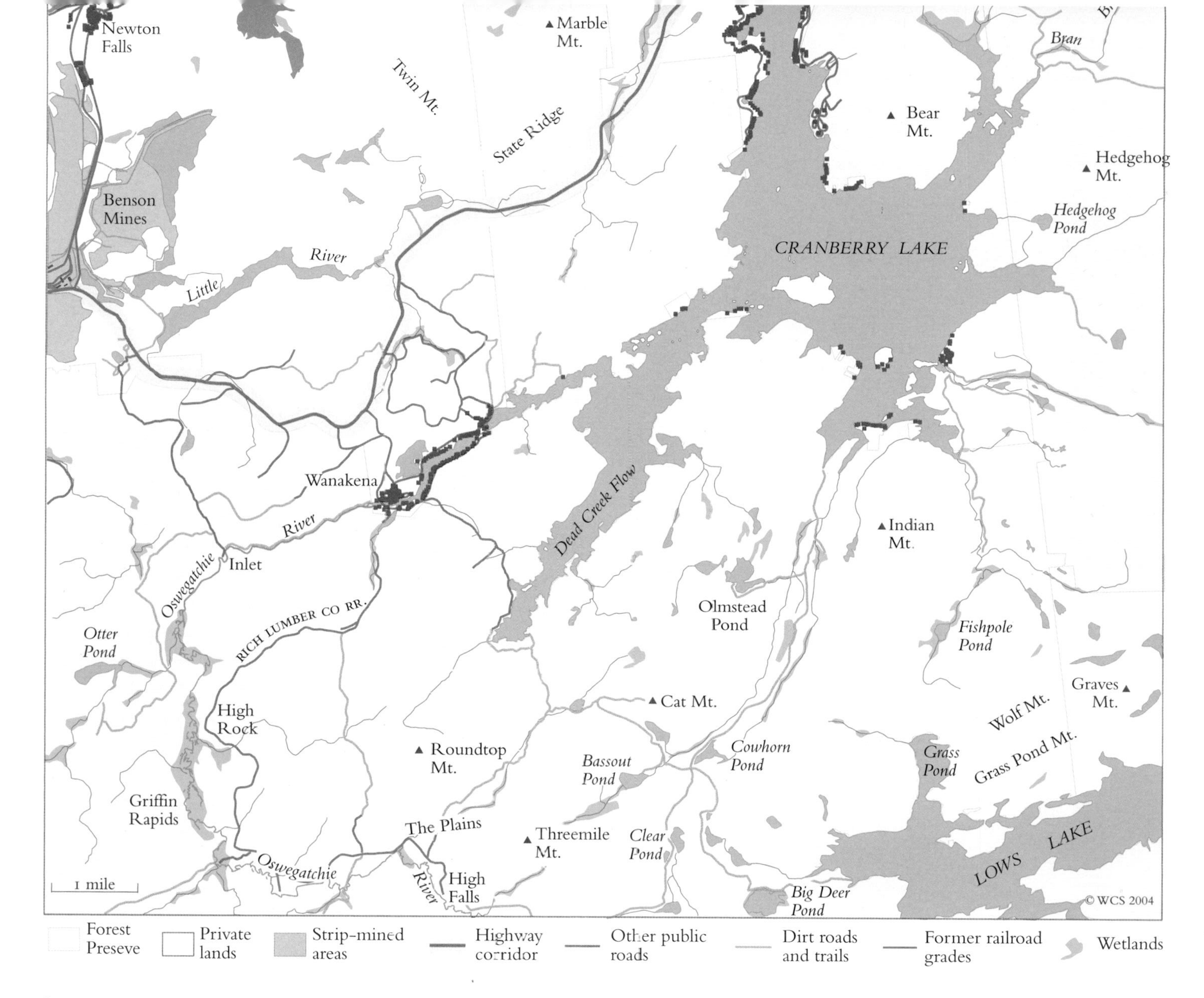

APPROXIMATELY a million and a half acres of the park consist of large private landholdings. Most of these lands are closed to the public and appear as empty spaces on road maps and guidebooks. But they are, of course, far from empty. Almost all are the workplaces of foresters and loggers and the operating areas of the skidders, feller-bunchers, and the other heavy machines used by contemporary logging crews. All have networks of private roads, and most have a scattering of camps and clubs. Some are private estates that are used only by their owners and their employees and guests. Many others are leased to clubs or individuals who build their own camps and have exclusive hunting and fishing rights. Some of the leased camps are rough, temporary, and comfortable-looking. Others are old, gracious, and dignified, really ownerships within ownerships, and clearly invested with the affections and traditions of many years of use.

The map shows a two-hundred square mile block of private timberlands near Cranberry Lake that is closed to the public and poorly known even to many long-time residents. We have omitted the ownership boundaries, for which no composite map exists, and the locations of the hunt clubs and the lands they lease, which the owners consider proprietary information.

This country is quite different from the Forest Preserve to the south. It is harder working, more used and mechanized, but also more private and personal. It lacks the spectacular features—the big mountains and cliffs, the large lakes—of the public lands. But it has beauties of its own, and some secrets too. Tracy and Long Ponds are backwoods gems, with beautiful wetlands and flowers. Pleasant Lake has a pair of loons that come up to greet visitors. Little Blue Pond has the loudest bullfrogs in the park. Albert Marsh has a mile or two of a lovely and very private stream, with white gentians on the banks. Just south of it, hidden within half-mile of alders, is a tiny bog pond with pink orchids. Buckthorn Ridge has some fine stands of young maples that Sam Parmalee remembers thinning when he was a young forester in the 1950s. And, most surprising of all, a few of the hills have some big maples and hemlocks that one or two foresters, stubbornly, unprofessionally, and tenderly, have kept hidden from their crews.

16-8 MORIAH

For two hundred years, from the opening of the mines near Cheever in the 1760s until Republic Steel turned off the pumps in the Harmony Mine in 1979, Moriah was an iron town.

Its development is a classic nineteenth-century story: the first furnaces near the shore in the 1820s; extravagantly rich discoveries and big open-pit mines in the fifties; immigrant workers, railroads, and anthracite furnaces in the sixties; mechanization and deep underground shafts in the seventies and eighties. In 1861 Port Henry and Crown Point supplied the iron that the Griswold Company in Troy made into the plates and machinery for the *Monitor;* in 1875 they produced 220,000 tons of iron ore, 8.8% of the total U.S. production.

Its subsequent decline is, sadly, a classic twentieth-century story: big expansions during the wars and shutdowns and layoffs afterwards; increasing labor costs with the switch to an native-born work force; competition from new beds in Canada and South America that were shallower and cheaper to work; buy-outs and consolidations and finally a takeover by a national company. The new owners, as any good managers would have, closed the furnaces, ran the Mineville mines though the war and the boom years of the 1950s when they were profitable, and closed them in 1971, after a hundred and fifty years, when it was clear they were not.

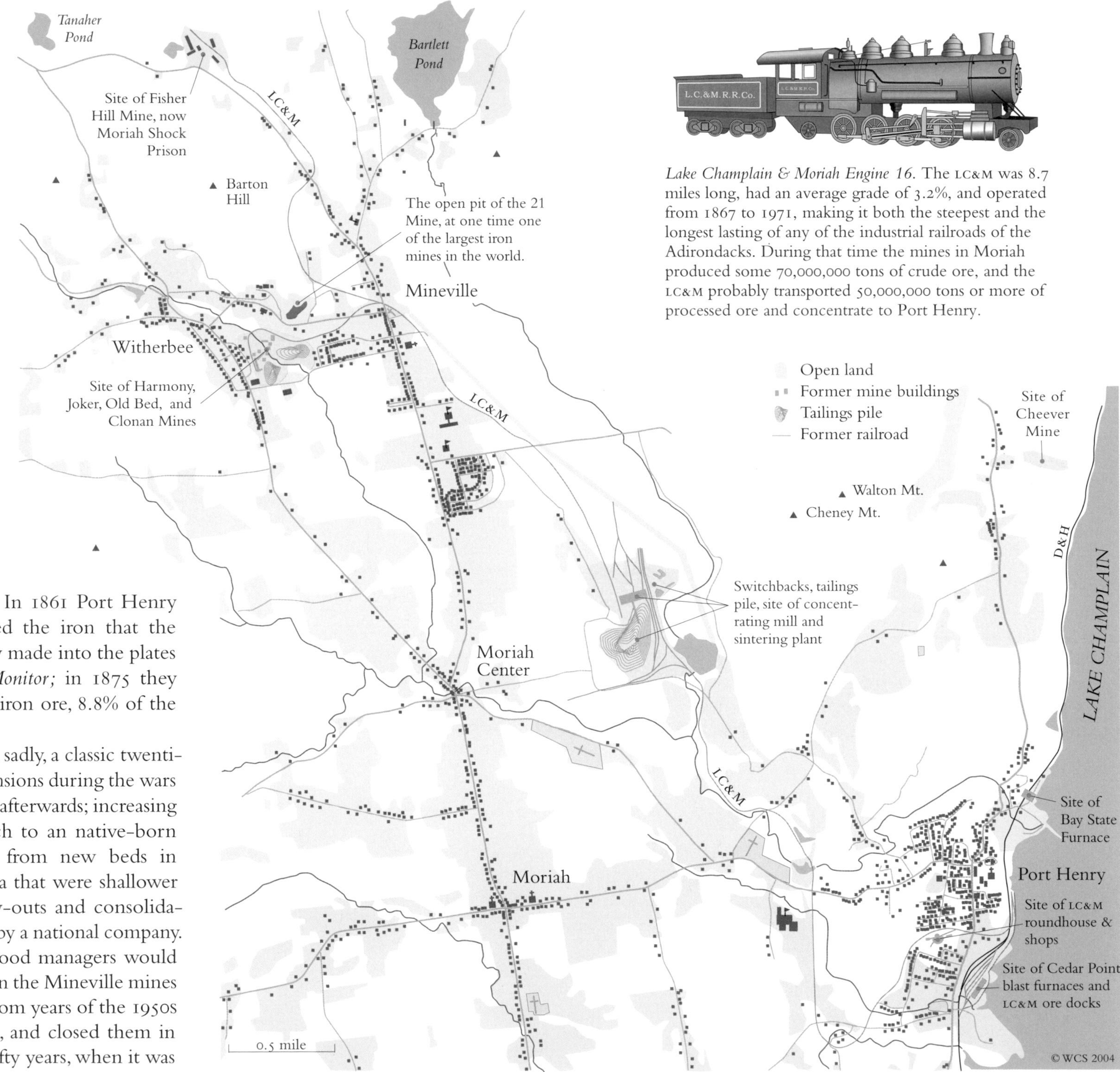

Lake Champlain & Moriah Engine 16. The LC&M was 8.7 miles long, had an average grade of 3.2%, and operated from 1867 to 1971, making it both the steepest and the longest lasting of any of the industrial railroads of the Adirondacks. During that time the mines in Moriah produced some 70,000,000 tons of crude ore, and the LC&M probably transported 50,000,000 tons or more of processed ore and concentrate to Port Henry.

MORIAH had grown steadily during the nineteenth century and by 1875 was the largest and probably the most prosperous town on the west side of Lake Champlain. The expansion and mechanization of the mines in the next forty years tripled employment and output and triggered a building boom. By the First World War Moriah was the major iron-mining town east of the Great Lakes and one of the largest iron-making towns north of Pennsylvania.

Prosperity in mining towns is often cyclical. The twenties and thirties were uncertain and hard, the forties and fifties boom years, the sixties difficult and fearful. In 1965, at the end of the good times, Moriah had thirteen grocers, eight clothing stores, six car dealers, six restaurants, three movie theaters, nine churches, and 1,450 students in six public schools. A decade later Moriah had lost five hundred jobs and a several million dollar payroll. Twelve grocers, six car dealerships, five churches, four restaurants, two parochial schools and all the movies eventually closed. The six public schools were combined into a smaller central school. School enrollment fell below nine hundred, and, despite scattered residential growth over the last twenty-five years, remains below nine hundred today.

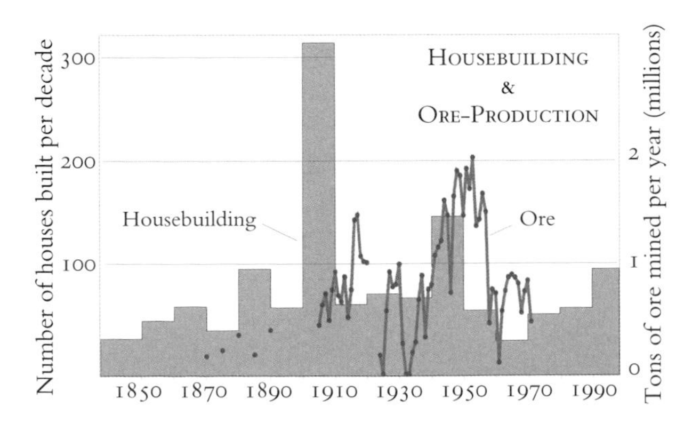

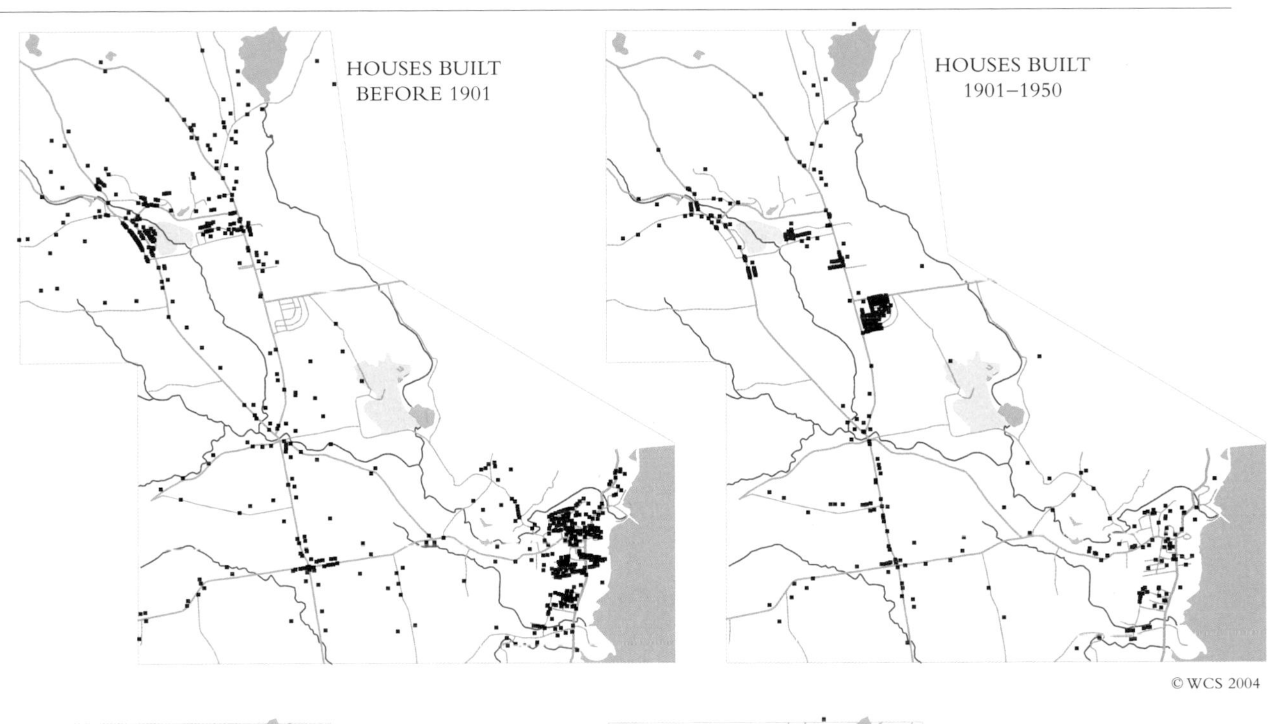

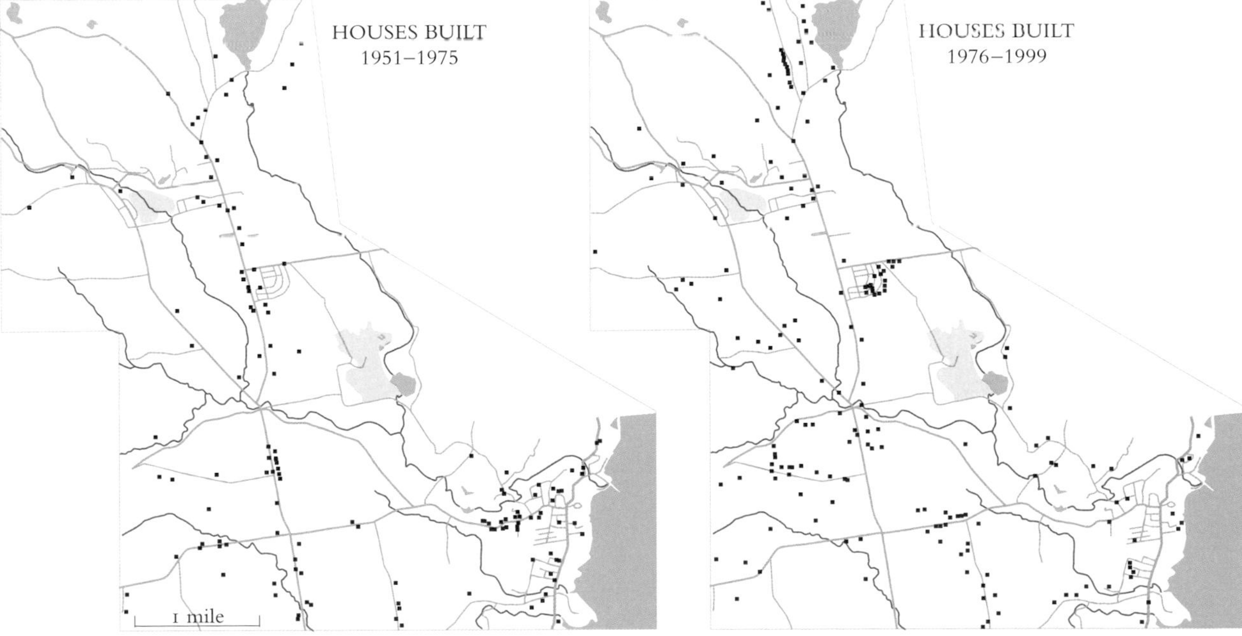

In 1920, Mineville was a modern, multicultural mining town. Witherbee & Sherman, who had run the mines for nearly seventy years, had built a school, hospital, and five hundred houses. Wages were, by industry standards, fairly high but not high enough to attract many local workers. The work force of 878 men was largely foreign: 44% were Polish, 18% Hungarian, and the rest mostly French, Irish, Italian, and Swedish. Only 14% were American-born. The workers were segregated by nationality, and safety and direction signs in the mines were in several languages.

The mines lay in a flat-bottomed bowl between the villages of Witherbee and Mineville. The most distinctive structures were the hundred-foot-high headframes over each of the active shafts. Around the headframes were the facilities that served the mine: the power houses, the huge electric hoists, the changing-houses for the miners, the mills that crushed and concentrated the ore, and the railroads that moved it. Below the headframes were the mines themselves: the shafts and levels that followed the veins of ore; the stopes where the mining took place; the scrapers and locomotives and power shovels, many built in the shops above, that moved the ore; and, all-important, the pumps that kept the mine dry and the skips that carried the men into the mines and the ore out of them.

When the mines were abandoned in 1979, the shafts were sealed and almost all the buildings were razed. Nothing remains above the ground except the tailing piles, the Harmony change-house, and the fenced-off pit of the original 21 Mine. The shafts and galleries and much of the machinery that was in them still exist but are now filled with water. Daring archaeologists may someday dive into them and, as we wonder at temples and pyramids, wonder at the people who made them.

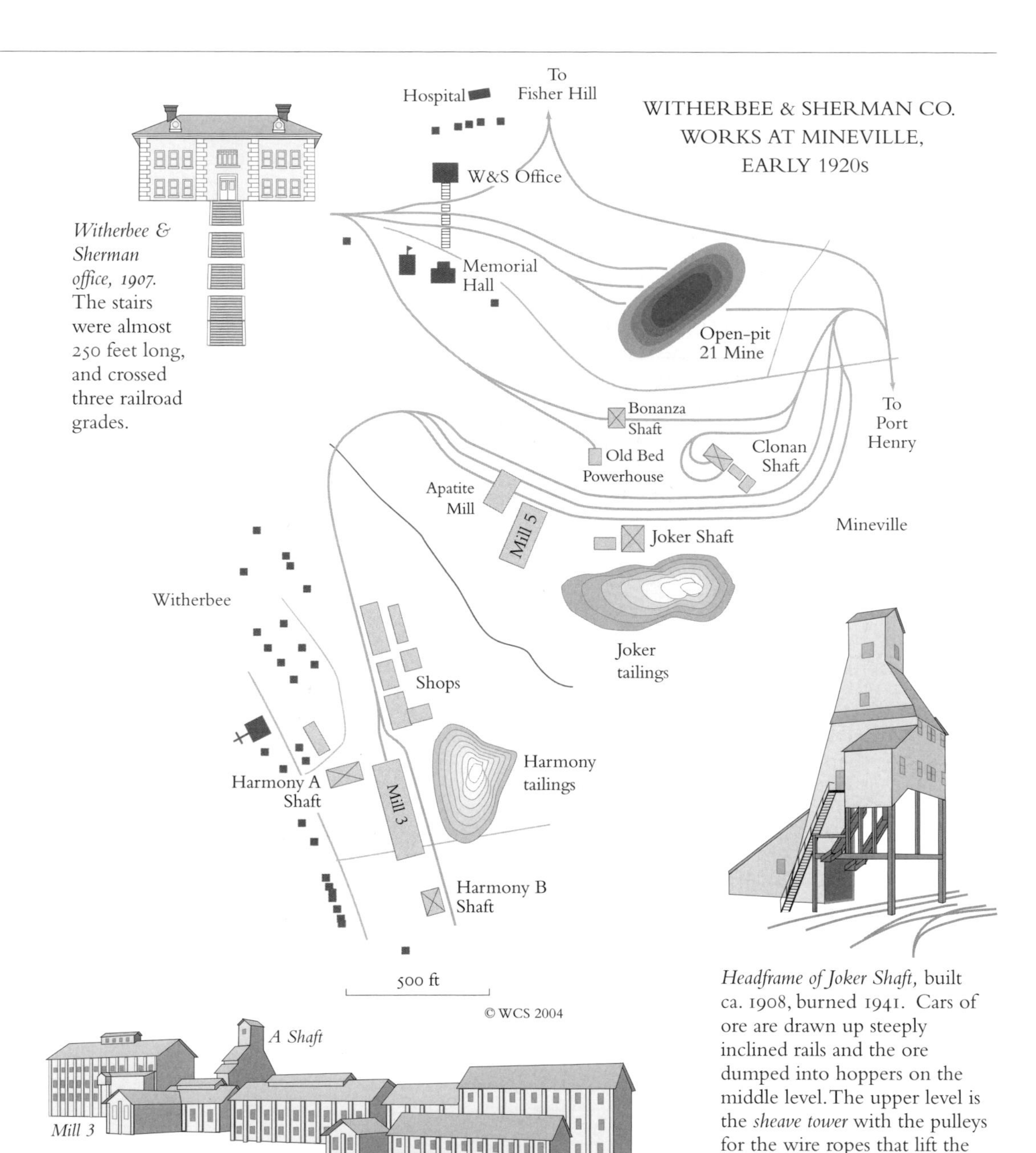

Witherbee & Sherman office, 1907. The stairs were almost 250 feet long, and crossed three railroad grades.

Mill 3, Harmony A Shaft, with associated shops and warehouses, ca. 1915. The mill dried, ground, sifted, and magnetically separated the ore, producing a concentrate used in blast furnaces. It was built in 1909, and, though "completely fireproof," burned in 1923. Five major structures at the mine burned between 1910 and 1941.

Headframe of Joker Shaft, built ca. 1908, burned 1941. Cars of ore are drawn up steeply inclined rails and the ore dumped into hoppers on the middle level. The upper level is the *sheave tower* with the pulleys for the wire ropes that lift the cars. The hoisting motors are in a separate building.

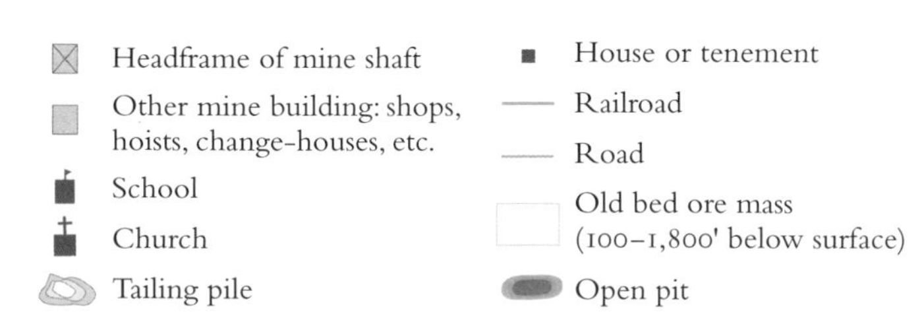

THE OLD-BED MINE

The Old Bed at Mineville was one of the richest and most famous iron deposits in North America. It was a thin sheet of ore, 10 feet thick and 500 to 2,000 feet wide, starting near the surface and dipping at 20–30 degrees to the southwest. It was divided into several units by trap-rock dikes and faults and required auxiliary shafts to follow the ore in each unit. In the 1920s it had been mined to a depth of about 2,000 feet. Eventually it descended 4,800 feet and the miners had to commute for forty-five minutes underground to reach the mining sites. The Harmony shafts served a parallel ore bed (not shown), which lay 700 to 800 feet above and somewhat to the southwest of the Old Bed.

THE OLD BED, FROM ABOVE

Surface

NE

-800

-1,330

-1100

-1,300

-1,800

-1,900

The whole mine was powered from a coal-burning steam engine on the surface; the hoists, locomotives and pumps were electric, the drills operated by compressed air. There were some electric lights, but miners still wore carbide lamps to work in the stopes.

The *Joker Shaft*, the original route to the Old Bed, was built about 1880. It is 17' x 8' and descends 1100' to the Old Bed ore veins shown here. At 500' it passed through the egg-shaped ore body shown at lower right that had been reached through the original 21-Mine and Bonanza shafts; by this time the ore in this body, except for the pillars, had been mined out.

Mined-out portions of bed, with the roof supported by ore pillars; these were salvaged in the 1940's

Ore was blasted from the ends of rising tunnels, the stopes, loaded by hand, power shovel, or scraper into trams, and then pushed by hand to the haulage levels. The trammers who filled and pushed the cars were the lowest paid workers in the mine, making 32 cents per hour base pay, with an extra 30 cents per car if they loaded more than eleven 1.5-ton cars in an 8-hour shift.

Loading ore from a rising stope with a scraper

Snatch block

Holes for blasting

Scraper

Loading chute

Hoist

3.5-ton tram

Surface

-1330

Depth

Country rock

Mined-out veins

Current mining area

Hoist

Haulage tracks

Tramway

Scraper

Hoisting engines

Pillar of ore

Traprock dike & fault

OLD-BED AND 21 MINES, CA. 1921

Joker Headframe

Hoist house

"21" open pit mine

Joker Shaft

Mined out Bonanza orebody

Pillars of ore left to support shafts

Skip (ore car) in shaft

Mined out raise

-1000' haulage level

Ore deposit ends at traprock dike and fault

Actively mined tunnel (stope)

-1330' haulage level

Main hoisting shaft, supported on steel frames and cedar beams

-1900' haulage level

Cleanout below haulage level

3000'

Ore was carried horizontally on the *haulage levels* in 4-ton, side-dump cars hauled by electric trolley locomotives. Eleven locomotives were left underground when the mine closed.

The *cleanouts* below the shafts had pumps that ran 24 hours a day for nearly 100 years. When the pumps were stopped in 1979 the mines began to flood and are now estimated to contain 4 billion gallons of water.

THE public lands of the Adirondacks are often described as economic liabilities and hence as a burden that local residents bear for the benefit of the visitors. The commonest complaints are that towns with large amounts of state land have lower tax revenues, fewer jobs, less commercial forestry, and less developable land.

This view is somewhere between disingenuous and dishonest. It ignores a number of important economic facts: the taxes the state pays on public lands, the jobs and revenues generated by the visitors who are drawn by the public lands, the benefits the residents derive from state lands, and the unfortunate fact that the profits from forestry rarely accrue to the towns that grow and cut the trees. And worse, its main assumption, that public lands are an economic burden, is simply untrue. There is no relation between the amount of public land and *any* index of prosperity that we have examined. Towns with much state land, though differing much among themselves, are no different on average—neither richer, nor poorer, nor more or less heavily taxed, nor more or less employed—than towns with little state land.

The reason for this is not that state lands have no economic significance but rather that they are only one of many factors that have economic significance. To show this, the maps compare the geography of five structural factors and six economic indices. Each of the eleven maps has a different pattern, showing than none of the variables correlate well with any other. Even poverty and unemployment, the most closely correlated, have important differences: there are some towns with high unemployment but low poverty, and some towns with low unemployment but high poverty. There are real patterns in the *individual* graphs, and so geography is clearly related to economics. But there are no real correlations *between* the graphs, and so no evidence that the amount of state land, or for that matter any other single structural factor, can control the overall prosperity of a town.

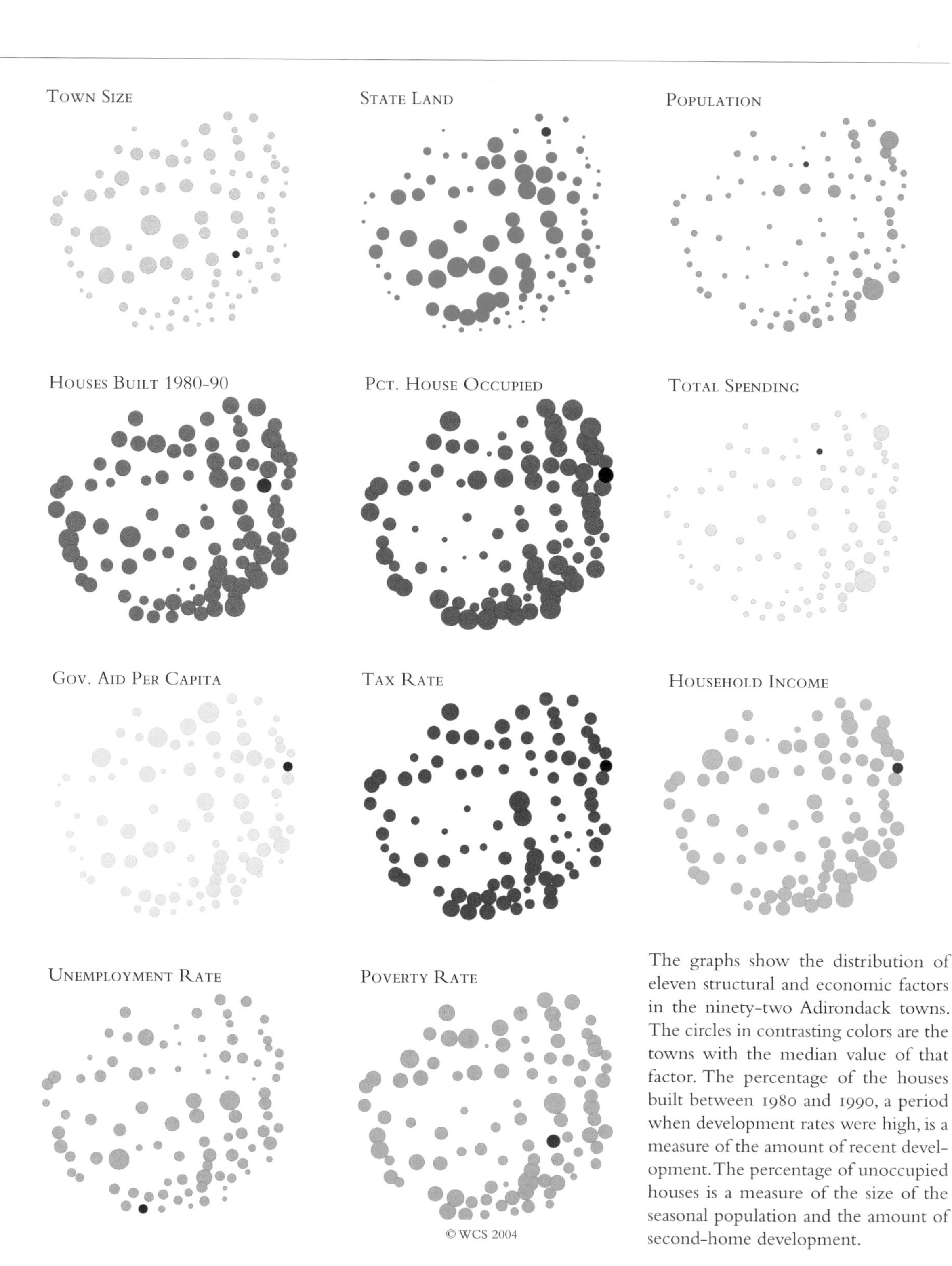

The graphs show the distribution of eleven structural and economic factors in the ninety-two Adirondack towns. The circles in contrasting colors are the towns with the median value of that factor. The percentage of the houses built between 1980 and 1990, a period when development rates were high, is a measure of the amount of recent development. The percentage of unoccupied houses is a measure of the size of the seasonal population and the amount of second-home development.

HERE are the statistical profiles of twelve Adirondack towns with large amounts—36 to 90 percent—of public land. If public land controlled prosperity, we would expect the profiles to be similar. The graphs suggest that this is not so and that these towns differ among themselves as much or more than they differ from other towns.

Harrietstown, for example, is both populous and prosperous. It has a substantial resident population (high percentage of occupied houses), needs a relatively high town budget to support them, but can, because of a lot of valuable commercial property, keep its tax rate low. North Hudson, with about the same percentage of state land, is the economic opposite. It has a small population, many houses owned by nonresidents, low town spending and lots of government aid, and some of the highest poverty and unemployment in the park.

The more closely you look at the graphs, the more contrasts you see. Newcomb, like North Hudson, has a small population, a lot of state land and a lot of unoccupied houses. But it spends twice as much per capita and has twice the tax rate but less than half the unemployment and poverty. Inlet and Arietta are two adjacent summer colonies with large amounts of state land and small resident populations. Inlet has had over twice as much recent development as Arietta and spends 1.5 times as much per capita but still has a lower tax rate.

The conclusion we draw from these graphs is negative but important. Local economic patterns are intricate and individual. Geography matters but does not dominate. If the wilderness towns are in some ways unusual it is because all towns are in some ways unusual and not because of the amount of Forest Preserve that they do, or do not, contain.

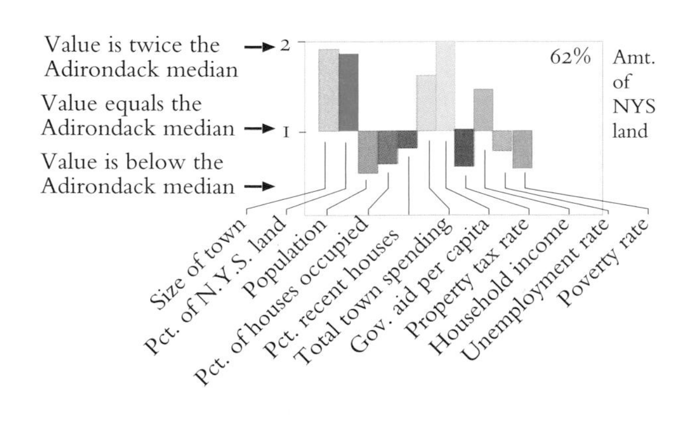

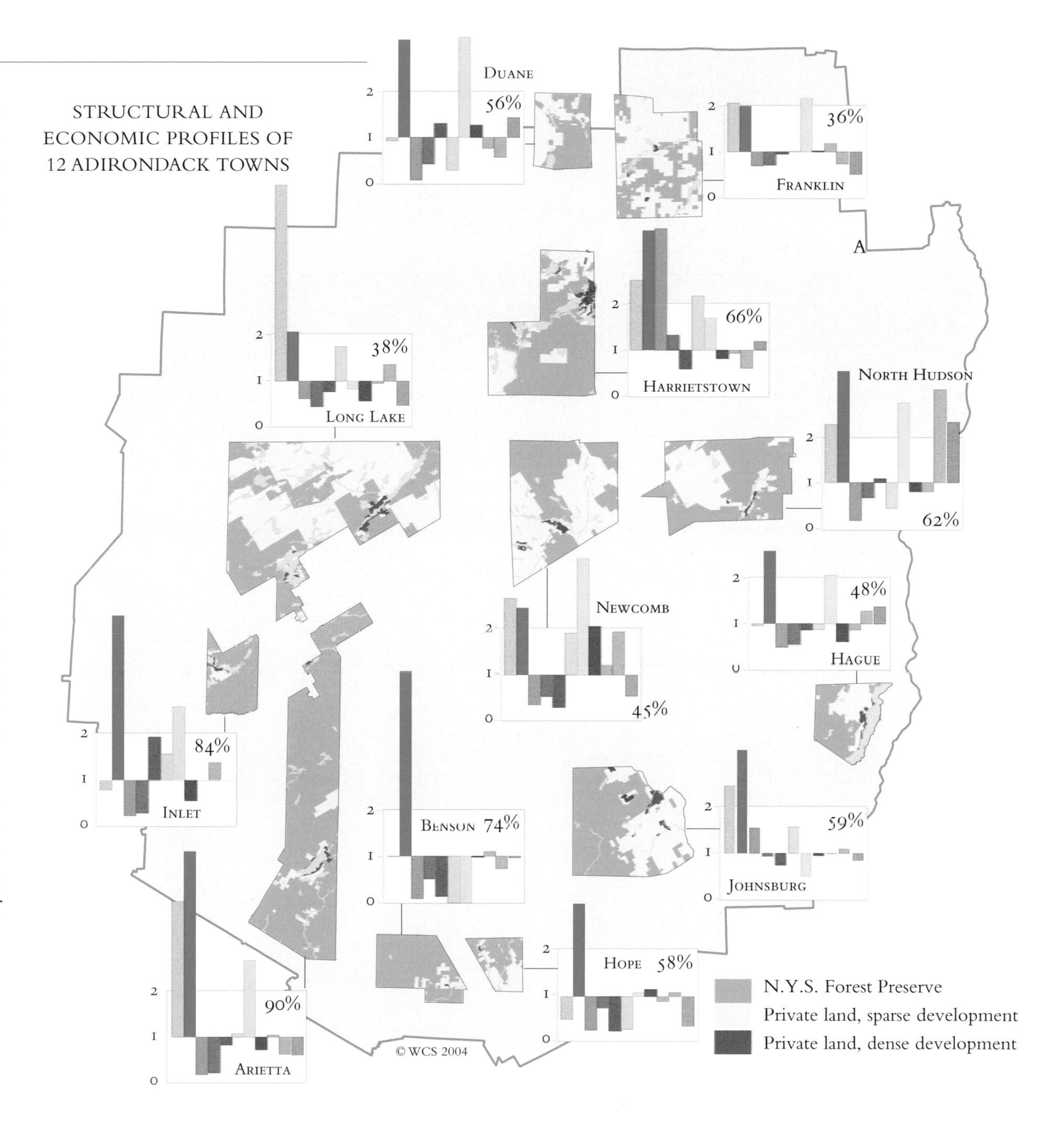

ENERGY POLLUTION

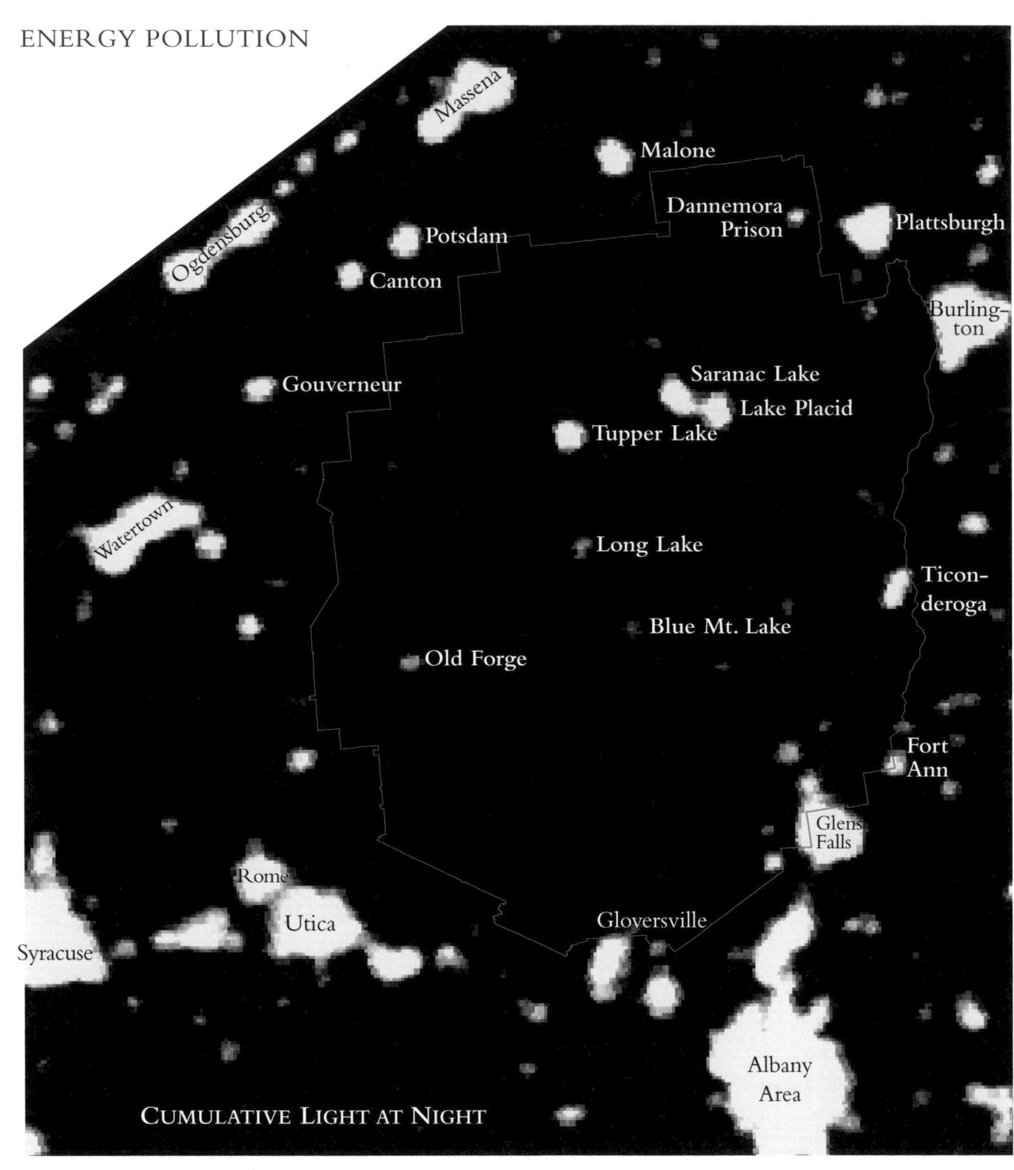

Because sight and hearing are sensitive to tiny amounts of energy, they are easily polluted by excess energy. Mild levels of energy pollution are a distraction or an annoyance; medium levels are a stress or an impairment; high levels are painful and disabling. The Adirondacks have some of the lowest levels of light pollution and thus some of the darkest skies and brightest stars in the eastern United States. But because they have large numbers of motor vehicles and watercraft, and because they are continually overflown by civilian aircraft and used as a training area by military aircraft, all parts of them are at least moderately polluted by noise.

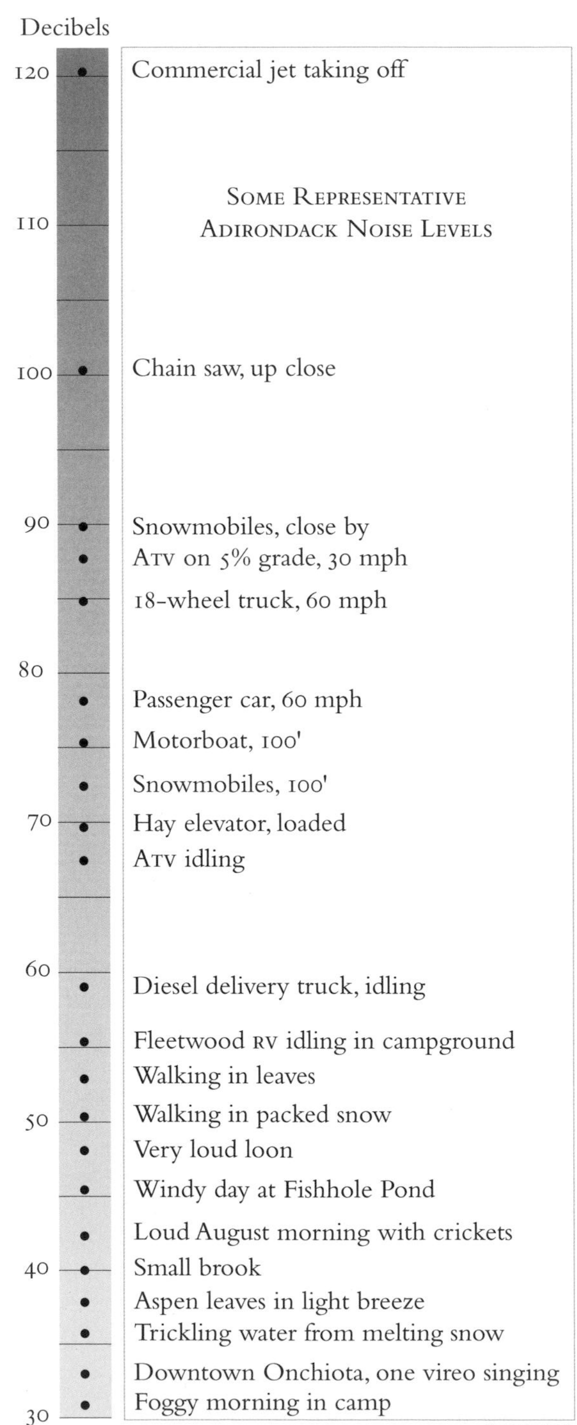

Noise levels vary greatly with the distance, speed and orientation of the source, and with the air temperature and weather. These are some average values measured in the Adirondacks in 2001–2002.

17 Pollution & Wastes

POLLUTANTS are the things we produce that subsequently become a burden to us. Some, like septic and toxic wastes, are the unwanted by-products of useful activities. Others, like old newspapers and junked cars, are trash, which is to say useful things that we no longer use. A few, like asbestos and plutonium, are very dangerous things that perhaps should never have been produced at all. And a few, like light and noise, are not things at all but rather forms of energy.

This chapter has two themes. The first is that, after a century of municipal and industrial negligence, we have finally gotten a lot better at managing the old pollutants like trash, sewage and PCBs. This has been a major achievement and the environment is much better for it. The second is that the growing human population is making growing amounts of new pollutants like phosphorus, carbon dioxide, and mercury, and right now we are not very good at managing them at all.

Though books rarely mention it, neither the old industrial Adirondacks of 1880 nor the happy cottage-and-camp Adirondacks of 1950 were particularly clean. The old Adirondacks generated large amounts of human and animal wastes, thought sewage treatment effete, and placed no restrictions on the emissions of spectacularly dirty industries like blast furnaces and tanneries. The busy Adirondacks of 1950 still barely treated most municipal sewage, discharged paper mill wastes to most main rivers, and were filled with cars, trucks, and motorboats whose exhaust, even by today's inadequate standards, was a lead-hydrocarbon soup.

As the maps in this chapter show, much progress has been made in dealing with old pollutants. Much waste is now recycled, and much of the rest is put in sealed landfills. All industrial discharges are registered and regulated, and the disposal of hazardous wastes is, in theory, strictly controlled. All sewage discharged to surface waters gets secondary treatment, and in a number of Adirondack watersheds there are few if any surface discharges to treat.

While this is promising, it does not mean that the old pollutants are gone. We still have PCBs in the sediments of Lake Champlain and the Hudson River, unremediated hazardous waste dumps throughout the state, and toxic discharges in all the old industrial cities. We have phosphorus and nitrogen, which are not removed by normal sewage treatment, accumulating in many lakes. And we have, of course, the pollutants from millions of internal combustion engines. Some of these engines, like those in the new hybrid vehicles, are remarkably clean. Others, like those in SUVs, many trucks, and the tens of thousands of ATVs, snowmobiles, and watercraft with two-stroke engines, are hardly better than their 1950 ancestors.

The old pollutants were, to a large extent, obvious, local, and avoidable. The new ones are more subtle, more widespread, and more deeply rooted in our way of life than the old ones. The most important new ones are *mercury*, produced by burning coal and trash; *sulfuric* and *nitric acids,* produced by burning coal and oil; and the greenhouse gases, particularly *carbon dioxide* and *methane*, produced from fossil fuels and by animal agriculture.

The struggle to control the new pollutants will be one of the great environmental challenges of this century. They are formidable adversaries. Because they act in small concentrations and often on ecosystems rather than individual organisms, they are hard to trace and their effects hard to isolate. Because they are generated by the burning of fossil fuels, they are ubiquitous. And because we are energy addicts and fossil fuels are our dominant way of obtaining energy, they will be with us for a long time to come.

We deal with the atmospheric acids and mercury in this chapter. We have not mapped greenhouse gas emissions, not because they are unimportant but because the data for a regional map do not exist. As a substitute, we provide a small discouraging graph, showing that, despite everything that has been learned about the political and environmental costs of fossil fuels, in the last eighteen years there has been no change in our dependence on them at all.

SOURCES OF ENERGY USED IN NEW YORK STATE

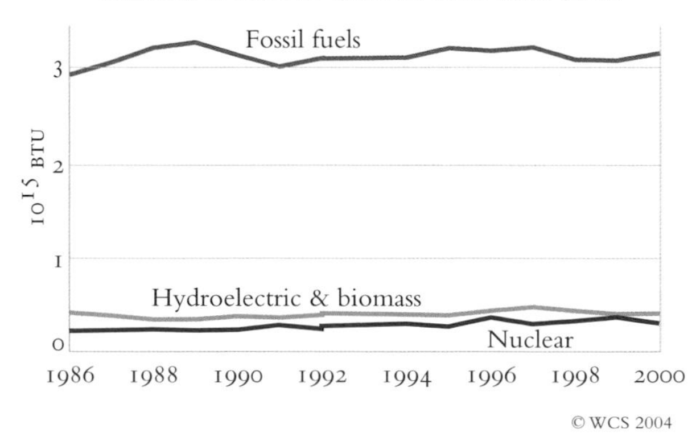

For many years, most New York trash went into unlined landfills, and many of the toxics from these landfills went straight into the local groundwater. In the late 1980s the Environmental Protection Agency prohibited unlined landfills, and New York State began replacing them with incinerators, recycling sheds, and new regional landfills with liners and caps and systems to collect and treat the wastes that leaked from them.

By 2000 all the old landfills had been closed, and 42% of the New York waste stream was being recycled. Thirty-two percent of the wastes went into lined landfills, 12% was burned, and 14%, mostly from New York City, was exported to other states. Sixty-one Adirondacks landfills had been closed, and the Adirondack counties were now recycling a meritorious 300,000 tons of waste each year. Thrifty Warren and Washington Counties, tied for second in the state, recycled 63% of their wastes; tourist-filled Hamilton County only 16%, and apathetic Franklin County, last in the state, less than 3%.

Recycling in New York is entirely voluntary, which sounds nice until we remember that governments and businesses rarely volunteer for things. Most of the of the wastes that are now recycled are recycled because individual households take the time to sort their trash. Relatively few businesses, fewer schools and universities, and almost no government offices recycle at all. Because they don't recycle their trash they have, unlike their citizens, no incentive to regulate how much of it they make or what it does and does not contain.

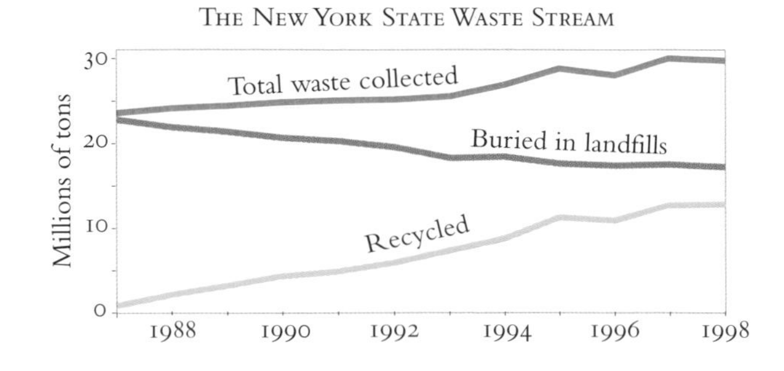

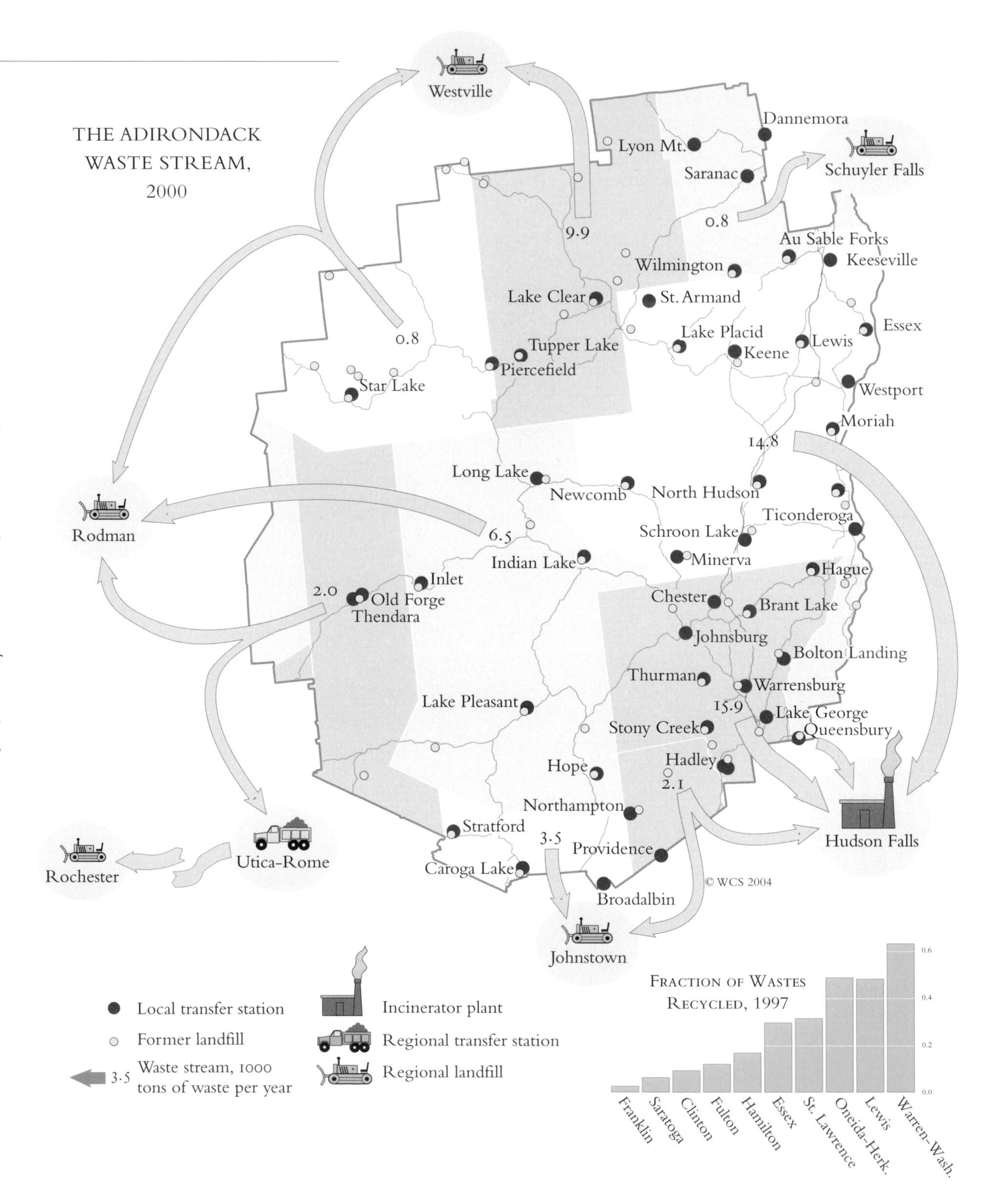

THIS map has both good and bad news. The good news is that the waters of the park are probably cleaner than they have been for a hundred and fifty years. The old tanneries and pulp mills and straight pipes from the town sewers to the river are gone. Every town and industry that discharges to surface waters now treats its wastes, and most have relatively little to treat. Several important watersheds have no discharges at all, and a number of others have only small ones.

The bad news is that while the rivers have been getting better, Lake Champlain has been getting worse. It currently receives about 86 tons of phosphorus from sewage per year, in addition to 400 tons from other sources. Three-quarters of the total phosphorus comes from Vermont and a quarter from New York. Some, but not enough, flushes out. The rest remains in the lake, enriching the water and sediments and changing their biology. In consequence, while nowhere near as altered as, say, Lake Oneida, Lake Champlain is now a very different lake than it was fifty years ago: greener, more opaque, and more fertile smelling. New fertility-dependent weeds abound: the southern bays are filled with water chestnut; the northern ones are being invaded by alien species of floating heart and water plantain; and Eurasian milfoil and zebra mussels, those master exploiters of surplus nutrients, are now throughout.

SOURCES OF PHOSPHORUS REACHING LAKE CHAMPLAIN

VERMONT, 75.8%
Agriculture, 44.6%
Sewage, 11.7%
Forests, 9.1%
Urban, 10.4%

NEW YORK, 24.2%
Sewage, 9.9%
Agriculture, 7.7%
Forests, 3.4%
Urban, 3.2%

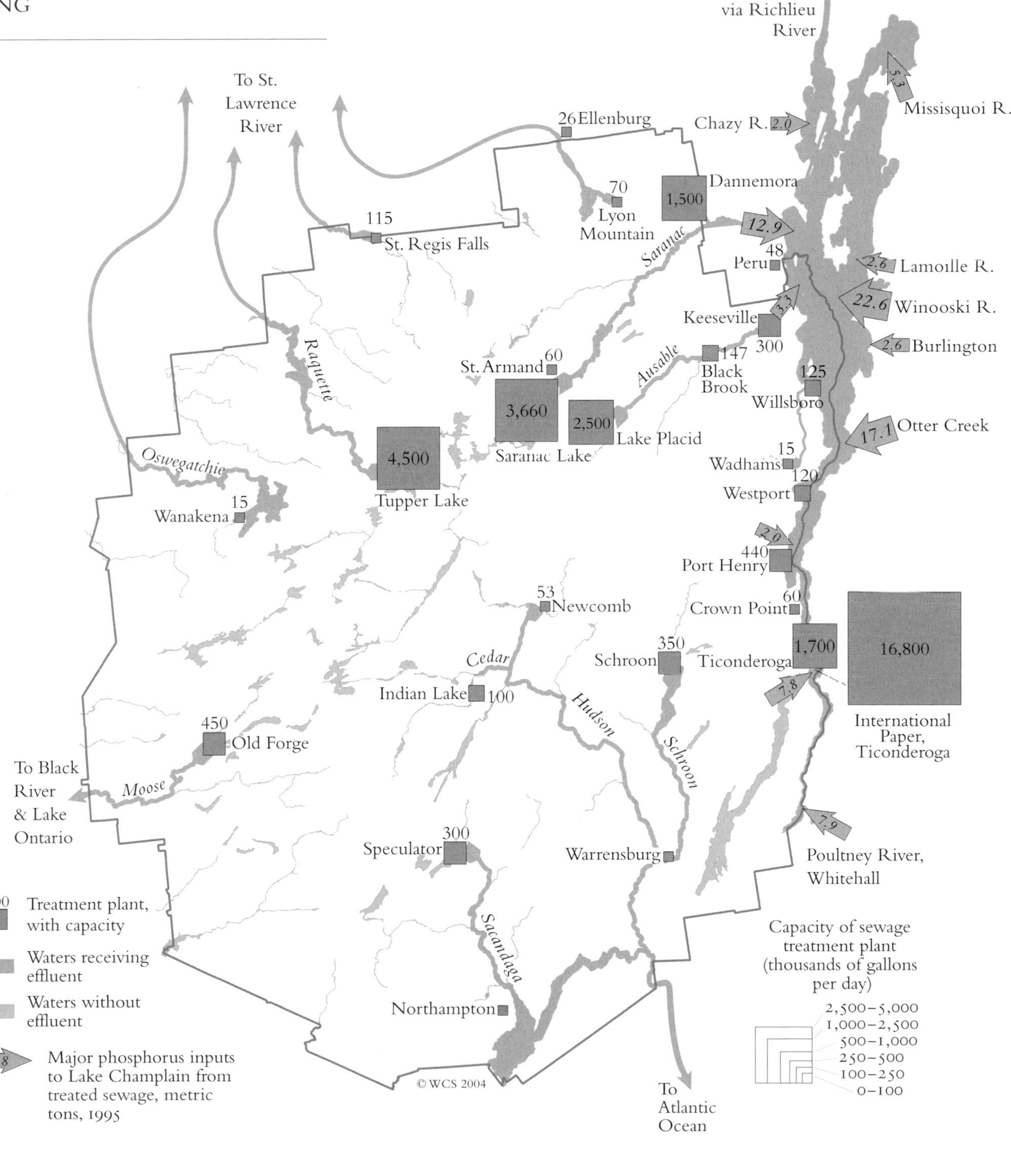

HERE is a snapshot of the toxic wastes that private industries produced in New York in 1995. It shows both how many toxic substances are still produced by industry and how far we have come from the old open-the-valve-and-off-she-goes days of forty years ago.

Note two things about these maps. First, because they are based on self-reporting rather than independent monitoring, they may underrepresent actual releases. And second that, as is too often the case, government has exempted itself from the reporting requirements it imposes on others. If a private industry releases chlorine or phosphorus, the releases will be in a public database. But the releases of chlorine from the city wastewater plant and phosphorus from in the munitions used at Fort Drum will not be. "Who oversees the overseers?" a Roman asked.

With these reservations, the maps show that in 1955, New York industries produced 383 million pounds of toxic wastes. They disposed of or treated about half of this in landfills and wastewater plants, recycled about a third, and released the remainder, about 10%, to the environment. The commonest materials released to water were metals, sulfuric acid, and n-butyl alcohol, an industrial solvent used in the manufacture of everything from brake fluid to antibiotics. The commonest materials released to air were acids, organic solvents, and, disturbingly, nickel and some other heavy metals.

The fate of these toxics after they are released is largely unknown. Some drift away, some accumulate in lakes and soils, some probably enter humans and animals. What harm, if any, they then do, we do not know. We regulate toxics because they are dangerous, but just how great the danger really is and whether our regulating is preventing it, no one seems able to say.

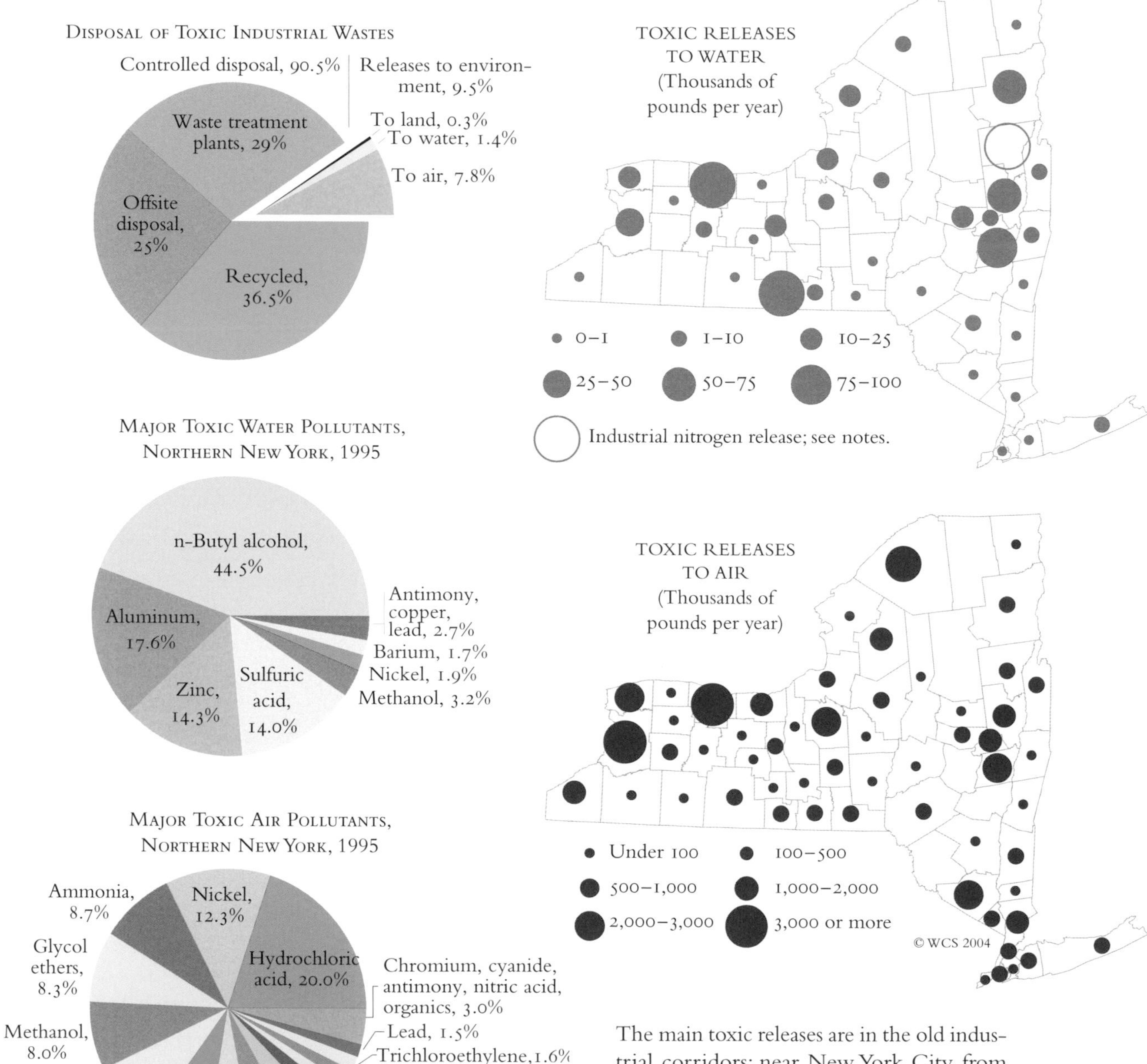

The main toxic releases are in the old industrial corridors: near New York City, from Albany to Glens Falls, and in the western canal towns from Syracuse to Buffalo. Many are in the Adirondack airshed and could easily affect the park.

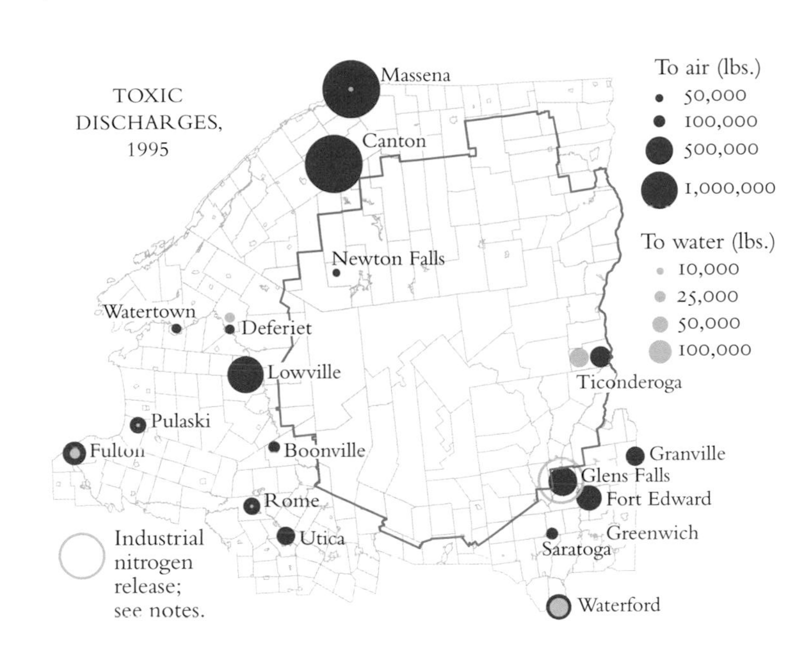

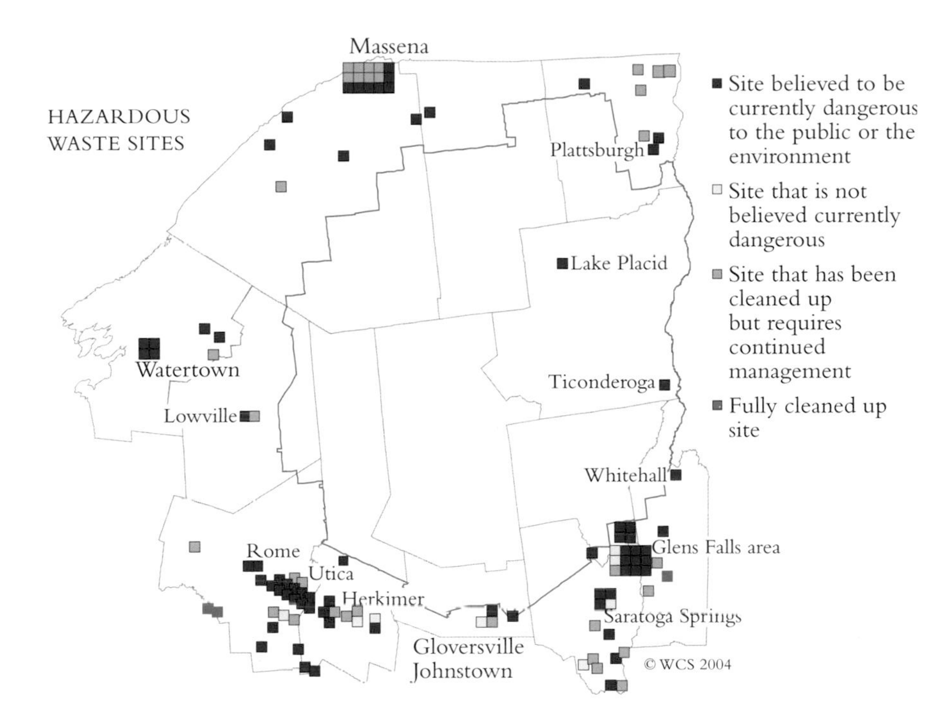

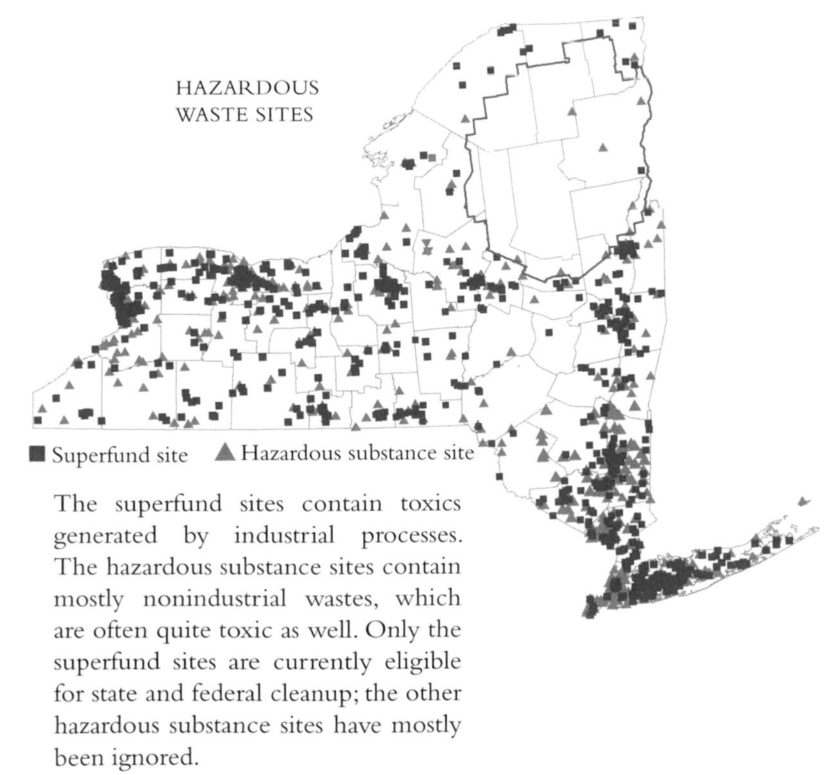

The superfund sites contain toxics generated by industrial processes. The hazardous substance sites contain mostly nonindustrial wastes, which are often quite toxic as well. Only the superfund sites are currently eligible for state and federal cleanup; the other hazardous substance sites have mostly been ignored.

CURRENTLY there are relatively few heavy industries and so relatively few toxic discharges in northern New York. The largest air discharges are from aluminum plants in Massena and a glassware manufacturer in Canton. The main water discharges are from paper mills in Ticonderoga and Glens Falls and from a silicone plant in Waterford.

Former industries, both dirtier and less regulated than now, produced large amounts of toxic waste. Some of these wastes are now gone. Others, dumped into lakes or sent to local landfills or just poured on the ground behind the plant, are still very much with us.

Because many of the wastes that persist in soils are quite dangerous, both the federal and state governments have *superfund* programs for cleaning up sites contaminated by toxic wastes.

Unfortunately, there are many, many contaminated sites, and cleaning them up turns out to be very difficult. Of the 1,024 New York hazardous waste sites known when the superfund program began in 1979, by 2001 only 16% were fully cleaned up and closed and another 22% partly cleaned up and being monitored. In northern New York, of the 106 potentially dangerous sites identified in Regions 5 & 6, by 1998 only three were fully cleaned up and closed and another thirty-two (30%) were closed but still being monitored.

ACID precipitation is rain or snow that contains sulfuric and nitric acids. The acids are produced by the hydration of sulfur and nitrogen oxides, which in turn are produced during combustion. The sulfur ultimately comes from the sulfur in fossil fuels, the nitrogen from the nitrogen in the atmosphere.

Acid precipitation began with the first large-scale use of fossil fuels in the 1800s. It intensified greatly in the 1950s and 1960s and then began to abate after the passage of the Clean Air Act in 1970.

The atmospheric acids affect both animals and plants. They damage green plants directly by depleting calcium from their membranes and making them less resistant to cold, disease, and other stresses. They alter forest soils by mobilizing aluminum, which is toxic to both plants and animals, and by depleting the reserves of the *base cations*, particularly calcium and magnesium, that are needed for plant growth. In small amounts they decrease fish and invertebrate diversity in ponds and streams; in larger amounts they kill fish and invertebrates directly.

Because forests can absorb nitrate and because early analyses of acidified lakes found more sulfates than nitrates, the federal Clean Air Act regulations have thus far concentrated on sulfur. Sulfur emissions have fallen 38% since 1978; nitrogen oxide emissions are largely unchanged and now exceed sulfur emissions.

Recent studies have shown that nitrogen emissions are more important than had been thought and that decreases in sulfur emissions will not be enough, by themselves, to control the effects of acid rain. The ability of forests to absorb nitrates turns out to be limited, especially when they are already acidified by sulfur. Nitrogen-loaded forests leak nitrates in snowmelt, and this leakage—the sawtooth nitrate peaks in the Big Moose graph—create highly acid *episodes* that can be even more damaging than the chronic acidification caused by sulfates.

PRECIPITATION CHEMISTRY, HUNTINGTON FOREST, 1978-1997

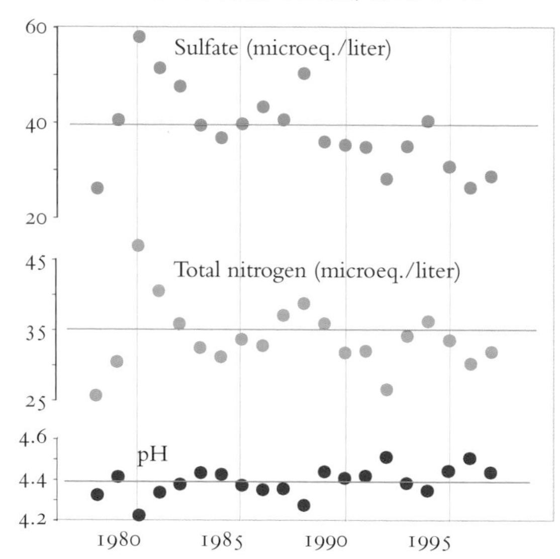

Sulfate concentrations in northeastern rain and snow reached a peak in the 1960s and have declined steadily as the emission controls required by the 1970 and 1990 Clean Air Act Amendments took effect. Nitrogen concentrations in the same period have been constant or shown slight increases; average pHs, which are largely controlled by sulfate concentrations, have risen slightly.

The chemistry of acidified streams and lakes has followed these changes closely, showing that precipitation chemistry is in fact controlling surface water chemistry. Sulfate concentrations have fallen and pHs risen; nitrates have largely stayed the same. Acid neutralizing capacities, a measure of the water's resistance to acidification, has shown at most a slight recovery thus far, and base cation concentrations, a measure of the reserve of mineral nutrients in soils, are steady or show continuing declines.

LAKE WATER CHEMISTRY, BIG MOOSE LAKE, 1992-2002

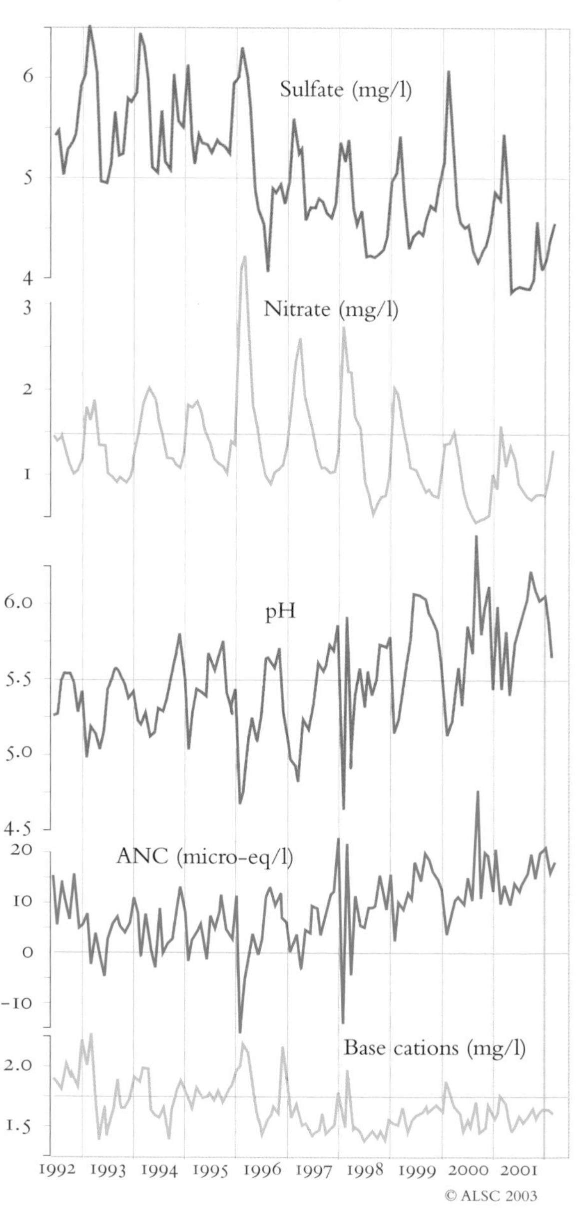

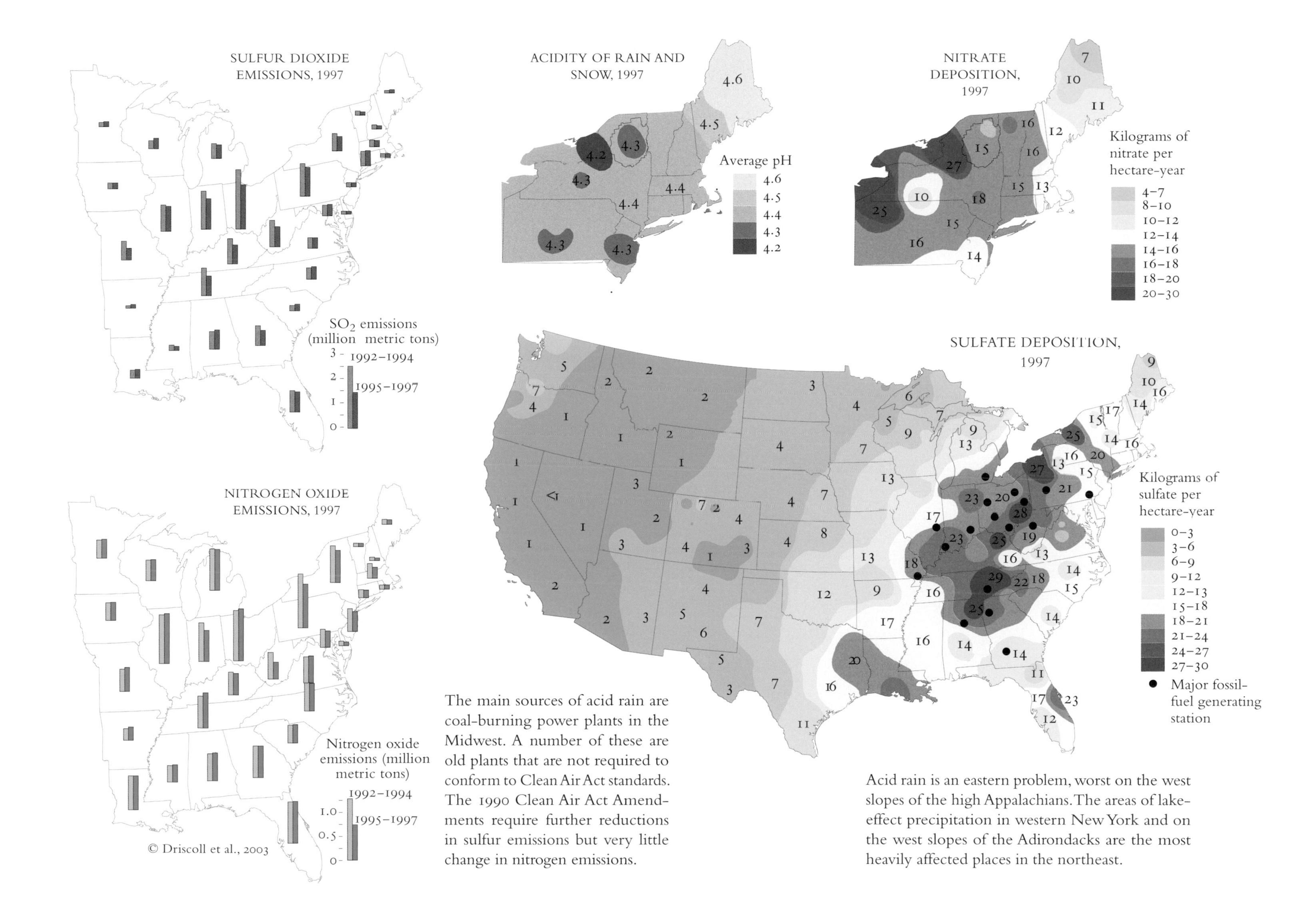

The main sources of acid rain are coal-burning power plants in the Midwest. A number of these are old plants that are not required to conform to Clean Air Act standards. The 1990 Clean Air Act Amendments require further reductions in sulfur emissions but very little change in nitrogen emissions.

Acid rain is an eastern problem, worst on the west slopes of the high Appalachians. The areas of lake-effect precipitation in western New York and on the west slopes of the Adirondacks are the most heavily affected places in the northeast.

ROCKS and soils contain bases that can neutralize the acids in acid rain. If the bases are carbonates and bicarbonates, the reaction is a true neutralization. If, as is more commonly the case, the bases are base cations like calcium or magnesium, the reaction is an *exchange reaction* in which the hydrogen binds to the soil and the base cation is released.

Either way, as the water running out of the soil becomes less acid the soil itself becomes more acid. If the soil has large pools of bases, like the soils derived from the Paleozoic rocks shown in blue on the map, this can go on indefinitely. But if it has only limited pools of bases, like the soils derived from the granitic rocks (1) shown in cream, the exchange reactions can deplete the base pools rapidly.

The map shows the approximate neutralizing capacities of the different Adirondack rocks. The best neutralizers within the park are marbles and dolomites (4), followed by anorthosites, amphibolites, and some metasediments (3), followed by a variety of granitics (1,2) and glacial sands and gravels (G). But bear in mind that Adirondack soils are not necessarily derived from the rocks below them, and that a thick glacial soil can have an excellent neutralizing capacity even when the rocks under it have very little.

In watersheds with a high neutralizing capacity, like those of the northeast Adirondacks, acid rain has had little effect on either the soils or the surface waters. The soils still are fertile, plant growth is good, and almost no lakes have been acidified. In watersheds with a low neutralizing capacity—as in the west Adirondacks where acid deposition is most intense—the story is different: there acid rain has depleted the base reserves of the soils and acidified lakes. The trees on the depleted soils seem to be reproducing poorly (*Map 17-5*); acid-related toxics, particularly aluminum and mercury, are moving through the acidified watersheds; and both fish and fish-eating animals are accumulating these toxics and suffering from them (*Map 17-6*).

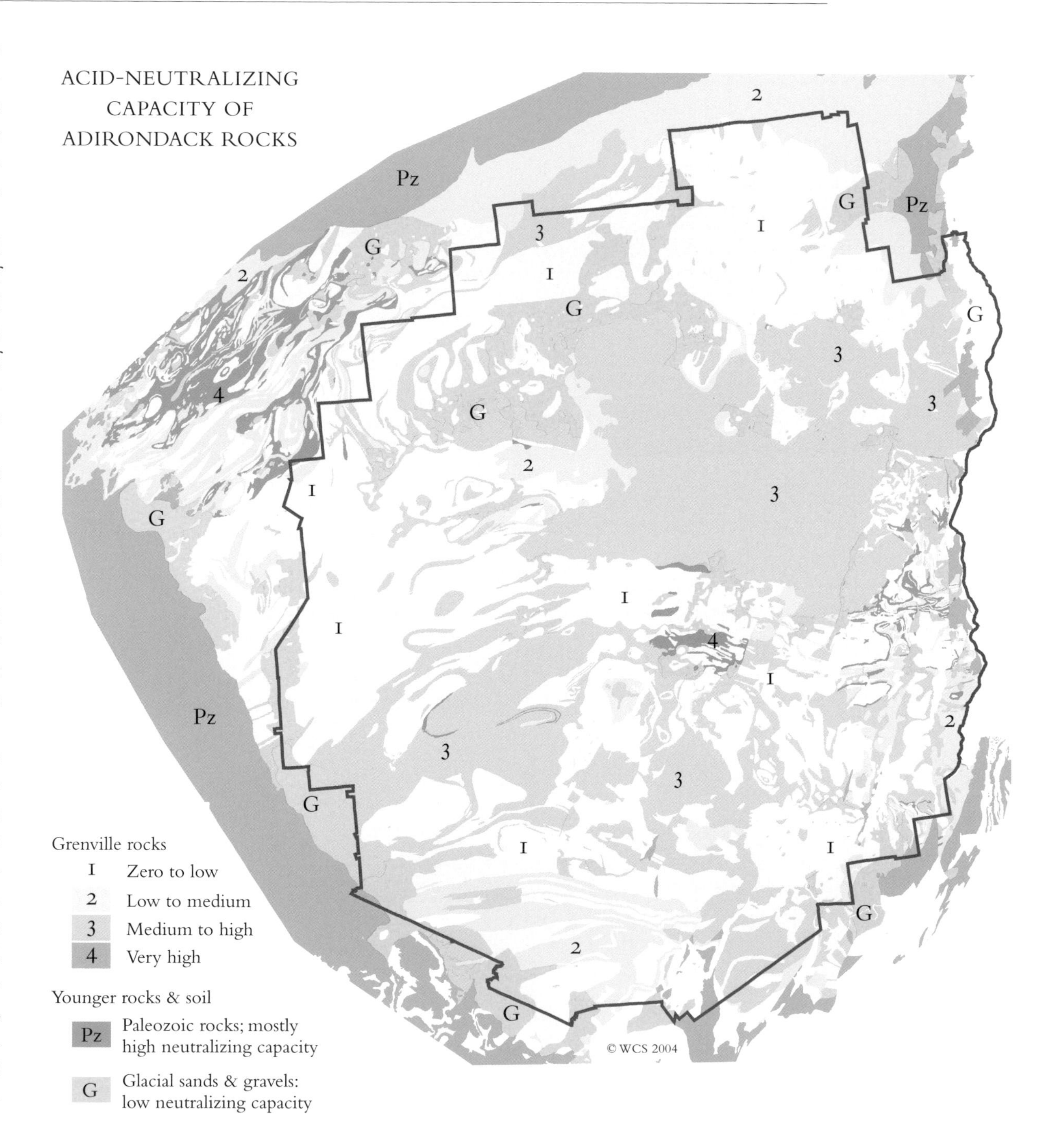

The Adirondack Lakes Survey, a large-scale study of lake chemistry and biology that began in the 1980s, proved that there was a consistent relation between acid rain, lake acidity and fish. They found that about a third of the 1,469 lakes they sampled had pHs below 5.6 and that while some of these lakes were naturally acid many others had been acidified by acid rain. The most acid lakes were found in the western part of the park where acid rain was most intense and the neutralizing capacity of the soils the lowest. In these lakes the numbers of fish, plant, and invertebrate species were always lower than in unacidified lakes.

Subsequent work, and especially several studies that reconstructed the chemical histories of lakes from the types of diatoms found in sediment cores, has confirmed all these conclusions. The best current estimates are that the western Adirondacks have more recently-acidified lakes than anywhere in the United States. In the last hundred years the pHs of almost all Adirondack lakes over four acres in size and with current pHs under 6.5 have decreased, and the number of highly acid lakes, with pHs under 5.6, has doubled.

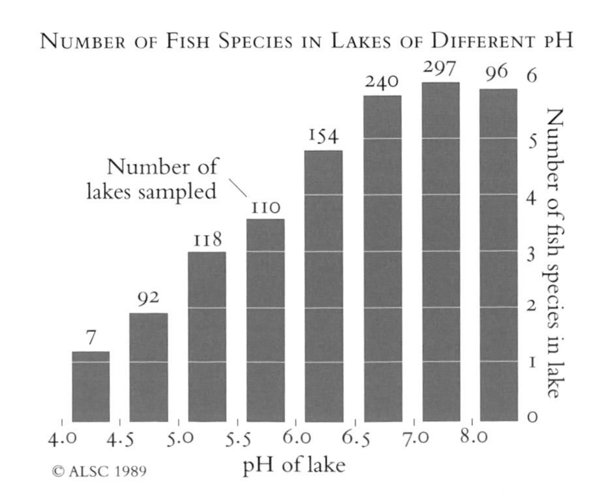

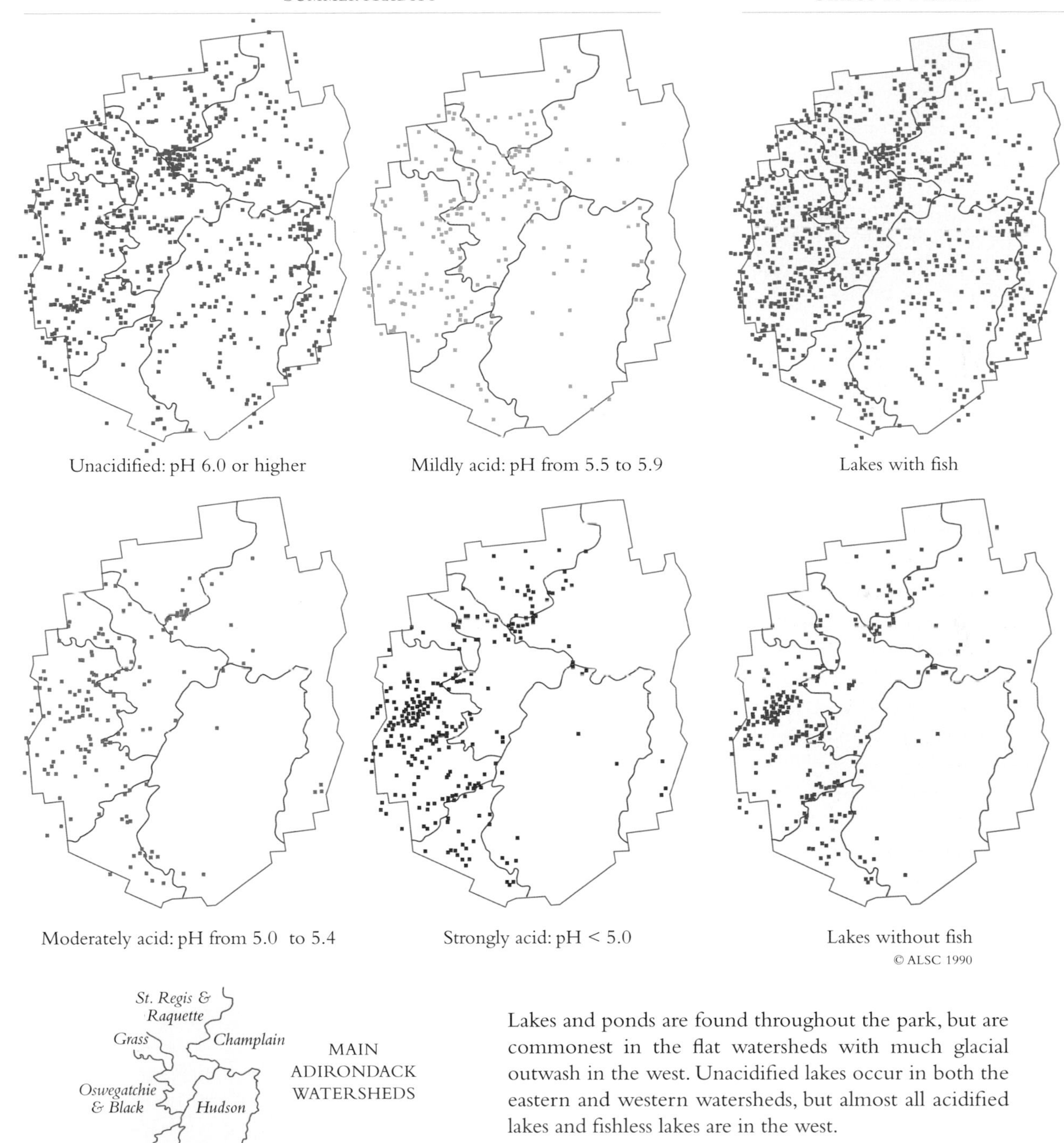

Lakes and ponds are found throughout the park, but are commonest in the flat watersheds with much glacial outwash in the west. Unacidified lakes occur in both the eastern and western watersheds, but almost all acidified lakes and fishless lakes are in the west.

SUGAR maple, one of the commonest forest trees at middle elevations, is no longer reproducing successfully in much of the western Adirondacks.

Healthy maple stands, like those in the northeastern Adirondacks, reproduce through a *seedling bank*. Maple seeds germinate on the forest floor. The seedlings, which are highly shade-tolerant, persist for many years, growing slowly and waiting for the light from a canopy opening. The layer of preestablished seedlings, which is often quite dense, allows maple forests to regenerate rapidly after storms and harvests.

Most maple stands in the western Adirondacks, do not seem to be producing seedling banks. Their understories are dominated by young beeches. Almost no maple seedlings appear in test plots, and almost no saplings under forty years old occur in the understory. Tree-ring counts suggest that sixty or seventy years ago many of these stands were reproducing normally; then something changed, and by forty years ago many of them were not reproducing at all.

The timing and geography of this failure, added to what we know of maple biology, suggest that soil depletion, caused or exacerbated by acid rain, is a likely cause. Maples require moderate amounts of soil calcium; trees on calcium-deficient soils are less vigorous, less healthy, less shade-tolerant, and less capable of producing flowers and seeds. Acid rain depletes soil calcium and other bases (*Maps 17-3,4*). Maples in parts of the eastern Adirondacks where acid deposition is relatively lower and the acid-neutralizing capacity higher are still reproducing well and producing dense seedling banks. Maples in the western Adirondacks, where the lakes are most acidified and the soils poorest in calcium and other bases, stopped reproducing just about the time acid rain reached its peak intensity and have not reproduced since.

BEECH TAKES OVER THE FOREST UNDERSTORY

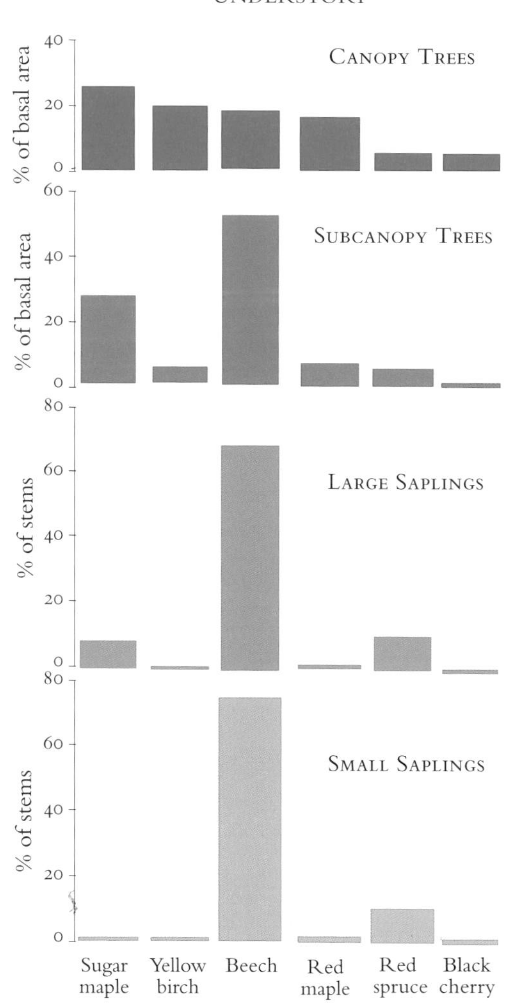

Sugar maple forests in the western Adirondacks tend to be dominated by a mixture of sugar and red maples, yellow birch, and beech. Beech is dominant among the smaller canopy trees and almost the only species occurring in the understory.

MAPLE REGENERATION FAILS

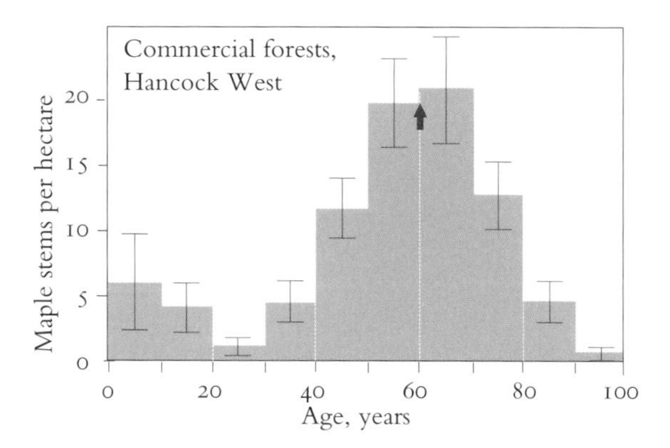

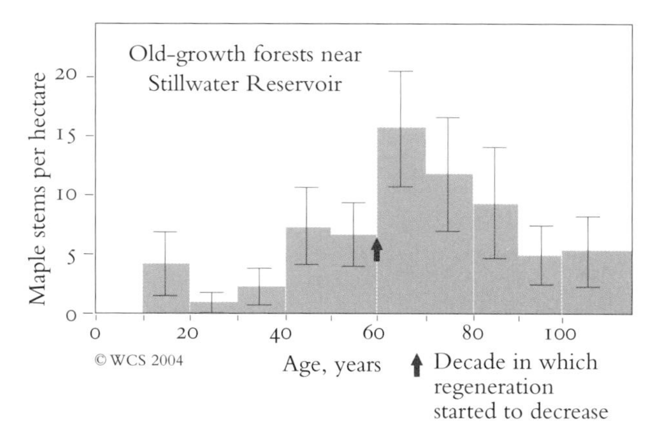

The bars show the numbers of sugar maples of different ages in two west Adirondack forests. Trees less than a foot high or more than eight inches in diameter were not sampled, and thus some of the trees younger than five years or older than about eighty years are missing from the curves. In a normally reproducing forest there are more young trees than old trees; in these forests, where regeneration is failing, there are fewer twenty- and forty-year-old trees than sixty-year-old trees.

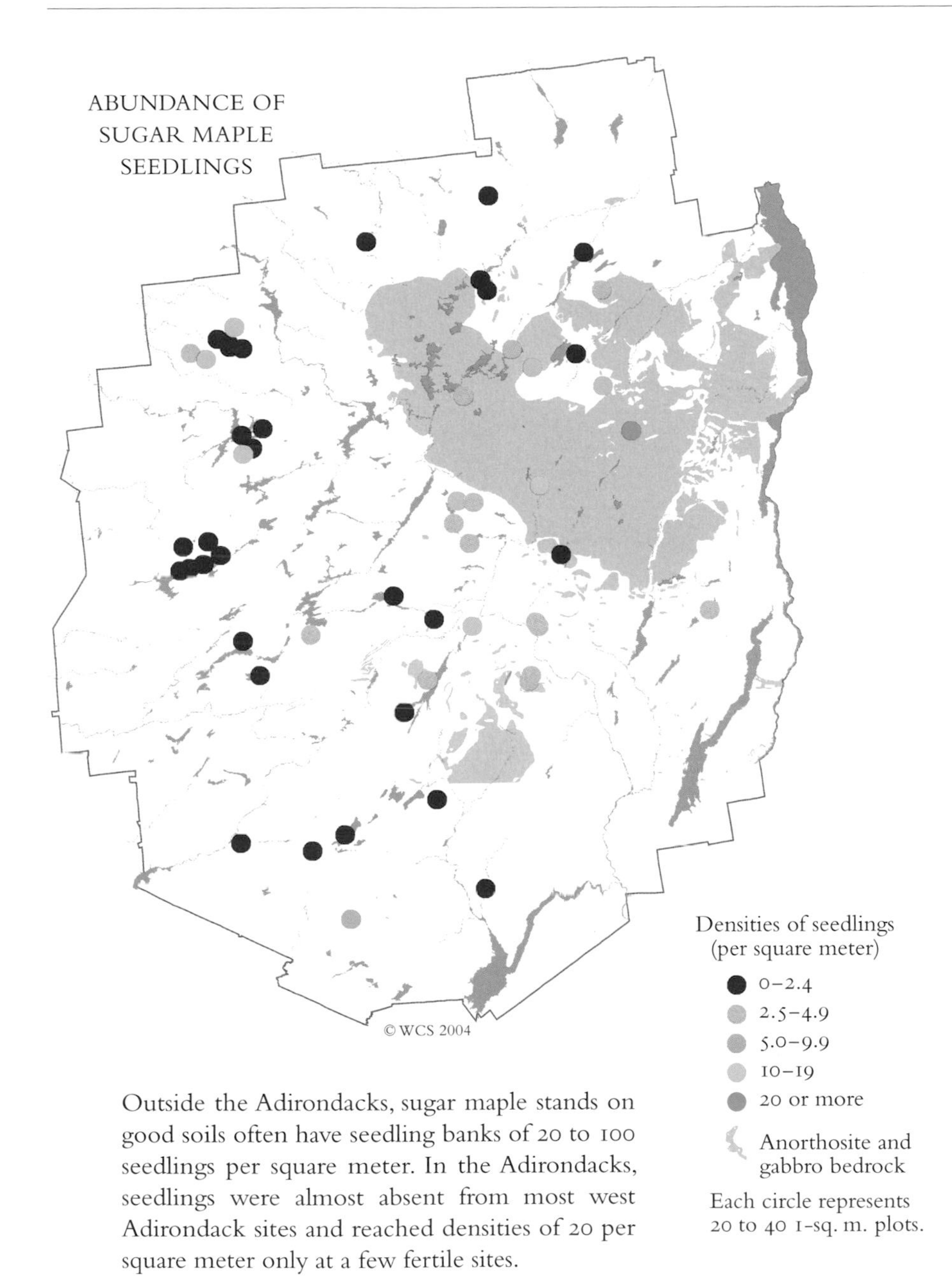

Outside the Adirondacks, sugar maple stands on good soils often have seedling banks of 20 to 100 seedlings per square meter. In the Adirondacks, seedlings were almost absent from most west Adirondack sites and reached densities of 20 per square meter only at a few fertile sites.

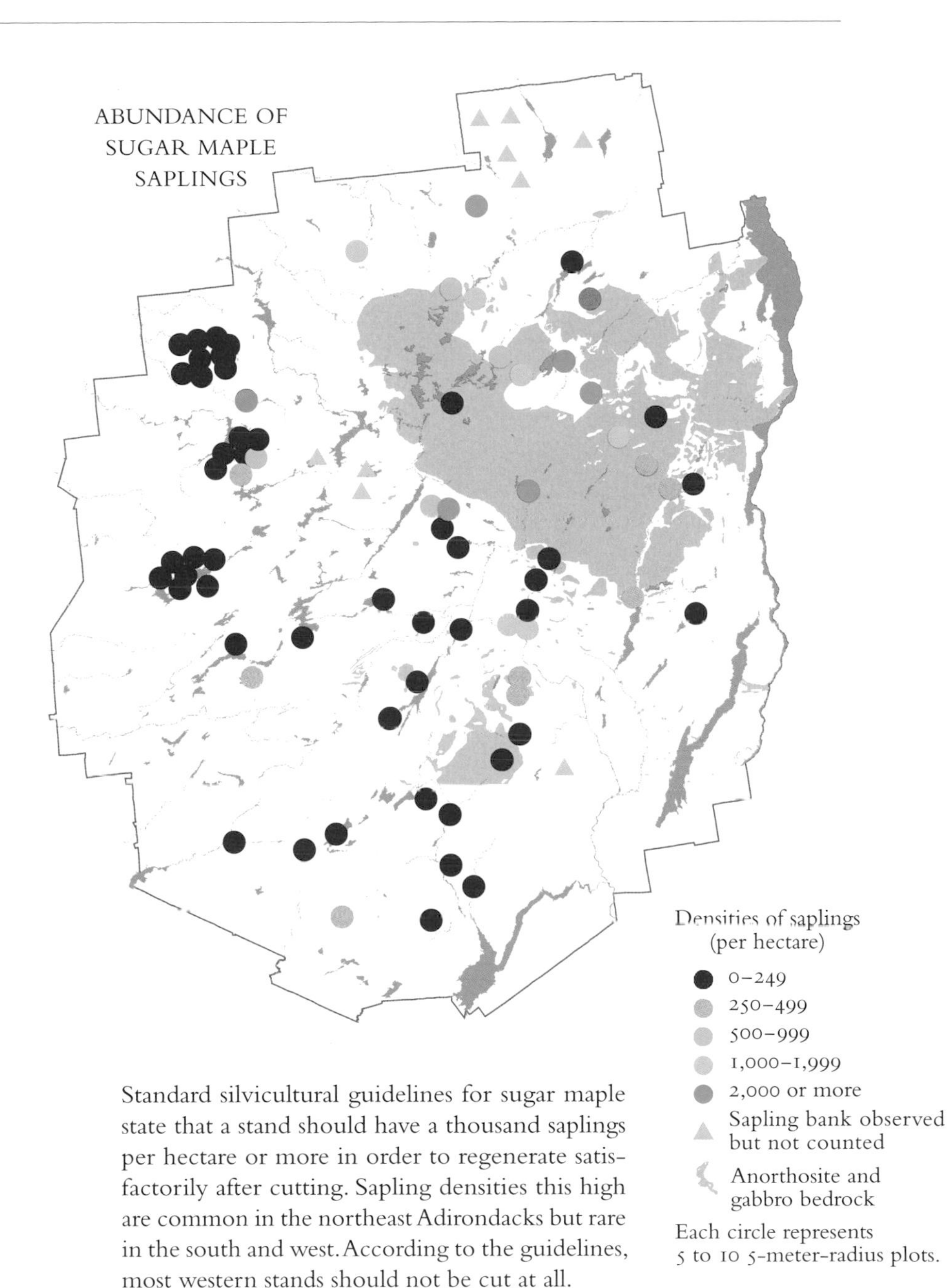

Standard silvicultural guidelines for sugar maple state that a stand should have a thousand saplings per hectare or more in order to regenerate satisfactorily after cutting. Sapling densities this high are common in the northeast Adirondacks but rare in the south and west. According to the guidelines, most western stands should not be cut at all.

METHYLMERCURY, an atom of mercury bound to a carbon atom and three hydrogen atoms, is one of the most toxic substances known in nature. Take a gram of mercury, the amount in one clinical thermometer, allow the bacteria in a wetland to convert it to methylmercury, and the result is more than enough methylmercury to contaminate a twenty-five-acre lake. Concentrate that gram of methylmercury in living tissue and it could make 2,000 fish unsafe for human consumption or decrease the reproductive success of thirty pairs of loons.

Methylmercury is an unusual poison because it is both physiologically and ecologically dangerous. Because it can pass through the blood-brain barrier and affect the brain directly, very tiny amounts can affect behavior or damage the nervous system, especially in young animals. Because it can bind to proteins it accumulates in tissue and can move up the food chain from prey to predators. The concentrations of methylmercury in fish are typically a million times higher than in lake water; the concentrations in the loons that eat the fish typically ten times higher still.

Traditionally, mercury contamination was thought of as an urban problem, restricted to the areas near mercury-using industries. Then in the 1980s it was discovered that many remote lakes in Ontario and the upper Midwest had high levels of mercury. Mercury, it seemed, was a ubiquitous component in acid rain, and the acids from in the rain were facilitating its movement through watersheds and making it more available to animals.

By the mid 1990s we had learned that the Northeast had serious mercury problems as well. One study found yellow perch exceeding the Canadian limit for safe consumption on fourteen out of sixteen lakes and fish exceeding the U.S. limit in nine of sixteen lakes. Fishermen were advised not to eat more than one meal a month of fish from Minnesota, inland Maine, and many lakes in the Adirondacks. Sadly, for the eagles and otters and loons on these lakes that was not an option.

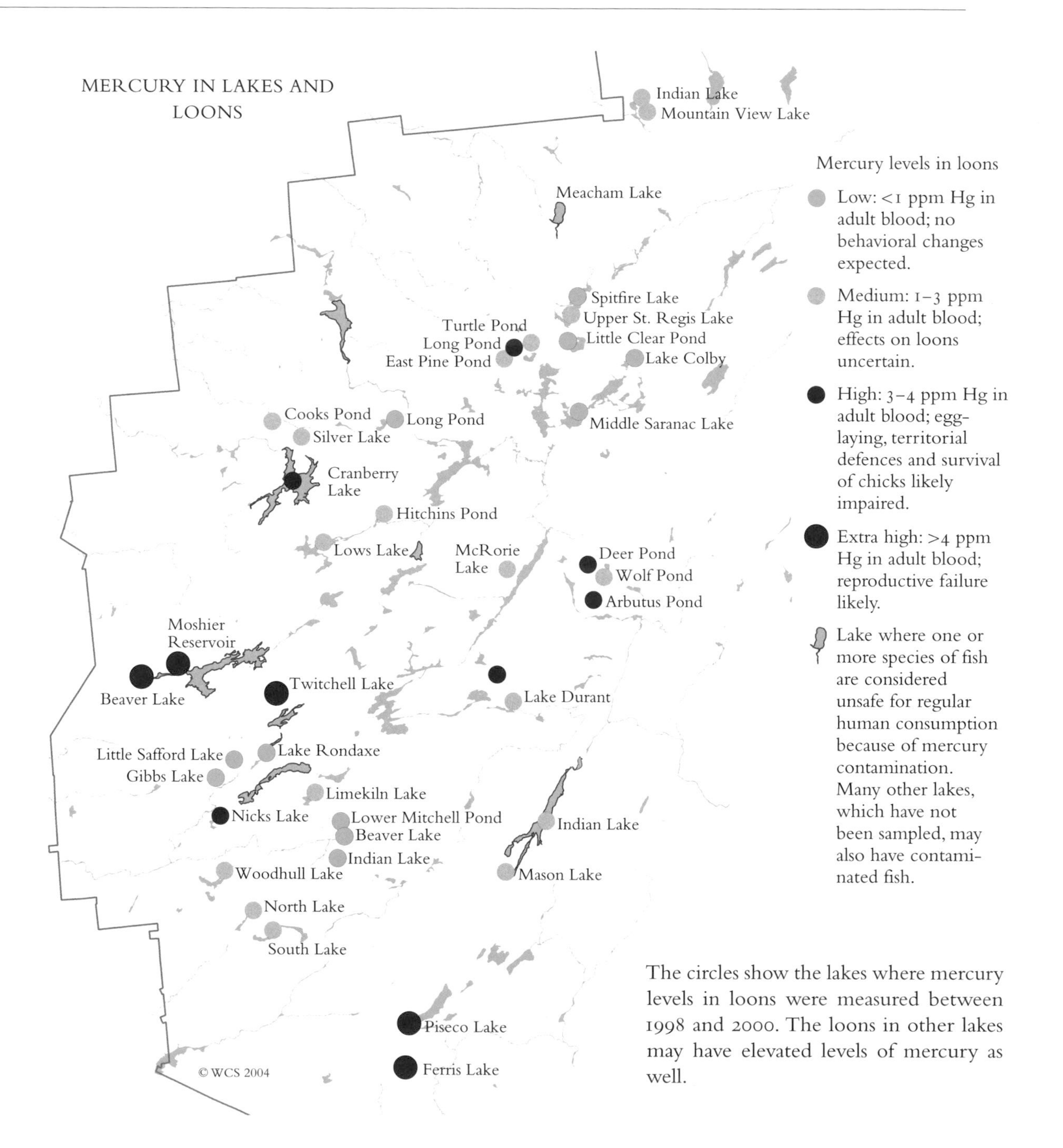

The circles show the lakes where mercury levels in loons were measured between 1998 and 2000. The loons in other lakes may have elevated levels of mercury as well.

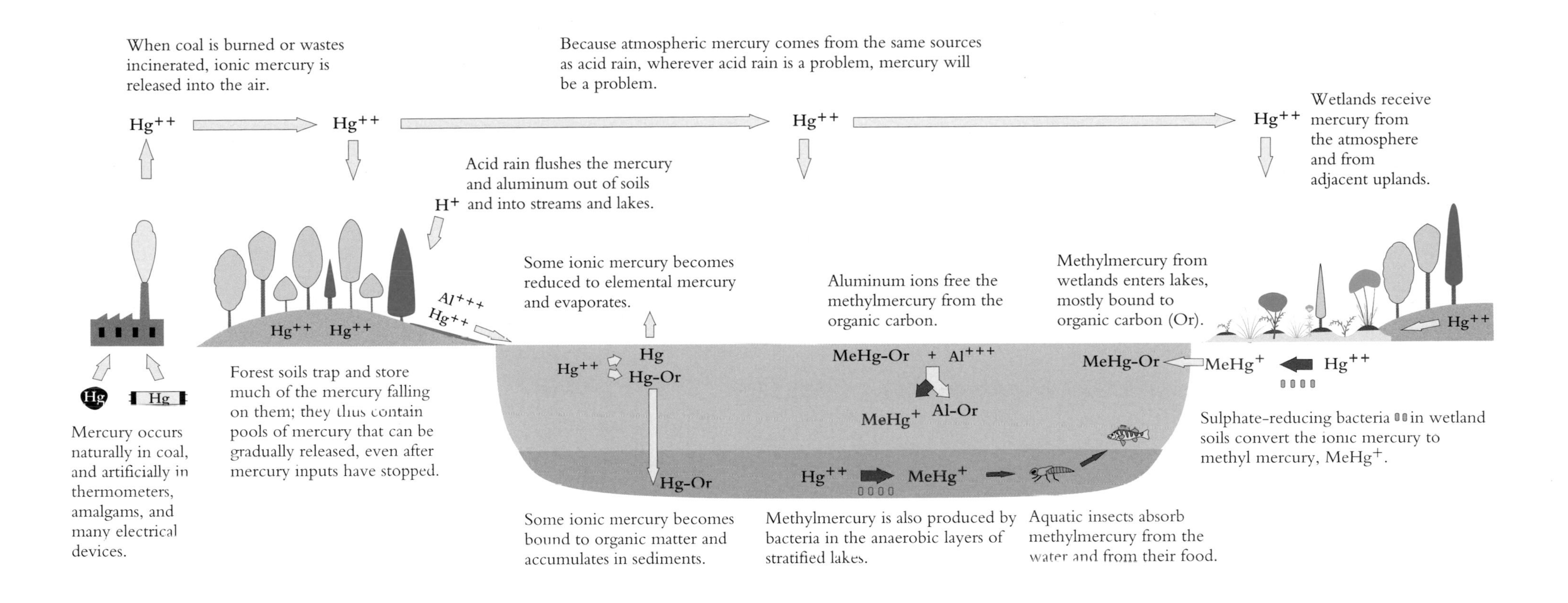

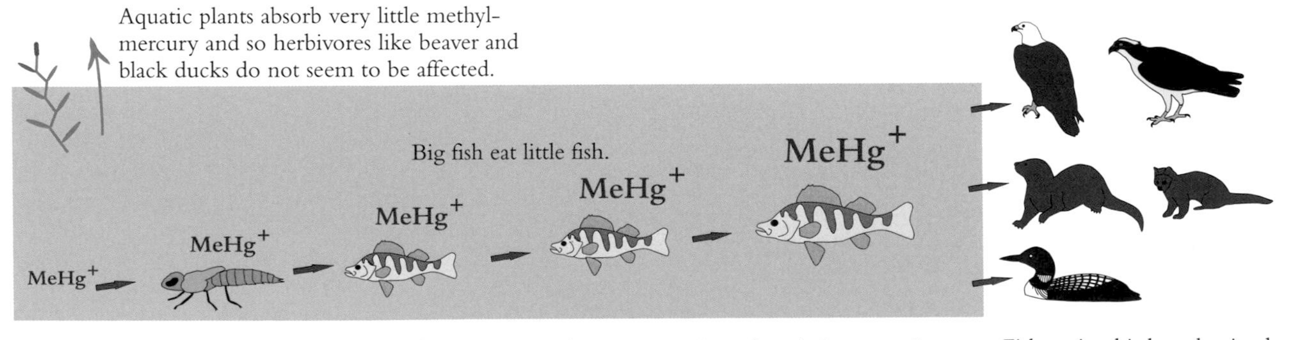

Little fish eat insects and accumulate methylmercury in their tissues. Ninety-five percent of the mercury in fish tissue is methylmercury.

The concentration of methylmercury in large yellow perch may be one million to twenty million times its concentration in the water.

Fish-eating birds and animals accumulate the methyl-mercury in their tissues, typically at concentrations ten times greater than those in their prey.

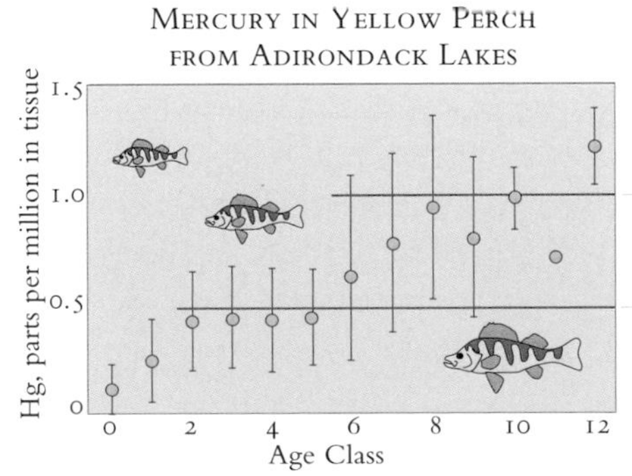

Mercury concentrations in perch increase as they move up the food chain. At about age two, they start eating small fish; at about age eight they start eating smaller perch.

HOW FAST IS THE CLIMATE WARMING?

Quite fast, but thus far almost entirely in the winter. Trend graphs for northern New York, which parallel those for elsewhere in the Northeast, show a gradual warming early in the century, a cooling in mid-century, and a rapid warming from 1970 on. Both the warming and the cooling are believed to be caused by the burning of fossil fuels. The mid-century cooling is believed to have been caused by sulfuric acid, which reflects sunlight. (It also causes acid rain, and the onset of the cooling, both in America and Europe, coincides with the onset of acid rain.) The warming of the last thirty years is part an unprecedented warming of the Northern Hemisphere. It is thought to be the result of increases in carbon dioxide and other greenhouse gases, mostly from fossil fuels, which trap heat that would have otherwise escaped from the earth (*Map 2-7*).

The current warming rates are very significant. If they continue they will give Albany the climate of Washington, D.C. by the year 2100. If the current rate accelerates, as all the climate models predict it will, Albany could have the climate of Washington, D.C. by 2050, and that of Richmond or Atlanta by the end of the century.

A winter warming of 5–10°F will have great effects on the Adirondacks. In the short term, when the Adirondacks are the only place in the state with persistent snow cover, winter sports and winter tourism may prosper. But in the long term there is little prosperity, natural or human, in sight. It is very likely that neither the major northern forest trees nor the boreal forest animals associated with them will survive much more than one forest generation of rapid climate changes (*Maps 4-2,3*). Fifty years from now we may have Adirondack winters without snow and ice and forests that are the biological analogues of the dying coral reefs seen in the tropics today: stressed, structurally altered, not reproducing, and unable to support the birds and animals that once lived in them.

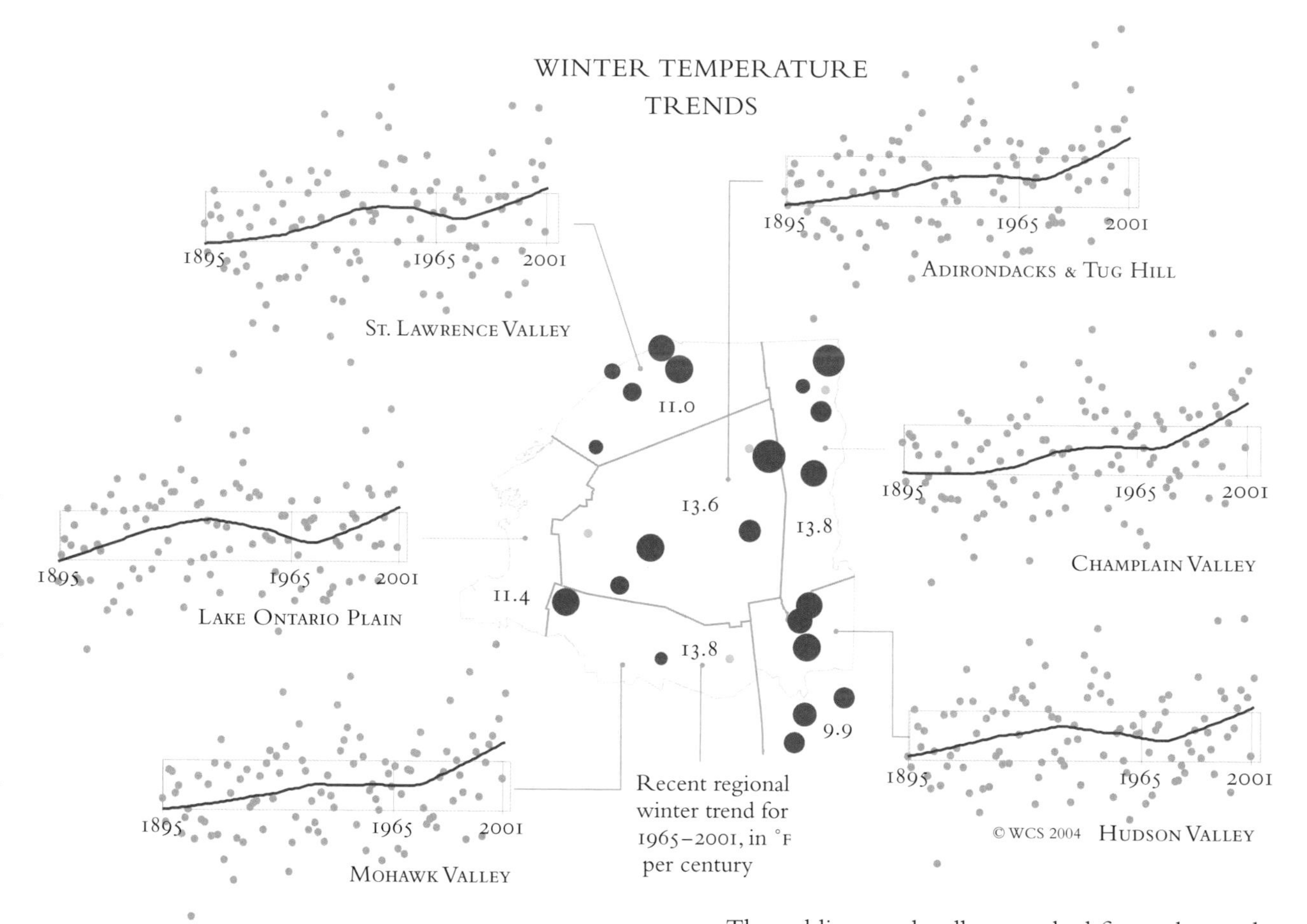

Regional trends in winter (DJF) temperature, 1895-2000

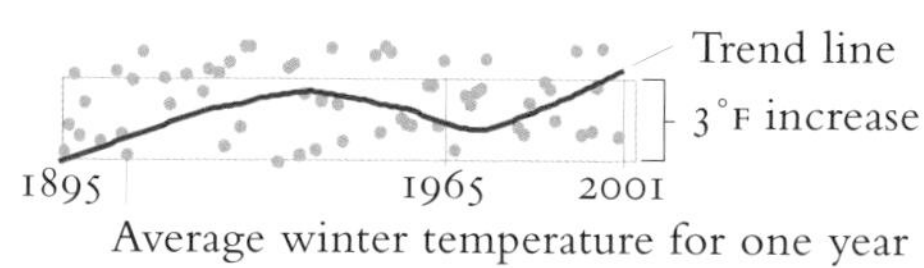

Average winter temperature for one year

Recent (1966–2001) winter temperature trends at individual stations,

- 20°F per century
- 10°F per century
- Trend not significant

The red lines,are locally smoothed fits to the yearly data (the brown dots) and show the trends in winter temperatures for each of the seven climate divisions of northern New York. The red circles give the recent winter temperature trends for 1965–2001, calculated by linear regression at the twenty-one stations for which there are thirty or more years of continuous data and for which the trend is significant at the 10% level. The green circles are four other stations for which the trends were not significant. The numbers give the recent trend for each climate division and were calculated from the average of all the stations in the division, not just those with thirty-year records.

18 Seven Questions about Change

In the intense debates about the management of the park in the 1960s and 1970s, both sides tended to see the park and its future in very simple ways. Adirondack land was either settled or wild, private or public, wilderness or nonwilderness. Conservationists tried to move as many parcels of land as they could into a box marked "public-wilderness," while keeping as many lands as possible from moving into a box marked "private-developed." A future in which the right lands were in the right box was a good future.

These debates accomplished many important things that we now take for granted: a major expansion of the Forest Preserve, a system of wildernesses, a parkwide land-use planning act. But in retrospect they have a curiously static quality. Land was either protected or not; once protected it was deemed safe, and conservationists turned their attention elsewhere.

Land, of course, is not static, and protected lands are not necessarily safe. The park has changed enormously in the last two centuries, and continues to change now. Many people are aware of these changes. Fewer people are aware of how deep and thorough some of them are, or how indifferent they are to our hard-won distinctions between protected and unprotected lands, or how they may combine to produce a park very unlike the one we currently know.

This final chapter is a primer on change. We present a number of graphs showing the trends we think most important to the future of the park. We speculate on what they may mean to us and the Adirondacks and then, at the very end, on some basic questions that they pose.

IS ACID RAIN STILL A PROBLEM?

YES it is. Its effects are subtle, but, as we are learning, cumulative and persistent. The Clean Air Act has decreased sulfate pollution but has done very little to change nitrate pollution (*Map 17-3*). Average lake and stream pHs have increased slightly, but the rain is still very acid, and our streams and lakes are still subject to the highly acid, nitrate-dominated episodes that kill fish. Many forest soils are acidified and nutrient-depleted. They will recover eventually, but recovery may take hundreds of years and may not begin while there are continuing inputs of acid.

All these factors point to continuing or increasing biological problems. Nitrate peaks during storms and spring run-off will continue to kill young fish, and thus many acid-damaged fish populations are unlikely to recover (*Map 17-4*). Nitrate inputs to forests will remain high, further acidifying soils and making trees more vulnerable to pests and diseases. Acidified soils will continue to affect the health of acid-sensitive trees (*Map 17-5*), limiting the ability of forests to cope with climate change. Acids stored in soils and wetlands will continue to leach into lakes for many years. Methylmercury, acid-deposition's most deadly associate, will continue to enter food chains and threaten wildlife and humans (*Map 17-6*).

Each of these effects is bad by itself. Applied in combination to ecosystems that will be increasingly stressed by climate change, there is every likelihood that they will produce substantial, long-continued, and unpredictable change.

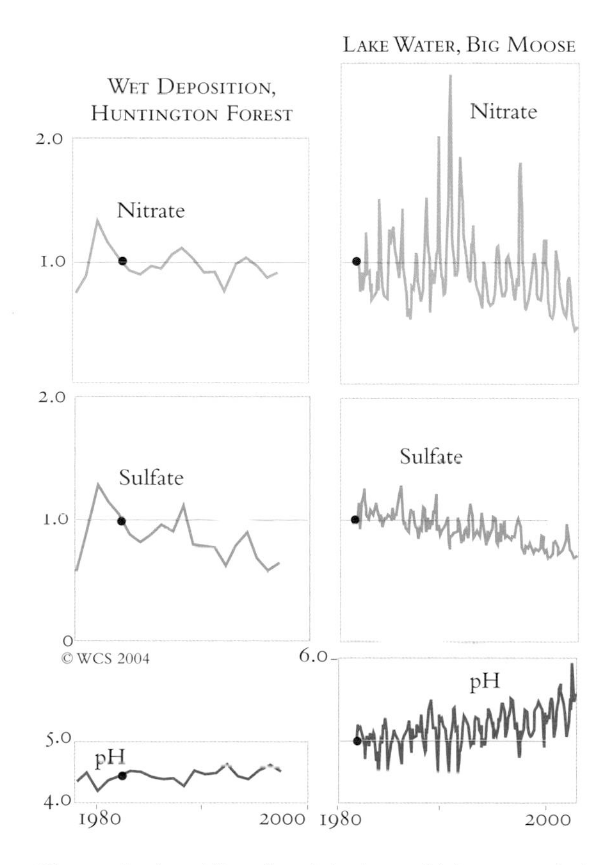

Changes in the acidity of precipitation and lake water relative to 1985. The graphs of sulfate and nitrate are *trend graphs*, in which the values of several different variables are set equal to 1 at some reference year, indicated by the black dots. Trend graphs compare *rates of change* but not absolute values. They tell you that sulfate has decreased more rapidly than nitrate, but not whether sulfate levels are higher or lower than nitrate levels. The pH values, which because they are logarithms already show fractional changes accurately, are given in absolute units. Note that an increase in pH is a decrease in acidity.

FOR the past fifty years remarkably well. All the species for which we have reasonably good information are holding their own or increasing. Many have made impressive recoveries. Beaver, mountain lion, and moose have returned from local extinction; fisher, otter, and bear have recovered from greatly reduced populations. Of the large animals that were formerly widespread in the Adirondacks, only the wolf is currently missing.

The most spectacular increases have been in the deer populations near the edge of the park. Deer populations all over northern New York crashed in the deep-snow winters of 1968 to 1972. Those on the edges of the park recovered by the early 1990s and are now two to eight times more abundant than before the crash (*Map 4-9*).

In small numbers deer are an amenity or a resource. In large numbers they are a problem. They lower forest diversity, disperse the Lyme disease tick, damage crops and gardens, and, perhaps most significantly, prevent forest reproduction. At the edge of the park their current reproductive capacity, augmented by abundant food, mild winters, and the ten-to-one ratio of females to males maintained by hunting, is extremely high. The coyotes and hunters are doing their best, but the graphs suggest that neither are currently able to control the population.

It is tempting to ask whether the return of wolves might solve our deer problems and head off the moose problems that we may soon encounter as well. If ecology were simply a matter of getting the right species lists—restocking the empty cupboard—it might. But geography matters too, and if wolves return they will find a very different geography than the one they left a hundred years ago. The deer are no longer a wilderness species but rather a rural and suburban one. Wolves, so far as we know, are not a good suburban species, and it may not be in either their or our interest to encourage them to become one. And even if they were to penetrate rural areas, it is quite possible that the well-fed and highly fecund deer populations they will find there are nowhere near as easy to control as their wilderness ancestors were.

The return of the wolf would be an event of great biological and emotional significance. But if it happens it is not likely to be a return to the protected center of a wild mountain range. Instead it will be a return to a populated edge where human-wildlife interactions are complex and intense and their outcomes almost completely unpredictable.

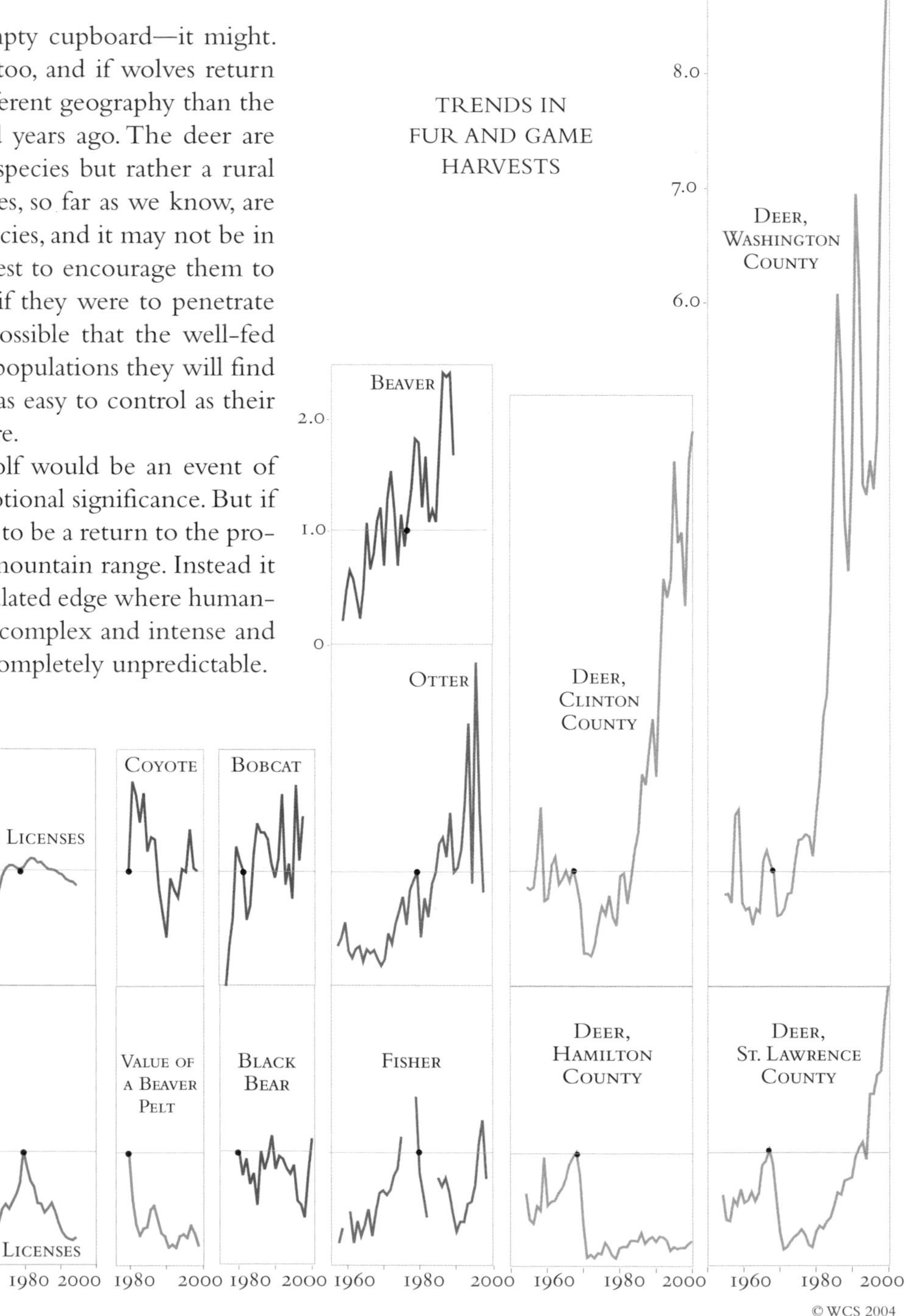

NEW lands are still being added to the Forest Preserve, but both the rate and the pattern are different from what they were in the past.

A hundred years ago, Adirondack land was cheap and available. The state, which was trying to create a large continuous public wilderness, simply bought everything it could. By 1919, twenty-seven years after the park was created, New York owned over half of the original park. By then, however, land was more expensive and, as the park became more settled and ownerships more permanent, it became less available. When it became clear that the Adirondacks would never become a continuous wilderness, the state slowed its purchases and switched its strategy, focusing on extending the wilderness areas that it had rather than creating new ones.

In the 1960s, growing concerns over large-scale land development combined with a growing wilderness movement to create a major acquisition program that added over 280,000 acres of land in the next thirty years. This program, though effective, generated much local opposition. In the 1990s the acquisition rate fell again and is now less than a third of what it was a decade ago.

With the fall in the acquisition rate, there has come a change in acquisition strategy, driven partly by higher land costs and partly by the desire to keep commercial forest lands in commercial use. When lands of high recreational or biological value, like Santanoni or Whitney Park, appear on the market, the state still tries to acquire them outright. But when lands of lower public value, such as large industrial timberlands, are offered for sale, the state now carves them up into recreational and commercial portions and keeps only the recreational portions for the Forest Preserve. The rest are sold to commercial buyers, with the state holding a conservation easement to limit or prevent development.

Thus far this strategy has worked very well. Some choice lands have become public, some less-than-choice lands have remained commercial, and no major tracts of forest lands have been developed. But the last ten years have been a period of relatively stable ownerships. There are signs, discussed on the next page, that the next twenty may be very different. If so, acquisition policy may have to change again and quite possibly in ways that no one can yet anticipate.

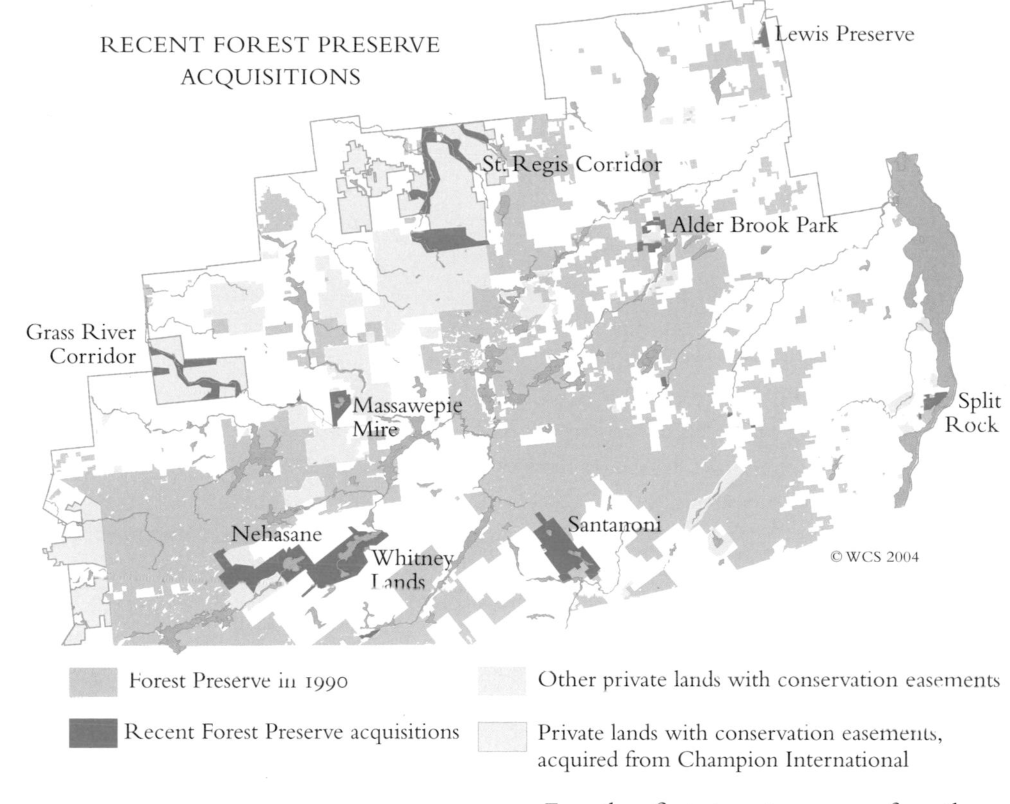

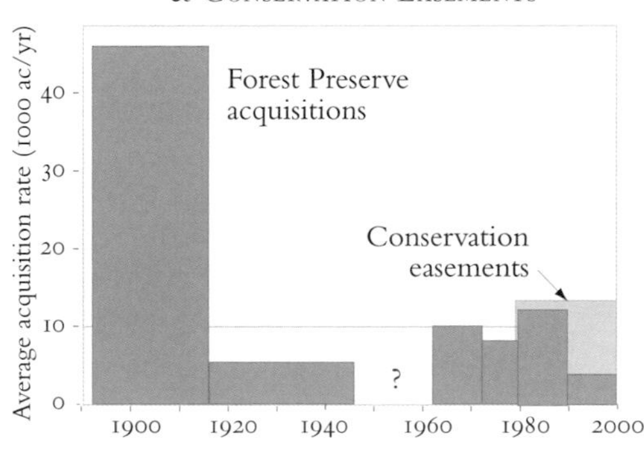

For the first twenty years after the Adirondack Park was created, Forest Preserve land was acquired at an average rate of over 46,000 acres per year. This slowed to about 5,000 acres per year for the next twenty years, doubled when a series of bond acts were passed in the 1960s and 1970s, and slowed again to about 3,500 acres per year in the 1990s. The first conservation easements were purchased by the state about 1978. Since then an average of 13,000 acres of conservation easements have been purchased every year, over three times the rate at which lands have been added to the Forest Preserve.

YES, but with significant losses. Farming and forestry are still the largest users of commercial land in the park. Both contribute substantially to local economies, and both protect scenic lands—wild lakeshores in the Adirondack interior and open lands in the Champlain Valley—with high development potential. The loss of either the revenues they generate or the land-protection they provide could be a very significant change.

Farming and forestry are both seriously troubled. Farming's troubles are old and simple: farmers are small players in large noncompetitive markets and are often paid less than their production costs. Our dairy farmers sell milk at 1975 prices and buy feed, fuel, and equipment at 2003 prices. This is, to put it mildly, an almost impossible economic handicap. Since 1970 the total real earnings of northern New York farmers have declined 70%, and farms are currently going out of business at an average rate of 100 per year. The drought in 2002 and the hard winters in 2003 and 2004 were very costly, and the rate may well accelerate.

Forestry's troubles are more recent and more complicated. Adirondack forests, because of a mixture of biological and historical problems, have much low-grade wood, for which there is at present almost no market, and little high-grade wood, for which there is a good market (*Map 13-2*). Paper-mill closings have depressed the pulp market, and widespread forest health problems, exacerbated by climate change and growing deer populations, will limit the amount of good wood which can be produced in the near future. The somber result is that many Adirondack forests will have little commercial value once the current canopy trees are harvested.

Unless there are some basic changes in the markets for logs and milk, it is likely that large amounts of Adirondack farm and forest land will appear on the market in the next few decades. This will pose two important problems for the Adirondack Park. In the short term the challenge will be to prevent the development of farm and forest lands; in the long term it will be to hold them until the commodity markets improve or the forests recover and they can be returned to productive use. The present easement-acquisition process (p. 247) is an answer to the first, but because it can currently absorb only 10,000-20,000 acres per year, it is only a partial answer. The answer to the second could be some form of land bank, but currently none exists. Very soon, we may need one badly.

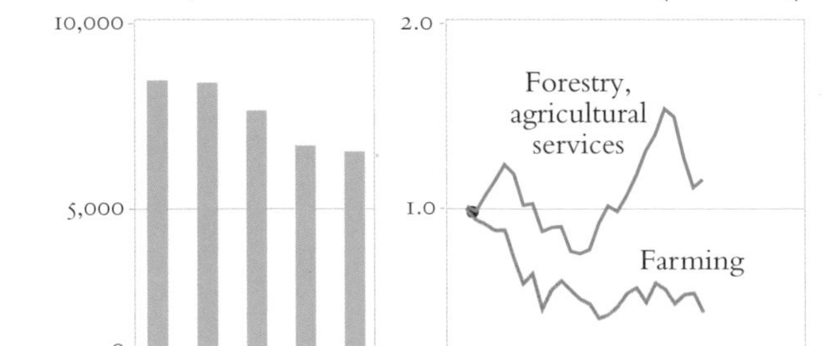

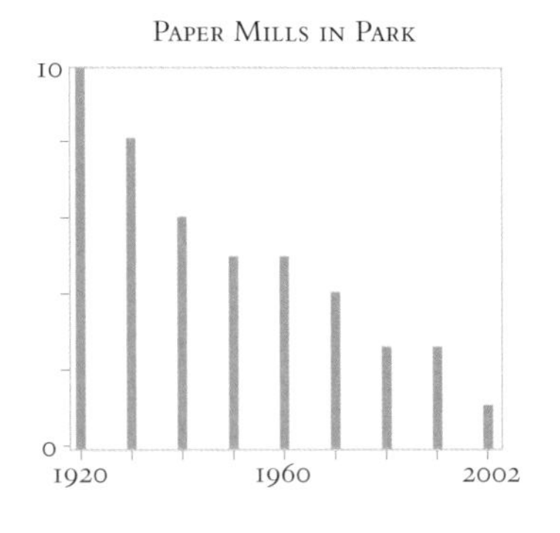

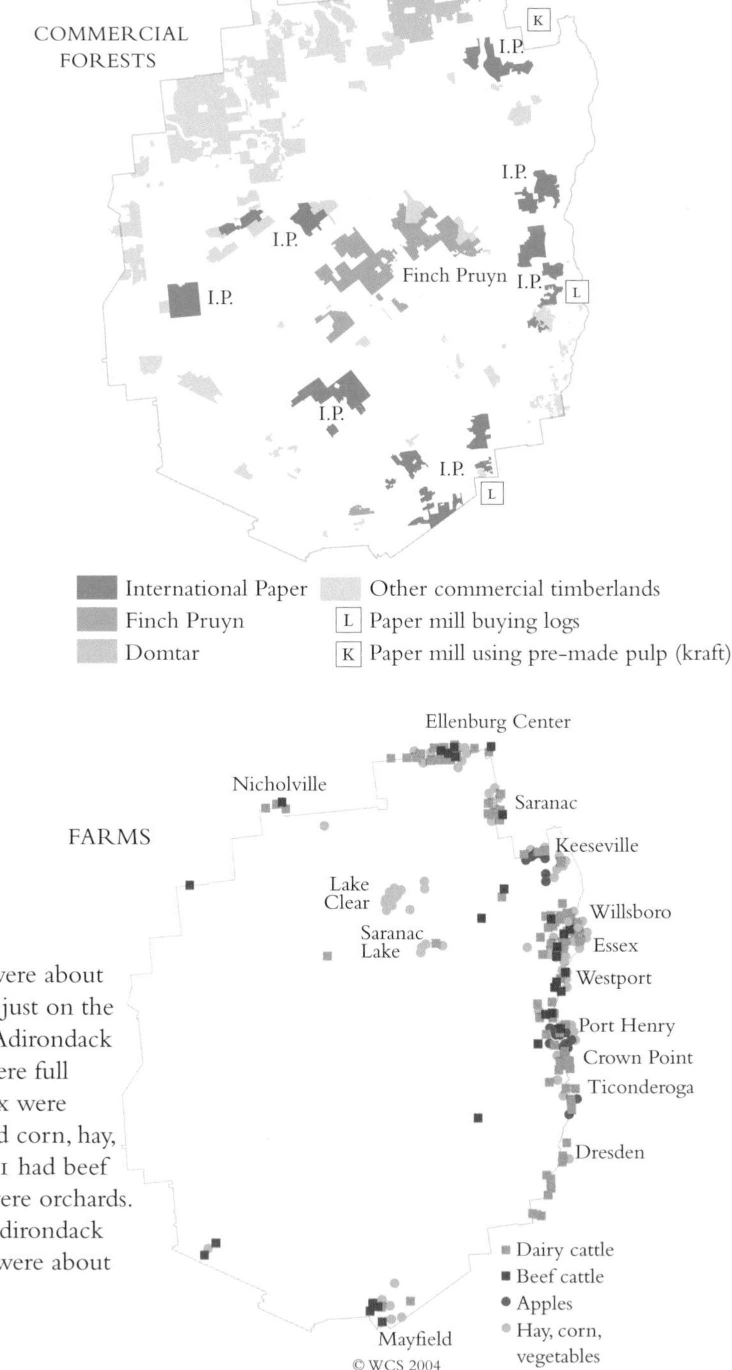

In 2000 there were about 216 farms in or just on the borders of the Adirondack Park. Not all were full time. Ninety-six were dairies, 76 raised corn, hay, or vegetables, 31 had beef cattle, and 21 were orchards. In the twelve Adirondack counties there were about 6,000 farms.

NO, but it was not intended to. It has kept most development within previously developed areas and prevented large developments that were judged inappropriate for the park. But it has no mechanism for regulating the rate of development directly, and there is no evidence that it is doing so indirectly.

Currently the park is developing steadily. The map shows about 70,000 houses in the park; there are in addition another 13,000 commercial buildings which are not mapped. Most of these have been built since 1900, giving an average construction rate of about 800 structures per year.

Between 1990 and 2000, local permits were issued for about 860 structures per year, of which 830 were residences on previously undeveloped parcels and the remainder either commercial buildings or new structures on previously developed parcels. This is close to the average for the last hundred years but about 20% below the growth peak from 1960-1990. Forty-three percent of the new structures were reviewed by the Park Agency and thus were regulated by the Land Use and Development Act; the remaining 57% were either reviewed by local planning and zoning boards or not reviewed at all. Many of the new structures are summer cottages or second homes, but no accurate numbers are available.

The continuing development of the park has had significant effects. Almost every private lakeshore is now continuously developed (*Maps 16-1,2*), and most of the smaller public roads that go through private land now have substantial amounts of new development.

If development continues at its present rate, the character of the park will gradually but irrevocably change, becoming less and less like the wilderness its founders intended and more and more like the settled and suburban areas outside. Privately very few people want to see this happen. But public opposition to development is very difficult, and the crucial questions of when growth should stop and how we can find the public will to stop it are rarely discussed.

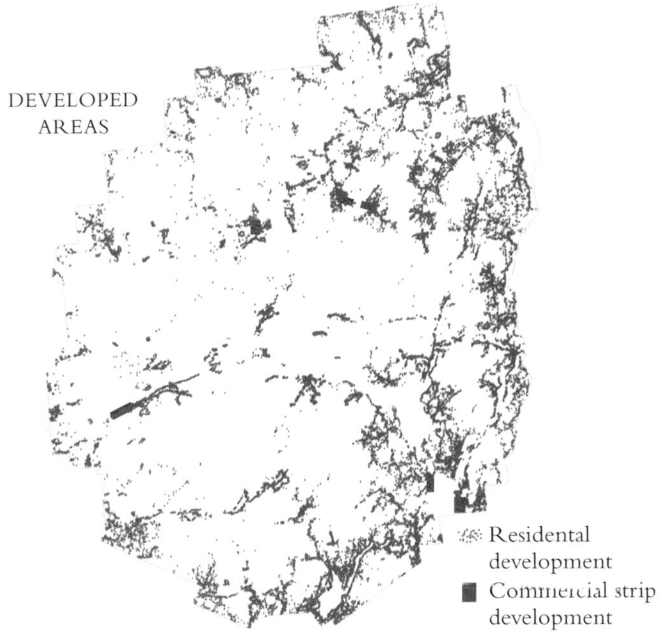

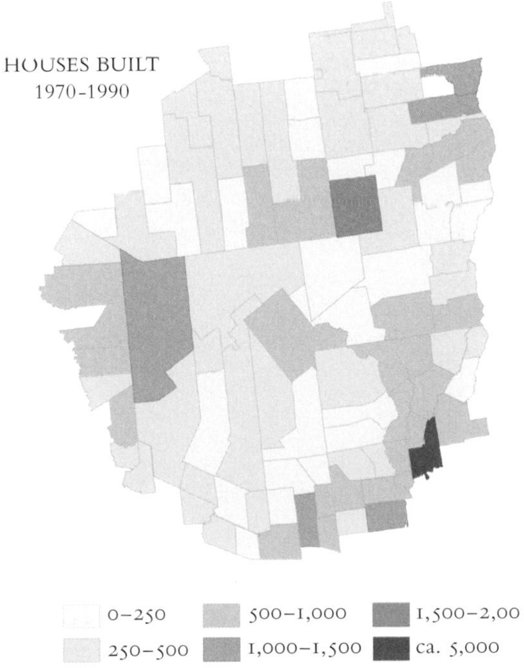

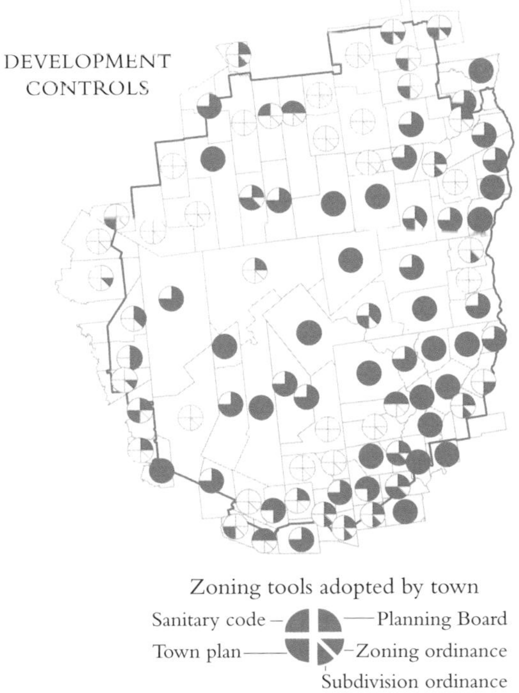

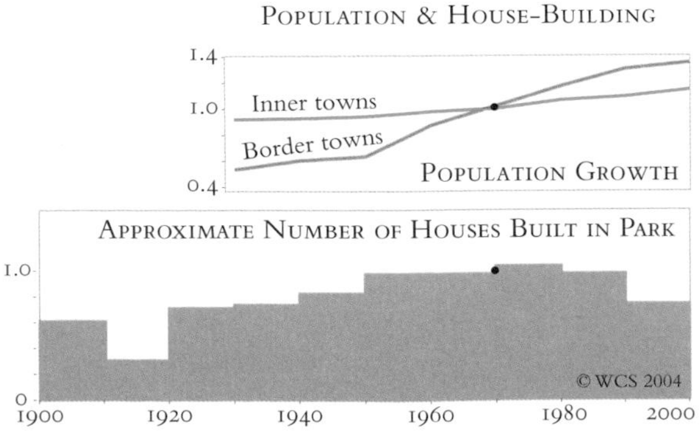

THE indicators we have suggest that in many areas they are.

Hiking, climbing, rafting, and registrations of recreational vehicles are up. Outfitters report substantial increases in boat rentals and in canoe and especially kayak sales. Informal observations suggest that *all* the major summer canoeing rivers now receive steady use and that most popular wilderness campsites are occupied most summer weekends. Ice-climbing and sales of ice-climbing gear have increased significantly; many outfitters are swelling more snowshoes than skis, and snowshoeing may be one of the fastest-growing Adirondack sports.

With other activities the picture is less clear. There have been no recent increases in the number of people using the state campgrounds, but most campgrounds are already at or near capacity on summer weekends and there may not be much room for change. Snowmobilers say that although the snowmobiling season is getting shorter and that the corridors open to them can be drastically limited when, as in 2002, the lake and river crossings aren't safe, participation is very high when conditions are good. Certainly the rapid increase in snowmobile registrations bears them out. Skiers say that the use of ski areas does not seem to have changed much in the last ten years; no official figures are available.

As recreation grows and user-groups overlap, there will be increasing competition among recreationists for access and privacy. To manage this competition we will have to answer two difficult questions: how wild should the park be, and what uses are consistent with its wildness. Answering the first means finding a way to limit use, which is, paradoxically, may be most unpopular with those who would like the park to be the most wild. Answering the second will pit user against user, and it is quite possible that the resulting conflict may be as damaging as any of the uses themselves.

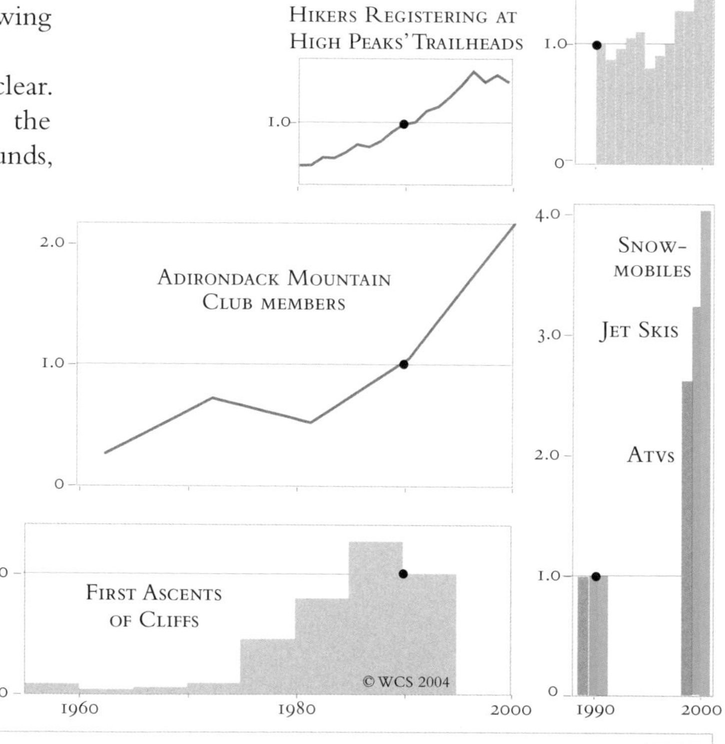

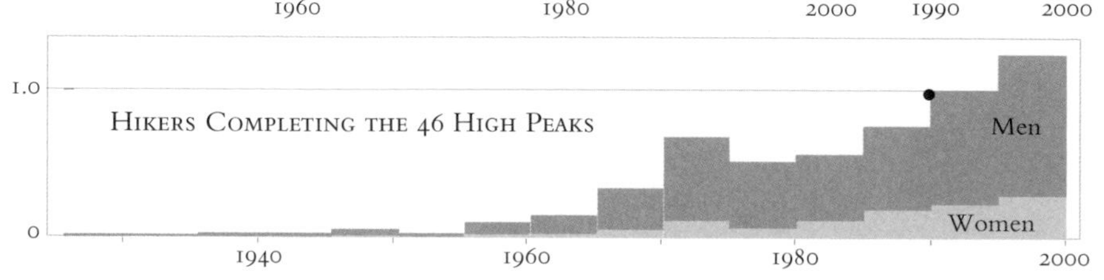

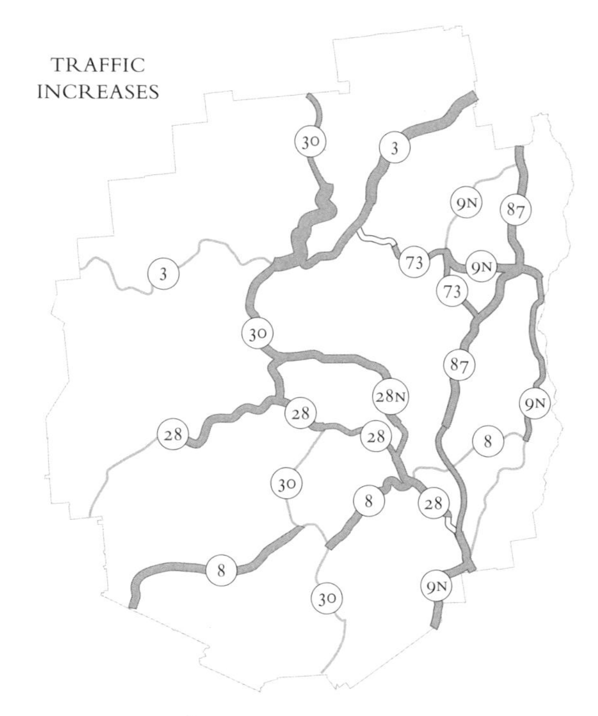

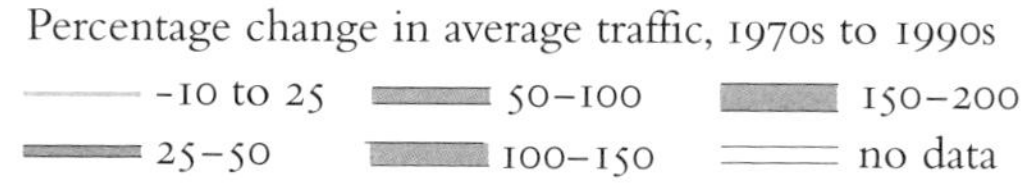

Although providing services to recreationists is one of the park's major industries, there are no overall estimates of the total number of recreationists, or what they do, or how fast their numbers are increasing. The numbers we have suggest increases of 30% to 100% per decade in overall car and truck traffic and in many outdoor sports, possibly excepting skiing. Hiker registrations in the High Peaks and Hudson Gorge rafting have increased by about 70% since 1990; ATV registrations in northern New York have increased by 150% since 1990, jet ski registrations by 200%, and snowmobile registrations by 300%.

THE *Adirondack Atlas* has no conclusion, but the questions and trends discussed in this chapter suggest a summary.

The Adirondack Park began as an attempt to create a protected wild place. It has done this with remarkable success, though not in the way or to the extent its founders envisioned.

After much controversy and nearly a century after the park was created, three critical questions—how to protect a core of public lands, how to regulate the *pattern* (though not the amount) of development on the private lands, and how and where to create wilderness areas—have now found broadly acceptable solutions. As a result we can now say that, as the founders intended, much of the Adirondacks is well protected.

But protected really only means *defended,* not *permanent* or *changeless.* All protected places still change. The Adirondacks are changing in climate, chemistry, ecology, economics, ownership, development, and usership. Some of these changes are slow, others rapid.

The changes may be thought of as taking us away, gradually but very steadily, from the Adirondacks we know to a new place which lies to the right of the lines in the graphs. The trends are too variable and too interrelated for us to give any picture of the place we are heading toward. Hence this page, alone in the *Atlas,* has no map.

Each of the types of change threatens things about the Adirondacks that we value and poses new questions about what the park should contain and how it should be managed.

If we hope to preserve the things about the Adirondacks that are important to us, we must find answers to these questions. In particular, by sometime in the park's second century we will need to know:

- How to produce energy without producing atmospheric acids, greenhouse gases, and toxics.
- How to assure that there are local markets for the food, wood, and fiber produced locally.
- How developed the park should be and how to stop development when it reaches that point.
- How to bank forest and farm lands which are ecologically damaged or which have no commercial use until they have recovered or until the market for their products has returned.
- What animals are appropriate to the changing climate and, in particular, what large carnivores are necessary to keep the herbivores under control.
- How crowded the woods should be, what uses should and should not be allowed in them, and how to regulate the people that use them.

These are very hard questions, both politically and intellectually. They involve not only the old political problem of balancing the common good against individual freedom but the new, and even harder, political problem of balancing the human good against the natural good.

Hard or not, they have to be faced and soon. Everything that we have learned in preparing the *Atlas* suggests that the character of the Adirondacks that some of us will know in thirty years and that our children and grandchildren will know in sixty years will depend on whether we can answer these questions in a way that makes human and ecological sense, and whether we can convince other people to accept our answer.

If we can't, loss and discord lie ahead and the new place we are traveling to will be unsettling and fractious. If we can, there is every chance that the new place, though different from the present Adirondacks, will have a new consistency and new beauties and will be as loved, and as rewarding to its lovers, as our Adirondacks are now.

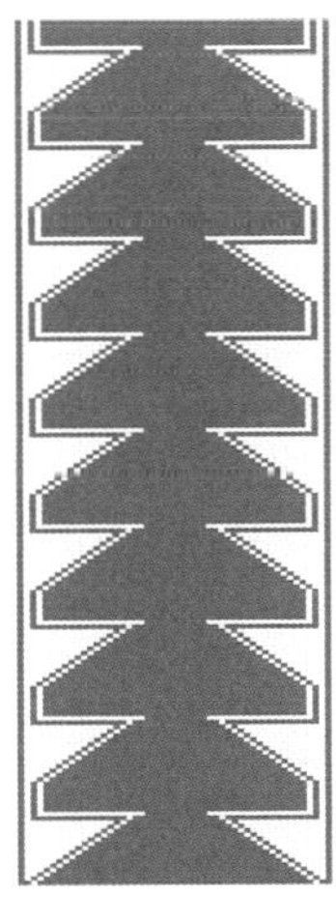

The *Evergrowing Tree,* symbolizing to the Iroquois the continuing life of their society and and its dependence on the forest and the land.

This section includes a brief bibliography of the most important and most frequently used sources followed by notes on the individual chapters and maps. The abbreviations preceding the bibliography entries are used in the notes.

Frequently Used Sources

Digital Sources

Many of the data in this atlas were taken from digital rather than print sources. The ones used most often are listed here. Others are cited in the notes for the individual maps.

Web Sites for New York State Agencies

DEC Department of Environmental Conservation (www.dec.state.ny.us)

DCJ Division of Criminal Justice (www.criminaljustice.state.ny.us)

DOH Department of Health (www.health.state.ny.us)

ED Department of Education (www.nysed.gov)

SL State Library (www.nysl.nysed.gov)

ORPS Office of Real Property Services (www.orps.state.ny.us)

DOT Department of Transportation (www.dot.state.ny.us)

Other Web Sites

EDTAC Economic Development Technical Assistance Center at the State University of New York at Plattsburgh (www .edtac.tacsuny.com)

On-line GIS Data

CGDR Cornell Geospatial Data Information Repository (www.cugir.mannlib.cornell.edu)

NYGCH New York State GIS Clearinghouse (www.nysl.nysed.gov/gis)

VGIS Vermont Geographical Information System (www.vcgi.org)

United States Census Data

Data from the 2000 census were obtained on line at www.census.gov. Older data were taken from the following CDs:

USCA U.S. Department of Agriculture, *1997 Census of Agriculture*. USDA AC97-CD-VOL1, 1999.

USAC U.S. Department of Commerce, U.S. Census Bureau, *USA Counties 1998*. USDC CD-COMP-USACOUNT98, 1999.

SAUS U.S. Census Bureau, *Statistical Abstract of the United States 1999*. CD-COMP-ABSTR99, 1999.

CCM Geolytics, *Census CD + Maps* (town-level data from the 1990 CENSUS IN GIS format)

Other Data Distributed on CDs

APGI Adirondack Lakes Survey Corp., Adirondack Park Agency, Northern Forest Lands Inventory, N.Y. Departments of Education, Environmental Conservation, Health, and Transportation, N.Y. Office of Real Property Services, U.S. Geological Survey, U.S. Environmental Protection Agency, *Shared Adirondack Park Geographic Information CD-ROM*. 2001.

ORPS N.Y. Office of Real Property Services, *2000 Real Property Data*. Albany, New York: 2000. On APGI.

AAM U.S. Department of Human Services, *Atlas of United States Mortality*. NCHS Atlas, CD-ROM NO. 1, 1997

TRI U.S. Environmental Protection Agency, *1987–1995 Toxics Release Inventory*. EPA-749-C-97-003, 1997.

Cartographic Sources

The county, state, and country outlines used in many maps were taken from *ESRI Data & Maps*, CDs 1–4, 1998, distributed by the Environmental Systems Research Institute, Redlands, Calif.; New York State hydrology, town boundaries and various administrative boundaries are from APGI. Larger scale maps, for instance *Map 14-4* or *Map 15-1*, are based on digital images of U.S.G.S. topographic sheets, obtained from NYGCH or Maptech Inc. (Greenland, N.H.). Additional information in some large-scale maps was taken from digital orthophoto quads supplied by the New York State Department of Transportation.

Major Adirondack & Regional References

The following are basic Adirondack references that were extensively consulted while preparing the *Atlas*.

GAP Bauer, Peter, Peter Sterling, and Todd Thomas, *Growth in the Adirondack Park: Analysis of Rates and Patterns of Development* (North Creek, N.Y.: Residents' Committee to Protect the Adirondacks, 2001)

TR Commission on the Adirondacks in the Twenty-first Century, *The Adirondack Park in the the Twenty-first Century: Technical Reports,* 2 vols. (Albany, N.Y.: State of New York, 1990)

AD Donaldson, Alfred L., *A History of the Adirondacks* (New York: Century Co., 1921)

LH Farrell, Patrick F., *Through the Light Hole: A Saga of Adirondack Mines and Men* (Utica, N.Y.: North Country Books, 1996)

GNY Isachsen, Y.W., E. Landing, J.M. Lauber, L.V. Rickard, and W.B. Rogers, eds., *Geology of New York: A Simplified Account* (Albany, N.Y.: New York State Museum, Geological Survey, 1991)

RA Kudish, Michael, *Railroads of the Adirondacks, A History* (Fleischmanns, N.Y.: Purple Mountain Press, 1996)

GF McMartin, Barbara, *The Great Forest of the Adirondacks* (Utica, N.Y.: North Country Books, 1994)

HH ———, *Hides, Hemlocks, and Adirondack History: How the Tanning Industry Influenced the Growth of the Region* (Utica, N.Y.: North Country Books, 1992).

PA ———, *Perspectives on the Adirondacks: A Thirty-Year Struggle by People Protecting Their Treasure* (Syracuse: Syracuse University Press, 2002).

RAN Richards, Frank E., *Richards Atlas of New York* (Phoenix, N.Y.: Frank E. Richards, 1959)

TA Schneider, Paul, *The Adirondacks: A History of America's First Wilderness* (New York, Henry Holt and Co., 1997)

CT Terrie, Philip G., *Contested Terrain: A New History of Nature and People in the Adirondacks* (Syracuse: Syracuse University Press, 1997)

FW ———, *Forever Wild: A Cultural History of Wilderness in the Adirondacks* (Syracuse: Syracuse University Press, 1994)

WW ———, *Wildlife and Wilderness: A History of Adirondack Mammals* (Fleischmanns, New York: Purple Mountain Press, 1993)

GNYS Thompson, John H., *The Geography of New York State* (Syracuse: Syracuse University Press, 1966)

VV VanValkenburgh, Norman J., *The Forest Preserve of New York State in the Adirondack and Catskill Mountains: A Short History* (Fleischmanns, N.Y.: Purple Mountain Press, 1996)

AF Williams, Michael, *Americans and Their Forests: A Historical Geography* (Cambridge: Cambridge University Press, 1989)

Front Matter

Title page. The image is a mosaic of side-looking airborne radar images acquired by the U.S. Geological Survey and originally published by the New York State Museum as their *Map and Chart Series Number 42*; © N.Y.S. Museum, 1996, and used by permission. The red double-H symbol is the log mark of Henry Hewit, a logger on the Raquette River in the mid nineteenth century. The log marks, which were stamped into the ends of the logs, allowed the logs to be sorted when they reached the mill.

The map on the copyright page is based on a reproduction of Louis Brion de la Tour's *Carte du Theatre de la Guerre Entre les Anglais et les Americans* (Paris, 1777) in J. Kevin Graffagnino, *The Shaping of Vermont: From the Wilderness to the Centennial, 1749–1877* (Rutland, Vt.: Vermont Heritage Press, 1983, p. 23).

The *Human Footprint* (p. ix) is adapted from Eric W. Sanderson et al., "The Human Footprint and the Last of the Wild," *Bioscience* 52, no. 10 (2002): 891–904, with permission. The model used to construct the map explicitly assumes that the effects of a road or a settlement extend over a mile into the woods; therefore, areas like the Adirondacks, with few public roads, are mapped as wilder than areas like the Green Mountains or Taconics, which have more. If the road-affected corridors were assumed to be narrower, or if the extensive system of private roads in the Adirondacks were taken into account, the Taconics and the Green Mountains would appear relatively wilder and the Adirondacks less so.

1 About the Adirondacks & the Atlas

Source data for maps from APGI and *ESRI Data & Maps*, CDs 1–4, 1998; the delimitation of the Adirondack interior in the map on p. 5 is based on the 1885 Sargent Commission map; see *Map 6-1*. The phrase "the limits of the possible" from Fernand Braudel's is the subtitle of *The Structure of Everyday Life,* which is vol. 1 of *Civilization and Capitalism: 15th–18th Century* (Berkeley: Univ. of California Press, 1992).

2 Environments

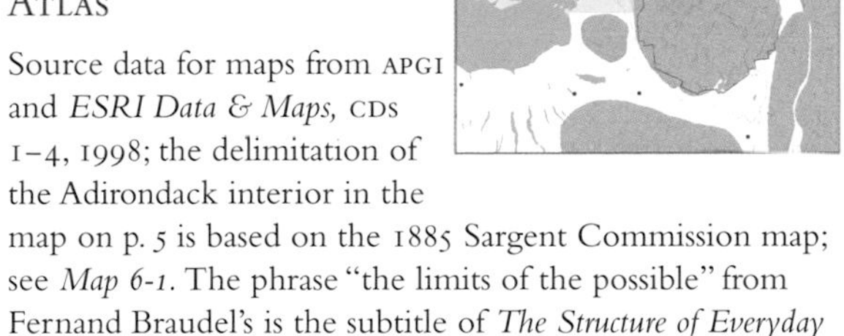

General. No comprehensive treatment of the park's environments exists. The most useful regional references are GNY and New England Regional Assessment Group, *Preparing for a Changing Climate: The Potential Consequences of Climate Variability and Change. New England Regional Overview* (Durham, N.H.: U.S. Global Change Research Program, University of New Hampshire, 2001). Older information on landforms, climate, and hydrology is found in GNYS.

Opening. APGI, GNY.

2-1 Redrawn, with permission, from drawings by Barbara Tewksbury in GNY.

2-2 APGI; Gary A. Nottis, ed., *Epicenters of Northeastern United States and Southern Canada, Onshore and Offshore; Time Period 1534–1980* (Albany, N.Y.: New York State Museum, Map and Chart Series Number 38, 1983); New York State Museum, *Bedrock Geology of New York* (Albany, N.Y., New York State Museum, 1999; on-line at CGDR, NGCH); used with permission. The modified Mercalli scale used for earthquake intensities is based on observed damage and ground motion; it is extensively used to map historical earthquakes for which quantitative measurements of intensity do not exist.

There is very extensive literature on the structural relations and geochemistry of the Adirondack rocks. Two recent reviews are D.W. Rankin et al., *Proterozoic Rocks East and Southeast of the Grenville Front,* in *The Geology of North America, Vol. C-2, Precambrian: Coterminous U.S.* (Geological Society of America, 1993); and James McClelland, J. Stephen Day, and Jonathan M. McLelland, "The Grenville Orogenic Cycle (ca. 1350-1000 MA): An Adirondack Perspective," *Tectonophysics* 265 (1996): 1–28.

2-3 Based on RAN, fig. 69, with additional information from the *USGS Mineral Resource Database* (http://mrdata.usgs.gov) and the files of the Adirondack Museum.

2-4 Soils from New York State Museum, *Surficial Geology of the Adirondack Park* (Albany, N.Y., New York State Museum, 1996, on-line at CGDR, NGCH), redrawn with permission. Locations of farms from ORPS database. The best statistics on farm decline are from USCA. For a useful summary see Guy H. Hutt and Michelle K. Hutt, *Agriculture in the Adirondack Park* (TR, vol. 2, 114–135, 1990).

2-5 Watersheds from APGI. Hydroelectric plants from James Sutherland et al., *Water Resources and Water Quality in the Adirondack Park* (TR, vol. 1, 1990), an excellent general reference. No full database of contemporary impoundments exists; the map, compiled from maps, observations, and personal communications, is likely incomplete. Because the original extent of many lakes was not accurately mapped and because a dam may raise the water level a little or a lot, the distinction between natural and artificial lakes is sometimes uncertain. Lake George, though dammed, is relatively little changed from its original state and thus considered natural. The current Indian Lake is much larger than the natural lake that preceded it and thus artificial. The original size of Meachem and St. Regis Lakes is uncertain: we treat them as natural lakes with dams, but may be wrong.

For three (among many) discussions of New York's hydraulic economy and the Adirondack log-transport system, see GNYM chap. 8; GF, chap. 4–5; and AF, 160–183. For the Setting Pole Dam (1870–85), see GF, 55–56. For the fights over the proposed Moose River and Hudson Gorge dams, an important and much retold part of Adirondack conservation history, see TA, chap. 25; CT, 150; and Jeff Jones, "Waterlogged," *Adirondack Life,* May/June (2002): 46.

2-6 Maps and diagrams original, based on USGS topographic maps and fieldwork by the author. Digital maps of the wetlands in the Adirondack portions of the Hudson and Oswegatchie-Black watersheds from the Adirondack Park Agency.

2-7 Regional climate data are from the National Oceanic and Atmospheric Administration, *Climate Atlas of the U.S.* (Washington: National Climatic Data Center, 2001). Adirondack station data are from the National Climatic Data Center (www.ncdc.noaa.gov); climate-change graphs redrawn, with permission, from *Preparing for a Changing Climate: The Potential Consequences of Climate Variability and Change, New England Regional Overview* (Durham, N.H.: U.S. Global Change Research Program, 2001), prepared by the New England Regional Assessment Group at the University of New Hampshire.

The map of regional climate was prepared from a data set of the monthly means for each climate division from 1895 to 2000, created by the National Climatic Data Center and supplied to us by Stephen Hale of the University of New Hampshire. The trends are the slope of a linear regression of the winter mean against the year; eight are significant at the 95% level, one at the 90% level, and one (marked n.s) at the 80% level.

The attempt to understand global warming may be the largest and most sophisticated international scientific program ever undertaken on this planet. Consensus reports on the state of our knowledge are issued every few years by the Intergovernmental Panel on Climate Change. The most recent is J.T. Houghton et al., eds., *Climate Change 2001: The Scientific Basis* (Cambridge: Cambridge University Press, 2001).

3 The Adirondack Park

The most comprehensive accounts of the creation and administration of the park are FW, VV, and

Eleanor Brown, *The Forest Preserve of New York* (Glens Falls, N.Y.: Appalachian Mountain Club, 1985).

Graham, Frank, Jr., *The Adirondack Park: A Political History* (Syracuse: Syracuse University Press, 1984).

Opening. Main map and graph from APGI. Because of the large number of parcels of state land in the Adirondacks and the uncertainties in their boundaries, the historical maps and the estimates of the amounts of state land at different times are often inconsistent with each other. Our estimates are taken from digital versions of historical maps and have no particular authority.

3-1 Historical information from the four sources listed above, plus Phillip G. Terrie, *A Park for the Adirondacks* (TG, vol 1, 10-23), and Thomas L. Cobb, *The Adirondack Park and the Evolution of its Current Boundary* (TR, vol. 3, 24–31). The 1885 map is based on the Sargent Committee's map of 1885; the 1893, 1938, and 1964 maps were published by the the N.Y. Forest Commission and its successors the Conservation Commission and the Department of Environmental Conservation; the 2001 map is from the current *Adirondack Park Agency State Land Map*. All are in the collection of the Adirondack Museum. Digital versions of the 1964 and 2001 maps are in APGI; the others were digitized for this project.

The maps are labeled *State Lands* rather than *Forest Preserve* because they include about 36,000 acres of lands acquired for other purposes (administration, corrections, sylviculture, transportation, etc.) that the state deems to be "inconsistent acquisitions" and therefore neither part of the Forest Preserve nor subject to the forever wild clause of the state constitution. Whether there is any statutory authority for this category of lands is politically and legally controversial. See Robert C. Glennon, *Non-Forest Preserve: Inconsistent Use* (TR, vol. 1, 74–111).

3-2 The maps are based on 2001 *Adirondack Park Agency Land Use and Development Plan Map and State Land Map* in APGI; for the development of these maps, see Charles W. Scrafford, *The Adirondack Park State Land Master Plan Origins and Current Status* (TR, vol. 1, 32–53) and Robert C. Glennon, *History of Private Land Use Regulation in the Adirondack Park* (TR, vol. 2, 136–183).

3-3 No comprehensive digital maps of ownership exist. The maps are adapted, with permission, from those in GAP and maps of easements maintained by the Adirondack Nature Conservancy; see Paul N. Miller, *An Analysis of the Use and Ownership of Private and Non-Forest Preserve Public Lands* (TR, vol. 2, 274–295).

3-4 Maps from APGI. The best picture of the overall complexity of managing the Forest Preserve and of the mix of issues and authorities involved in it is Eleanor Brown's *The Forest Preserve of New York,* cited above. The tensions between agency and agency and between local and state governments are constantly discussed but much less written about. Two frank accounts are James C. Dawson, *Forest Preserve Management* (TR vol. 1, 54–73) and PA, especially chap. 5–7.

4 Animals & Plants

Useful general references are:

Robert F. Anderle and Janet R. Carroll, eds., *The Atlas of Breeding Birds of New York State* (Ithaca: Cornell University Press, 1988)

John Bull, *Birds of New York State* (Garden City, N.Y.: Doubleday, 1974)

Michael Kudish, *Adirondack Upland Flora: An Ecological Perspective* (Saranac, N.Y.: Chauncy Press, 1992)

Peter Matthiessen, *Wildlife in America* (New York: Viking Press, 1959)

C.H. Merriam, *The Mammals of the Adirondack Region* (New York: L.S. Foster, 1884)

D. Andrew Saunders, *Adirondack Mammals* (Syracuse: State University of New York, College of Environmental Sciences and Forestry, n.d.)

John H. Whitaker Jr. and William J. Hamilton Jr., *Mammals of the Eastern United States,* 3rd ed. (Ithaca: Cornell University Press, 1998).

Opening. A Wildlife Scorecard is based on Bull, Anderle and Carroll, and Terrie. For a good regional discussion of the problems over overlapping human and animal populations, see William J. McShea et al., ed., *The Science of Overabundance: Deer Ecology and Population Management* (Washington, D.C.: Smithsonian Institution Press, 1997).

4-1 From fieldwork by the author. The forest-cover image is our recoloring of a 1982 satellite image computer-classified by the Adirondack Park Agency in 1996; each pixel represents a two-hundred-foot square. The shrub- and evergreen-covered wetlands common in the Adirondacks have not, thus far, been mapped accurately from satellite images; our map merges the wetland layers in the original image withother layers.

4-2 Continental distributions of trees from John Laird Farrar, *Trees in Canada* (Markham, Ont.: Fitzhenry & White-

side, 1997) and E.L. Little Jr., *Atlas of United States Trees,* vol. 1, 4 (Washington, D.C.: USDA Forest Service, 1971). Adirondack distributions from fieldwork by the author. Tree range shifts redrawn, with permission, from Louis R. Iverson et al., *Atlas of Current and Potential Distributions of Common Trees of the Eastern United States* (Radnor, Pa.: USDA Forest Service, Northeastern Research Station, General Technical Report NE-265, 1999; available on-line at www.fs.fed.us/ne/delaware/atlas).

4-3 Distribution maps from Anderle and Carroll. Boreal bird diversity map original, from their data. For more recent distribution information and changes in breeding bird abundance, see the *N.Y.S. Breeding Bird Atlas,* www.dec.state.us/apps/bba/results/index.htm, and the *North American Breeding Bird Survey,* ww.mbr-pwrc.usgs.gov/bbs/bbs.htm

4-4 Timeline and map derived from WW, chap. 3–7, and personal communications. Map of southern birds based on the *Breeding Bird Atlas.*

4-5 Information compiled from Merriam, *Mammals of the Adirondack Region;* Charles Eugene Johnson, "An Investigation of the Beaver in Herkimer and Hamilton Counties in the Adirondacks," *Roosevelt Wildlife Bulletin,* 1 no. 2 (1922): 117–186; Charles Eugene Johnson, "The Beaver in the Adirondacks: Its Economics and Natural History," *Roosevelt Wildlife Bulletin,* 4 no. 4 (1927); Harry V. Radford, "History of the Adirondack Beaver," *Annual Reports N.Y. State Forest, Fish, and Game Commission* (1904–1906): 389–418; other annual reports of the Forest, Fish, and Game Commission and later the Conservation Commission; and unpublished data supplied by the N.Y. Department of Environmental Conservation.

4-6 Wetland image from K.M. Roy et al., *Influences on Wetlands and Lakes in New York State: A Catalog of Existing and New GIS Layers for the 400,000-Hectare Oswegatchie/Black River Watershed* (Ray Brook, N.Y.: Adirondack Park Agency, 1997). Diagrams original.

4-7 Map redrawn with permission from Heather M. Fener, "Coyote *(Canis latrans)* Colonization of New York: The Influence of Human-Induced Landscape Changes," Masters thesis, Columbia University, 2001. For the eastward migration of the coyote and general coyote biology, see Peter Parker, *The Eastern Coyote, the Story of Its Success* (Halifax, N.S.: Nimbus, 1995) and Matthew Gompper, *The Ecology of Coyotes in Northeastern North America: Current Knowledge and Priorities for Future Research,* (Bronx, N.Y.: Wildlife Conservation Society, 2001). Moose data from the N.Y. Department of Environmental Conservation.

4-8 Radiolocations of lynx supplied by Kent Gustafson of the New Hampshire Division of Fish and Game; additional information from Ray Masters of the Adirondack Ecological Center; both were involved in the original releases. The rubbing-pad survey was conducted by John Weaver, using methods that had been quite successful in the western United States. The lynx sees a visual lure (a shiny pie-plate on a swivel) up in a tree, comes to investigate, finds the rubbing pad and rubs his head on it, leaving hairs from which DNA can be extracted. In two years of sampling, no lynx hair and almost no bobcat hair was found. The absence of bobcat hair is puzzling, as bobcat are certainly present in the study area. It is possible that the our forests are too dense for cats to find the lures or that our summers are too wet for the scent on the pads to persist.

Cougar sitings supplied by the N.Y. Department of Environmental Conservation. Like other conservation departments the DEC is discomfited by the return of the cougar, and emphasizes that they consider all the sightings, even ones by their own biologists, unverified.

4-9 Data from N.Y. Department of Environmental Conservation. For a recent study of deer distribution within the park, see Genevieve M. Nesslage and William F. Porter, "A Geostatistical Analysis of Deer Harvest in the Adirondack Park, 1954–1997," *Wildlife Society Bulletin* 29(3): 787–794, 2001.

4-10 Data from the unpublished *Histfur* database maintained by the N.Y. Department of Environmental Conservation.

4-11 Based on data from Justina Ray of the University of Toronto, Matthew Gompper of the University of Missouri, and Roland Kay of the New York State Museum. Each three-mile transect had six track boxes and three camera traps, and was walked once a month to collect scat. The transects did not include wetlands, and so mink and otter were not observed. For a review of recent work on the conservation biology of carnivores, see Justina C. Ray, *Mesocarnivores of Northeastern North America: Status and Conservation Issues* (Bronx, N.Y.: Wildlife Conservation Society, 2000).

4-12 The maps and account of wolf recovery in the west and midwest are based on information from the U.s. Fish and Wildlife Service (www.mexicanwolf.fws.gov, www.outer-banks.com/alligator-river), the International Wolf Center (www.wolf.org), the Wolf Recovery Foundation (www.forwolves.org), and other links to specific recovery programs available through these sites. The map of potential wolf habitat is adapted from Daniel J. Harrison and Theodore G. Chapin, *An Assessment of Potential Habitat for Eastern Timber Wolves in the Northeastern United States and Connectivity with Occupied Habitat in Southeastern Canada* (Bronx, N.Y.: Wildlife Conservation Society, 1997). For a general review of wolf biology and the possibility of a wolf recovery in the Adirondacks, see Angie Hodgson, *Wolf Restoration in the Adirondacks? The Questions of Local Residents* (Bronx, N.Y.: Wildlife Conservation Society, 1997).

4-13 Based on field work by the author, 1983–2003. Moss drawings © Sue Williams, 2002.

4-14 Based on data and maps from the Wadsworth Center of the N.Y. Department of Health, on-line at wadsworth.org/rabies. For the ecology of emergent diseases and the public health problems they pose, see two books by Laurie Garrett, *The Coming Plague: Newly Emerging Diseases in a World out of Balance* (New York: Penguin Books, 1994), and *Betrayal of Trust: The Collapse of Global Public Health* (New York: Hyperion, 2000).

4-15 Bear incident statistics from the N.Y. Department of Environmental Conservation; accounts of incidents from a log kept by the Adirondack Mountain Club at Adirondack Loj; animal injuries from unpublished data supplied by county health departments.

4-16 Fish-stocking information from records of the N.Y. Department of Environmental Conservation, Regions 5 and 6; BTI information from pesticide applications submitted by the towns to the Adirondack Park Agency. Since very little nonagricultural pesticide use is permitted in the park, the popularity of BTI raises some interesting regulatory problems. While as a general rule no pesticides may be applied to directly to Forest Preserve lands or any Adirondack wetlands or lakes, BTI is often used upstream from all three.

5 War, Settlement, & Industry

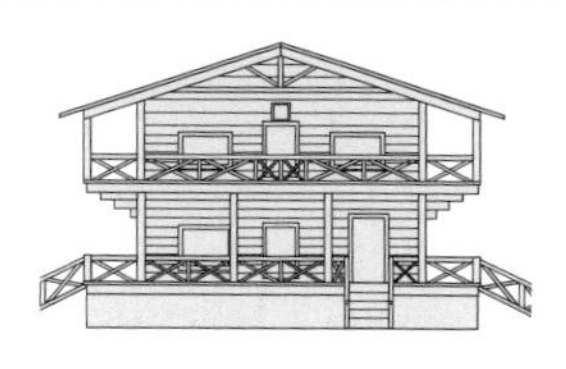

In preparing this chapter much use was made of AD, AF, CT, RA, LH, TA, GNYS and the following:

Mark C. Carnes and John A. Garraty, *Mapping America's Past: A Historical Atlas* (New York: Henry Holt & Co., 1996).

Colin G. Calloway, *New Worlds for All: Indians, Europeans and the Remaking of Early America* (Baltimore: Johns Hopkins University Press, 1997).

Guy Omeron Coolidge, *The French Occupation of the Champlain Valley from 1609 to 1759* (Harrison, N.Y: Harbor Hill Books, 1979).

William G. Dean et al., ed., *Concise Historical Atlas of Canada* (Toronto: University of Toronto Press, 1998).

William N. Fenton, *The Great Law and the Longhouse: A Political History of the Iroquois Confederacy* (Norman, Okla.: University of Oklahoma Press, 1988).

William Fox, "A History of the Lumber Industry in the State of New York," *Annual Report of the Forest, Fish, and Game Commission* (1900) 237–282.

John Keegan, *Fields of Battle: The Wars for North America* (New York: Random House, 1995).

D.W. Meinig, *The Shaping of America: A Geographical Perspective on 500 Years of History*, vol. 1, *Atlantic America, 1492–1800* (New Haven: Yale University Press, 1986).

John Richard Moravek, "The Iron Industry as a Geographic Force in the Adirondack-Champlain Region of New York State, 1800–1871," Ph.D. diss., University of Tennessee, Knoxville, 1976.

Samuel Eliot Morison, *The Oxford History of the American People* (New York: Oxford University Press, 1965).

Daniel K. Richter, *The Ordeal of the Longhouse, the Peoples of the Iroquois League in the Era of European Colonization* (Chapel Hill, University of North Carolina Press, 1992).

Opening. Large map compiled from *History of the Lumber Industry,* and the *Geography of New York State.* Sawmills from AF, 97; mines and forges from a ms. list of mining sites at the Adirondack Museum. Map of Iroquois and neighbors based on a map in the *Ordeal of the Longhouse.* On this map the *roquois* and *Algonquins* are specific tribes; the *Iroquoians* and *Algonquians* are groups of tribes with similar languages.

5-1 Based on information compiled from Harold A. Innis, *The Fur Trade in Canada* (New Haven: Yale University Press, 1962); Francis Jennings, *The Ambiguous Iroquois Empire: The Covenant Chain Confederation of Indian Tribes with English Colonies from Its Beginnings to the Lancaster Treaty of 1744* (New York, W.W. Norton & Co., 1985); Thomas Elliot Norton, *The Fur Trade in Colonial New York* (Madison: University of Wisconsin Press, 1974); *The Ordeal of the Longhouse;* and plates 43 and 60 in the *Historical Atlas of Canada.*

5-2 Based on the *French Occupation of the Champlain Valley* and a database of forts and military installations from the from N.Y. Department of Military Affairs, www.dmna.state.ny.us/forts.

5-3 Based on the *Fields of Battle,* chap. 2; Francis Jennings, *Empire of Fortune: Crowns, Colonies, & Tribes in the Seven Years War in America* (New York, W.W. Norton & Co., 1988); *Oxford History,* chap. 11; *Atlantic America,* 267–280, 284–288; *Mapping America's Past,* 52–59; the *Historical Atlas of Canada,* plate 35; and historical and chronological information from the Fort Ticonderoga Association, www.fort-ticonderoga.org."

"[Some] damned silly thing." The full quote, attributed to Bismarck, is "If there ever is another war in Europe, it will come out of some damned silly thing in the Balkans." It was later considered prophetic because the mobilizations that led to World War I were triggered by the assassination of the Archduke Franz Ferdinand in Sarajevo.

5-4 Based on the *Fields of Battle,* chap. 3; *Oxford History,* chap. 15-16; *Atlantic America,* 307–322; *Mapping America's Past,* 66-67; and Charles A. Jellison, *Ethan Allen: Frontier Rebel* (Syracuse: Syracuse University Press, 1983, chap. 6–7).

5-5 Maps based on the *Geography of New York State,* fig. 43; for the displacement of the Iroquois, see Barbara Graymount, *The Iroquois in the American Revolution* (Syracuse: Syracuse University Press, 1972); Lawrence M. Hauptman, *Conspiracy of Interests: Iroquois Dispossession and the Rise of New York State* (Syracuse: Syracuse University Press, 1999); Max M. Mintz, *Seeds of Empire: The American Revolutionary Conquest of the Iroquois* (New York: New York University Press, 1999); and *The Great Law,* parts 5-6.

An act of attainder nullifies the civil rights, and with them the property rights, of a person considered guilty of serious crimes, especially treason. The New York Act of Attainder of 1779, which attainted both Indians and a number of prominent Loyalists, was controversial because as a legislative act it was essentially a summary judgment: no evidence against those attainted was presented, and no possibility of judicial review was provided.

As an example of how sketchily the standard histories treat the annexation of Indian lands, *The Geography of New York State* devotes only five sentences in a thirty-three page chapter on the geography of expansion to explaining how New York became available to expand into.

5-6, 5-7, Maps based on the Adirondack Park Agency State Lands Map of 1983; this is the most detailed map of Adirondacks tracts and lotting ever produced; thus far there is no digital equivalent. For the early grants see *The Great Forest,* chap. 3; *The Adirondacks,* chap. 6; *A History of the Adirondacks,* chap. 9–13.

5-8 Based on the *Oxford History,* chap. 24; *Mapping America's Past,* 80–81; *Historical Atlas of Canada,* plate 38; D.W. Meinig, *The Shaping of America: A Geographical Perspective on 500 Years of History,* vol. 2, *Continental America, 1800–1867* (New Haven: Yale University Press, 1986, 41–52). The map of the Battle of Plattsburgh is redrawn, with permission, from a map at www.historiclakes.org, © Jim Millard, 2001. For the concentration of fire in naval engagements, see John Keegan, *The Price of Admiralty: The Evolution of Naval Warfare* (New York: Penguin Books, 1989, esp. 42, 66, 74–75).

5-9 Data for maps of populations and industries from GF, chap. 5; HH, part 2; *The Iron Industry;* and a ms. list of mining sites at the Adirondack Museum. The map of the interior is an original compilation of somewhat contradictory information from early maps and guidebooks in the library of the Adirondack Museum. The railroads are from AR. Additional information about this period from: Albert Fowler, *Cranberry Lake: From Wilderness to Adirondack Park* (Syracuse: Syracuse University Press, 1962); Harold K. Hochschild, *Lumberjacks and Rivermen in the Central Adirondacks, 1850–1950* (Blue Mountain Lake, N.Y.: Adirondack Museum, 1962) and *The MacIntyre Mine: From Failure to Fortune* (Blue Mountain Lake, N.Y.: Adirondack

Museum, 1962); Richard Plunzed, *Two Adirondack Hamlets in History: Keene and Keene Valley* (Fleischmanns, N. Y.: Purple Mountain Press, 1999); Louis Simmons, *Mostly Spruce and Hemlock* (Tupper Lake, N.Y., 1976); and Peter C. Welsh, *Jacks, Jobbers and Kings: Logging the Adirondacks, 1850–1950* (Utica, N.Y.: North Country Books, 1995). An excellent regional account of early iron-making is Victor R. Rolando, *200 Years of Soot and Sweat: The History and Archeology of Vermont's Iron, Charcoal, and Lime Industries* (Manchester, Vt.: Vermont Archeological Society, 1992)

Our map of industrial sites includes Dannemora Prison, which contained a mine, a forge, and a furnace and had been built in 1845 to use convict labor to produce iron. See Anne Mackinnon, "Welcome to Siberia," *Adirondack Life* Nov./Dec. (1997): 41–51.

5-10 Maps based on *Railroads of the Adirondacks;* see also Harold K. Hochschild, *Adirondack Railroads, Real and Phantom* (Blue Mountain Lake, N.Y.: Adirondack Museum, 1962) and *The Great Forest*, 128–138.

5-11 Based on Harvey H. Kaiser, *The Great Camps of the Adirondacks* (Boston: D. R. Godine, 1986); Craig Gilborn, *Durant: The Fortunes and Woodland Camps of a Family in the Adirondacks* (Blue Mountain Lake, N.Y.: Adirondack Museum, 1986); Seneca Ray Stoddard, *The Adirondacks Illustrated* (Glens Falls: Huth, 1890); Harold K. Hochschild, *Life and Leisure in the Adirondack Backwoods* (Blue Mountain Lake, N.Y.: Adirondack Museum, 1962), *An Adirondack Resort in the Nineteenth Century* (Blue Mountain Lake, N.Y.: Adirondack Museum, 1962), and *Blue Mountain Lake 1870-1900, Stage Coaches and Luxury Hotels* (Blue Mountain Lake, N.Y.: Adirondack Museum, 1962); Jeanne Winston Adler, *Early Days in the Adirondacks: The Photographs of Seneca Ray Stoddard* (New York: Harry N. Abrams, 1997); Mark Caldwell, *Saranac Lake: Pioneer Health Resort* (Saranac Lake, N.Y.: Historic Saranac Lake, 1995); *History of the Adirondacks*; *Mostly Spruce and Fir; The Adirondacks,* chap. 21–22; and a ms. inventory of great camps in the files of the Adirondack Museum.

The size and the locations of the private parks are from GF, 148–149, and the *Eighth and Ninth Reports of the Forest, Fish, and Game Commission,* for 1902–3; their boundaries are are very approximate.

5-12 The map of industries is compiled from GF, HH, LH and a ms. list of of mining sites at the Adirondack Museum; mill locations and pulp and lumber production from the *Annual Report of the Forest, Fish and Game Commission* (Albany, N.Y.: J.B. Lyon, 1900, 102–123). Railroads and post offices from *Railroads of the Adirondacks; Post Offices in the United States in 1890* (James Bennett, 1973); and a 1914 U.S. Postal Service map at the Adirondack Museum.

5-13 Locations and dates of industries from RA, LH, and files of the Adirondack Museum.

5-14 Text based on Norman Friedman, *The Fifty-year War: Conflict and Strategy in the Cold War* (Annapolis: Naval Institute Press, 2000); an on-line history of Plattsburgh Air Force Base and the 380th Bomber Wing by P. Colin at www.geocities.com/~pcolin/380thbw.html; and an on-line chronology of nuclear weapons at www.ask.ne.jp/~hanaku/english/np9y.html. Maps based on a list of cold war installations from the N.Y. Department of Military Affairs, www.dmna.state.ny.us/forts, and from a web site documenting the Nike missiles, http://alpha.fdu.edu/~bender. The drawing of the Atlas F is based on a number of historic photos; the silo is based on drawings from www.siloworld.com.

Useful general references are George F. Kennan, *Russia and the West under Lenin and Stalin* (Boston: Little, Brown & Co., 1960); Walter LaFeber, *American, Russia, and the Cold War, 1945–1980,* 4th ed. (New York: John Wiley & Sons, 1980); and Derek Leebaert, *The Fifty-Year Wound: The True Price of America's Cold War Victory* (Boston: Little, Brown & Co., 2002).

6 Forest Change

General. Two excellent general references on the forest history of the eastern U.S. are AF and

Gordon G. Whitney, *From Coastal Wilderness to Fruited Plain: A History of Environmental Change in Temperate North America, 1500 to the Present* (Cambridge: Cambridge University Press, 1994).

The most detailed treatment of Adirondack forests is GF; for a summary of Adirondack windstorm information, see

Jerry Jenkins, *Notes on the Adirondack Blowdown of July 15th, 1995: Scientific Background, Observations, and Policy Issues* (Bronx, N.Y.: Wildlife Conservation Society, 1995).

Opening. Presettlement forest composition maps from data supplied by Charles Cogbill and © Charles Cogbill, 2003. For methods see Charles V. Cogbill, John Burke, and G. Motzkin, "Forests of Presettlement New England: Spatial and Compositional Patterns Based on Town Proprietor Surveys," *Journal of Biogeography* 29 (2003):1–26. The land use history map is original; the areas on it are obtained from *Maps 6-1* to *6-4* in this chapter.

6-1 The author or authors of the Sargent Commission map are not known. Our digital version was prepared from a paper original supplied by Barbara McMartin. Estimates of cutting rates are compiled from GF, 31, 34–37, 49, 61.

6-2 The digital version of the 1916 Fire Map was prepared by Applied GIS Inc. for the Adirondack Park Agency and is on APGI. Areas burned from GF, 141. Because some areas burned several times, the total burned area on the map is less than the total of the separate burns in the graph. For the decline in fire frequency, see *Notes on the Adirondack Blowdown*, 41–44.

6-3 For a history of the tax sales and their relation to the creation of the park, see the GF, 73–100. The map is based on a ms. map prepared by Barbara McMartin from the original inventories of tax sale lands in the state archives. It was prepared, laboriously, by recreating the lotting grids from the 1983 *State Lands* map and then transferring the tax-sale informations from the manuscript map. The acreages of early acquisitions are difficult to establish accurately. The ones given in the bar graph are calculated directly from the map and are less than those given in the GF, 80, which are based on the state's original estimates of what it had purchased. Our map may be incomplete or the state's estimates incorrect or both.

6-4 Text compiled from the GF, the files of the Adirondack Museum, and conversations with foresters. The map of contemporary forests is based on the 2001 *Adirondack Park Agency Land Use and Development Plan Map and State Land Map* in APGI, with early acquisitions from *Map 6-3* and farmland from the ORPS database.

6-5 For the histories of the 1950 and 1995 storms and the legal issues connected with salvage on the Forest Preserve, see *Notes on the Adirondack Blowdown*, 3–4, 7–27, and 64–69. The map of the 1950 blowdown is based on a blueprint prepared by Gerald Rider of the DEC and digitized by the Center for Remote Sensing at SUNY Plattsburgh and Applied GIS; the map of the 1995 derecho was prepared by Kurt Schwartz and the staff of the GIS Section of the DEC Division of Lands and Forests and is based on air photos taken in July and August 1995. The map of radar reflectivities is based on archived computer data supplied by John Cannon and Jeffrey Waldstreicher of the Albany office of the National Weather Service. It is adapted from maps 8–10 in *Notes on the Adirondack Blowdown*.

Radar reflectivities are measured in decibels of radio frequency energy in the reflected signal, *dbz*. Rain and snow reflect the frequencies used for weather radar but clouds do not. Thus the intensity of the radar reflection is a measure of the instantaneous intensity of precipitation, and its frequency shift (Doppler shift) a measure of the horizontal velocity of the storm.

6-6 The map of the ice storm was prepared by GIS Section of the DEC Division of Lands and Forests. Information of the impacts of the ice storm comes from a summary on the Federal Emergency Management Agency website (www.fema.gov/reg-ii/1998/nyice2.htm), from the Niagara Mohawk Power Company, and from Charles Cogbill. For a discussion of the ecology of windstorms, see *Notes on the Adirondack Blowdown*, chap. 7–12, and the references cited there. An elegant modeling study of the long-term effects of windstorms using field data from the 1995 blowdown is Michael J. Papaik, "Modeling the Effects of a Wind-Dominated Disturbance Regime in Transition Oak-Northern Hardwoods Forests," Masters thesis, Bard College, 1998.

7 Vital Statistics

Most of the material in this chapter is from the 1990 and 2000 censuses. The contemporary vital statistics (*Maps 7-2, 3*) are from the New York State Department of Health (www.health.state.ny.us); historical birth and death rates from HSUS.

9-3 "Change the bride with the spring" From W.B. Yeats, *The Gift of Harun al-Rashid*.

8 Employers, Jobs, & Income

All the data are from U.S. Census except for the maps of jobs and job growth within the Adirondack Park, which are based on data from Alan Beideck, "Blue Line Labor: An Update of Employment and Payrolls in the Adirondack Park," *Adirondack Jour. of Environmental Studies* 6, no. 3 (1999): 18–22 and the maps of government facilities (*8-3*) and social services (*8-5*) which are from data compiled by the *Atlas* staff. A useful general reference is James Dunne, *Demographic and Economic Characteristics of the Adirondack Park* (TR, vol. 2, 10–29, 1990).

Opening. If all workers worked in the town or the county they lived in, the distinction between a map of workers and a map of jobs would be unimportant. In the Adirondacks, where many people commute out of the park, it is very important.

9 Death, Injury, Disease, & Crime

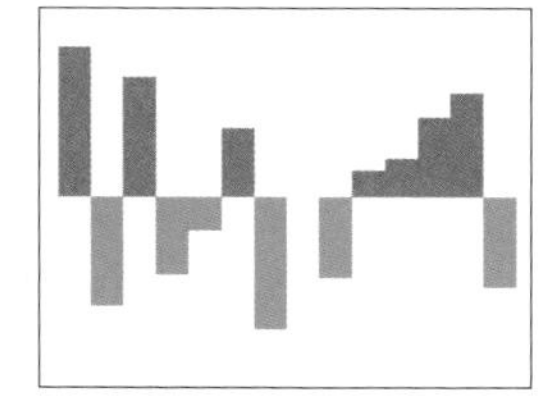

The state health data in this chapter are from DOH; the national data from AAM.

9-1 Maps from the New York State Department of Health Cancer Surveillance Improvement Initiative (www.health.state.us/nydoh/cancer/csii) and AAM. An excellent series of short papers on cancer incidence and cancer risk factors, produced by the Centers for Disease Control, are available at www.seer.cancer.gov/publications/raterisk.

9-2 AIDS data from DOH; national Lyme disease statistics from the American Lyme Disease Foundation (www.aldf.com/usmap.asp)

9-3 The map of risks and outcomes is original, based on data from DOH.

9-4,5 Crime data from the USAC and New York State Division of Criminal Justice (www.criminaljustice.state.ny.us/crimnet/data.html); maps of murders in U.S. from AAM.

9-6 General information on New York prisons and prisoners from State of New York, Department of Correctional Services, *The Hub System: Profile of Inmates under Custody on January 1, 2000* (Albany, N.Y.: Dept. of Correctional Services, 2000) and from information supplied by the Department of Correctional Services to the *Atlas* staff. The figures on U.S. imprisonment rates are from Mark Mauer, *Americans Behind Bars: The International Use of Incarceration, 1992–93* (Washington, D.C.: The Sentencing Project, 1994); the figures on the sentencing of blacks and Latinos for drug offenses are from a 2002 paper on the Rockefeller Drug Laws prepared by the New York Legal Aid Society, available on the web site of the Drug Policy Alliance (www.drugpolicy.org); also see *Racial Disparities in the War on Drugs* on the Human Rights Watch web site (www.hrw.org/campaigns/drugs/war).

For a general account of northern New York prisons, see Anne Mackinnon, "Welcome to Siberia," *Adirondack Life* Nov./Dec. (1997): 41-51.

At present there is widespread criticism of the Rockefeller drug laws. Besides the papers listed above, see a summary of the issues prepared by the New York State Defenders Organization (www.nysda.org/hot_topics/Rockefeller_drug_laws) and a criticism from an international perspective by Human Rights Watch (www.hrw.org/campaigns/drugs). A number of articles citing criticisms of the Rockefeller laws by judges and trial lawyers can be found by searching the files of the *Albany Times-Union* (www.timesunion.com). The statement that there are more blacks and Latinos in the state prisons than in the state universities comes from *New York State of Mind?: Higher Education vs. Prison Funding in the Empire State, 1988–1998* (Center on Juvenile and Criminal Justice, www.cjcj.org/pubs, 1999).

A *23-hr. lockdown* is a "maximum control" prison where inmates are locked in their cells twenty-three hours a day and allowed one hour a day of excercise in individual wire-enclosed recreation pens attached to each cell. Such facilities are not new—solitary confinement has been around a long time—but the number of prisoners now held in them in the U.S. and the lengths of time for which they being held are unprecedented and have raised questions about humaneness and constitutionality. See Paul Grondahl, "Lockdown," *Albany Times-Union*, March 26, 2000, and subsequent letters and related articles (www.timesunion.com).

10 Schools & Colleges

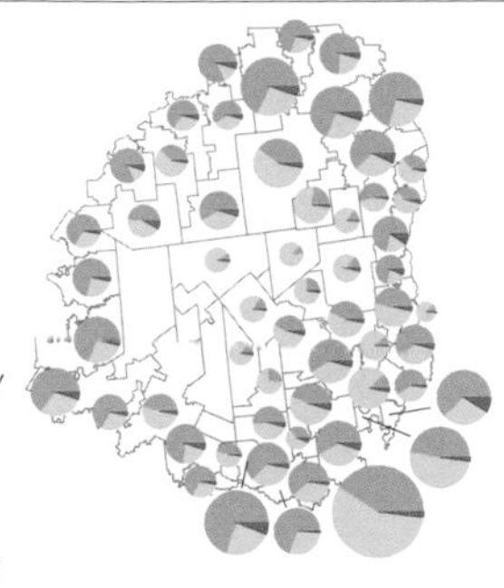

Opening. Map of schools compiled from the *Statistical Profiles of School Districts* prepared by the New York State Department of Education (www.nysed/chap655/index.html). Maps of educational attainment from U.S. Census.

10-1, 2 Data from *Statistical Profiles of School Districts.*

10-3 Both maps from data compiled by *Atlas* staff.

11 Town Budgets & Local Taxes

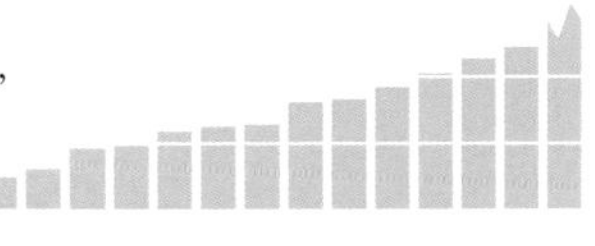

Two useful general references, both by James Dunne, are *Revenue and Expenditure Patterns of Adirondack Park Local Governments* in TR, vol. 2, 56–83; and *Significant Issues in Real Property Taxation in the Adirondack Park* in TR, vol. 2, 84–95.

Opening, 11-1 Based on data supplied by William Heidelmark of the New York State Office of Real Property Services (ORPS). Much valuable general information about tax law and tax rates can be found on the ORPS web site. Their discussion of equalization and equalized rates (www.orps.state.ny.us/ess/on-lineis/eistopics.htm) is especially valuable.

11-2 Based on William J. Heidelmark, *Exemptions from Real Property Taxation in New York State: 1998 County, City, and Town Assessment Rolls* (Albany: N.Y. Office of Real Property Services, n.d.); *Compensating Local Governments for Loss of Tax Base Due to State Ownership of Land* (Albany: N.Y. Office of Real Property Services, 1996); *Joint Report of the New York State Department of Environmental Conservation and the Board of Equalization and Assessment on the Forest Tax Laws* (Albany: Department of Environmental Conservation and the Board of Equalization and Assessment, 1993).

The Supreme Court case forbidding a lower jurisdiction to tax a higher one was *McCulloch v. Maryland*; the arguments, which lasted for nine days, pitted Daniel Webster against Luther Martin. See Peter Irons, *A People's History of the Supreme Court* (New York, Penguin Books, 1999).

12 Vital Services

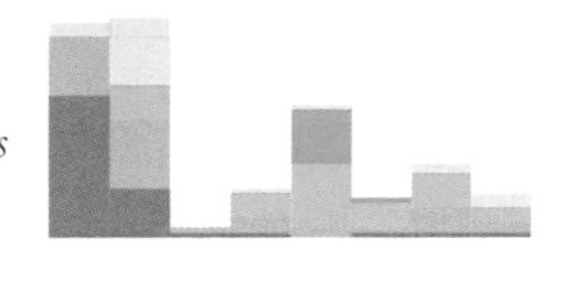

Opening. Maps compiled by *Atlas* staff, with information from the towns, counties, and New York State Police.

12-1 Maps of fires and wilderness emergencies based on records supplied by the DEC and transcribed by Ranger Steven Ovitt. Map of rangers based on rosters for DEC Regions 5 and 6, on-line at www.dec.state.ny.us/web site/protection/rangers, updated by Capt. John Streiff.

12-2 Map of health facilities based on the *Medical Directory of New York State* (Lake Success, N.Y.: Medical Society of the State of New York, 2001) and research by *Atlas* staff; information on the networks of clinics run by the Hudson Headwaters Health Network and the Adirondack Medical Center is at www.adirondackmedctr.org and www.hhhn.org.

12-3 Based on the *Medical Directory of New York State,* and the *American Dental Directory* (Chicago: American Dental Directory, 2001); information on nursing homes and adult care facilities from DOH and research by *Atlas* staff.

13 Business & Industry

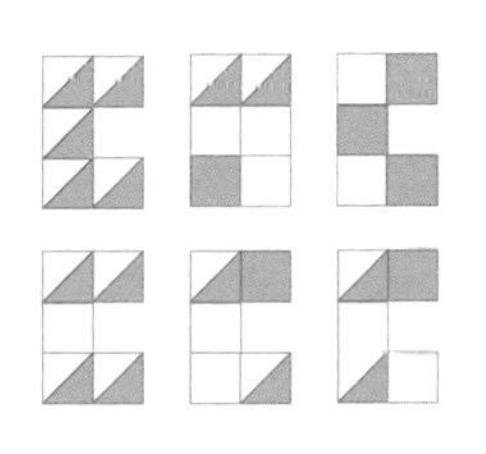

Much general information on business in northern New York can be found in a series of technical reports produced by the Economic Development and Technical Assistance Center (EDTAC) at SUNY Plattsburgh; see their web site, www.edtac.tacsuny.com.

Opening. Bank locations and assets from the *Thompson Bank Directory* (Skokie, Ill.: Thompson Financial Publishing, 2001); deposits from EDTAC; commercial property and number of businesses from ORPS. Fairs have at least a thousand-year history. Their great period, when they were the major wholesale centers, money markets, and currency exchanges in Europe, was perhaps from the twelfth through the sixteenth centuries. See Fernand Braudel *Civilization and Capitalism: 15th–18th Century*, vol. 2, *The Wheels of Commerce,* tr. Siân Reynolds (Berkeley: University of California Press, 1992, 82–93).

13-1 Employment figures from EDTAC, *New York: Manufacturing* (Twinsburg, Ohio: Harris Infosource, 2001), the *Dun and Bradstreet Million Dollar Directory* (Bethleham, Pa.: Dun and Bradstreet, 2001) and research by *Atlas* staff. Employment figures change frequently, and those given by different sources rarely agree.

13-2 Maps based on *New York: Manufacturing,* with mill locations from data supplied by the Northern Forest Lands Inventory and production figures and numbers of large mills from the *Directory of Primary Wood-Using Industries in New York State* (Albany, N.Y.: N.Y. Department of Environmental Conservation, 1996).

13-3 Information compiled by *Atlas* staff.

13-4,5 Compiled from the ORPS database, data supplied by the Northern Forest Lands Inventory and the Adirondack Regional Tourism Council.

13-6 Compiled by the *Atlas* staff.

14 Media & Culture

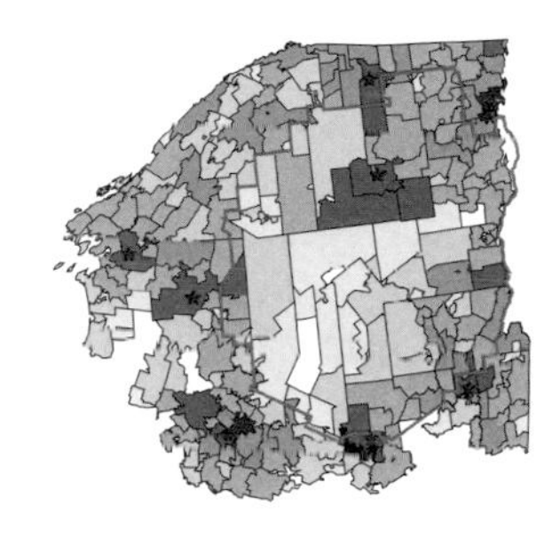

All the maps in this chapter are based on data compiled by the *Atlas* staff except the maps of radio and television stations, which use data from the Boston Radio Project (www.bostonradio.org), the Federal Communications Commission, and *2001 Radio Today: How America Listens Listens to Radio* (New York: Arbitron Inc., 2001).

14-1 Additional information from the New York State Newspaper Project (www.nysl.nysed.gov/nysnp). School and college newspapers and newspapers that are primarily advertising are not shown. Printers are only shown within the park.

For a detailed inventory of historical resources in Keene and Keene Valley, see Richard Plunz, ed., *Two Adirondack Hamlets in History: Keene and Keene Valley* (Fleischmanns, N.Y: Purple Mountain Press, 1999); for background information on Saranac Lake, see Mark Caldwell, *Saranac Lake: Pioneer Health Resort* (Saranac Lake: Historic Saranac Lake. 1993).

15 Outdoor Recreation

The best general general account of Adirondack recreation is Thomas L. Cobb, *A Profile of Outdoor Recreation Opportunties in the Adirondack Park*, in TR, vol. 1, 188–245.

Opening. Map compiled by *Atlas* staff using data from the ORPS, DEC, and Department of Transportation.

15-1 Contemporary trails from APGI; historical trails from the Adirondack Mountain Club *Guide to Adirondack Trails,* ed. Orra A. Phelps (Lake George, N.Y.: Adirondack Mountain Club, 1935); trail use from the DEC; 46ers from www.adk46ers.or. The bars give the number of people who finish climbing all forty-six peaks over 4, 000 feet in that year.

15-2 Data from Thomas R. Rosecrans, *Adirondack Rock and Ice Climbs*, (Lake George, New York: Rosecrans Outing and Climbing Klub, 1946); Don Mellor, *Climbing in the Adirondacks: A Guide to Rock and Ice Routes in the Adirondack Park,* 3rd ed. (Lake George, N.Y.: Adirondack Mountain Club, 1995), and Laura Waterman and Guy Waterman, *Yankee Rock and Ice: A History of Climbing in the Northeastern United States* (Mechanicsburg, Pa.: Stackpole Books, 1993). The drawings of Noonmark and the Beer Walls are based, with permission, on photographs in *Climbing in the Adirondacks;* the originals are © Adirondack Mountain Club, 1995. The climbs listed in the graphs are *free* climbs, where a rope is used for protection and not support. They are rated by their peak and not their average difficulty. A 5.1 climb is only a little more difficult than steep walking. A 5.13, the most difficult yet done in the Adirondacks, is near or at the limit of what a very good climber can do on a very good day.

15-3 Maps based on Paul Jamieson and Donald Morris, *Adirondack Canoe Waters: North Flow* (Glens Falls, N.Y.: Adirondack Mountain Club, 1988), Alec C. Proskine, *Adirondack Canoe Waters: South and West Flow* (Glens Falls, N.Y.: Adirondack Mountain Club, 1985), and unpublished maps of the Hudson River Gorge by John Berry and Pat Cunningham. The map of the areas that would have been flooded by Gooley Dam is redrawn from a map prepared by the Association for the Protection of the Adirondacks, one of several Adirondack organizations which played a major role in the fight to stop the dam.

15-4 Compiled by the *Atlas* staff.

15-5 Map of motorized lakes based on data from the Residents Committee to Protect the Adirondacks. The snowmobile trails from the New York State Department of Parks, Recreation, and Historic Preservation. For the legal history of motorboat use in the park, see Judith M. LaBelle, *Conflicts among Water-Based Recreational Uses* in TR, vol. 1, 246–257; for the regulatory history of snowmobiling in the park, see Dave Gibson, "A Chronology of Snowmobile Policies and Issues in the Adirondack Park" (unpublished ms. available from the Association for the Protection of the Adirondacks, 1999); also Cobb, *Outdoor Recreation Opportunities*, 203–6, and PA, 240–241. A recent article (Paul Grondahl, *Power vs. Paddle: Should More Lakes Be Motor Free?)* on motorless lakes is on the web site of the *Adirondack Explorer* at www.adirondackexplorer.org/decmotorless.htm). The de-motorization of lakes is highly divisive and controversial; see the letters to the editor in recent issues of the *Explorer.*

15-6,7 The map of roads and trails in wild forests was prepared from USGS topographic maps; data for the remaining maps from DEC, the New York State Department of Transportation, and the Adirondack Mountain Club. For the current status of the various Unit Management Plans see www.dec.state.ny.us/web site/dlf/publands/ump.

The roads in wild forests open to the disabled are listed in *Roads and Trails Open to Motor Vehicle Use by People with Mobility Impairment Disabilities* (Albany, N.Y.: New York State Department of Environmental Conservation, 2001). In theory, since the roads in wild forests that are open to general vehicle use have been officially designated by the DEC, there should be some master list of these roads. In practice, either a list has never been compiled or the DEC does not want to make it public; in any event, several requests for a full list have been refused.

For a personal narrative of recreation conflicts and the unit management planning process, see PA, especially 62–66, 249–277. For the legal, biological, and social issues connected with the use of ATVs in the park, see Leslie Karasin, *All-Terrain Vehicles in the Adirondacks: Issues and Options* (Bronx, N.Y.: Wildlife Conservation Society, 2003).

16 Changing Towns

Useful general references on Adirondack towns are GAP; William Johnston and James Hotaling, *Hamlet Revitalization,* TR, vol. 2, 30–47; Roger Trancik, *Hamlets of the Adirondacks: History, Preservation, and Investment* (Ithaca, N.Y.: R. Trancik, 1983); and Roger Trancik, *Hamlets of the Adirondacks: A Manual of Development Strategies* (Ithaca, N.Y.: R. Trancik, 1984).

Opening, 16-3,4,8. Data for maps showing the ages of houses and the progress of recreational development are from ORPS; the ORPS database only includes *existing* houses, and so our maps only show the houses built in a certain period that have survived to the present. The map of original industries is based on AD, GF, HH, RA, and research by the author.

16-1,2,5 Land classification from APGI; current houses from ORPS; former industries from RA.

16-3,4,6-8 Redrawn from USGS topographic maps, with additions by the author.

16-4 For John Brown and Gerrit Smith, see TA, 105–114.

16-5 For industrial history see GF, HH, RA, LH, Harold Hochschild, *The MacIntyre Mine: From Failure to Fortune* (Blue Mountain Lake, N.Y.: Adirondack Museum, 1962), and Henry Dornburgh, *Why the Wilderness Is Called Adirondack: The Earliest Account of the Founding of the MacIntyre Mine* (Fleischmanns, N.Y.: Harbor Hill Books, originally published 1885, reprinted 1999).

16-6 For the history of Cranberry Lake, see Albert Fowler, *Cranberry Lake from Wilderness to Adirondack Park* (Syracuse, N.Y.: Adirondack Museum/Syracuse University Press, 1968) and Susan Thomas Smeby, *Cranberry Lake and Wanakena* (Charleston, S.C.: Tempus, 2002); for the Rich Lumber Company and Emporium Forestry, see GF. The details of the mills at Cranberry Lake and Wanakena are based on drawings in RA.

16-7 Text based on fieldwork by the author, 1995-2002.

16-8,9 Text, maps, and drawings based on LH, and manuscript maps in the collection of Patrick Farrell's papers at the Adirondack Museum.

16-10 All maps original, based on geographic data from APGI and economic data from the ORPS.

17 Pollution & Wastes

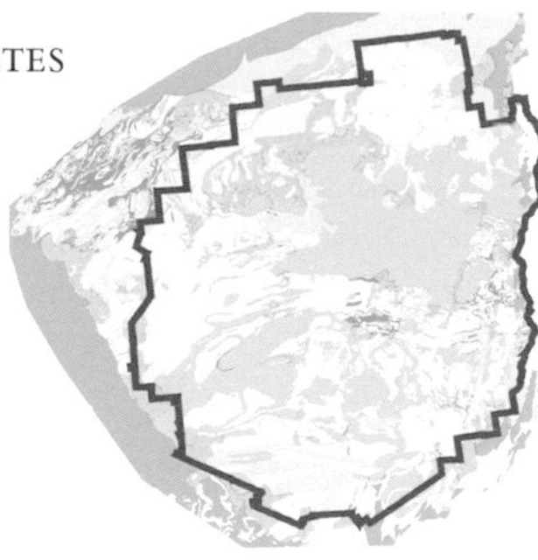

Opening. Noise level observations by the author using an Extech meter meeting ANSA standards. Nighttime light from an image collected by the Defense Meteorological Satellite Program. New York State energy sources from *Patterns and Trends: New York State Energy Profiles: 1986–2001* (Albany, N.Y.: New York Energy Research and Development Authority, 2001), available at www.nyserda.org/energy/info.html.

17-1 Data from DEC and Lake Champlain Basin Program, *The Lake Champlain Atlas* (South Burlington, Vt.: Northern Cartographic, 1999); for a detailed treatment of Adirondack hydrology and water quality, see James W. Sutherland et. al., *Water Resources and Water Quality in the Adirondack Park* in TR, vol. 1, 1990.

17-2 Data from TRI and the databases of waste sites maintained by the New York State Division of Environment Remediation (www.dec.state.ny.us/website/der/info).

17-3 Deposition data from the National Acid Deposition Program (www.nadp.sws.uiuc.edu/sites and www.nadp.sws.uiuc.edu/isopleths); stream chemistry redrawn, with permission, from the Adirondack Lakes Survey (www.adirondacklakessurvey.org); emission data redrawn, with permission, from from Charles T. Driscoll et. al., "Acidic Deposition in the Northeastern United States: Sources and Inputs, Ecosystem Effects, and Management Strategies," *Bioscience* 51 no. 3 (2003): 180–198, which is the most authoritative regional summary of acid deposition currently available.

17-4 Neutralizing capacity from a map of *Bedrock Neutralizing Capacity* prepared by the New York Geological Survey, available from NYGCH. Fish diversity redrawn from *Adirondack Lakes Survey: An Evaluation of Fish Communities and Water Chemistry* (Adirondack Lakes Survey Corporation, 1989, page 3-197); fish and pH distributions redrawn from Joan P. Baker and Steven A. Gherini, *Adirondack Lakes Survey: An Interpretive Analysis of Fish Communities and Water Chemistry 1984–1987,* (Ray Brook, N.Y.: Adirondack Lakes Survey, 1990, 3–30,31), both with permission. These are the classic works on Adirondack lake acidification. For a more recent summary, see Jerry Jenkins, Karen Roy, and Charles T. Driscoll, *Adirondack Lakes Survey: Lake and Stream Acidification, 1990–2000* (Ray Brook, N.Y.: Adirondack Lakes Survey, 2004).

17-5 Based on research by the author. See Jerry Jenkins, *Hardwood Regeneration Failure in the Adirondacks* (Bronx, N.Y.: Wildlife Conservation Society, 1998) and *Sugar Maple Regeneration Failure in the Adirondacks: Geography, History, & Demography* (Bronx, N.Y.: Wildlife Conservation Society, 2004).

17-6 Loon data from the Northeastern Loon Study Workgroup, *Monitoring Mercury in Common Loons: New York Field Report, 1998–2000,* available from the Adirondack Cooperative Loon Program (www.adkscience.org/loons); fish advisories from DOH. The diagram on the mercury cycle based on Charles T. Driscoll et al., "The Mercury Cycle and Fish in Adirondack Lakes," *Environmental Science & Technology* 28 (1994). 136A 143A; Charles T. Driscoll et al., "The Role of Dissolved Organic Carbon in the Chemistry and Bioavailability of Mercury in Remote Adirondack Lakes," *Water, Air, & Soil Pollution* 80 (1995): 499–508; Charles T. Driscoll et. al., "The Chemistry and Transport of Mercury in a Small Wetland in the Adirondack Region of New York, U.S.A.," *Biogeochemistry* 40 (1998): 137–146; and Peter Lorey and Charles T. Driscoll, "Historical trends in the Deposition of Mercury in Adirondack Lakes," *Environmental Science and Technology* 33 (1999): 718–722. The association between acid deposition and raised mercury levels in lakes is now very well established in the U.S. and Canada; to what extent this results from a relation between the *sources* of acids and mercury and to what extent it reflects the increased *transport* and *bioavailability* of mercury in acid environments is not known.

For entrance points to the rapidly growing literature on the effects of mercury on loons see N.M. Burgess et al., "Mercury and Reproductive Success of Common Loons Breeding in the Maritimes" in *Mercury in Atlantic Canada: A Progress Report* (Sackville, New Brunswick: Environment Canada, 104–109); D.C. Evers et al., "Geographic Trends in Mercury Measured in Common Loon Feathers and Blood," *Toxicological Chemistry* 17, no.2 (1998): 173–183; K.E. Parker, "Common Loon Reproduction and Chick Feeding on Acidified Lakes in the Adirondack Park, New York," *Canadian Journal of Zoology* 66 (1988) 804–810). Two on-line sources are A.M. Scheulhammer, "Methylmercury Exposure and Effects in Piscivorous Birds" in the proceedings of a 1995 workshop organized by the Canadian Mercury Network (www.eman-rese.ca/eman/reports/publications/mercury95) and an anonymous paper on the web site of the Department of Conservation Biology at the University of Minnesota, "Mercury Levels and Effects on Common Loon (*Gravia immer*) Behavior in the Upper Midwestern United States" (www.consbio.umn.edu/waterbirds/loon.html).

18 Seven Questions about Change

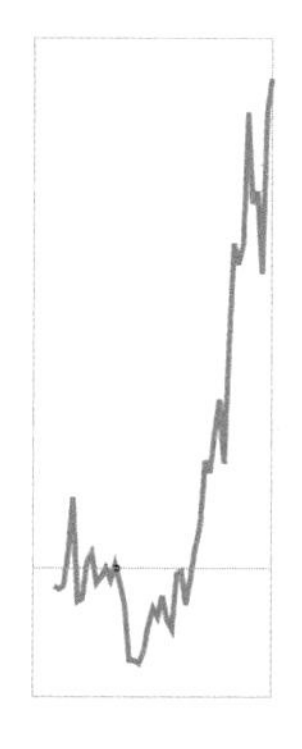

Many of the data in the chapter have been presented elsewhere in the *Atlas,* though often in a different form. For sources see the notes to *Map 2-7* (climate trends by division); *Map 3-3* (ownership); *Maps 4-7,8,9,10* (wildlife); *Map 8-3* (farm and forest income); *Maps 15-1,2,3,7* (recreation); and *Map 17-2* (acid rain). The climate trends for individual stations were calculated from daily weather summaries compiled on a CD, *Cooperative Summary of the Day: Eastern U.S.* prepared by the National Climatic Data Center. The locations of farms came from the ORPS database; information about the numbers of farms is from USCA. The location of the houses in the park and the number of houses built in each decade is from the ORPS database on APGI. Information about the local control of development is from GAP. Information about changes in traffic is based on data from the DOT, compiled by *Atlas* staff.

The symbol of the Evergrowing Tree that ends this chapter and that of the Iroquois League and the Tree of Peace on page x are based on photographs of eighteenth-century wampum belts in William N. Fenton, *The Great Law and the Longhouse: A Political History of the Iroquois Confederacy* (Norman, Okla.: University of Oklahoma Press, 1988).

The sixty-two New York counties are the primary subdivisions of the state. Legally, they are public corporations, chartered by the state, and allowed to govern, tax, and judge their residents. Historically, they are the enlarged descendents of ten original counties chartered by Charles II in 1683. Unlike the relatively weak counties of New England, the New York counties retain much of the power and autonomy of their royal ancestors and dominate the towns and cities they contain.

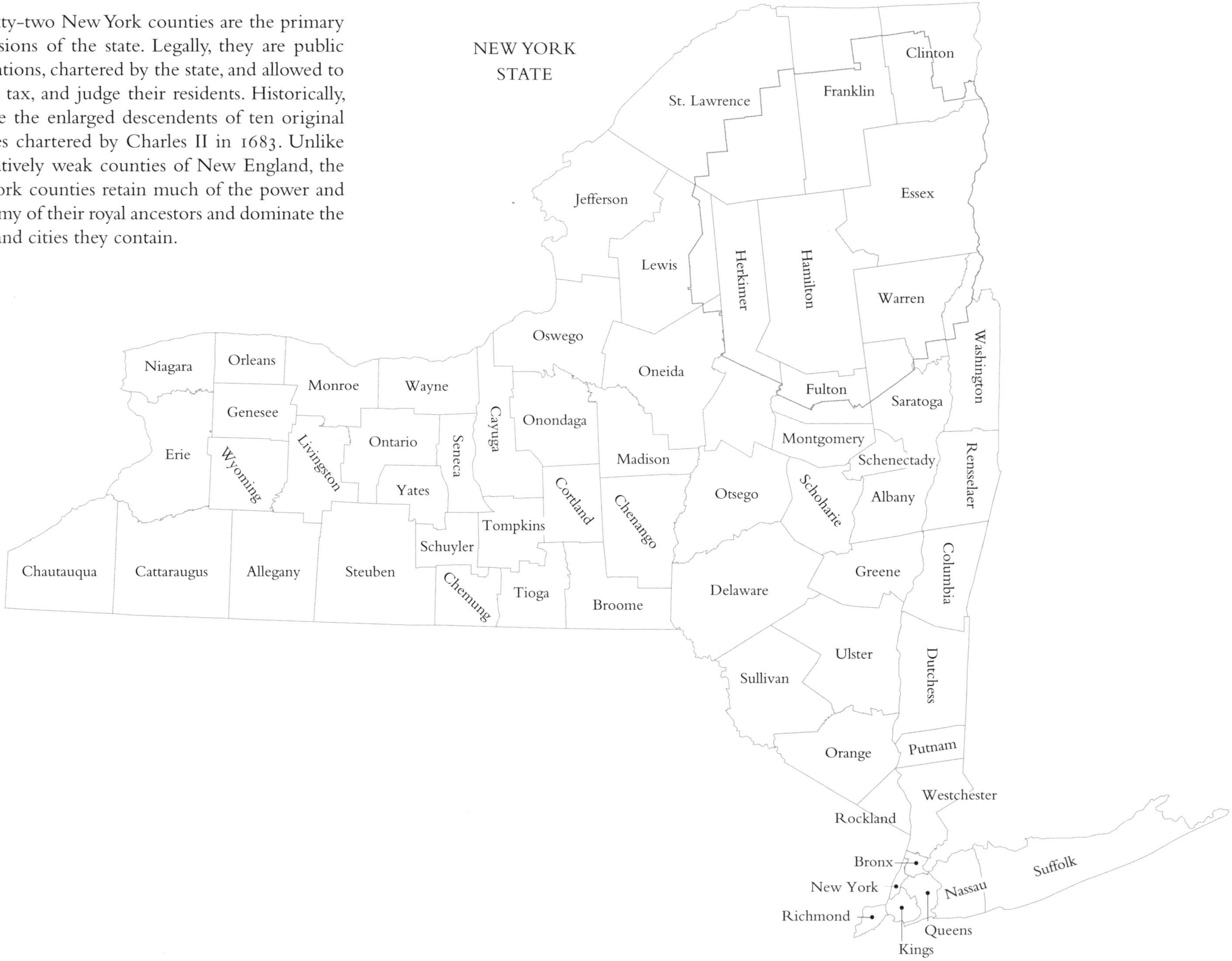

Northern New York has fifteen counties. Twelve are partly or wholly within the Adirondack Park. Hamilton and Essex are completely included in the park, and Warren almost completely included. Oneida contains only small pieces of the park. Washington, Lewis, St. Lawrence, and Saratoga are unequally divided and contain more non-Adirondack land than Adirondack land. Fulton, Herkimer, Clinton, and Franklin are more equally and are half to two-thirds Adirondack land.

New York counties are subdivided into *towns, cities,* and *villages.* All these are what the U.S. Census calls incorporated places: state-chartered public corporations with prescribed boundaries, powers, and functions. In New York, again unlike New England, this subdivision is hierarchical: villages are subservient to the towns they are in and cities and towns subservient to their counties.

The northern New York cities, shown in dark blue, are all river and canal towns in the industrial corridors outside the Adirondacks. The largest Adirondack settlements, like Tupper Lake and Speculator, are *incorporated villages* and thus legally distinct from the towns that contain them. The smaller settlements, like Johnsburg, are *hamlets*. Geographically, they have identities and histories; legally, they have no powers or boundaries and are simply parts of the towns in which they lie.

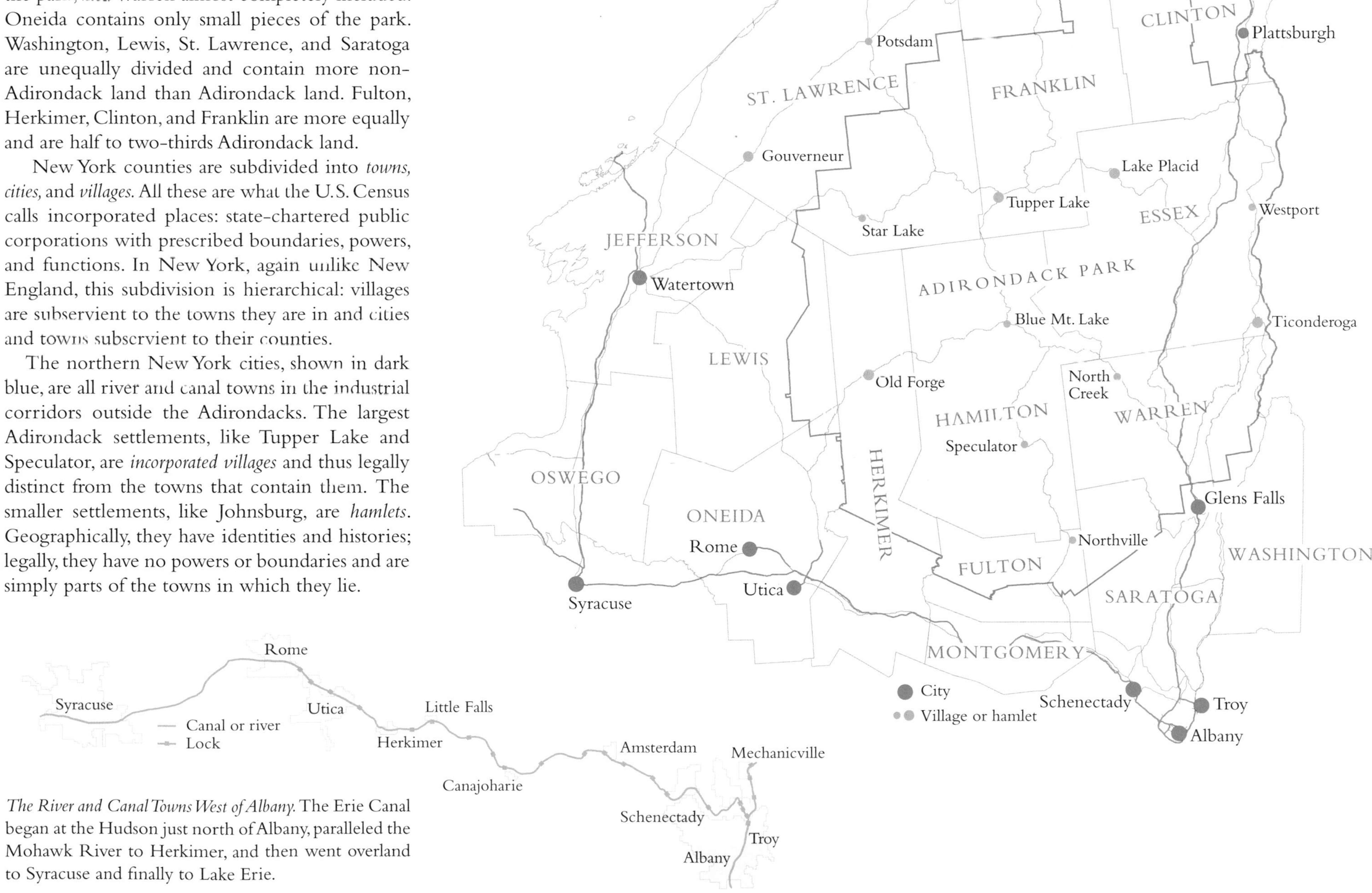

The River and Canal Towns West of Albany. The Erie Canal began at the Hudson just north of Albany, paralleled the Mohawk River to Herkimer, and then went overland to Syracuse and finally to Lake Erie.

Legally, a New York *town* is a portion of a county with an elected government that collects taxes and administers a piece of ground of reasonable size. Towns may, like Queensbury, be densely settled or may, like Arietta, have almost no settlement at all. A *village* is a similar entity, nested within and partially governed by one or more towns, that is always settled and administers a somewhat smaller piece of ground. A *settlement* is any recognizable group of houses and businesses, with or without a government and a legal identity.

The Adirondack Park contains parts of ninety-two towns. Sixty-two are completely within the Blue Line. The remaining thirty have some land outside the Blue Line and some inside. Some of the border towns, like Belmont and Lyonsdale, are largely inside the park; a few others, like Lawrence, Plattsburgh, and Remsen (with arrows showing the Adirondack portions), are almost completely outside. Most, however, have substantial amounts of land and population on both sides of the Blue Line. This creates, in many cases, a complicated kind of dual governance and makes the compilation of accurate Adirondack park statistics somewhere between daunting and impossible (*Map 3-4*, p. 113).

Town completely within park
Portion of border town within park
Blue Line
Portion of border town outside park

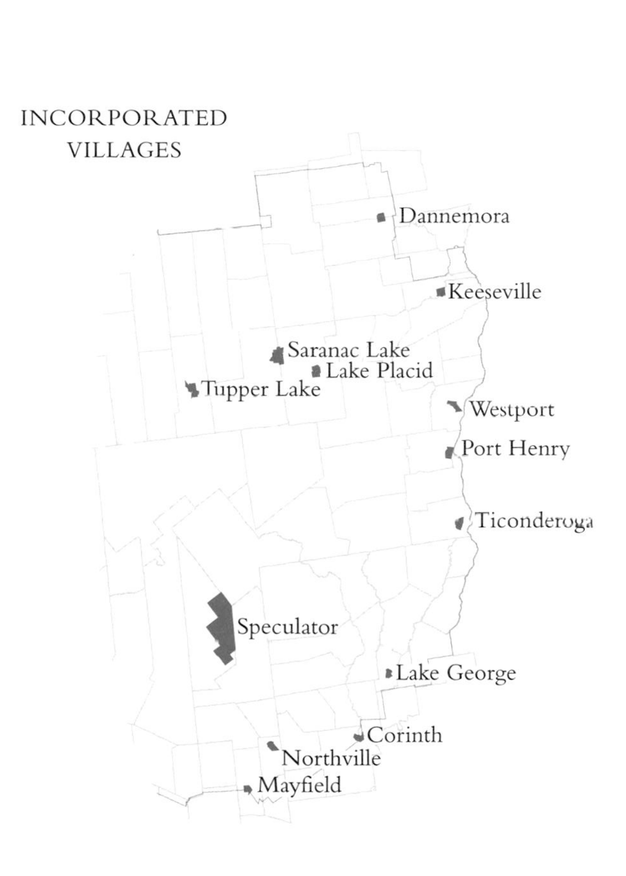

The maps show thirteen incorporated villages and ninety-four other unincorporated settlements of various sizes. Many smaller settlements might also be recognized. Some are shown in *Maps 5-9* and *5-12*.

The Adirondacks are estimated to contain at least 2,700 lakes a half acre or more in size, plus an unknown number of smaller ponds. The maps show approximately a hundred of the largest ones, grouped by watersheds. Many of these lakes are artificial or artificially enlarged: see *Map 2-5* to separate them from the natural ones.

The traditional division of the Adirondacks into five large watersheds is slightly confusing. The Hudson and its two major tributaries, the Schroon and the Sacandaga are a natural unit. The West Canada Creek watershed, which contains one of the headwaters of the Mohawk River is also a part of Hudson watershed, but joins the Hudson south of the park. The Oswegatchie and Black River watersheds, grouped together here, are at least as separate as the West Canada and the Hudson. The Oswegatchie and its West Branch flow northwest to the Saint Lawrence; the Black river and its four northern tributaries, the Beaver, Independence, Otter, and Moose, turn west and flow into Lake Ontario.

The north and northwestern flows of the Adirondacks are flatter than the southern ones and have many lakes and a complicated drainage pattern. Three long rivers, the Ausable, Saranac, and Raquette, reach the flat center of the park and meet in a the lake district west of the High Peaks. Three smaller ones, the Boquet, St. Regis, and Grass, drain the steeper sides of the dome and have fewer lakes.

The southern part of the Lake Champlain watershed, not shown on this map, has no major rivers and, except for Lake George and the South Bay of Lake Champlain, no major lakes.

This is an index to the subjects of the maps and to the subjects, places, animals, plants, and people referred to in the text. It is not a full gazetteer or concordance and does not include the titles of the maps, the labels on the maps, or the material in the notes and sources. It also omits casual references, like those to overexertion and gravity on page 133, and references to most places outside the Adirondacks, like those to the Ohio and Mississippi Valleys on the same page. For compactness, each topic in a two-page spread is indexed only once; thus, information about the growth of the park, indexed on page 26, will be found on page 27 as well.

Navita de ventis, de tauris narrat arator,
Enumerat miles vulnera, pastor oves.

We speak of what we know, Propertius said:
the sailor of winds, the plowman of bulls; the
soldier counts wounds, the shepherd sheep.

The *Adirondack Atlas*, our record of our
winds and sheep, was designed and composed
by Jerry Jenkins in White Creek, New York and
printed by Friesens in Altona, Manitoba.
The type is Monotype Bembo. The
paper is Jenson Satin.

• White Creek